INTEGRATION

INTEGRATION

By

Edward James McShane

PRINCETON

PRINCETON UNIVERSITY PRESS

1944

London: Oxford University Press

Second Printing 1947
Third Printing 1950
Fourth Printing 1957

Printed in the United States of America

Preface

The swift development of analysis in the twentieth century, beginning with the theory of the Lebesgue integral, has been of tremendous mathematical importance. No mathematician today can afford to be ignorant of the modern theories of integration, and it is to the profit of a student of mathematics that he become acquainted with these ideas early in his graduate studies. On the other hand, most of the writings on integration are written by mature mathematicians for mature mathematicians, often in an admirably concise form which is not appreciated by a beginner. This book is written with the hope that it will open a path to the Lebesgue theory which can be travelled by students of little maturity.

It is for the sake of such readers that details are explicitly presented which could ordinarily be regarded as obvious. An experienced mathematician may regard many proofs as verbose. Probably some of them are unnecessarily wordy, even for the veriest beginners; equally probably there are details omitted as obvious which will not be obvious to all readers. In view of the audience to whom this is addressed, the latter must be considered the graver fault.

The scheme of introducing the Lebesgue and Lebesgue-Stieltjes integral here adopted is a modification of that of Daniell, the integral appearing as the result of a two-stage generalization of the Cauchy (or Stieltjes) integral. Perhaps this manifestation of a connection between continuous functions and summable functions may help the beginner to feel at home in the newer theory.

There are few historical remarks on the theorems and methods here used and there is practically no bibliography. These are not usually of great interest to a beginner, and a student who wishes to continue further into the subject will necessarily read treatises—above all, Saks' *Theory of the Integral*—which will furnish bibliographical and historical references.

In only a few features can this book make claims to novelty. An expert will usually recognize known proofs used in assorted combinations and modifications. One acknowledgement must however be made. The latter part of the chapter on differential equations owes much to a mimeographed set of lecture notes on differential equations by Professor G. A. Bliss.

Part of the material in this book has been used in teaching graduate classes at the University of Virginia, and in several respects the choice of subject matter and of forms of proof has been guided by the comments of the students, especially by those of Dr. B. J. Pettis.

Shortly after the manuscript reached the Editors of the Princeton Mathematical Series I was called to the Aberdeen Proving Ground to help with the work in exterior ballistics. As a result, I lacked the time to perform the usual final tasks. I am most grateful to the Editors for their kindness in taking over duties which properly should have devolved upon the author, and thereby advancing the date of publication by many months. In particular, I owe thanks to Dr. Paco Lagerstrom, who worked long and efficiently over the manuscript.

In the correction of proof I have been greatly assisted by Miss Mary Jane Cox, who not only read all proof-sheets but pointed out a number of places in which rewording was needed for the sake of clarity.

Finally, I wish to thank Princeton University Press for its cooperativeness and efficiency.

E. J. McSHANE.

CHARLOTTESVILLE, VIRGINIA,
November 21, 1943.

Contents

CHAPTER I

Some Theorems on Real-valued Functions

The entire subject-matter of this book rests upon the properties of real numbers, with which we assume the reader to be familiar. No appeal is made to geometric intuition. Nevertheless, it is often convenient to use the language of geometry, and this is permissible if we define all our geometric expressions in terms of number.

If q is a positive integer, we shall say that each ordered q-tuple $(x^{(1)}, \cdots, x^{(q)})$ of real numbers is a "point in q-dimensional space," or a "point in R_q." For convenience in notation we prefer to put the indices $^{(1)}, \cdots, ^{(q)}$ up instead of down; the lower position will be reserved for subscripts distinguishing different points from each other. Usually we abbreviate by writing x for $(x^{(1)}, \cdots, x^{(q)})$. A standing notational convention will be the following. If any letter, with or without affixes, is used to denote a point in q-dimensional space, the q numbers defining the point will be denoted by the same symbol (with the same affixes if any) with superscripts $^{(1)}, \cdots, ^{(q)}$. Thus if we speak of a point y_0 in q-dimensional space R_q we mean the ordered q-tuple

$$(y_0^{(1)}, y_0^{(2)}, \cdots, y_0^{(q)}).$$

Two points of a space R_q are identical if and only if corresponding numbers in the two q-tuples are equal; that is, if x and y are both in R_q the equation $x = y$ has the same meaning as the q equations

$$x^{(1)} = y^{(1)}, \qquad \cdots, \qquad x^{(q)} = y^{(q)}.$$

An ordered q-tuple and an ordered p-tuple $(p \neq q)$ will never be regarded as identical.

Having this system of abbreviation, it is reasonable to proceed a step further and define sums $x + y$ and products cx, where x and y are points in R_q and c is a real number. The definitions are

$$\begin{aligned} x + y &= (x^{(1)} + y^{(1)}, \cdots, x^{(q)} + y^{(q)}), \\ cx &= (cx^{(1)}, \cdots, cx^{(q)}). \end{aligned}$$

We have little use for these symbols until the later chapters.

In order that a collection E of points of R_q shall be called a *point set* in R_q we require only that, given any point x of R_q, it must be possible to determine whether or not x belongs to the collection E.

However, there is a great deal of trouble concealed in this statement. The difficulty lies in giving a precise meaning to the word "determine." Clearly it is not possible to list all the infinitely many points of R_q one by one, marking each as belonging to E or not belonging to E. Some rule must be given. This leads to a further question. What is a rule? Now we have begun to enter the domain of foundations of mathematics; and, without denying the importance of such studies, we shall turn back again to the narrower study of the points of our spaces R_q. We shall assume that the reader has some reasonably adequate concept of a rule; if he has doubts of this, as we all may well have, we can only refer him to the various publications on mathematical logic and the foundations of mathematics.

A simple example of a point-set in R_q is R_q itself; for, given any x in R_q, we know at once that it belongs to R_q. The "empty" set Λ, which contains no points whatever, is also a point-set in R_q; for, given any x in R_q, we know that it does *not* belong to Λ. Two point-sets E_1, E_2 in R_q are identical if and only if each point x which belongs to E_1 belongs also to E_2 and each point x which belongs to E_2 belongs to E_1.

The set of all x for which the statement S holds will sometimes be denoted by $\{x \mid S\}$. E.g. $\{x \mid 0 \leqq x \leqq 1\}$ would be the closed interval consisting of all real numbers between 0 and 1. (Cf. also §4.)

Given any set E in R_q, the set of all points of R_q which do not belong to E is called the *complement* of E, and is denoted by CE. Thus the complement of the whole space R_q is the empty space Λ, and conversely; in symbols, $CR_q = \Lambda$ and $C\Lambda = R_q$. It is easy to see that for every point-set E in R_q the equation $C(CE) = E$ holds. For if x is in E, it is not in CE, and is therefore in $C(CE)$; and if x is in $C(CE)$, it is not in CE, and is therefore in E.

If E_1 and E_2 are point sets in R_q, we say that E_1 *is contained in* E_2 (in symbols, $E_1 \subset E_2$) or that E_2 *contains* E_1 (in symbols, $E_2 \supset E_1$) in case every point x which belongs to E_1 also belongs to E_2. Thus in particular $E \subset E$ and $\Lambda \subset E$ for every set E. Further, we define $E_1 \cup E_2$ to be the set of all points x belonging to one or both of E_1, E_2; we define* $E_1 \cap E_2$ or E_1E_2 to be the set of all points x belonging to both E_1 and E_2; and we define $E_1 - E_2$ to be the set of all points x which belong to E_1 but not to E_2.†

* $E_1 \cap E_2$ is sometimes called the *product*, sometimes the *intersection* of E_1 and E_2, $E_1 \cup E_2$ is called the sum or the union of E_1 and E_2.

† In defining the set theoretical operations we might of course have considered any collection of elements instead of R_q.

EXAMPLE. In R_1, put $E_1 = \{x \mid 0 \leqq x \leqq 2\}$ and $E_2 = \{x \mid 1 \leqq x \leqq 3\}$. Then $E_1 \cup E_2$ is the set $\{x \mid 0 \leqq x \leqq 3\}$, E_1E_2 is the set $\{x \mid 1 \leqq x \leqq 2\}$, $E_1 - E_2$ is $\{x \mid 0 \leqq x < 1\}$, $E_2 - E_1$ is $\{x \mid 2 < x \leqq 3\}$.

If $\{E_\alpha\}$ is any (finite or infinite) collection of sets in R_q, we define the sum (union) $\bigcup E_\alpha$ to be the set of all x contained in at least one of the sets E_α, and we define the product (intersection) $\bigcap E_\alpha$ to be the set of all x belonging to all the sets E_α.*

EXAMPLE. In one dimensional space R_1, let E_n be $\{x \mid 0 \leqq x \leqq 1/n\}$ $(n = 1, 2, 3, \cdots)$. Then $\bigcup E_n$ is $\{x \mid 0 \leqq x \leqq 1\}$, $\bigcap E_n$ is the single point 0. If E_n is $\{x \mid 0 < x < 1/n\}$, then $\bigcup E_n$ is $\{x \mid 0 < x < 1\}$, while $\bigcap E_n = \Lambda$.

The following relationships are easily verified:

$$C(E_1 \cup E_2) = CE_1 \cap CE_2, \qquad C(\bigcup E_\alpha) = \bigcap (CE_\alpha),$$
$$C(E_1 \cap E_2) = CE_1 \cup CE_2, \qquad C(\bigcap E_\alpha) = \bigcup (CE_\alpha),$$
$$E_1 - E_2 = E_1 \cap CE_2,$$
$$C(E_1 - E_2) = CE_1 \cup E_2.$$

For instance, a point x is in $C(\bigcup E_\alpha)$ if and only if it is not in $\bigcup E_\alpha$, which is true if and only if it is in no one of the sets E_α, which is true if and only if it is in every set CE_α and therefore in $\bigcap (CE_\alpha)$. Again, applying this equality to the sets CE_α, we have

$$C(\bigcup CE_\alpha) = \bigcap (CCE_\alpha),$$

whence by taking complements

$$\bigcup CE_\alpha = C(\bigcap E_\alpha).$$

If $\{E\}$ is a collection of sets, the sets of the collection $\{E\}$ are *disjoint* if no point x belongs to more than one of the sets of the collection.

A useful tool in studying properties of point sets E is the characteristic function.

1.1. *The characteristic function $K_E(x)$ of the set E is that function whose value is 1 if x is in E and whose value is 0 if x is not in E.*

In the next theorem we assemble some simple properties of characteristic functions. However, it is desirable first to define sums and products of characteristic functions. This is trivial for finite sums and products. Given an infinite aggregate of symbols α, and

* Some authors write $E_1 + E_2$, $E_1 \cdot E_2$, ΣE_α, ΠE_α for $E_1 \cup E_2$, $E_1 \cap E_2$, $\bigcup E_\alpha$, $\bigcap E_\alpha$.

corresponding to each α a number t_α which is either 0 or 1, we define the product of all t_α to be 0 if any one of them is 0 and to be 1 if all the t_α are 1. The sum of all t_α is the number n if exactly n of the t_α have the value 1, and is ∞ (>1) if an infinite number of the t_α have the value 1. Concerning this symbol ∞ we shall have more to say shortly.

1.2. (*a*) *For any collection* $\{E_\alpha\}$ *of sets,* $K_{\bigcap E_\alpha}(x) = \Pi K_{E_\alpha}(x)$.

(*b*) *For any collection* $\{E_\alpha\}$ *of sets,* $K_{\bigcup E_\alpha}(x)$ *is the smaller of the numbers* 1 *and* $\Sigma K_{E_\alpha}(x)$.

(*c*) *The sets* E_α *are disjoint if and only if* $\Sigma K_{E_\alpha}(x) \leqq 1$.

(*d*) *If the sets* E_α *are disjoint, then* $K_{\bigcup E_\alpha} = \Sigma K_{E_\alpha}$.

(*e*) $K_{CE}(x) = 1 - K_E(x)$.

(*f*) *If* $E_1 \subset E_2$, *then* $K_{E_1}(x) \leqq K_{E_2}(x)$.

To prove (*a*), we observe that if the left member has the value 1, then x is in $\bigcap E_\alpha$, so it is in every E_α, so $K_{E_\alpha}(x) = 1$ for every α, and the product of the characteristic functions is 1. Otherwise the left member has the value zero, x is not in $\bigcap E_\alpha$, it is therefore lacking from some E_α; for this E_α we have $K_{E_\alpha}(x) = 0$, and the product of the characteristic functions is 0.

To prove (*b*), if x is in $\bigcup E_\alpha$ it is in at least one E_α, so at least one term of the sum $\Sigma K_{E_\alpha}(x)$ is 1. Hence $K_{\bigcup E_\alpha}(x) = 1$, and the smaller of 1 and $\Sigma K_{E_\alpha}(x)$ is also 1. If x is not in $\bigcup E_\alpha$, then $K_{\bigcup E_\alpha}(x)$ is 0 and so is every term of the sum $\Sigma K_{\bigcup E_\alpha}(x)$. So the sum is 0, which is the smaller of 0 and 1.

In (*c*), if the sets E_α are disjoint, each x belongs to at most one set E_α, so at most one term in the sum is 1, the others all being 0. So the sum is 0 or 1. Conversely, if the sum is never more than 1, there is no x for which two or more of the characteristic functions have the value 1. That is, no x belongs to more than one of the sets E_α, and the E_α are disjoint.

Statement (*d*) follows at once from (*b*) and (*c*). Or we can prove it directly. If x is in $\bigcup E_\alpha$, it is in exactly one of the sets E_α, so both members of the equation have the value 1. If x is not in $\bigcup E_\alpha$, it is not in any E_α, so both members of the equation have the value 0.

If x is in CE, it is not in E, so $K_E(x) = 0$ and $1 = K_{CE}(x) = 1 - K_E(x)$. If x is not in CE it is in E, so $K_E(x) = 1$ and $0 = K_{CE}(x) = 1 - K_E(x)$. This proves (*e*).

For (*f*) we observe that if x is not in E_1, then

$$0 = K_{E_1}(x) \leqq K_{E_2}(x),$$

while if x is in E_2 it is also in E_1, so

$$1 = K_{E_1}(x) = K_{E_2}(x).$$

2. Next we proceed to investigate some properties of sets in R_q which depend, at least in part, on the concept of distance. If x is a point in R_q (or, as an alternative name, a vector in R_q) we define its distance from the origin (or, alternatively, the length of the vector) to be the quantity $||x||$ defined by the equation

$$||x|| = \left[\sum_{i=1}^{q} (x^{(i)})^2\right]^{\frac{1}{2}}.$$

If x and y are points of R_q, we define their distance $||x, y||$ by the equation

$$||x, y|| = ||x - y|| = \left[\sum_{i=1}^{q} (x^{(i)} - y^{(i)})^2\right]^{\frac{1}{2}}.$$

We now establish the four fundamental properties of this distance, which are the following.

(1) For all points x, y of R_q,

$$||x, y|| \geqq 0.$$

(2) If x and y are points of R_q, $||x, y|| = 0$ if and only if $x = y$.
(3) For all points x, y of R_q,

$$||x, y|| = ||y, x||.$$

(4) For all points x, y, z of R_q,

$$||x, y|| + ||y, z|| \geqq ||x, z||.$$

Properties (1) and (3) are evident from the definition. Also, $||x, y|| = 0$ if and only if each difference $x^{(i)} - y^{(i)}$ has the value zero, which establishes (2). Property (4) is called the "triangle inequality"; in geometric language, it states that the sum of two sides of a triangle is at least equal to the third side. In order to prove it, it is convenient first to establish the highly useful Cauchy inequality.

If $a_1, \cdots, a_q, b_1, \cdots, b_q$ are real numbers, then

$$\left[\sum_{i=1}^{q} a_i^2\right]^{\frac{1}{2}} \left[\sum_{i=1}^{q} b_i^2\right]^{\frac{1}{2}} \geqq \sum_{i=1}^{q} a_i b_i.$$

It is evident that

$$\sum_{i,j=1}^{q} [a_i b_j - a_j b_i]^2 \geqq 0;$$

that is,

$$\sum_{i,j=1}^{q} a_i^2 b_j^2 - 2 \sum_{i,j=1}^{q} a_i b_j a_j b_i + \sum_{i,j=1}^{q} a_j^2 b_i^2 \geqq 0.$$

In the first double sum we first collect all the terms containing a_1, then those containing a_2, and so on. We find

$$\begin{aligned}\sum_{i,j=1}^{q} a_i^2 b_j^2 &= a_1^2(b_1^2 + \cdots + b_q^2) + a_2^2(b_1^2 + \cdots + b_q^2) + \cdots \\ &\qquad\qquad + a_q^2(b_1^2 + \cdots + b_q^2) \\ &= (a_1^2 + \cdots + a_q^2)(b_1^2 + \cdots + b_q^2) \\ &= \left(\sum_{i=1}^{q} a_i^2\right)\left(\sum_{i=1}^{q} b_i^2\right).\end{aligned}$$

A similar process can be applied to each of the other two double sums; we thus find

$$\left(\sum_{i=1}^{q} a_i^2\right)\left(\sum_{i=1}^{q} b_i^2\right) - 2\left(\sum_{i=1}^{q} a_i b_i\right)\left(\sum_{i=1}^{q} a_i b_i\right) + \left(\sum_{i=1}^{q} a_i^2\right)\left(\sum_{i=1}^{q} b_i^2\right) \geqq 0.$$

If we transpose the middle term and divide by 2, we obtain

$$\left(\sum_{i=1}^{q} a_i^2\right)\left(\sum_{i=1}^{q} b_i^2\right) \geqq \left(\sum_{i=1}^{q} a_i b_i\right)^2.$$

The left member of the Cauchy inequality is non-negative. If the right member is also non-negative, the Cauchy inequality follows from the preceding inequality by taking the square roots of both members; if the right member is negative, the inequality is evidently satisfied.

EXERCISE. Let us say that the q-tuples $(a_1, \cdots, a_q)$ and $(b_1, \cdots, b_q)$ are proportional if there are numbers h, k not both zero such that $ha_i = kb_i$, $i = 1, \cdots, q$. Show that the absolute values of the two members of the Cauchy inequality are equal if and only if the q-tuples are proportional. (If they are proportional and, say, $h \neq 0$, we can substitute kb_i/h for a_i and verify equality. If equality holds, show that it also holds in the first inequality in the proof. From this deduce proportionality of the q-tuples.)

Returning to the proof of the triangle inequality, we first observe that by the Cauchy inequality

$$\left[\sum_{i=1}^{q} (x^{(i)} - y^{(i)})^2\right]^{\frac{1}{2}} \left[\sum_{i=1}^{q} (y^{(i)} - z^{(i)})^2\right]^{\frac{1}{2}} \geqq \sum_{i=1}^{q} (x^{(i)} - y^{(i)})(y^{(i)} - z^{(i)}).$$

We add the same quantity to both members of this inequality to obtain

$$\tfrac{1}{2}\sum_{i=1}^{q} (x^{(i)} - y^{(i)})^2 + \left[\sum_{i=1}^{q} (x^{(i)} - y^{(i)})^2\right]^{\frac{1}{2}} \left[\sum_{i=1}^{q} (y^{(i)} - z^{(i)})^2\right]^{\frac{1}{2}} + \tfrac{1}{2}\sum_{i=1}^{q} (y^{(i)} - z^{(i)})^2$$

$$\geqq \tfrac{1}{2}\sum_{i=1}^{q} (x^{(i)} - y^{(i)})^2 + \sum_{i=1}^{q} (x^{(i)} - y^{(i)})(y^{(i)} - z^{(i)}) + \tfrac{1}{2}\sum_{i=1}^{q} (y^{(i)} - z^{(i)})^2,$$

or

$$\tfrac{1}{2}\left\{\left[\sum_{i=1}^{q} (x^{(i)} - y^{(i)})^2\right]^{\frac{1}{2}} + \left[\sum_{i=1}^{q} (y^{(i)} - z^{(i)})^2\right]^{\frac{1}{2}}\right\}^2 \geqq \tfrac{1}{2}\left[\sum_{i=1}^{q} [(x^{(i)} - y^{(i)}) + (y^{(i)} - z^{(i)})]^2\right].$$

Multiplying both members by 2 and changing notation, we find

$$[||x, y|| + ||y, z||]^2 \geqq ||x, z||^2,$$

whence the triangle inequality follows at once.

If x_0 is any point of the space R_q, and ϵ is any positive number, we define the *ϵ-neighborhood* $N_\epsilon(x_0)$ of the point x_0 to be $\{x \mid ||x, x_0|| < \epsilon\}$. Thus in space of one dimension the ϵ-neighborhood of x_0 consists of $\{x \mid x_0 - \epsilon < x < x_0 + \epsilon\}$; in three-space, $N_\epsilon(x_0)$ consists of the points inside of the sphere of radius ϵ with center at x_0.

EXERCISE. If x is in R, and h and k are both in $N_\epsilon(x)$, so is every number y between h and k.

EXERCISE. Given two points x_1, x_2 of R_q, we say that a point x is on the line-segment joining x_1 and x_2 if there is a number t between 0 and 1 such that $x^{(i)} = tx_1^{(i)} + (1 - t)x_2^{(i)}$. Show that if x_1 and x_2 are both in $N_\epsilon(y)$, so is every point on the line-segment joining x_1 and x_2. (Use the equation $||cx|| = |c| \cdot ||x||$ and the triangle inequality.)

A point x is *interior* to a set E if it is possible to find a neighborhood* $N_\epsilon(x)$ every point of which belongs to E. A point-set E is *open* if every point x which belongs to E is interior to E. For example,

* In such a case as this, in which we merely wish to state that there is *some* neighborhood $N_\epsilon(x)$ with a given property, and the size of ϵ is of no importance, we shall sometimes write merely $N(x)$ instead of $N_\epsilon(x)$.

let us give the name "open interval" to a point set consisting of $\{x \mid a^{(1)} < x^{(1)} < b^{(1)}, \cdots, a^{(q)} < x^{(q)} < b^{(q)}\}$ where the $a^{(i)}$ and $b^{(i)}$ are finite constants for which $a^{(i)} < b^{(i)}$. Then every open interval is an open set. For if x belongs to the interval, each of the numbers $x^{(i)} - a^{(i)}$ and $b^{(i)} - x^{(i)}$ is positive. Denote by 2ϵ the smallest of them, and consider the neighborhood $N_\epsilon(x)$. If x_0 is in this neighborhood, then

$$\begin{aligned} a^{(i)} < x^{(i)} - \epsilon < x^{(i)} - \| x, x_0 \| \leqq x^{(i)} - | x^{(i)} - x_0^{(i)} | \leqq x_0^{(i)} \\ \leqq x^{(i)} + | x^{(i)} - x_0^{(i)} | \leqq x^{(i)} + \| x, x_0 \| < x^{(i)} + \epsilon < b^{(i)}, \end{aligned}$$

so x_0 is also in the open interval. Hence every open interval is an open set.

If we give the name "closed interval" to $\{x \mid a^{(i)} \leqq x^{(i)} \leqq b^{(i)}\}$, where the $a^{(i)}$ and $b^{(i)}$ are finite constants such that $a^{(i)} \leqq b^{(i)}$, we see that a closed interval is *not* an open set. For the point a belongs to the closed interval; but every neighborhood of a contains points x_0 with $x_0^{(i)} < a^{(i)}$, so that x_0 can not belong to the closed interval.

A point x is called an *accumulation point* of a set E if every neighborhood of x contains infinitely many points of E. The point x itself may or may not belong to E. For example, every point of an open interval is an accumulation point of the interval. If in one-space we take E to be the set of points $1, \frac{1}{2}, \frac{1}{3}, \cdots, 1/n, \cdots$, then 0 is an accumulation point of E; and in fact we easily verify that it is the only accumulation point of E.

The set of all points x which are accumulation points of a set E is called the *derived set* of E, and is denoted by E'. The set $E \cup E'$ is the *closure* of E, and is denoted by $\bar{E}$.

A set E is *closed* in case every accumulation point of E is itself a point of E; in symbols, if $E' \subset E$. Thus the open interval

$$\{x \mid a^{(i)} < x^{(i)} < b^{(i)}\}$$

is *not* a closed set; for the point a is an accumulation point of the interval, but does not belong to the interval. The closed interval $\{x \mid a^{(i)} \leqq x^{(i)} \leqq b^{(i)}\}$ is a closed set. For suppose that a point x is not in the interval. Then the above inequalities do not all hold. Suppose, to be specific, that $x^{(1)} < a^{(1)}$. Define $\epsilon = \frac{1}{2}(a^{(1)} - x^{(1)})$. For every point x_0 in $N_\epsilon(x)$ we have $x_0^{(1)} \leq x^{(1)} + | x_0^{(1)} - x^{(1)} | < x^{(1)} + \epsilon < a^{(1)}$, so x_0 is not in the interval. Hence x can not be an accumulation point of the interval; and since no point outside of the interval is an accumulation point of the interval, the interval is a closed set.

It is by no means true that every set is either open or closed. In one-space, the "half-open interval" $\{x \mid a \leqq x < b\}$ is neither open nor closed; for no neighborhood of a lies in the set, while b is an accumulation point which does not belong to the set.

We here introduce a notation for intervals in R_1. We define

$$\begin{aligned}[a, b] &\equiv \{x \mid a \leqq x \leqq b\},\\ [a, b) &\equiv \{x \mid a \leqq x < b\},\\ (a, b] &\equiv \{x \mid a < x \leqq b\},\\ (a, b) &\equiv \{x \mid a < x < b\}.\end{aligned}$$

Thus a square bracket at either end connotes that that end is included, a round bracket connotes that it is excluded. By an interval in R_q we shall mean any non-empty set defined either by inequalities $a^{(i)} \leqq x^{(i)} \leqq b^{(i)} (i = 1, \cdots, q)$ or by the inequalities obtained by replacing some or all of the signs "$\leqq$" by the sign "$<$." In R_q there will be 4^q different types of intervals. Sometimes it is convenient to use special symbols for four of these, in analogy with the preceding; thus an interval $\{x \mid a^{(i)} \leqq x^{(i)} < b^{(i)} (i = 1, \cdots, q)\}$ can be designated by the abbreviation $[a, b)$.

Let I be an interval defined either by inequalities $a^{(i)} \leqq x^{(i)} \leqq b^{(i)}$ or by inequalities obtained by replacing some or all of the signs $\leqq$ by $<$. It is easy to verify that

$$\bar{I} = I' = \{x \mid a^{(i)} \leqq x^{(i)} \leqq b^{(i)}\},$$

which is a closed interval; and that the set of interior points of I is the set $\{x \mid a^{(i)} < x^{(i)} < b^{(i)}\}$, which is either an open interval or the empty set (the latter in case $a^{(i)} = b^{(i)}$ for some i).

If we consider R_q as a subset of itself, we easily see that it is both open and closed, and the same is true of the empty set Λ. We could show that no other sets in R_q are both open and closed.

It sometimes happens that we are interested only in subsets of some given set S, and wish to neglect the rest of the space R_q. For instance, let us anticipate slightly and consider a function $f(x)$ defined and continuous on the set S of all rational numbers. Let E be $\{x \mid f(x) > 0, x \text{ in } S\}$. This set is surely not open; it can not contain any neighborhood of any point, because it contains no irrational numbers. Yet for every x_0 in E there is an $\epsilon > 0$ such that $|f(x) - f(x_0)| < \frac{1}{2}f(x_0)$ if x is any *rational* for which $|x - x_0| < \epsilon$. So for such x we have $f(x) > 0$. Thus $N_\epsilon(x_0)$ is not contained in E, but all points of S in $N_\epsilon(x_0)$ are in E. This suggests the following definition. If a set E is contained in a set S, and for each x in E there is an $\epsilon > 0$ such that $N_\epsilon(x) \cap S \subset E$, then E is said to be *open relative to S*. The corresponding closure concept is this. If a set E is contained in a set S, and all accumulation points of E *which belong to* S also belong to E, then E is closed relative to S. That is, $E \subset S$ is closed relative to S if $E'S \subset E$.

In particular, if S is R_q itself, these properties take the form of ordinary openness and closedness, since the factor S in $N_\epsilon(x_0)S$ and $E'S$ may be omitted if it is the whole space under consideration.

EXAMPLE. Let E be the set $(0, 1]$; that is, $\{x \mid 0 < x \leqq 1\}$. This is neither open nor closed. But if S is the set $(0, 2)$, then E is closed relative to S. For $E' = [0, 1]$, and $E'S = (0, 1] \subset E$. Also, if S_0 is the set $[-1, 1]$, E is open relative to S_0. For let x be any point of E. If $x \neq 1$, there is a neighborhood $N_\epsilon(x)$ contained in E, so $N_\epsilon(x)S \subset E$. If $x = 1$, take $\epsilon = 1$. The set $N_\epsilon(x)$ is then $(0, 2)$, and $N_\epsilon(x)S$ is $(0, 1]$, which is contained in E.

EXERCISE. If S is any set, and E and E_0 are respectively closed and open, then ES and E_0S are respectively closed and open relative to S. Also, S is both open and closed relative to S.

An immediate consequence of the definition of accumulation point is the following theorem.

2.1. *Let E be any set. In order that a point x_0 shall be an accumulation point of E, it is necessary and sufficient that for every positive ϵ the neighborhood $N_\epsilon(x_0)$ shall contain at least one point of E different from x_0.*

If x_0 is an accumulation point of E, every neighborhood $N_\epsilon(x_0)$ contains infinitely many points of E, hence contains points of E different from x_0. If x_0 is not an accumulation point of E, some $N_\epsilon(x_0)$ contains only a finite number of points of E. Let $x_1, \cdots, x_n$ be the points of E different from x_0 and in $N_\epsilon(x_0)$. Let δ be the smallest of the positive numbers $\|x_0, x_1\|, \cdots, \|x_0, x_n\|$. Then $N_\delta(x_0)$ contains none of the points $x_1, \cdots, x_n$, and therefore contains no point of E except perhaps x_0 itself.

Another trivial theorem is

2.2. *Let E be any set. In order that a point x_0 shall belong to $\bar{E}$, it is necessary and sufficient that for every positive ϵ the neighborhood $N_\epsilon(x_0)$ contain at least one point of E.*

If x_0 is in $\bar{E}$, either it is in E, in which case $N_\epsilon(x_0)$ always contains the point x_0 of E; or else it is in E', in which case $N_\epsilon(x_0)$ contains a point of E by 2.1. If x_0 is not in $\bar{E}$ it is not in E', so by 2.1 some $N_\epsilon(x_0)$ contains no point of E except perhaps x_0. But x_0 too is not a point of E, since it is not in $\bar{E}$. So $N_\epsilon(x_0)$ contains no point of E.

The justification for the name "closure" given to the set $\bar{E} = E \cup E'$ is contained in the following theorem.

2.3. *Let E be any set. Then E' and $\bar{E}$ are closed sets.*

Let x_0 be an accumulation point of $\bar{E}$; we must show that x_0 is in $\bar{E}$. For every positive ϵ, the neighborhood $N_\epsilon(x_0)$ contains at least one point x_1 of $\bar{E}$ different from x_0. Since x_1 is in $N_\epsilon(x_0)$ and is different from x_0 we have $0 < ||x_0, x_1|| < \epsilon$. Let δ be the smaller of the positive numbers $||x_0, x_1||$ and $\epsilon - ||x_0, x_1||$. Then $N_\delta(x_1)$ contains a point x_2 of E by 2.2. The point x_2 is different from x_0, since $||x_1, x_2|| < \delta \leqq ||x_1, x_0||$; and it is in $N_\epsilon(x_0)$, since $||x_0, x_2|| \leqq ||x_0, x_1|| + ||x_1, x_2|| < ||x_0, x_1|| + \delta \leqq \epsilon$. So every $N_\epsilon(x_0)$ contains a point of E different from x_0, and x_0 is in E' by 2.1. Therefore x_0 is in $\bar{E}$. A similar proof applies to E'.

As an immediate corollary we have

2.4. *E is a closed set if and only if $E = \bar{E}$.*

If E is closed, then $E' \subset E$, so $E = E \cup E' = \bar{E}$. If $E = \bar{E}$, then E is closed by 2.3.

EXERCISE. If $E \subset E_1$, then $E' \subset E_1'$ and $\bar{E} \subset \bar{E}_1$.

EXERCISE. If $E \subset G$ and G is closed, then $\bar{E} \subset G$.

The interrelation between open sets and closed sets is given in the following theorem.

2.5. *Let E be contained in S. If E is closed relative to S, then $S - E$ is open relative to S, and if E is open relative to S, then $S - E$ is closed relative to S.*

Let E be closed relative to S, and let x belong to $S - E$. Then x is not in E, and therefore not in $E'S$, which is contained in E. Now x is in S, but not in $E'S$, so it is not in E'. Being neither in E nor in E', it is not in $\bar{E}$. By 2.2, there is a positive ϵ such that $N_\epsilon(x)$ contains no point of E. Hence $N_\epsilon(x) \subset CE$, and so $N_\epsilon(x)S \subset S \cap CE = S - E$. This proves that $S - E$ is open relative to S.

Suppose now that E is open relative to S, and let x be any point of E. There is an ϵ such that $N_\epsilon(x)S \subset E$, hence $N_\epsilon(x)S$ contains no points of $S - E$. But the rest of $N_\epsilon(x)$, namely $N_\epsilon(x) \cap CS$, surely contains no points of $S - E$, for it contains no points of S. Therefore no point of $S - E$ is in $N_\epsilon(x)$, and x is not an accumulation point of $S - E$. We have thus proved that E and $(S - E)'$ have no points in common. Now the set $S \cap (S - E)'$ is in S, and has no point in common with E, so it must be entirely in $S - E$; that is, $S \cap (S - E)' \subset S - E$. This proves that $S - E$ is closed relative to S.

2.6. Corollary. *If E is closed, CE is open; and if E is open, CE is closed.*

In 2.5 we take $S = R_n$; then 2.5 takes the form 2.6.

EXERCISE. If S is any set, and E is closed relative to S, it is the product of S and a closed set (namely $\bar{E}$). If S is any set and E is open relative to S, it is the product of S and an open set.

2.7. *If E is contained in a closed set S and is closed relative to S, then E is closed. If E consists of interior points of S (in particular if S is open and $E \subset S$) and E is open relative to S, then E is open.*

To prove the first statement, we observe that every accumulation point of E is also an accumulation point of S, as follows at once from the definition. (Cf. exercise after 2.4.) Hence $E' \subset S'$. But S is closed, so $S' \subset S$, from which we have $E' \subset S$. Therefore $E'S = E'$. By hypothesis, E is closed relative to S, so $E'S \subset E$. This gives $E' = E'S \subset E$, so E is closed.

To prove the second part, let x_0 be a point of E. Since E is open relative to S, there is an $\alpha > 0$ such that $N_\alpha(x_0)S \subset E$. But x_0 is an interior point of S, so there is a $\delta > 0$ such that the neighborhood $N_\delta(x_0)$ is contained in S. Now let ϵ be the smaller of α and δ. Then $N_\epsilon(x_0) \subset N_\delta(x_0) \subset S$, so $N_\epsilon(x_0)S$ is the same as $N_\epsilon(x_0)$. Also, $N_\epsilon(x_0) = N_\epsilon(x_0)S \subset N_\alpha(x_0)S \subset E$. Thus there is a neighborhood $N_\epsilon(x)$ of x_0 completely contained in E, and E is open.

2.8. *If $\{E_\alpha\}$ is any collection of sets lying in a set S and open relative to S, their sum $\bigcup E_\alpha$ is open relative to S. If $E_1, \cdots, E_p$ is a finite collection of sets lying in S and open relative to S, their product $\bigcap E_i$ is open relative to S.*

Let x be a point in $\bigcup E_\alpha$. Then it belongs to some E_α. Since this E_α is open relative to S, there is an ϵ such that $N_\epsilon(x)S$ is contained in E_α, and hence surely contained in $\bigcup E_\alpha$. Thus $\bigcup E_\alpha$ is open relative to S.

Let x be a point in $\bigcap E_i$. For each i the point x is in the set E_i, so for each i we can find a positive number ϵ_i such that $N_{\epsilon_i}(x)S$ is contained in E_i. Let ϵ be the smallest of the numbers $\epsilon_1, \cdots, \epsilon_p$. Then for each i we see that $N_\epsilon(x) \cap S \subset N_{\epsilon_i}(x) \cap S \subset E_i$. Since $N_\epsilon(x) \cap S$ is contained in each E_i, it is contained in the product $\bigcap E_i$. Hence $\bigcap E_i$ is open relative to S.

2.9. *If $\{E_\alpha\}$ is any collection of sets each contained in a set S and closed relative to S, $\bigcap E_\alpha$ is closed relative to S. If $E_1, \cdots, E_p$ is a finite collection of sets closed relative to S, then $\bigcup E_i$ is closed relative to S.*

Since each E_α is closed relative to S, each CE_α is open relative to S by 2.5, so $\bigcup CE_\alpha$ is open relative to S by 2.8, so $\bigcap E_\alpha = C(\bigcup CE_\alpha)$ is closed relative to S by 2.5. Similarly, since each E_i is closed relative to S, each CE_i is open relative to S, so $\bigcap CE_i$ is open relative to S by 2.8, so $\bigcup E_i = C(\bigcap CE_i)$ is closed relative to S.

2.10. Corollary. *The product of any number of closed sets is closed; so is the sum of a finite number of closed sets. The sum of any number of open sets is open; so is the product of a finite number of open sets.*

EXAMPLE. If to each x in $[0, 1]$ we assign the open interval $(x - 1, x + 1)$, then the sum of all these intervals is the open set $(-1, 2)$.

The product of infinitely many open sets may not be open, and the sum of infinitely many closed sets may not be closed. For example, for each n let the set E_n be $(-1/n, 1 + 1/n)$. These are open; but their product is the closed interval $[0, 1]$. Again, for each $n > 3$ let E_n be $[1/n, 1 - 1/n]$. These sets are closed; but their sum is the open interval $(0, 1)$.

A point set E is *bounded* if there is a finite ϵ such that E is entirely in the ϵ-neighborhood of the origin $0 \equiv (0, 0, \cdots, 0)$: $E \subset N_\epsilon(0)$. This is equivalent to requiring that there is an $M > 0$ such that $|x^{(i)}| < M$ for all x in E and $i = 1, \cdots q$. For if this last condition holds then $||x, 0|| = [(x^{(1)})^2 + \cdots + (x^{(q)})^2]^{\frac{1}{2}} \leqq [M^2 + \cdots + M^2]^{\frac{1}{2}} = M\sqrt{q}$, so x is in $N_\epsilon(0)$ if $\epsilon > M\sqrt{q}$. Conversely, if $E \subset N_\epsilon(0)$, then for every x in E we have $|x^{(i)}| \leqq ||x, 0|| < \epsilon$.

A fundamental theorem in the theory of functions of a real variable is the Bolzano-Weierstrass Theorem:

2.11. *If E is a bounded set containing infinitely many points, there is a point x_0 which is an accumulation point of E.*

The proof of this theorem can be found in almost any book on the theory of functions of a real variable. For $n = 1$ it is given in Courant's Differential and Integral Calculus (English edition), p. 58; this proof is easily extended to any value of n.

Let us say that $x_n \to x_0$ (or $\lim\limits_{n\to\infty} x_n = x_0$) if $\lim\limits_{n\to\infty} ||x_n, x_0|| = 0$; that is, if to each positive ϵ there corresponds an N such that $||x_n, x_0|| < \epsilon$ whenever $n > N$. We then have a simple corollary of the Bolzano-Weierstrass theorem, as follows.

2.12. *If $x_1, x_2, x_3, \cdots$ is an infinite sequence of points of R_q, and the set $\{x_1, x_2, \cdots\}$ is bounded, then there exists a subsequence x_{n_1}, $x_{n_2}, \cdots$ which converges to a limit x_0.*

Suppose first that some point occurs infinitely many times among the x_n. Call the point x_0, and let $x_{n_1}, x_{n_2}, \cdots$ be the terms of the sequence which coincide with x_0. Then clearly $x_{n_i} \to x_0$.

If no infinite repetitions occur, there must be infinitely many different points among the x_n. By 2.11, these have an accumulation point x_0. In $N_1(x_0)$ there are infinitely many x_n; choose one, and call it x_{n_1}. In $N_{\frac{1}{2}}(x_0)$ there are infinitely many x_n; some of these must have subscripts greater than n_1. Choose one such; this is x_{n_2}. In $N_{\frac{1}{3}}(x_0)$ there are infinitely many x_n; choose one, x_{n_3}, with subscript greater than n_2. We continue the process; by an obvious induction, we obtain a sequence $x_{n_1}, x_{n_2}, \cdots$ such that $n_1 < n_2 < n_3 < \cdots$ and $|| x_0, x_{n_i} || < 1/i$. This is the subsequence sought.

2.13. *Let E be a closed set. If $x_1, x_2, \cdots$ is a sequence of points of E converging to a limit x_0, then x_0 is in E.*

If x_0 is not in E, it is not in $\bar{E}$, by 2.4. By 2.2, for some positive ϵ the neighborhood $N_\epsilon(x_0)$ contains no point of E, hence contains none of the x_n. This is impossible if $x_n \to x_0$.

We now establish a very useful property of sequences of closed sets.

2.14. *Let $E_1, E_2, \cdots$ be a shrinking sequence of closed sets (that is, $E_1 \supset E_2 \supset E_3 \supset \cdots$), and let E_1 be bounded. Then either there is a k such that the sets $E_k, E_{k+1}, E_{k+2}, \cdots$ are all empty, or else there is a point x_0 contained in all the sets E_i, that is x_0 is in $\bigcap_1^\infty E_i$.*

If any one set E_k is empty, then from the hypothesis $E_k \supset E_{k+1} \supset E_{k+2} \supset \cdots$ we have all succeeding sets empty. Otherwise, from each E_n we choose a point x_n. The points $\{x_1, x_2, \cdots\}$ form a bounded set, for they are all in E_1, which is bounded by hypothesis. So by 2.12 there is a subsequence $x_{n_1}, x_{n_2}, \cdots$ which converges to a limit point x_0. Choose any integer n. Only a finite number of the subscripts n_i are less than n; say $n \leqq n_k < n_{k+1} < \cdots$. But if $n_i \geqq n$, then $E_{n_i} \subset E_n$, and in particular x_{n_i} is in E_n. Hence the sequence $x_{n_k}, x_{n_{k+1}}, \cdots$ is a sequence of points of the closed set E_n, and by 2.13 the point x_0 is in E_n. Since this is true for each n, the theorem is proved.

3. A class E of objects $\{P\}$ is *denumerable* if its elements can be put into one-to-one correspondence with the positive integers; that is, if its elements can be arranged in an infinite sequence $P_1, P_2, P_3, \cdots$, in such a way that every element of E occurs in just one place in the sequence. Thus the set of positive integers is denumerable, being automatically in one-to-one correspondence with itself. Likewise the set of numbers $1, \frac{1}{2}, \frac{1}{3}, \cdots 1/n, \cdots$ is denumerable.

The next three theorems state important properties of denumerable sets.

3.1. *Every subset of a denumerable set is finite or denumerable.*

Let E be the set consisting of $P_1, P_2, \cdots$, and let E^* be a subset. Beginning with P_1, we examine each P_i until we come to an element P_i

belonging to E^*. We re-name this element P_1^*. Continuing from this element, we proceed until we reach another which belongs to E^*; this we re-name P_2^*. By repeating the process indefinitely, each element of E^* will be reached after a finite number of steps. If the process terminates, E^* is finite. If not, the elements of E^* are exhibited in a sequence $P_1^*, P_2^*, \cdots$, and E^* is denumerable.

3.2. *The sum of a finite or denumerable collection of sets, each of which is finite or denumerable, is itself finite or denumerable.*

We consider first the case of a denumerable collection $E_1, E_2, \cdots$ of sets, each of which is denumerable. Let the elements of E_i be $P_{i,1}$, $P_{i,2}$, $P_{i,3}, \cdots$. We now form the sequence

$$(*) P_{1,1}; P_{2,1}, P_{1,2}; P_{3,1}, P_{2,2}, P_{1,3}; \cdots ; P_{n,1}, P_{n-2,3}, \cdots, P_{1,n}; \cdots .$$

Schematically we can exhibit this in the form

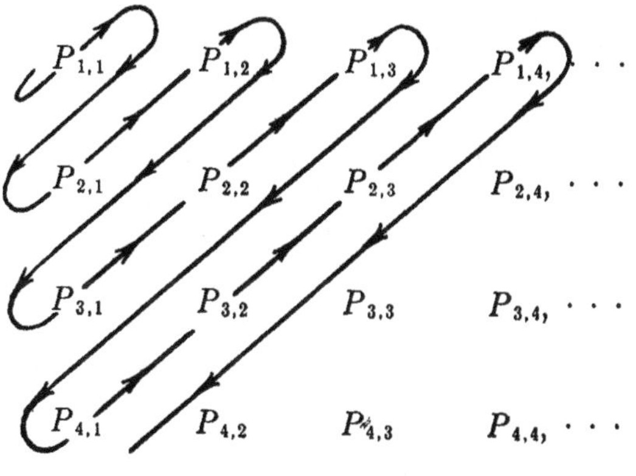

Clearly each element $P_{n,k}$ of $\bigcup E_i$ will be reached after a finite number of steps. However, it is possible that the same element P may have occurred in two different sequences, so that repetitions may occur in the sequence (*). We eliminate this by examining the elements of (*) in order, and removing any element which has previously occurred. Thus all the elements of $\bigcup E_i$ are exhibited without repetition in a sequence. This sequence can not be finite, for it contains all the elements of E_1, so by 3.1 $\bigcup E_i$ is denumerable.

This scheme can be applied with simple modifications to the cases of a finite set of denumerable sets and of a denumerable set of finite sets (the case of finite sets of finite sets is trivial). However, no additional proof is really needed, for we can fill out each finite sequence $P_{i,1}, P_{i,2}, \cdots P_{i,k}$ to an infinite sequence by repeating the last term infinitely many times, and if there are only a finite number of sets we repeat the last set infinitely many times. The extra terms thus introduced are later removed when we strike repetitions out of the sequence (*).

3.3. *Let $E_1, \cdots, E_k$ be sets each of which is finite or denumerable. Then the collection of ordered k-tuples $(x_1, \cdots, x_k)$ with x_1 in $E_1, \cdots,$ x_k in E_k, is finite or denumerable.*

We prove this first for $k = 2$. For each fixed x_1, the pairs (x_1, x_2) are in one-to-one correspondence with the elements x_2 of E_2; that is, to each x_1 corresponds a finite or denumerable collection of pairs (x_1, x_2). There are as many such collections of pairs as there are elements x_1 of E_1, that is, there are finitely or denumerably infinitely many collections. By 3.2, the total collection of pairs is finite or denumerable.

For general k, we proceed by induction. Suppose the theorem proved for $k - 1$; the $(k - 1)$-tuples $(x_1, \cdots, x_{k-1})$ form a finite or denumerable collection. The k-tuples $(x_1, \cdots, x_k)$ can be regarded as pairs $([x_1, \cdots, x_{k-1}], x_k)$ in which the first element is a $(k - 1)$-tuple and the second element is x_k. By the preceding paragraph, the collection of such pairs (i.e., of k-tuples $(x_1, \cdots, x_k)$) is finite or denumerable.

3.4. *The set of all rational numbers is denumerable.*

For each positive integer n, let E_n be the set consisting of the numbers $0, 1/n, -1/n, 2/n, -2/n, 3/n, -3/n, \cdots$. Each set E_n is denumerable, being exhibited as an infinite sequence without repetitions. Hence by 3.2 the set $\bigcup E_n$ is finite or denumerable. The set of all rational numbers is contained in $\bigcup E_n$, so by 3.1 it is finite or denumerable. It is not finite, since it contains all the integers; so it is denumerable.

EXAMPLE 1. In ν-dimensional space R_ν let us give the name "rational points" to the points $(x^{(1)}, \cdots, x^{(\nu)})$ such that each coordinate $x^{(i)}$ is a rational number. By 3.4 and 3.3, the rational points form a denumerable subset of R_ν.

EXAMPLE 2. In R_ν, the spheres $N_r(x_0)$ with rational points x_0 as centers and rational radii r form a denumerable set. For they are in one-to-one correspondence with the $(\nu + 1)$-tuples $(x_0^{(1)}, \cdots, x_0^{(\nu)}, r)$, where the first ν numbers are the (rational) coordinates of x_0 and r is the (positive rational) radius.

We can use the result in this last example to help in proving the *Borel theorem.* Given a collection of open sets $\{U\}$ and a set E, we say that the collection $\{U\}$ *covers* E if each point of E belongs to at least one set U of the collection: $E \subset \bigcup U$. The Borel theorem is the following.

3.5. *Let $\{U\}$ be a collection of open sets, and let E be a bounded closed set. If the collection $\{U\}$ covers E, it is possible to select a finite number $U_1, \cdots, U_n$ of sets of the collection $\{U\}$ in such a way that $\{U_1, \cdots, U_n\}$ covers E.*

We begin with the collection of all spheres with rational centers and radii; this is denumerable, by Example 2 above. A sphere of this collection will be *selected* if it is entirely within some single set of the collection $\{U\}$. By 3.1, the selected spheres form a finite or denumerable collection, and it is easily seen that the collection is not finite. We can then arrange the selected spheres in a sequence $S_1, S_2, \cdots$.

Our next step is to show that $\{S_1, S_2, \cdots\}$ covers E. Let x_0 be a point of E. By hypothesis x_0 is in a set U_0 of the collection $\{U\}$; since U_0 is open, there is a positive ϵ such that $N_\epsilon(x_0)$ is contained in U_0. There is a rational point $\bar{x}$ such that $\|\bar{x}, x_0\| < \epsilon/3$, and there is a rational number r such that $\epsilon/3 < r < 2\epsilon/3$. Clearly $N_r(\bar{x}) \subset N_\epsilon(x_0) \subset U_0$, since for each point x in $N_r(\bar{x})$ we have $\|x_0, x\| < \|x_0, \bar{x}\| + \|\bar{x}, x\| < \epsilon/3 + r < \epsilon$. Hence $N_r(\bar{x})$ is a selected sphere, and is one of the S_n. But x_0 is in this S_n, for $\|\bar{x}, x_0\| < \epsilon/3 < r$, so that x_0 is in $N_r(\bar{x})$. Thus $\{S_1, S_2, \cdots\}$ covers E.

Now we define

$$\begin{aligned} E_1 &= E, \\ E_2 &= E \cap CS_1, \\ &\cdots \\ E_n &= E \cap C[S_1 \cup \cdots \cup S_{n-1}], \\ &\cdots \end{aligned}$$

Each set $S_1 \cup \cdots \cup S_{n-1}$ is open by 2.10, hence E_n is closed by 2.6 and 2.10. It is easily seen that $E_1 \supset E_2 \supset E_3 \supset \cdots$. No point x_0 is in all the E_n; for then it would be in E but not in any S_n, which we have seen to be impossible. By 2.14, some set E_{k+1} is empty. That is, no point in E is in $C[S_1 \cup \cdots \cup S_k]$, whence every point of E is in $S_1 + \cdots + S_k$.

By the manner of selecting the S_i, each selected sphere S_i is completely contained in some set (we denote it by U_i) of the collection $\{U\}$. Hence

$$E \subset S_1 \cup \cdots \cup S_k \subset U_1 \cup \cdots \cup U_k,$$

establishing the theorem.

Our definition of a closed interval was such as to permit each $a^{(i)}$ to be equal to the corresponding $b^{(i)}$, in which case the interval $\{x \mid a^{(i)} \leqq x^{(i)} \leqq b^{(i)}\}$ shrinks to the single point $(a^{(1)}, \cdots, a^{(n)})$. *With* *the exception of this special case, every interval (open or closed) contains a non-denumerable infinity of points.* It is enough to prove this for closed intervals, since every open interval $\{x \mid a^{(i)} < x^{(i)} < b^{(i)}\}$ contains a closed interval $\{x \mid a^{(i)} + \epsilon \leqq x^{(i)} \leqq b^{(i)} - \epsilon\}$ (ϵ less than half the smallest $| b^{(i)} - a^{(i)} |$). It is enough to prove it for one dimension, for every closed interval $\{x \mid a^{(i)} \leqq x^{(i)} \leqq b^{(i)}\}$ contains the set of points $\{x \mid a^{(1)} \leqq x^{(1)} \leqq b^{(1)}, x^{(j)} = a^{(j)} \ (j = 2, \cdots, n)\}$. It is enough to prove it for the single interval [0, 1], for every interval $[a, b]$ can be mapped on this by the one-to-one mapping $x = (x' - a)/(b - a)$. Suppose now that all the numbers in [0, 1] were a denumerable set $a_1, a_2, \cdots$. Expand each a_i in a decimal fraction $a_i = .a_1^{(i)}a_2^{(i)}a_3^{(i)} \cdots$. For each n, define $b_n = 5$ if $a_n^{(n)} \neq 5$, $b_n = 6$ if $a_n^{(n)} = 5$. Thus $b_n \neq a_n^{(n)}$. Then the number $.b_1b_2 \cdots$ does not occur among the a_i; it is different from each a_n, since it has a different digit in the n-th decimal place.* Hence the assumption that *all* the numbers of the interval [0, 1] can be arranged in a sequence leads to a contradiction, and the set is not denumerable.

We now return to the subject of point sets, and prove that in a certain sense open sets are the next simplest sets after intervals. It is convenient to define a new term now. A collection of intervals is *non-overlapping* if no point which belongs to any interval of the collection is interior to any other interval of the collection. Thus in R_1 the intervals [0, 1] and [1, 2] are non-overlapping, although they are not disjoint because they have the point 1 in common.

3.6. *If G is an open set, it is the sum of a denumerable collection $I_1, I_2, \cdots$ of disjoint intervals I_n: $\{x \mid a_n^{(i)} \leqq x^{(i)} < b_n^{(i)}\}$. It is also the sum of a denumerable number of non-overlapping closed intervals $\bar{I}_n$.*

Consider first all intervals of the form $\{x \mid n_k^{(i)} \leqq x^{(i)} < n_k^{(i)} + 1\}$, where the numbers $n_k^{(i)}$ are integers. From these we select the ones whose closures are contained in G; these selected intervals we will call "the intervals selected at the first stage." Next consider all those intervals of the form $\{x \mid 2^{-1}n_k^{(i)} \leqq x^{(i)} < 2^{-1}(n_k^{(i)} + 1)\}$, the $n_k^{(i)}$ being integers. We select those whose closures are contained in G, but which are not contained in any interval selected at the first stage. These new intervals we will call "the intervals selected at the second stage." Proceeding by induction, for each integer p we consider the intervals $\{x \mid 2^{-p}n_k^{(i)} \leqq x^{(i)} < 2^{-p}(n_k^{(i)} + 1)\}$, the $n_k^{(i)}$ being integers. From these

* We must remember that numbers with different digits may be equal, for $.a_1a_2 \cdots a_n9999 \cdots = .a_1a_2 \cdots (a_n + 1)00000 \cdots$. But the decimal expression $.b_1b_2 \cdots$ can not contain any 9's or 0's, so this exceptional possibility for equality between $.b_1b_2 \cdots$ and $.a_1a_2 \cdots$ is ruled out.

we select intervals ("the intervals selected at the $(p + 1)$st stage") whose closures are contained in G but which are not contained in any interval selected at a previous stage.

All the intervals constructed at each stage form a denumerable collection by 3.3, so the intervals selected at each stage form a finite or denumerable set by 3.1, and the aggregate of all intervals selected at any stage whatever is finite or denumerable. We shall shortly see that it cannot be finite.

Let the collection of all selected intervals be denoted by $I_1, I_2, \cdots$. From the construction it is clear that they are disjoint; for two intervals of the type constructed are disjoint unless one contains the other. Moreover, each I_n is contained in G, so $\bigcup I_n$ is contained in G. It remains to show that G is contained in $\bigcup I_n$; that is, if x_0 is not in $\bigcup I_n$ it is not in G.

Let x_0 not belong to $\bigcup I_n$. For each positive integer, p, it is in some interval (call it I_p^*) of the p-th stage of subdivision, and this is not contained in any selected interval, since x_0 is not in $\bigcup I_n$. Since I_p^* is not in any interval selected at a previous stage and is not itself selected, its closure fails to be contained in G. That is, its closure must contain a point x_p of CG. By the definition of distance in §2 and the assumption made about the size of I_p^*, $\| x_p, x_0 \| \leqq [q \cdot (2^{-p+1})^2]^{\frac{1}{2}}$, which tends to 0 as p increases. Thus x_p tends to x_0 by the definition preceding 2.12. But CG is closed by 2.6, so by 2.13 the point x_0 is in CG. Therefore $\bigcup I_n$ contains G, and being also contained in G must be the same as G.

If we define $\bar{I}_n$ to be the closure of I_n, it is evident from the construction that no point of any $\bar{I}_n$ is interior to any other $\bar{I}_m$, and so the intervals $\bar{I}_n$ are non-overlapping. Since $\bar{I}_n$ contains I_n, we have $\bigcup \bar{I}_n \supset \bigcup I_n = G$. But the I_n were so chosen that their closures $\bar{I}_n$ were contained in G, so $\bigcup \bar{I}_n$ is contained in G, and must be identical with G.

Incidentally, this shows that the I_n cannot be finite in number. For suppose the I_n finite in number. Their sum contains a point for which $x^{(1)}$ has the least value on $\bigcup I_n$. This point cannot be interior to $\bigcup I_n$, and yet lies in the open set $G = \bigcup I_n$.

As a corollary, we have the following theorem.

3.7. *If G is open relative to a closed interval I, it is the sum of a finite or denumerable set of non-overlapping closed intervals $I_1, I_2, \cdots$.*

The set $I - G$ is closed by 2.5 and 2.7. So by 2.6 its complement $C(I - G)$ is open. But $C(I - G) = CI \cup G$. By 3.6, the open set $G \cup CI$ is the sum of denumerably many non-overlapping closed intervals I_n. Then

$$G = (G \cup CI) \cap I = (\bigcup I_n) \cap I = \bigcup (I_n \cap I).$$

From the sum in the last member we omit all sets $I_n \cap I$ which are empty; the rest form a finite or denumerable collection of non-overlapping closed intervals whose sum is G.

EXERCISE. If E is contained in S and is open relative to S, there is a sequence $I_1, I_2, \cdots$ of intervals I_n: $\{x \mid a_n^{(i)} \leqq x^{(i)} < b_n^{(i)}\}$ such that $E = \bigcup (S \cap I_n)$.

4. According to the definition of E. H. Moore, a function is a system consisting of

(a) a class $\mathfrak{X}$ of elements x;

(b) a class $\mathfrak{Y}$ of elements y;

(c) a law of correspondence by which to each element x of $\mathfrak{X}$ there is assigned exactly one element y of the class $\mathfrak{Y}$. (Such a function is sometimes also called a mapping of $\mathfrak{X}$ into $\mathfrak{Y}$.) The element y corresponding to the element x is denoted by $f(x)$, or $g(x)$, or some other similar symbol. Properly, the symbol $f(x)$ denotes the functional value y corresponding to the particular element x. However, it is common usage to use the symbol $f(x)$ also to denote the function itself. Thus $\sin x$ may mean either the function in which $\mathfrak{X}$ is the class of real numbers and $\mathfrak{Y}$ is also the class of real numbers, and to each x there is assigned the number $\sin x$; or else it may mean the particular number $\sin x$ which corresponds to some specific number x. However, this ambiguity does not often lead to confusion. The set of all values of $f(x)$ when x varies over $E \subset \mathfrak{X}$ will be denoted by $f(E)$. In symbols $f(E) = \{y \mid y = f(x), x \text{ in } E\}$. In particular, if E is the set $\{x \mid S\}$ of points for which statement S holds, the set

$$f(E) = f(\{x \mid S\})$$

will be alternatively denoted by $\{f(x) \mid S\}$. Thus, for example, if $f(x) = x^2$ we have $\{x^2 \mid -1 \leqq x \leqq 1\} = \{x \mid 0 \leqq x \leqq 1\}$. The set of all x such that $f(x)$ lies in E (the "inverse image of E") will be denoted by $f^{-1}(E)$; thus $f^{-1}(E) = \{x \mid f(x) \text{ in } E\}$. We write $f \leqq g$ if $f(x) \leqq g(x)$ for all x in $\mathfrak{X}$. Also $c \cdot f$ is the function whose value at the point x is $c \,.\, f(x)$, and so on (cf. the similar definitions for points in R_q at the beginning of §1).

We shall be chiefly (though not exclusively) concerned with functions in which the class $\mathfrak{X}$ is a point set in R_q. For the class $\mathfrak{Y}$ we shall always use a slight enlargement (which we denote by R^*) of the class of real numbers; $\mathfrak{Y}$ shall consist of all real numbers plus two elements which we denote by ∞ (or $+\infty$) and $-\infty$. We shall use the following calculation rules for these new symbols:

$$-\infty < a < \infty \text{ for every real number } a,$$
$$\infty \cdot a = a \cdot \infty = \infty \quad \text{if} \quad 0 < a \leqq \infty,$$
$$\infty \cdot a = a \cdot \infty = -\infty \quad \text{if} \quad -\infty \leqq a < 0,$$
$$(-\infty) \cdot a = a \cdot (-\infty) = -\infty \quad \text{if} \quad 0 < a \leqq \infty,$$
$$(-\infty) \cdot a = a \cdot (-\infty) = \infty \quad \text{if} \quad -\infty \leqq a < 0,$$
$$\frac{a}{\infty} = \frac{a}{(-\infty)} = 0 \quad \text{if} \quad a \text{ is real},$$
$$\infty + a = a + \infty = \infty \quad \text{if} \quad a > -\infty,$$
$$-\infty + a = a + (-\infty) = -\infty \quad \text{if} \quad a < \infty,$$
$$|\infty| = |-\infty| = \infty,$$
$$\infty \cdot 0 = 0 \cdot \infty = (-\infty) \cdot 0 = 0 \cdot (-\infty) = 0.$$

The last rule may seem strange, but functions with values $\pm\infty$ will usually occur in connection with limiting processes of such a nature that the rule will prove convenient.

4.1. *For all numbers a, b in R^* the sum $a + b$ is defined (finite or infinite) unless $a = \infty$ and $b = -\infty$ or $a = -\infty$ and $b = \infty$.*

A simple and frequently used result is

4.2. *If a, b, c, d are in R^*, and $a \geqq b$ and $c \geqq d$, then $a + c \geqq b + d$, provided that the sums are defined.*

This is clear if a, b, c, d are all finite. Otherwise either a or c is $+\infty$, in which case $a + c = \infty \geqq b + d$; or else b or d is $-\infty$, in which case $a + c \geqq -\infty = b + d$.

In order to avoid verbosity it is convenient to define neighborhoods of ∞ and $-\infty$. We define $N_\epsilon(\infty) = \{x \mid x > 1/\epsilon\}$ and $N_\epsilon(-\infty) = \{x \mid x < -1/\epsilon\}$. With this definition we see that

4.3. *If x, y, z are any numbers (finite or infinite) in R^*, then (a) if x is in $N_\epsilon(y)$, then $|x|$ is in $N_\epsilon(|y|)$; (b) if x is in $N_\alpha(y)$ and y is in $N_\beta(z)$, and α and β are $\leqq \frac{1}{2}$, then x is in $N_{\alpha+\beta}(z)$.*

To prove (*a*), if y is finite then $|x - y| < \epsilon$, so $||x| - |y|| \leqq |x - y| < \epsilon$. If $y = \infty$, then $x > 0$, so $|x| = x$ and $|y| = y$, and $|x|$ is in $N_\epsilon(|y|)$. If $y = -\infty$, then $x < -1/\epsilon$, so $|x| > 1/\epsilon$, and $|x|$ is in $N_\epsilon(|y|) = N_\epsilon(\infty)$.

To prove (*b*), we observe that it is obvious if any two of the numbers x, y, z are equal. Suppose therefore that they are different. If z is finite, then $|y - z| < \beta$, and y is finite; so $|x - y| < \infty$. Therefore $|x - z| \leqq |x - y| + |y - z| < \alpha + \beta$, and x is in $N_{\alpha+\beta}(z)$. If $z = \infty$, then $y > 1/\beta$, and $|x - y| < \alpha$. Therefore $x > (1/\beta) - \alpha$. A simple calculation shows that $(1/\beta) - \alpha > 1/(\alpha + \beta)$ if α and β are positive numbers not greater than $\frac{1}{2}$. Hence $x > 1/(\alpha + \beta)$, and x is in $N_{(\alpha+\beta)}(\infty)$. A like discussion applies to the case $z = -\infty$.

In accordance with this definition, ∞ is an accumulation point of a set E if E contains arbitrarily large numbers x; that is, if for every $\epsilon > 0$ there are real numbers x in E such that $x > 1/\epsilon$.

Suppose now that $f(x)$ is defined on a point set E in R_q or in R^*, and that x_0 is an accumulation point of E. Then as usual we say that $f(x)$ approaches the number k as x tends to x_0 (or $f(x) \to k$ as $x \to x_0$) if for every $\epsilon > 0$ there is a $\delta > 0$ such that $f(x)$ is in $N_\epsilon(k)$ for all points x of E in $N_\delta(x_0)$, with the possible exception of x_0 itself. More compactly,

4.4. *Let $f(x)$ be defined on a set E, and let x_0 be a point of E'. Then $\lim_{x \to x_0} f(x) = k$ is defined to mean that to every $\epsilon > 0$ there corresponds a $\delta > 0$ such that $f(x)$ is in $N_\epsilon(k)$ for all x in* $N_\delta(x_0)E - (x_0)$.*

EXAMPLE. Let $f(x) = x^{-2}$ for all $x \neq 0$. Then $\lim_{x \to \infty} f(x) = 0$; for $f(x)$ is in $N_\epsilon(0)$ (i.e., $|f(x)| < \epsilon$) whenever x is in the $\sqrt{\epsilon}$-neighborhood of ∞ (i.e., $x > 1/\sqrt{\epsilon}$). Likewise, $\lim_{x \to 0} f(x) = \infty$.

In particular, if E is the set of positive integers, it is customary to denote the independent variable by n instead of x and to write f_n instead of $f(n)$. For this case definition 4.4 takes the following form. *In order that*

$$\lim_{n \to \infty} f_n = k$$

it is necessary and sufficient that to every $\epsilon > 0$ there shall correspond an integer n_ϵ such that f_n is in $N_\epsilon(k)$ whenever $n > n_\epsilon$. If we write $\delta = 1/n_\epsilon$, this differs from 4.4 only in that δ is required to be the reciprocal of an integer; and this restriction on δ is easily seen to be immaterial. Thus our definition includes the familiar definition of the limit of a sequence.

Both 4.4 and the re-phrasing for sequences can be applied at once to functions in which the functional values $f(x)$ and the limit k are in the q-dimensional space R_q. Thus, for instance, we have incidentally defined limits such as

$$\lim_{n \to \infty} x_n = x_0,$$

where x_n and x_0 are in R_q; the definition reduces exactly to that in §2 (the paragraph preceding 2.12).

* In accordance with §1, $N_\delta(x_0)E - (x_0)$ is the set of all points x which belong both to E and to $N_\delta(x_0)$, but not to the set (x_0) consisting of the single point x_0; i.e. $N_\delta(x_0)E - (x_0) = \{x \mid x \text{ in } N_\delta(x_0),\ x \text{ in } E,\ x \neq x_0\}$.

We can use the notion of convergent sequences to give another formulation to 4.4 which is of frequent usefulness.

4.5. *Let $f(x)$ be defined on E and let x_0 be in E'. Then* $\lim_{x \to x_0} f(x)$ *exists and has the value k if and only if for every sequence $x_1, x_2, \cdots$ of points of E distinct from x_0 and tending to x_0 the limit* $\lim_{n \to \infty} f(x_n)$ *exists and has the value k.*

Suppose $\lim_{x \to x_0} f(x) = k$, and let ϵ be a positive number. There is a $\delta > 0$ such that $f(x)$ is in $N_\epsilon(k)$ if x is in $N_\delta(x_0)E - (x_0)$. Let $x_1, x_2, \cdots$ be a sequence of the type described. Since $x_n \to x_0$, there is a $\gamma > 0$ such that x_n is in $N_\delta(x_0)$ if n is in $N_\gamma(\infty)$; that is, if $n > 1/\gamma$. But $x_n \neq x_0$ and x_n is in E, so for $n > 1/\gamma$ the point x_n is in $N_\delta(x_0)E - (x_0)$, and $f(x_n)$ is in $N_\epsilon(k)$. Therefore $\lim_{n \to \infty} f(x_n) = k$.

Suppose it is not true that $\lim_{x \to x_0} f(x) = k$. Then there is an $\epsilon > 0$ such that no δ serves; that is, in every set $N_\delta(x_0)E - (x_0)$, no matter what δ is, there is a point x for which $f(x)$ is not in $N_\epsilon(k)$. Let δ take on the values $1, \frac{1}{2}, \frac{1}{3}, \cdots$. For each of these values of δ we choose an x_n in $N_\delta(x_0)E - (x_0)$ such that $f(x_n)$ is not in $N_\epsilon(k)$. Then $x_n \to x_0$, $x_n \neq x_0$, and $f(x_n)$ does not converge to k.

5. Let S be a set (*not* the empty set) of real numbers. A number* M is an *upper bound* for the numbers x of S in case $x \leqq M$ for every x in S. M is the *supremum* or *least upper bound* (abbreviated sup) of the numbers of S if: (*a*) *for every x in S the inequality $x \leqq M$ holds;* (*b*) *for every number $m < M$, there exists a number x in S such that $x > m$.*

Property (*a*) asserts that M is an upper bound, property (*b*) that no smaller number is an upper bound. Analogously, N is a *lower bound* for the numbers S in case $x \geqq N$ for all x in S; and N is the *infimum* or *greatest lower bound* (abbreviated inf) if (*a*) *$x \geqq N$ for every x in S;* (*b*) *for every number $n > N$, there exists a number x in S such that $x < n$.*†

A fundamental property, "*Dedekind continuity,*" of the system of real numbers is that *every non-empty set of numbers which has a* (*finite*) *upper bound has a* (*finite*) sup. By change of sign, a similar statement holds for lower bounds. Since we allow ∞ and $-\infty$ as numbers, this becomes: *Every non-empty set of numbers in R^* has a* sup *and an* inf.

* We remember that ∞ and $-\infty$ are allowed as numbers.

† Many authors use the notation least upper bound and greatest lower bound (abbreviated l.u.b. and g.l.b.) for supremum and infimum. If the collection S is finite the terms maximum and minimum (abbreviated max and min) are sometimes used.

From the definitions it is clear that if M_0 is an upper bound for S, then $M_0 \geqq \sup S$; likewise, if m_0 is a lower bound for S, then $m_0 \leqq \inf S$.

A simple but important property of these bounds is

5.1. *Let S, S^* be two non-empty sets of numbers in R^* and $S^* \subset S$. Put $m = \inf S$, $m^* = \inf S^*$, $M^* = \sup S^*$, $M = \sup S$. Then $m \leqq m^* \leqq M^* \leqq M$.*

(Observe that this contains the statement that for the arbitrary non-empty set S the relation $M \geqq m$ holds.) We prove first $M^* \geqq m^*$. Let x be any number in S^*; then $M^* \geqq x$ and $x \geqq m^*$. For all x in S, and a fortiori for all x in S^*, the inequalities $x \leqq M$ and $x \geqq m$ hold. Hence M, m are upper and lower bounds for the x in S^*. Therefore the *least* upper bound M^* can not exceed M, and likewise $m^* \geqq m$.

EXAMPLES. If S consists of all rational numbers x such that $1 < x < 3$, then $\inf S$ and $\sup S$ are 1 and 3 respectively. If S is the class of all positive integers, then $\inf S$ is 1 and $\sup S$ is ∞. If S consists of all integers, then $\inf S$ is $-\infty$ and $\sup S$ is ∞. If S consists of the single number ∞, then $\inf S$ and $\sup S$ are both ∞.

5.2. *If of two numbers a, b we know that $a \leqq x$ for all numbers $x > b$, then $a \leqq b$. Likewise, if $a \geqq x$ for all $x < b$, then $a \geqq b$.*

We prove the first statement; the second is similar. This is really a special case of 5.1; for the set S^* of numbers $x > b$ is contained in the set S of numbers $x \geqq a$, so $\inf S^* = b$ is $\geqq \inf S = a$. However, we need not refer to 5.1. For if the theorem were false and $a > b$, there would be an h such that $a > h > b$. Then $h > b$ but $a > h$, contrary to hypothesis.

Let $f(x)$ be a function defined on a set E. By a lower (upper) bound for f on E we shall mean a lower (upper) bound of the set $f(E)$. We shall also sometimes write $\sup f$ (on E) or $\sup f(x)$ for $\sup f(E) = \sup \{f(x) \mid x \text{ in } E\}$ and $\inf f$ for $\inf f(E)$. When there are several variables involved, for instance when we are studying a set of functions $f_1(x), f_2(x), \cdots$ we shall occasionally write a subscript on "sup" and "inf" to indicate the collection whose bounds we are seeking. Thus $\sup_i \{f_i(x)\}$ denotes the supremum of the numbers $f_1(x), f_2(x), \cdots$ at a particular point x, while $\sup_x f_i(x)$ denotes $\sup \{f_i(x) \mid x \text{ in } E\}$ (for a fixed i). The symbol $\sup \{f_1(x), f_2(x), \cdots, f_n(x)\}$ denotes that function whose value at each x is the greatest of the numbers $f_1(x), \cdots, f_n(x)$.

5.3. *If $f(x)$ is defined on E and $E^* \subset E$, then* $\inf f(E) \leqq \inf (fE^*) \leqq \sup f(E^*) \leqq \sup f(E)$.

Since $f(E^*) \subset f(E)$ our theorem follows at once from 5.1.

5.4. *If f and g are both defined on E and $f \leqq g$, then* $\sup f \leqq \sup g$ *and* $\inf f \leqq \inf g$.

For all x in E we have

$$\inf f \leqq f(x) \leqq g(x) \leqq \sup g.$$

So $\inf f$ is a lower bound for g and hence $\inf f \leqq \inf g$. Similarly $\sup g$ is an upper bound for f and hence $\sup g \geqq \sup f$.

5.5. *If $f(x)$ is defined on E, and $0 \leqq c < \infty$, then:* (*a*) $\sup cf = c \sup f$; (*b*) $\inf cf = c \inf f$; *while if* $-\infty < c < 0$, *then* (*c*) $\sup cf = c \inf f$; (*d*) $\inf cf = c \sup f$.

For $c = 0$ both (*a*) and (*b*) reduce to $0 = 0$. If $c < 0$, then for all x we have $f(x) \leqq \sup f$, so $cf(x) \geqq c \sup f$. That is, $c \sup f$ is a lower bound for $cf(x)$. If $h > c \sup f$, then $h/c < \sup f$. Therefore there is an x in E such that $f(x) > h/c$, whence $cf(x) < h$. Hence if $h > c \sup f$, then h is not a lower bound for cf, and so (*d*) is proved. To prove (*c*) we use (*d*):

$$c \inf f = c \inf c^{-1}[cf(x)] = c \cdot c.^{-1} \sup cf = \sup cf.$$

Suppose $c > 0$. We write $c = (-1)(-c)$; then by (*c*) and (*d*)

$$\sup cf = \sup [(-c)(-1)f] = -c \inf (-1)f = (-c)(-1) \sup f,$$
$$\inf cf = \inf [(-c)(-1)f] = -c \sup (-1)f = (-c)(-1) \inf f,$$

establishing (*a*) and (*b*).

5.6. *If the functions $f_1(x)$ and $f_2(x)$ and their sum $f_1(x) + f_2(x)$ are defined* on a set E, then*

$$\inf [f_1 + f_2] \geqq \inf f_1 + \inf f_2,$$
$$\sup [f_1 + f_2] \leqq \sup f_1 + \sup f_2,$$

*provided that the right members of these inequalities are defined.**

For each x in E we have

$$f_1(x) \leqq \sup f_1, \qquad f_2(x) \leqq \sup f_2;$$

hence by 4.2

$$f_1(x) + f_2(x) \leqq \sup f_1 + \sup f_2.$$

Thus $\sup f_1 + \sup f_2$ is *an* upper bound for $f_1(x) + f_2(x)$, and can not be less than the least upper bound. This establishes the second

* Cf. 4.1.

inequality. The first is obtained from the second by replacing f_1 and f_2 by $-f_1$ and $-f_2$ and using 5.5.

EXAMPLE. Let E be any (non-empty) set. For each x the distance $||x, \bar{x}||$ is a function defined on E. Define $d(x)$ (the distance from x to the set E) to be $\inf\{||x, \bar{x}|| \mid \bar{x} \text{ in } E\}$. Since 0 is a lower bound for $||x, \bar{x}||$, we have at once $d(x) \geqq 0$. This function is continuous. For if x_1, x_2 are any two points, we have by §1

$$||x_1, \bar{x}|| \leqq ||x_1, x_2|| + ||x_2, \bar{x}_,||$$

for all $\bar{x}$ in E. Then by 5.4 $d(x_1) \leqq ||x_1, x_2|| + d(x_2)$. But we may interchange x_1 and x_2 and thus get $d(x_2) \leqq ||x_1, x_2|| + d(x_2)$; that is, $|d(x_1) - d(x_2)| \leqq ||x_1, x_2||$. So $|d(x)_1 - d(x_2)| < \epsilon$ whenever $||x_1, x_2|| < \epsilon$. This proves in fact that $d(x)$ is uniformly continuous.*

This function $d(x)$ has the value 0 if and only if x is in $\bar{E}$. If x is in $\bar{E}$, for every $\epsilon > 0$ there is a point $\bar{x}$ of E in $N_\epsilon(x)$, by 2.2. Then $||x, \bar{x}|| < \epsilon$, so $d(x) = \inf ||x, \bar{x}|| < \epsilon$. Since this is true for every $\epsilon > 0$, by 5.2 we have $d(x) \leqq 0$. But $d(x)$ is not negative, so $d(x) = 0$. Conversely, if $d(x) = 0$, for every $\epsilon > 0$ there is a point $\bar{x}$ of E such that $||x, \bar{x}|| < \epsilon$. That is, for every positive ϵ the neighborhood $N_\epsilon(x)$ contains a point of E, so by 2.2 the point x is in $\bar{E}$.

6. One of the chief concerns of analysis is the behavior of a function at the points which are near a given point. If we are interested in the upper bound of a function from this point of view, we shall first of all not be particularly concerned with the value of the function $f(x)$ *at* the given point x_0, and second we shall not care about the supremum of $f(x)$ for distant points. Suppose to be specific that $f(x)$ is defined on a set E, and that x_0 is an accumulation point of E. Whether or not x_0 is in E is immaterial; if it is, we simply disregard the number $f(x_0)$. For each ϵ we find $\sup\{f(x) \mid x \text{ in } N_\epsilon(x_0) \cap E, x \neq x_0\}$; we denote this by $M_\epsilon(f; x_0)$. If we choose any one ϵ, this still is not adequate to describe the behavior of f near x_0, for points at a distance greater than $\epsilon/2$ may enter into the definition of $M_\epsilon(f; x_0)$. But if we reduce ϵ, we observe that the set of points x in $N_\epsilon(x_0)E - (x_0)$ is reduced, so by (5.3) $M_\epsilon(f; x_0)$ is reduced. Since we are interested only in $M_\epsilon(f; x_0)$ for small ϵ, it follows that only the smaller values of $M_\epsilon(f; x_0)$ are of importance. This suggests to us that the number which will be the local substitute for the $\sup f$ will be $\inf_\epsilon M_\epsilon(f; x_0)$. This number will be named the *upper limit* of $f(x)$ as x tends to x_0; it is denoted by $\limsup_{x \to x_0} f(x)$. Summing up, (and defining the lower limit analogously):

* Cf. Courant, Differential and Integral Calculus, vol. I, p. 51.

6.1. *If $f(x)$ is defined on a set E and x_0 is an accumulation point of E, then*

$$\limsup_{x\to x_0} f(x) \equiv \inf \{M_\epsilon(f; x_0) \mid \epsilon > 0\},$$

where

$$M_\epsilon(f; x_0) \equiv \sup f(N_\epsilon(x_0) \cap E - (x_0));$$

and

$$\liminf_{x\to x_0} f(x) \equiv \sup \{m_\epsilon(f; x_0) \mid \epsilon > 0\},$$

where

$$m_\epsilon(f; x_0) \equiv \inf f(N_\epsilon(x_0) \cap E - (x_0)).$$

These limits are always defined (finite, $+\infty$ or $-\infty$), but they are not always equal.

EXAMPLE. Let $f(0) = 2, f(x) = 1$ if x is rational and $\neq 0, f(x) = 0$ if x is irrational. Take $x_0 = 0$. Then $M_\epsilon(f; 0) = 1$ and $m_\epsilon(f; 0) = 0$ for all $\epsilon > 0$, and so $\limsup_{x\to x_0} f(x) = 1$, $\liminf_{x\to x_0} f(x) = 0$.

Since we frequently have to use sequences of numbers, we shall now observe the special form of these definitions which applies to sequences $a_1, a_2, \cdots$ of numbers. A sequence is a function in which the independent variable ranges over the set E of the positive integers, and we are interested in the behavior of the function as $x \to \infty$. So in 6.1 we set $x_0 = \infty$. Then $N_\epsilon(x_0)E$ consists of all integers greater than $1/\epsilon$, so $M_\epsilon(f; x_0) \equiv \sup \{a_i \mid i > 1/\epsilon\}$, $m_\epsilon(f; x_0) \equiv \inf \{a_i \mid i > 1/\epsilon\}$. Hence

$$\limsup_{i\to\infty} a_i = \inf_\epsilon \sup \{a_i \mid i > 1/\epsilon\}.$$

It is easily seen that this is the same as

$$\limsup_{i\to\infty} a_i = \inf_n \sup \{a_i \mid i \geqq n\}.$$

Likewise,

$$\liminf_{i\to\infty} a_i = \sup_n \inf \{a_i \mid i \geqq n\}.$$

Before we start to prove any theorems, we shall collect several obvious consequences of the definitions:

6.2. *Let $f(x)$ be defined on a set E, and let x_0 be an accumulation point of E.*

(*a*) *If $h > \limsup_{x\to x_0} f(x)$, there exists an $\epsilon > 0$ such that for all positive numbers $\alpha \leqq \epsilon$ the inequalities*

$$\limsup_{x\to x_0} f(x) \leqq M_\alpha(f; x_0) \leqq M_\epsilon(f; x_0) < h$$

hold.

(b) *If* $h < \liminf_{x \to x_0} f(x)$, *there exists an* $\epsilon > 0$ *such that for all positive numbers* $\delta \leqq \epsilon$ *the inequalities*

$$\liminf_{x \to x_0} f(x) \geqq m_\delta(f; x_0) \geqq m_\epsilon(f; x_0) > h$$

hold.

(c) *If* $h < M_\epsilon(f; x_0)$, *there exists a point* x *in* $N_\epsilon(x_0)E - (x_0)$ *such that* $h < f(x) \leqq M_\epsilon(f; x_0)$.

(d) *If* $h > m_\epsilon(f; x_0)$, *there exists a point* x *in* $N_\epsilon(x_0)E - (x_0)$ *such that* $m_\epsilon(f; x_0) \leqq f(x) < h$.

To prove (a), we observe that by 6.1 the number h is greater than the inf of $M_\epsilon(f; x_0)$ for all $\epsilon > 0$, so by the definition of inf there is an $\epsilon > 0$ such that $M_\epsilon(f; x_0) < h$. If $\alpha \leqq \epsilon$, then $N_\alpha(x_0)E - (x_0)$ is contained in $N_\epsilon(x_0)E - (x_0)$. So the sup of $f(x)$ on the former set is not greater than its sup on the latter set, by 5.3. Hence $M_\alpha(f, x_0) \leqq M_\epsilon(f; x_0)$. Finally, by 6.1 the upper limit of $f(x)$ as $x \to x_0$ is the inf of all $M_\epsilon(f; x_0)$, so it is not greater than any one of them. This proves (a). We can prove (b) analogously, or by replacing $f(x)$ by $-f(x)$ and using 5.5.

Statements (c) and (d) are immediate consequences of the definitions of $M_\epsilon(f; x_0)$ and $m_\epsilon(f; x_0)$.

From 5.5 we readily obtain

6.3. *Let* $f(x)$ *be defined on* E, *and let* x_0 *be an accumulation point of* E. *If* $0 \leqq c < \infty$, *then*

(a) $\limsup_{x \to x_0} cf(x) = c \limsup_{x \to x_0} f(x)$,

(b) $\liminf_{x \to x_0} cf(x) = c \liminf_{x \to x_0} f(x)$;

while if $-\infty < c < 0$, *then*

(c) $\limsup_{x \to x_0} cf(x) = c \liminf_{x \to x_0} f(x)$,

(d) $\liminf_{x \to x_0} cf(x) = c \limsup_{x \to x_0} f(x)$.

If $c = 0$, (a) and (b) reduce to $0 = 0$. If $c < 0$, by 5.5 we have for every $\epsilon > 0$

$$\begin{aligned} m_\epsilon(cf; x_0) &= \inf \{cf(x) \mid x \text{ in } N_\epsilon(x_0)E - (x_0)\} \\ &= c \cdot \sup \{f(x) \mid x \text{ in } N_\epsilon(x_0)E - (x_0)\} = cM_\epsilon(f; x_0), \end{aligned}$$

and again by 5.5

$$\begin{aligned} \liminf_{x \to x_0} cf(x) &= \sup m_\epsilon(cf; x_0) = \sup cM_\epsilon(f; x_0) = c \inf M_\epsilon(f; x_0) \\ &= c \limsup_{x \to x_0} f(x). \end{aligned}$$

Hence (*d*) is established. By (*d*),

$$c \liminf_{x\to x_0} f(x) = c \liminf_{x\to x_0} c^{-1}\cdot cf(x) = c\cdot c^{-1}\cdot \limsup_{x\to x_0} cf(x),$$

which is (*c*).

If $c > 0$, then by (*c*) and (*d*)

$$\limsup_{x\to x_0} cf(x) = \limsup_{x\to x_0} (-1)(-c)f(x) = (-1) \liminf_{x\to x_0} (-c)f(x)$$
$$= c \limsup_{x\to x_0} f(x),$$

and (*a*) is established. Interchanging sup and inf in the proof of (*a*) gives the proof of (*b*).

The next four theorems are designed to set forth the relationship between upper and lower limits as defined in 6.1 and the (unique) limit as defined in 4.4. The first of these, Theorem 6.4, is the analogue of the sequential test for convergence established in 4.5. Theorem 6.6 shows that the equality of the upper and the lower limit is a necessary and sufficient condition for the existence of a limit in the ordinary sense.

6.4. *Let $f(x)$ be defined on a set E, and let x_0 be an accumulation point of E. If $h = \limsup_{x\to x_0} f(x)$ or if $h = \liminf_{x\to x_0} f(x)$, then there exists a sequence $x_1, x_2, \cdots$ of points of E distinct from x_0 and tending to x_0 such that $\lim_{n\to\infty} f(x_n) = h$. But if $h > \limsup_{x\to x_0} f(x)$ or if $h < \liminf_{x\to x_0} f(x)$, no such sequence $x_1, x_2, \cdots$ can exist.*

We shall prove the statements about the upper limit; the statements about the lower limit can be similarly proved, or they can be obtained by use of 6.3.

Suppose that h is equal to the upper limit of $f(x)$ as $x \to x_0$. We first show that to every positive number n there corresponds a positive number $\alpha \leqq 1/n$ such that $M_\alpha(f; x_0)$ is in the neighborhood $N_{1/n}(h)$. If $h = \infty$, this is evident; for then $M_\alpha(f; h) \geqq \limsup_{x\to x_0} f(x) = h = \infty$, so $M_\alpha(f; x_0) = h$, and is in $N_{1/n}(h)$ no matter what n and α we choose. If $h < \infty$, the neighborhood $N_{1/n}(h)$ contains a number $k > h$. Since $h = \inf_\epsilon M_\epsilon(f; x_0)$, by 6.2 there is a positive number $\alpha < 1/n$ such that

$$h \leqq M_\alpha(f; x_0) < k.$$

Therefore $M_\alpha(f; x_0)$ is between two numbers of $N_{1/n}(h)$, and must itself belong to $N_{1/n}(h)$.

Next we show that there is a point x_n in $N_\alpha(x_0)E - (x_0)$ for which $f(x_n)$ is in $N_{1/n}[M_\alpha(f; x_0)]$. If $M_\alpha(f; x_0) = -\infty$, this is evident; for then $f(x) \leqq M_\alpha(f, x_0) = -\infty$ for all x in $N_\alpha(x_0)E - x_0$. Thus if x_n

is any point of $N_\alpha(x_0)E - x_0$ we have $f(x_n) = -\infty$, which is in $N_{1/n}(-\infty)$. If $M_\alpha(f; x_0) > -\infty$, there are numbers $k < M_\alpha(f; x_0)$ in $N_{1/n}[M_\alpha(f; x_0)]$. By 6.2(c), there is then a point x_n in $N_\alpha(x_0)E - (x_0)$ for which $k < f(x_n) \leqq M_\alpha(f; x_0)$. Thus $f(x_n)$ is between two numbers of $N_{1/n}[M_\alpha(f; x_0)]$, and is therefore in that neighborhood.

Summing up, x_n is in $N_\alpha(x_0)E - (x_0)$. So $x_n \neq x_0$, and $\| x_n, x_0 \| < \alpha \leqq 1/n$, whence $x_n \to x_0$. Moreover $f(x_n)$ is in $N_{1/n}[M_\alpha(f; x_0)]$ and $M_\alpha(f; x_0)$ is in $N_{1/n}(h)$, so by 4.3b (if $n \geqq 2$) $f(x_n)$ is in $N_{2/n}(h)$. Therefore $f(x_n) \to h$.

Thus we have shown that it is always possible to pick a sequence $x_1, x_2, \cdots$ of points of E distinct from x_0, tending to x_0, and such that $f(x_n) \to \limsup_{x \to x_0} f(x)$. Suppose now that $h > \limsup_{x \to x_0} f(x)$; we must show that no such sequence exists. Let k be a number such that

$$h > k > \limsup_{x \to x_0} f(x).$$

If there is a sequence of the type specified with $f(x_n)$ tending to h, then for all but a finite number of these x_n we have $f(x_n) > k$. For every positive ϵ the neighborhood $N_\epsilon(x_0)E - (x_0)$ contains infinitely many x_n, so it contains some x_n with $f(x_n) > k$. Hence $M_\epsilon(f; x_0) = \sup \{f(x) \mid x \text{ in } N_\epsilon(x_0)E - (x_0)\}$ is greater than k for all ϵ, and

$$\limsup_{x \to x_0} f(x) = \inf_\epsilon M_\epsilon(f; x_0) \geqq k.$$

But k was a number greater than the upper limit of $f(x)$ as $x \to x_0$. This contradiction proves that no sequence $\{x_n\}$ with $x_n \to x_0$ and $x_n \neq x_0$ and $f(x_n) \to h > \limsup_{x \to x_0} f(x)$ can exist.

This theorem justifies the names "upper limit" and "lower limit"; for the upper and lower limits of $f(x)$ as $x \to x_0$ are respectively the greatest and the least numbers which are limits of sequences $f(x_n)$, where x_n is in E, $x_n \neq x_0$ and $x_n \to x_0$.

A direct consequence of 6.4 is

6.5. Corollary. *Let $f(x)$ be defined on E, and let x_0 be an accumulation point of E. Then*

$$\limsup_{x \to x_0} f(x) \geqq \liminf_{x \to x_0} f(x).$$

Let h be the upper limit of $f(x)$ as $x \to x_0$. By 6.4, there is a sequence $x_1, x_2, \cdots$ such that x_n is in E, $x_n \neq x_0$, $x_n \to x_0$, and $f(x_n) \to h$. Again by 6.4, this makes it impossible that $h < \liminf_{x \to x_0} f(x_0)$, which establishes our theorem.

6.6. *Let the function $f(x)$ be defined on a set E, and let x_0 be an accumulation point of E. Then the limit $\lim_{x \to x_0} f(x)$ exists and has the value k if and only if*

$$\limsup_{x \to x_0} f(x) = \liminf_{x \to x_0} f(x) = k.$$

Suppose that the limit of $f(x)$ as $x \to x_0$ exists and is equal to k. By 4.5, if $x_1, x_2, \cdots$ is any sequence of points of E distinct from x_0 and tending to x_0, then $f(x_n) \to k$. But by 6.4, the points $x_1, x_2, \cdots$ can be so chosen that

$$\lim_{n \to \infty} f(x_n) = \limsup_{x \to x_0} f(x).$$

This is only possible if the upper limit of $f(x)$ is k. Replacing sup by inf, we find that $\liminf_{x \to x_0} f(x)$ must also be equal to k.

Conversely, suppose that the upper and lower limits both have the value k. Let ϵ be an arbitrary positive number. If k is finite, by 6.2 there are positive numbers α, β such that

$$k - \epsilon < m_\alpha(f; x_0) \leqq k \leqq M_\beta(f; x_0) < k + \epsilon.$$

We define γ to be the smaller of α and β; then

$$k - \epsilon < m_\gamma(f; x_0) \leqq M_\gamma(f; x_0) < k + \epsilon.$$

By definition of m_γ and M_γ, this implies that $f(x)$ is in $N_\epsilon(k)$ whenever x is in $N_\gamma(x_0)E - (x_0)$. Hence in this case $\lim_{x \to x_0} f(x)$ exists and is equal to k. If $k = +\infty$, by 6.2 there is a positive number α such that $m_\alpha(f; x_0) > 1/\epsilon$. That is, $f(x)$ is in $N_\epsilon(\infty)$ whenever x is in $N_\alpha(x_0)E - (x_0)$, so in this case also the limit $\lim_{x \to x_0} f(x)$ exists and has the value $k (= \infty)$. The case $k = -\infty$ can be discussed in a like manner.

An immediate consequence of 6.6 which happens to be in a form convenient for applications is the following.

6.7. *Let $f(x)$ be defined on E; let x_0 be an accumulation point of E, and let k be a number in R^*. In order that $\lim_{x \to x_0} f(x)$ shall exist and have the value k, it is necessary and sufficient that the inequalities*

$$\liminf_{x \to x_0} f(x) \geqq k \quad \text{and} \quad \limsup_{x \to x_0} f(x) \leqq k$$

both hold.

Suppose that both inequalities hold. By 6.5

$$\liminf_{x \to x_0} f(x) \geqq k \geqq \limsup_{x \to x_0} f(x) \geqq \liminf_{x \to x_0} f(x).$$

Hence both the upper limit and the lower limit of $f(x)$ as $x \to x_0$ have the value k; so by 6.6, the limit exists and is equal to k. The converse is obvious from 6.6.

An important special class of functions for which we can be sure that limits exist (finite, $+\infty$ or $-\infty$) is the class of monotonic functions. A function $f(x)$ of a single real variable x, defined on a set E, is *monotonic increasing* if $f(x_2) \geqq f(x_1)$ whenever $x_2 \geqq x_1$ and both x_1 and x_2 are in E; it is *monotonic decreasing* if $f(x_2) \leqq f(x_1)$ whenever $x_2 \geqq x_1$ and both x_1 and x_2 are in E. Then

6.8. *If $f(x)$ is defined and monotonic increasing on the $\{x \mid a < x < b\}$, then*

$$\lim_{x\to a} f(x) = \inf f, \qquad \lim_{x\to b} f(x) = \sup f.$$

If $f(x)$ is defined and monotonic decreasing on $\{x \mid a < x < b\}$, then

$$\lim_{x\to a} f(x) = \sup f, \qquad \lim_{x\to b} f(x) = \inf f.$$

(Observe that a and b are permitted to have the values $-\infty$, ∞ respectively.) We prove the first statement; the second follows from the first if we replace f by $-f$. Also we prove only that $\lim_{x\to a} f(x) = \inf f$; the statement about the limit as $x \to b$ can be proved analogously, or else can be derived by replacing f by $-f(-x)$.

If we give ϵ any particular value, say $\epsilon = 1$, by 5.1 we have

$$m_1(f; a) = \inf f(N_1(a) - (a)) \geq \inf f.$$

Hence

$$(A) \qquad \liminf_{x\to a} f(x) = \sup m_\epsilon(f; a) \geqq m_1(f; a) \geqq \inf f.$$

On the other hand, let h be any number greater than $\inf f$. For some number $\bar{x}$ in $a < \bar{x} < b$ we have $f(\bar{x}) < h$, by the definition of inf. Whether a is finite or $-\infty$, we see that for sufficiently small ϵ the neighborhood $N_\epsilon(a)$ is entirely to the left of $\bar{x}$; that is, $x < \bar{x}$ for all x in $N_\epsilon(a)$. For all such x we have $f(x) \leqq f(\bar{x}) < h$, so

$$M_\epsilon(f; a) \leqq h.$$

It follows that

$$\limsup_{x\to a} f(x) = \inf M_\epsilon(f; a) \leqq h.$$

This holds for all $h > \inf f$, so by 5.2 we deduce

$$(B) \qquad \limsup_{x\to a} f(x) \leqq \inf f.$$

By 6.7, inequalities (*A*) and (*B*) imply the desired conclusion.

EXERCISE. If $a_1, a_2, \cdots$ is a monotonic increasing sequence of numbers, $\lim\limits_{n\to\infty} a_n$ exists and is equal to $\sup a_n$. An analogous statement holds for decreasing sequences.

We have already had an example of a monotonic function in the definition of upper and lower limits; for we saw that $M_\epsilon \geqq M_\delta$ if $\epsilon \geqq \delta$. Hence M_ϵ is a monotonic increasing function on the range $0 < \epsilon < \infty$, and so by 5.6 we have the following theorem.

6.9. *If $f(x)$ is defined on E, and x is in E', then*

$$\limsup_{x\to x_0} f(x) = \inf M_\epsilon(f; x_0) = \lim_{\epsilon\to 0} M_\epsilon(f; x_0),$$
$$\liminf_{x\to x_0} f(x) = \sup m_\epsilon(f; x_0) = \lim_{\epsilon\to 0} m_\epsilon(f; x_0).$$

From 6.8 we can deduce a corollary concerning the right and left limits of monotone functions.

Let $f(x)$ be defined on a subset E of one-dimensional space R_1, and let x_0 be a point of E. We put $E_+ = \{x \mid x > x_0, x \text{ in } E\} = (x_0, \infty) \cap E$ and $E_- = \{x \mid x < x_0, x \text{ in } E\} = (-\infty, x_0) \cap E$. Then

6.10. *If $f(x)$, considered as defined only in E_+, has a limit as $x \to x_0$, we call this limit the right limit of $f(x)$ as $x \to x_0$, and denote it by $f(x_0+)$. If $f(x)$, considered as defined only on E_-, has a limit as $x \to x_0$, we call this the left limit of $f(x)$ at x_0, and denote it by $f(x_0-)$. Also, the upper and lower limits of $f(x)$, considered as defined only on E_+, as $x \to x_0$, are denoted by*

$$\limsup_{x\to x_0+} f(x), \qquad \liminf_{x\to x_0+} f(x)$$

respectively; and the corresponding upper and lower limits when $f(x)$ is regarded as defined only on E_- are denoted by

$$\limsup_{x\to x_0-} f(x), \qquad \liminf_{x\to x_0-} f(x)$$

respectively.

We now prove

6.11. *Let $f(x)$ be monotonic on an interval I. If x_0 is in $\bar{I}$ and is not the left end-point of I, $f(x_0-)$ exists, and if x_0 is in $\bar{I}$ and is not the right end-point of I, $f(x_0+)$ exists. Moreover, if $f(x)$ is monotonic increasing [monotonic decreasing] the inequalities*

$$f(x_0-) \leqq f(x_0), \qquad f(x_0) \leqq f(x_0+)$$
$$[f(x_0-) \geqq f(x_0), \qquad f(x_0) \geqq f(x_0+)]$$

are valid whenever their terms are defined.

Suppose, for example, that $f(x)$ is monotonic increasing, and that $\bar{I}$ is $[\alpha, \beta]$. If $\alpha < x_0 \leqq \beta$, the set E_- is either $[\alpha, x_0)$ or (α, x_0), and if α happens to belong to E_- we can throw it out without changing $f(x_0-)$ or $\sup f(x)$ for x on E_-. Then by 6.8 $f(x_0-)$ exists and

$$f(x_0-) = \sup f(E_-).$$

If furthermore x_0 is in I, then $f(x) \leqq f(x_0)$ for all x in E_-, so

$$\sup f(E_-) \leqq f(x_0).$$

Hence $f(x_0-) \leqq f(x_0)$. The other parts of the theorem can be established in a like manner.

EXERCISE. If $f(x_0-)$ and $f(x_0+)$ exist and are equal, then $\lim\limits_{x \to x_0} f(x)$ exists and is equal to both of the one-sided limits.

EXERCISE. If $f(x)$ is bounded and monotonic on an interval I, its discontinuities are at most denumerable. (Suggestion: Show that for each n there can be only a finite number of points at which

$$|f(x-) - f(x+)| > 1/n).$$

It is actually enough to assume that $f(x) \neq \pm\infty$, without assuming it bounded.

The next theorem is of considerable importance.

6.12. *If $f_1(x)$ and $f_2(x)$ and their sum $f_1(x) + f_2(x)$ are defined on a set E, and x_0 is an accumulation point of E, then*

$$\limsup_{x \to x_0} (f_1 + f_2) \leqq \limsup_{x \to x_0} f_1 + \limsup_{x \to x_0} f_2,$$

provided that the sum on the right is defined (cf. 4.1).

If either of the terms on the right is $+\infty$, the theorem is obviously true. We assume then that neither is $+\infty$. Remembering that $M_\epsilon(f; x_0)$ is a sup, by 5.6

$$M_\epsilon(f_1 + f_2; x_0) \leqq M_\epsilon(f_1; x_0) + M_\epsilon(f_2; x_0).$$

Now let h, k be any pair of numbers which are greater than $\limsup f_1$, $\limsup f_2$ respectively. By 6.2*a* we can find $\epsilon > 0$ small enough so that $M_\epsilon(f_i; x_0) < h$ and $M_\epsilon(f_2; x_0) < k$. Then for this ϵ

$$M_\epsilon(f_1 + f_2; x_0) < h + k,$$

and a fortiori

$$\limsup_{x \to x_0} (f_1 + f_2) < h + k.$$

But here $h + k$ can be any number greater than $\limsup f_1 + \limsup f_2$; so by 5.2 our inequality is true.

The inequality established in 6.12 can be written in many different forms. In the next theorem we collect some of these variant forms of 6.12.

6.13. *Let $f_1(x)$ and $f_2(x)$ be defined on a set E, and let x_0 be an accumulation point of E. Then any of the following statements which do not involve undefined expressions* (i.e. *sums of the form* $\infty + (-\infty)$) *are true:*†

$$(\alpha) \quad \limsup [f_1(x) + f_2(x)] \leqq \limsup f_1(x) + \limsup f_2(x),$$
$$(\beta) \quad \liminf [f_1(x) + f_2(x)] \geqq \liminf f_1(x) + \liminf f_2(x),$$
$$(\gamma) \quad \limsup [f_1(x) + f_2(x)] \geqq \limsup f_1(x) + \liminf f_2(x),$$
$$(\delta) \quad \liminf [f_1(x) + f_2(x)] \leqq \liminf f_1(x) + \limsup f_2(x),$$
$$(\epsilon) \quad \limsup [f_1(x) - f_2(x)] \leqq \limsup f_1(x) - \liminf f_2(x),$$
$$(\zeta) \quad \limsup [f_1(x) - f_2(x)] \geqq \limsup f_1(x) - \limsup f_2(x) \geqq \liminf [f_1(x) - f_2(x)],$$
$$(\eta) \quad \liminf [f_1(x) - f_2(x)] \geqq \liminf f_1(x) - \limsup f_2(x),$$
$$(\vartheta) \quad \liminf [f_1(x) - f_2(x)] \leqq \liminf f_1(x) - \liminf f_2(x) \leqq \limsup [f_1(x) - f_2(x)].$$

Here (α) is the inequality of 6.12. (β) follows from (α) if we replace $f_1(x)$, $f_2(x)$ by $-f_1(x)$, $-f_2(x)$ and use 6.3. (ϵ) follows from (α), and (η) from (β), if we replace $f_2(x)$ by $-f_2(x)$.

To establish (γ), we first notice that if $f_2(x)$ takes on the value $-\infty$ on every set $N_\epsilon(x_0) \cap E - (x_0)$, then $m_\epsilon(f_2; x_0) = -\infty$ for every ϵ, and the right member of (γ) is $-\infty$. If f_2 takes on the value $+\infty$ on every such set, then so does $f_1 + f_2$, and $M_\epsilon(f_1 + f_2; x_0)$ is always $+\infty$. The left member of (γ) is then $+\infty$. There remains the principal case, in which there is a positive ϵ such that $f_2(x)$ is finite on $N_\epsilon(x_0)E - (x_0)$. We may then consider f_2 as always finite, since its values outside of $N_\epsilon(x_0)$ do not affect either member of (γ). Now to prove (γ) we need only replace $f_1(x)$ by $f_1(x) + f_2(x)$ in (ϵ), noting that if $\liminf f_2(x) = +\infty$, by hypothesis $\limsup f_1(x) > -\infty$, which by (ϵ) implies $\limsup [f_1(x) + f_2(x)] = \infty$. From (γ) we obtain (δ) by replacing $f_1(x)$ by $-f_1(x)$ and $f_2(x)$ by $-f_2(x)$. If we replace $f_2(x)$ by $-f_2(x)$ in (γ) and (δ), we obtain the first inequalities in (ζ) and (ϑ) respectively. If we replace $f_1(x)$ by $-f_2(x)$ and $f_2(x)$ by $f_1(x)$ in (γ) and (δ), we obtain the second inequalities in (ϑ) and (ζ) respectively.

From theorem 6.13 we readily obtain

6.14. *Let the functions $f_1(x)$ and $f_2(x)$ be defined on a set E, and let x_0 be an accumulation point of E. If*† $\lim f_2(x)$ *exists, then*

† To save printing, the symbol $x \to x_0$ is understood, instead of printed, under each lim sup, lim inf and lim.

(a) $$\limsup [f_1(x) + f_2(x)] = \limsup f_1(x) + \lim f_2(x),$$
(b) $$\liminf [f_1(x) + f_2(x)] = \liminf f_1(x) + \lim f_2(x),$$

provided that the sums are defined.

For by 6.13α, 6.6 and 6.13 γ,

$$\begin{aligned}\limsup [f_1(x) + f_2(x)] &\leqq \limsup f_1(x) + \limsup f_2(x) \\ &= \limsup f_1(x) + \lim f_2(x) = \limsup f_1(x) + \liminf f_2(x) \\ &\leqq \limsup [f_1(x) + f_2(x)].\end{aligned}$$

Thus (a) is proved. If we replace $\leqq$ by $\geqq$ and interchange inf and sup, we obtain (by 6.13β, δ) the proof of (b).

EXERCISE. Prove the following theorem.

Let $f(x)$ be defined on E, and let x_0 be an accumulation point of E. The equation $\limsup_{x \to x_0} f(x) = k$ *holds if and only if:*

(a) *for every number* $h > k$ there is an $\epsilon > 0$ such that $f(x) < h$ on $N_\epsilon(x_0)E - (x_0)$; and*

(b) *for every number* $h < k$ and every neighborhood $N_\epsilon(x_0)$ there is a point x in $N_\epsilon(x_0)E - (x_0)$ such that $f(x) > h$.*

Likewise, $\liminf_{x \to x_0} f(x) = k$ *if and only if:*

(c) *for every number* $h < k$ there is an $\epsilon > 0$ such that $f(x) > h$ on $N_\epsilon(x_0)E - (x_0)$;*

(d) *for every number* $h > k$ and every neighborhood $N_\epsilon(x_0)$ there is a point x in $N_\epsilon(x_0)E - (x_0)$ such that $f(x) < h$.*

It is now easy to establish a generalized form of the Cauchy convergence criterion.

6.15. *Let $f(x)$ be defined and finite on E, and let x_0 be an accumulation point of E. In order that* $\lim_{x \to x_0} f(x)$ *shall exist and be finite, it is necessary and sufficient that the following condition be satisfied. For every positive number ϵ, there is a $\delta > 0$ such that $| f(x') - f(x'') | < \epsilon$ for every pair of points x', x'' in $N_\delta(x_0)E - (x_0)$.*

If the limit exists and is finite, for every positive ϵ there exists by definition a positive δ such that $f(x)$ is in $N_{\epsilon/2}$ $(\lim_{x \to x_0} f(x))$ whenever x is in $N_\delta(x_0)E - (x_0)$. It follows that for every pair x', x'' of points of this set we have $| f(x') - f(x'') | < \epsilon$.

Conversely, let the condition be satisfied. Let ϵ be an arbitrary positive number, and let δ be the number specified in the condition. Choose a point x'' in $N_\delta(x_0)E - (x_0)$; such a point exists, by 2.1.

* If no such number exists, this condition is considered to be automatically satisfied. This happens in (a) and (d) if $k = +\infty$, and in (b) and (c) if $k = -\infty$.

Then by hypothesis, for every x in $N_\delta(x_0)E - (x_0)$ we have $|f(x) - f(x'')| < \epsilon$, whence

$$f(x'') - \epsilon < f(x) < f(x'') + \epsilon.$$

Thus $f(x'') - \epsilon$ and $f(x'') + \epsilon$ are respectively a lower bound and an upper bound for $f(x)$ on $N_\delta(x_0)E - (x_0)$. By definition,

$$f(x'') - \epsilon \leqq m_\delta(f; x_0) \leqq \liminf_{x \to x_0} f(x)$$
$$\leqq \limsup_{x \to x_0} f(x) \leqq M_\delta(f; x_0) \leqq f(x'') + \epsilon.$$

Hence the upper and lower limits of $f(x)$ as $x \to x_0$ are both finite, and differ by at most 2ϵ. Since 2ϵ is an arbitrary positive number, the upper and lower limits are equal, and by 6.6 $\lim_{x \to x_0} f(x)$ exists and is finite.

Suppose that $f(x, y)$ is defined for all x in a set E contained in the space R_q or R^* and for all y in a set Y, concerning which we make no assumptions. Let $f_0(y)$ be a function defined on Y. If x_0 is an accumulation point of E, the statement that

$$\lim_{x \to x_0} f(x, y) = f_0(y)$$

for all y in Y has by 4.4 the following meaning. To each y in Y and each positive ϵ there corresponds a positive δ, depending on y and ϵ, such that $f(x, y)$ is in $N_\epsilon(f_0(y))$ whenever x is in $N_\delta(x_0)E - (x_0)$. In general we cannot hope that given ϵ, one single δ can be chosen which will serve for all y in the set Y. Whenever this happens to be true, the convergence is called *uniform.*

6.16. *If $f(x, y)$ is defined for all x in a set E in R_q or in R^* and for all y in a set Y, and $f_0(y)$ is defined on Y, and x_0 is an accumulation point of E, then the statement*

$$\lim_{x \to x_0} f(x, y) = f_0(y) \text{ uniformly on } Y$$

is defined to mean that to each positive ϵ there corresponds a positive δ such that $f(x, y)$ is in $N_\epsilon(f_0(y))$ whenever x is in $N_\delta(x_0)E - (x_0)$. (The δ is independent of y.)

For uniform convergence there is a criterion quite analogous to Cauchy's criterion (6.15). This is the following.

6.17. *Let $f(x, y)$ be defined and finite for all x in a set E in R_q or in R^* and all y in a set Y. Let x_0 be an accumulation point of E. In order that $f(x, y)$ shall converge uniformly on Y to some finite limit as $x \to x_0$,* *it is necessary and sufficient that the following condition be satisfied. To each positive number* ϵ *there corresponds a positive number* δ *such that* $|f(x', y) - f(x'', y)| < \epsilon$ *for every pair of points* x', x'' *in* $N_\delta(x_0)E - (x_0)$ *and every* y *in* Y.

If $f(x, y)$ converges uniformly on Y to a finite limit $f_0(y')$, and ϵ is positive, there is a δ such that $|f(x, y) - f_0(y)| < \epsilon/2$ whenever x is in $N_\delta(x_0)E - (x_0)$. So if x' and x'' are both in this set we have

$$|f(x', y) - f(x'', y)| < \epsilon.$$

Conversely, let the condition be satisfied. Then for each individual y in Y the condition in 6.15 is satisfied, so for each such y the function $f(x, y)$ approaches a finite limit $f_0(y)$ as $x \to x_0$. Let ϵ be a positive number. By hypothesis, there is a $\delta > 0$ such that

$$|f(x', y) - f(x'', y)| < \frac{\epsilon}{2}$$

whenever x' and x'' are in $N_\delta(x_0)E - (x_0)$. For each particular y there is a $\delta(y)$ such that

$$|f(x'', y) - f_0(y)| < \frac{\epsilon}{2}$$

whenever x'' is in $N_{\delta(y)}(x_0)E - (x_0)$. Let $\gamma(y)$ be the smaller of δ and $\delta(y)$. If x' is in $N_\delta(x_0)E - (x_0)$, we choose an x'' in $N_{\gamma(y)}(x_0)E - (x_0)$. By the two inequalities above,

$$|f(x', y) - f_0(y)| < \epsilon$$

for all x' in $N_\delta(x_0)E - (x_0)$. But δ is independent of y, so $f(x, y)$ converges uniformly to $f_0(y)$ as x tends to x_0.

7. From the concept of limit we can proceed in the usual way to the concept of continuity. Let $f(x)$ be defined on a set E, and let x_0 be a point of E which is also an accumulation point of E, so that x_0 is in EE'. We say that $f(x)$ is *continuous at* x_0 if $f(x_0)$ is finite, and $\lim f(x)$ exists and is equal to $f(x_0)$. Furthermore, $f(x)$ is *continuous* on E if it is finite-valued on E and is continuous at each point of EE'. (Notice that nothing other than finiteness is required of $f(x)$ at points of E which are not accumulation points of E.)

EXERCISE. If $f(x)$ is continuous on a bounded closed set E, it is uniformly continuous; that is, to each positive ϵ there corresponds a positive δ such that $|f(x_1) - f(x_2)| < \epsilon$ whenever x_1 and x_2 are in E and $||x_1, x_2|| < \delta$. (To each x corresponds $\delta(x)$ such that

$|f(x) - f(x_0)| < \epsilon/2$ if x_0 is in $N_{2\delta(x)}(x)E$. A finite number of the neighborhoods $N_{\delta(x)}(x)$ cover E, say those with $x = x_1, x_2, \cdots, x_k$. Then $\delta = \inf \{\delta_1, \cdots, \delta_k\}$ serves.)

In a similar way, from the ideas of upper and lower limits we can form two concepts related to continuity:

Let $f(x)$ be defined on a set E, and let the point x_0 of E be an accumulation point of E. We define $f(x)$ to be lower semi-continuous at x_0 if $\liminf_{x \to x_0} f(x) \geqq f(x_0)$, and we define $f(x)$ to be upper semi-continuous at x_0 if $\limsup_{x \to x_0} f(x) \leqq f(x_0)$.

We shall say that $f(x)$ is upper [lower] semi-continuous on E in case it is upper [lower] semi-continuous at each point x_0 of EE'. In these last definitions there is no requirement that $f(x)$ be finite-valued.

EXERCISE. If $f(x)$ is lower semi-continuous on a bounded closed set E, it attains its greatest lower bound on E; that is, there is an x_0 in E such that $f(x_0) = \inf f(E)$. If in addition $f(x) \neq -\infty$ on E, it has a finite lower bound on E. Also, if $f(x)$ is upper semi-continuous on E, it attains its least upper bound on E. (By definition of inf, there is a sequence x_n in E for which $\lim_{n \to \infty} f(x_n) = \inf f(E)$. Use 2.12, 2.13, 6.4.)

7.1. *If $f(x)$ is defined and finite-valued on E, then $f(x)$ is continuous on E if and only if it is both upper and lower semi-continuous on E.*

The function $f(x)$ is both upper and lower semi-continuous on E if and only if $\liminf_{x \to x_0} f(x) \geqq f(x_0) \geqq \limsup_{x \to x_0} f(x)$ for every x_0 in EE'. By 6.7, this is true if and only if $\lim_{x \to x_0} f(x) = f(x_0)$. By definition, this last is true if and only if $f(x)$ is continuous at x_0.

7.2. *If $f(x)$ is upper semi-continuous on E and $g(x)$ is lower semi-continuous on E, then if $0 \leqq c < \infty$ the function $cf(x)$ is upper semi-continuous on E and $cg(x)$ is lower semi-continuous on E; while if $-\infty < c \leqq 0$, the function $cf(x)$ is lower semi-continuous on E and $cg(x)$ is upper semi-continuous on E.*

For $c = 0$ this is trivial, as 7.1 shows. Suppose $c < 0$ and let x_0 be any point of EE'. Then by 5.4

$$\liminf_{x \to x_0} cf(x) = c \limsup_{x \to x_0} f(x) \geqq cf(x_0),$$

so $cf(x)$ is lower semi-continuous. By replacing f by g and interchanging inf and sup we prove that $cg(x)$ is upper semi-continuous.

If $c > 0$, then by the proof just completed $(-c)f(x)$ is lower semi-continuous, and therefore $cf(x) \equiv (-1)(-c)f(x)$ is upper semi-continuous. Likewise $(-c)g(x)$ is upper semi-continuous, so $cg(x) = (-1)(-c)g(x)$ is lower semi-continuous.

7.3. *Let $f(x)$ be defined on E, and let x_0 be a point of E. Then $f(x)$ is lower semi-continuous at x_0 if and only if for every number $h < f(x_0)$ there is a positive number ϵ such that $f(x) > h$ for all x in $N_\epsilon(x_0)E$. Also, $f(x)$ is upper semi-continuous at x_0 if and only if for every number $h > f(x_0)$ there is a positive number ϵ such that $f(x) < h$ for all x in $N_\epsilon(x_0)E$.*

We shall prove the first statement; the second follows by change of sign. Suppose first that x_0 is in E but not in E'. Then $f(x)$ is lower semi-continuous at x_0, since our definition of lower semi-continuity requires nothing of f at points of $E - E'$. Also, by 2.1 there is an ϵ such that $N_\epsilon(x_0)E$ consists of x_0 alone, so that if $h < f(x_0)$ the equation $f(x) > h$ holds at every point (i.e., x_0 alone!) of $N_\epsilon(x_0)E$. So the two statements are always true, and are therefore equivalent to each other.

We still have to consider the principal case, in which x_0 is in EE'. Suppose $f(x)$ lower semi-continuous at x_0, and let h be a number less than $f(x_0)$. Then

$$h < f(x_0) \leqq \liminf_{x \to x_0} f(x),$$

so by 6.2 there is a positive ϵ for which $m_\epsilon(f; x_0) > h$. Therefore by definition of m_ϵ we have $f(x) \geqq m_\epsilon(f; x_0) > h$ for all x in $N_\epsilon(x_0)E - (x_0)$. But we do not need to except the point x_0, for the relation $f(x) > h$ holds at x_0 too by choice of h. Hence $f(x) > h$ at all points of $N_\epsilon(x_0)E$.

To prove the converse, let h be a number less than $f(x_0)$. By hypothesis, there is a positive ϵ such that $f(x) > h$ on $N_\epsilon(x_0)E$. Hence $m_\epsilon(f; x_0) \geqq h$, and by 6.1

$$\liminf_{x \to x_0} f(x) \geqq h.$$

Here h is any number less than $f(x_0)$, so by 5.2

$$\liminf_{x \to x_0} f(x) \geqq f(x_0).$$

This completes the proof.

7.4. *If $f(x)$ is lower semi-continuous on E and $g(x)$ is upper semi-continuous on E, then the set G consisting of those points x of E at which $g(x) < f(x)$ is open relative to E, and the set C consisting of those points x of E at which $f(x) \leqq g(x)$ is closed relative to E.*

We shall prove G open relative to E; then, since $C = E - G$, it will follow by 2.6 that C is closed relative to E. Let x_0 be a point of G; we must show that there is a positive number γ such that $N_\gamma(x_0)E \subset G$.

Since x_0 is in G, we have $f(x_0) > g(x_0)$, so there is a number h such that $f(x_0) > h > g(x_0)$. By 7.3, there are positive numbers ϵ, δ such that $f(x) > h$ for all x in $N_\epsilon(x_0)E$ and $h > g(x)$ for all x in $N_\delta(x_0)E$. Let γ be the smaller of ϵ and δ. For every x in $N_\delta(x_0)E$ we then have $f(x) > h > g(x)$, so all such x are in G. That is, $N_\gamma(x_0)E \subset G$, completing the proof.

7.5. *If K is a collection of functions $f(x)$ defined and lower semi-continuous on E, and for each x in E we define $g(x)$ to be* $\sup \{f(x) \mid f \text{ in } K\}$, *then $g(x)$ is lower semi-continuous on E. Likewise, if K is a collection of functions $f(x)$ defined and upper semi-continuous on E, and for each x in E we define $h(x)$ to be* $\inf \{f(x) \mid f \text{ in } K\}$, *then $h(x)$ is upper semi-continuous on E.*

We prove the first part; the second part is obtained by changing signs and using 7.2. Let x_0 be in E, and let h be any number less than $g(x_0)$. By the definition of $g(x_0)$ there is a function $f(x)$ in the collection K such that $f(x_0) > h$. By 7.3 there is a positive ϵ such that $f(x) > h$ for all x in $N_\epsilon(x_0)E$. Since $g(x) \geqq f(x)$ by definition, it is also true that $g(x) > h$ for all x in $N_\epsilon(x_0)E$. By 7.3, this proves that $g(x)$ is lower semi-continuous at x_0. Since x_0 is an arbitrary point of E, this implies that $g(x)$ is lower semi-continuous on E.

EXERCISE. In order that $f(x)$ shall be lower semi-continuous on E, it is necessary and sufficient that for every constant c, the set $\{x \mid f(x) \leqq c\}E$ be closed relative to E.

7.6. *If $f_1(x), \cdots, f_n(x)$ are defined and lower semi-continuous on E, and $f(x) = \inf \{f_1(x), \cdots, f_n(x)\}$, then $f(x)$ is lower semi-continuous on E. If $g_1(x), \cdots, g_n(x)$ are upper semi-continuous on E, and $g(x) = \sup \{g_1(x), \cdots, g_n(x)\}$, then $g(x)$ is upper semi-continuous on E.*

We prove the first statement; the second follows by change of sign. Let x_0 be a point of E, and let h be a number less than $f(x_0)$. Then $h < f_i(x_0)$ for $i = 1, \cdots, n$, so for each i there is (by 7.3) an $\epsilon_i > 0$ such that $f_i(x) > h$ on $N_{\epsilon_i}(x_0)E$. If ϵ is the smallest of the ϵ_i, then $f_i(x) > h$ for all x in $N_\epsilon(x)E$ and all i. One of these $f_i(x)$ (the least) is equal to $f(x)$, so $f(x) > h$ for x in $N_\epsilon(x_0)E$. This is true for all $h < f(x_0)$, so by 7.3 $f(x)$ is lower semi-continuous.

EXERCISE. There is a strong resemblance between Theorems 7.5 and 7.6 on semi-continuous functions and Theorems 2.8 and 2.9 on relatively open and closed sets. With the help of the preceding exercise, show that 7.5 and 7.6 can be deduced from 2.8 and 2.9.

7.7. *If $f(x)$ and $g(x)$ are both upper semi-continuous on E, and neither one takes on the value $+\infty$, then $f(x) + g(x)$ is upper semi-*

continuous on E. *If* $f(x)$ *and* $g(x)$ *are both lower semi-continuous on* E, *and neither one takes on the value* $-\infty$, *then* $f(x) + g(x)$ *is lower semi-continuous on* E.

We prove the first statement; the second follows by change of sign. Let x_0 be any point in EE'. We can not have $\limsup_{x\to x_0} f(x) = +\infty$, for this would imply $f(x_0) \geqq \limsup_{x\to x_0} f(x) = \infty$, contrary to hypothesis. A similar statement applies to $g(x)$. Therefore we may use 4.2 and 6.12 to obtain

$$\begin{aligned} f(x_0) + g(x_0) &\geqq \limsup_{x\to x_0} f(x) + \limsup_{x\to x_0} g(x) \\ &\geqq \limsup_{x\to x_0} (f(x) + g(x)). \end{aligned}$$

This proves that $f(x) + g(x)$ is upper semi-continuous.

7.8. *If* $f(x)$ *and* $g(x)$ *are both lower semi-continuous and non-negative, then* $f(x)g(x)$ *is also lower semi-continuous. If* $f(x)$ *and* $g(x)$ *are both upper semi-continuous and non-negative, and there is no point* x *in* E *at which one of them is* ∞ *and the other* 0, *then* $f(x)g(x)$ *is upper semi-continuous.*

Suppose them both lower semi-continuous and $\geqq 0$, and let x_0 be a point of EE'. If either $f(x_0)$ or $g(x_0)$ is 0, then $f(x_0)g(x_0) = 0$. Then $f(x)g(x)$ is necessarily lower semi-continuous at x_0, for $f(x)g(x) \geqq 0$, so that

$$\liminf_{x\to x_0} f(x)g(x) \geqq 0 = f(x_0)g(x_0),$$

If neither one is 0, then $f(x_0) > 0$ and $g(x_0) > 0$. Let h be any number less than $f(x_0)g(x_0)$. We can find positive numbers $h_1 < f(x_0)$ and $h_2 < g(x_0)$ such that $h_1h_2 > h$. By 7.3, there are positive numbers ϵ, δ such that $f(x) > h_1$, for all x in $N_\epsilon(x_0)E$ and $g(x) > h_2$ for all x in $N_\delta(x_0)E$. If γ is the smaller of ϵ and δ, then for all x in $N_\gamma(x_0)E$ both inequalities hold, so that

$$f(x)g(x) > h_1h_2 > h$$

for all such x. By 7.3, this implies that $f(x)g(x)$ is lower semi-continuous.

Suppose now that $f(x)$ and $g(x)$ are upper semi-continuous, and that we never have $f(x) = \infty$ where $g(x) = 0$ or vice versa. Let x_0 be any point of EE'. If one of the factors in $f(x_0)g(x_0)$ is ∞, the other is positive, so

$$f(x_0)g(x_0) = \infty \geqq \limsup_{x\to x_0} f(x)g(x).$$

Otherwise, both $f(x_0)$ and $g(x_0)$ are finite. We can now repeat the proof of the first part of the theorem, reversing all inequalities.

EXERCISE. A function is sometimes called "continuous in the generalized sense" if it satisfies the definition at the beginning of §7 except for the requirement of finiteness. Let $\varphi(t)$ be continuous in the generalized sense and monotonic increasing on the set $-\infty \leqq t \leqq \infty$. If $f(x)$ is defined and lower [upper] semi-continuous on E, so is $\varphi(f(x))$. If $\varphi(t)$ is continuous and monotonic decreasing on $-\infty \leqq t \leqq \infty$, then if $f(x)$ is defined and lower [upper] semi-continuous on E, $\varphi(f(x))$ is defined and upper [lower] semi-continuous on E. From this we can deduce 7.2. Also, if we let $\varphi(t) = \log t$ $(0 < t < \infty)$, $\varphi(t) = -\infty$ $(t \leqq 0)$, $\varphi(\infty) = \infty$, we can deduce 7.8, except for the case in which one factor is 0 and the other is ∞.

7.9. *If $f(x)$ is defined and lower semi-continuous on E, and there is a constant M such that $f(x) \geqq M$ for all x in E, then it is possible to find a sequence of functions $\varphi_1(x), \varphi_2(x), \cdots$, each defined and continuous on the whole space R_q, such that $M \leqq \varphi_1(x) \leqq \varphi_2(x) \leqq \varphi_3(x) \leqq \cdots$ for all x in R_q and $\lim_{n\to\infty} \varphi_n(x) = f(x)$ for all x in E. Likewise, if $f(x)$ is upper semi-continuous and $f(x) \leqq M$, there are functions $\psi_n(x)$ continuous on R_q such that $M \geqq \psi_1(x) \geqq \psi_2(x) \geqq \cdots$ for all x in R_q and $\lim_{n\to\infty} \psi_n(x) = f(x)$ for all x in E.*

Again the second statement follows from the first by change of sign. For $n = 1, 2, \cdots$ and all x in R_q we define

$$\varphi_n(x) = \inf \{f(\bar{x}) + n \,||\, x, \bar{x} \,||\ \ |\, \bar{x} \text{ in } E\}.$$

Clearly $f(\bar{x}) + n \,||\, x, \bar{x} \,|| \leqq f(\bar{x}) + (n+1) \,||\, x, \bar{x} \,||$ for all $\bar{x}$, so by 5.4 $\varphi_n(x) \leqq \varphi_{n+1}(x)$. Also, $\varphi_n(x) \geqq \inf f(E) \geqq M$. Let x_1 and x_2 be any points of R_q. For every $\bar{x}$ in E we have

$$\begin{aligned} f(\bar{x}) + n \,||\, x_1, \bar{x} \,|| &\leqq f(\bar{x}) + n(\,||\, x_1, x_2 \,|| + ||\, x_2, \bar{x} \,||\,) \\ &= (f(\bar{x}) + n \,||\, x_2, \bar{x} \,||\,) + n \,||\, x_1, x_2 \,|| \end{aligned}$$

So by 5.4

$$\varphi_n(x_1) \leqq \varphi_n(x_2) + n \,||\, x_1, x_2 \,||.$$

Interchanging x_1 and x_2, we get $\varphi_n(x_2) \leqq \varphi_n(x_1) + n \,||\, x_1, x_2 \,||$. Therefore $|\, \varphi_n(x_1) - \varphi_n(x_2) \,| \leqq n \,||\, x_1, x_2 \,||$, which proves that $\varphi_n(x)$ is continuous on all of R_q.

Finally, let x_0 be a point in E; we must show that $\varphi_n(x_0) \to f(x_0)$. Let h be any number less than $f(x_0)$. By 7.3 there is an $\epsilon > 0$ such that $f(x) > h$ for all x in $N_\epsilon(x_0)E$. Now choose a number n_0 large enough so that $M + n_0\epsilon > h$. If $n > n_0$, then in the expression $f(\bar{x}) + n \| x_0, \bar{x} \|$ either $\| x_0, \bar{x} \| \geqq \epsilon$ or $\bar{x}$ is in $N_\epsilon(x_0)E$. In the first case, $f(\bar{x}) + n \| x_0, \bar{x} \| \geqq M + n\epsilon > M + n_0\epsilon > h$. In the second case, $f(\bar{x}) + n \| x_0, \bar{x} \| \geqq f(\bar{x}) > h$ by the choice of ϵ. Thus h is a lower bound for $f(\bar{x}) + n \| x_0, \bar{x} \|$, and by definition $\varphi_n(x_0) \geqq h$. Since n was any number greater than n_0, this implies* $\lim_{n\to\infty} \varphi_n(x_0) \geqq h$. But h was any number $< f(x_0)$, so by 5.2 $\lim_{n\to\infty} \varphi_n(x_0) \geqq f(x_0)$.

On the other hand, in the definition of $\varphi_n(x_0)$ we can take $\bar{x} = x_0$, because x_0 is in E. So one possible value of $f(\bar{x}) + n \| x_0, \bar{x} \|$ is $f(x_0)$, and $\varphi_n(x_0) \leqq f(x_0)$. Since this is true for all n, we have $\lim_{n\to\infty} \varphi_n(x_0) \leqq f(x_0)$. This, with the preceding inequality, proves $\varphi_n(x_0) \to f(x_0)$, and completes the proof of our theorem.

7.10. Corollary. *If $f(x)$ is defined and lower semi-continuous on E, and there is a constant M such that $f(x) \geqq M$ for all x in E, there is a function $g(x)$ defined and lower semi-continuous on the whole space such that $g(x) = f(x)$ for all x in E.*

Let $\{\varphi_n(x)\}$ be the sequence of functions of 7.9, and define $g(x) = \lim_{n\to\infty} \varphi_n(x)$. The limit exists, because $\varphi_1(x) \leqq \varphi_2(x) \leqq \cdots$; it is lower semi-continuous, by 7.5; and it coincides with $f(x)$ for all x in E, by 7.9.

8. There are several classes of functions which distinguish themselves by their usefulness in one branch or another of analysis. One such class, important in the present connection, is the class of monotonic functions (cf. §6). A peculiarly important property of these functions will later turn out to be that (in a sense which can be made quite precise) they are almost everywhere continuous and almost everywhere have a derivative, the latter property not being shared by all continuous functions.

We often wish to add and subtract functions; but the difference of two monotonic functions (e.g., x^3 and x) may not be monotonic. Hence another class assumes importance—the class of functions which are sums or differences of monotonic functions. The principal object of this section is to show that this class is the same as the class of functions of *bounded variation* which we now define.

Let us suppose that $f(x)$ is defined and finite-valued on an interval $I = [a, b]$. We subdivide I by means of a finite number of points α_i, where $a = \alpha_0 < \alpha_1 < \cdots < \alpha_n = b$, and form the sum

* The limit exists; see the exercise after 6.8.

$\sum_{i=0}^{n-1} | f(\alpha_{i+1}) - f(\alpha_i) |$. The sup of this sum for all subdivisions of I is called the *total variation* of f over I, and is denoted by $T_f[I]$. If $T_f[I]$ is finite we say that $f(x)$ is of *bounded variation* (abbreviated "$f(x)$ is BV") on I.

Before proceeding with the theory we shall give an example of a function which is continuous but has not limited total variation. Such a function is $f(x) = x \cos (\pi/x)$, $0 < x \leqq 1$; $f(0) = 0$. For if we choose the α_i to be 0, $1/n$, $1/(n-1)$, $\cdots$ 1, we find

$$\sum | f(\alpha_{i+1}) - f(\alpha_i) | = \left| \frac{\cos n\pi}{n} - 0 \right| + \left| \frac{\cos (n-1)\pi}{n-1} - \frac{\cos n\pi}{n} \right| + \cdots + \left| \frac{\cos 2\pi}{2} - \frac{\cos \pi}{1} \right| = \frac{2}{n} + \frac{2}{n-1} + \cdots \frac{2}{2} + 1;$$

and since this is greater than a partial sum of the divergent series $\Sigma 1/n$, the sup is ∞.

We verify at once that every function which is finite-valued and monotonic on an interval $I = [a, b]$ is of BV. Suppose to be specific that $f(x)$ is monotonic increasing; then for every subdivision $a = \alpha_0 < \cdots < \alpha_n = b$ we have

$$\begin{aligned} \Sigma | f(\alpha_{i+1}) - f(\alpha_i) | &= \Sigma(f(\alpha_{i+1}) - f(\alpha_i)) \\ &= f(b) - f(a). \end{aligned}$$

Hence $f(b) - f(a)$ is the common value of all the sums, and is therefore their sup, $T_f[I]$.

An easy consequence of the definition is

8.1. *If $f(x)$ and $g(x)$ are of BV on I, and μ and γ are finite constants, then $h(x) \equiv \mu f(x) + \gamma g(x)$ is of BV on I.*

For if $a = \alpha_0 < \alpha_1 < \cdots < \alpha_n = b$ is a subdivision of I, then

$$\begin{aligned} \Sigma | h(\alpha_{i+1}) - h(\alpha_i) | &= \Sigma | \mu f(\alpha_{i+1}) + \gamma g(\alpha_{i+1}) - \mu f(\alpha_i) - \gamma g(\alpha_i) | \\ &= \Sigma | \mu[f(\alpha_{i+1}) - f(\alpha_i)] + \gamma[g(\alpha_{i+1}) - g(\alpha_i)] | \\ &\leqq | \mu | \Sigma | f(\alpha_{i+1}) - f(\alpha_i) | + | \gamma | \Sigma | g(\alpha_{i+1}) - g(\alpha_i) | \\ &\leqq | \mu | T_f[I] + | \gamma | T_g[I]. \end{aligned}$$

In particular, the difference of two monotonic functions is of BV.

8.2. *If $\alpha = \alpha_0 < \alpha_1 < \cdots < \alpha_n = \beta$ is a subdivision of the interval I: $\alpha \leqq x \leqq \beta$, and $\alpha = \beta_0 < \beta_1 < \cdots < \beta_m = \beta$ is a finer subdivision (that is, a subdivision which includes all the points α_i among the β_j), then*

$$\Sigma | f(\beta_{j+1}) - f(\beta_j) | \geqq \Sigma | f(\alpha_{i+1}) - f(\alpha_i) |.$$

For each interval $\alpha_i \leqq x \leqq \alpha_{i+1}$ is subdivided into a number of intervals $\alpha_i = \beta_k < \beta_{k+1} < \cdots < \beta_h = \alpha_{i+1}$; and from the equation

$$(f(\beta_{k+1}) - f(\beta_k)) + (f(\beta_{k+2}) - f(\beta_{k+1})) + \cdots + (f(\beta_h) - f(\beta_{h-1})) \\ = f(\alpha_{i+1}) - f(\alpha_i)$$

we have the inequality

$$\sum_{j=k}^{h-1} |f(\beta_{j+1}) - f(\beta_j)| \geqq |f(\alpha_{i+1}) - f(\alpha_i)|.$$

Adding the inequalities for $i = 0, 1, \cdots, n-1$, we obtain the desired result.

If I is the interval $\alpha \leqq x \leqq \beta$, it is sometimes convenient to use the symbol $T_f[\alpha, \beta]$ to denote $T_f[I]$. For completeness, we define $T_f[\alpha, \alpha] = 0$. With this notation, we state

8.3. *If $a \leqq \alpha \leqq \gamma \leqq \beta \leqq b$, then $T_f[\alpha, \gamma] + T_f[\gamma, \beta] = T_f[\alpha, \beta]$.*

If $\alpha = \gamma$ or $\gamma = \beta$, the result reduces to an obvious identity. Otherwise, let h, k, l, be numbers less than $T_f[\alpha, \gamma]$, $T_f[\gamma, \beta]$, $T_f[\alpha, \beta]$ respectively. Then we can subdivide (α, γ) by points β_i, and (γ, β) by points γ_i, and (α, β) by points δ_i, in such a way that

(a) $$h < \Sigma |f(\beta_{i+1}) - f(\beta_i)| \leqq T_f[\alpha, \gamma],$$
(b) $$k < \Sigma |f(\gamma_{i+1}) - f(\gamma_i)| \leqq T_f[\gamma, \beta],$$
(c) $$l < \Sigma |f(\delta_{i+1}) - f(\delta_i)| \leqq T_f[\alpha, \beta].$$

Now take all the points $\beta_i, \gamma_i, \delta_i$ together, and re-name them $\alpha_0(= \alpha)$, $\alpha_1, \cdots, \alpha_k(= \gamma), \cdots, \alpha_n(= \beta)$, from left to right. Then these points provide a finer subdivision than those in (a), (b), and (c); and so by 8.2

(e) $$h < \sum_{i=0}^{k-1} |f(\alpha_{i+1}) - f(\alpha_i)| \leqq T_f[\alpha, \gamma],$$

(f) $$k < \sum_{i=k}^{n-1} |f(\alpha_{i+1}) - f(\alpha_i)| \leqq T_f[\gamma, \beta],$$

(g) $$l < \sum_{i=0}^{n-1} |f(\alpha_{i+1}) - f(\alpha_i)| \leqq T_f[\alpha, \beta].$$

But the sums in (e) and (f) add to give the sum in (g). Hence

$$h + k < T_f[\alpha, \beta] \qquad \text{and} \qquad l < T_f[\alpha, \gamma] + T_f[\gamma, \beta].$$

Replacing h, k, l by their upper bounds, as we may by 5.2,

$$T_f[\alpha, \gamma] + T_f[\gamma, \beta] \leqq T_f[\alpha, \beta]$$

and

$$T_f[\alpha, \beta] \leqq T_f[\alpha, \gamma] + T_f[\gamma, \beta],$$

establishing the equality.

This theorem can be extended readily, by induction, to any finite number of intervals. One of its consequences is

8.4. *If $f(x)$ is of BV on I, it is also of BV on every interval contained in I.*

For let I be the interval $\alpha \leqq x \leqq \beta$. Then

$$T_f[\alpha, \beta] \leqq T_f[a, \alpha] + T_f[\alpha, \beta] + T_f[\beta, b] = T_f[a, b].$$

A trivial consequence of the definition of total variation is the inequality

8.5. $|f(\beta) - f(\alpha)| \leqq T_f[\alpha, \beta)]$.

For $T_f[\alpha, \beta]$ is by definition not less than $\Sigma\, |f(\alpha_{i+1}) - f(\alpha_i)|$ for any subdivision of the interval $[\alpha, \beta]$, and in particular for the subdivision in which $\alpha_0 = \alpha$ and $\alpha_1 = \beta$ and there are no other α_i.

We can now prove

8.6. *If $f(x)$ is of BV on I, then there exist monotonic increasing functions $p(x)$, $n(x)$ such that $f(x) = p(x) - n(x)$.*

Define $p(x) = \frac{1}{2}(T_f[a, x] + f(x))$, $n(x) = p(x) - f(x)$. That $f(x) = p(x) - n(x)$ is obvious. We must show that p and n are monotonic increasing. Consider any two numbers α and $\beta > \alpha$ in I. Then by 8.3 and 8.5

$$\begin{aligned} p(\beta) - p(\alpha) &= \tfrac{1}{2}(T_f[a, \beta] - T_f[a, \alpha] + f(\beta) - f(\alpha)) \\ &\geqq \tfrac{1}{2}\{T_f[\alpha, \beta] - |f(\beta) - f(\alpha)|\} \geqq 0. \end{aligned}$$

Also,

$$\begin{aligned} n(\beta) - n(\alpha) &= p(\beta) - p(\alpha) - f(\beta) + f(\alpha) \\ &= \tfrac{1}{2}(T_f[a, \beta] - T_f[a, \alpha] - f(\beta) + f(\alpha)) \\ &\geqq \tfrac{1}{2}\{T_f[\alpha, \beta] - |f(\beta) - f(\alpha)|\} \geqq 0. \end{aligned}$$

Hence both p and n are monotonic increasing, and the theorem is proved.

Taking this in conjunction with the remark after 8.1, we see that a function $f(x)$ is of BV on I if and only if it is the difference of two monotonic functions.

9. As we mentioned in the first paragraph of §8, the function of BV will later be shown to have a derivative at all points with relatively few exceptions. But it does not follow that such a function is the integral of its derivative. For this to be the case, we shall later find that the function must satisfy a certain condition called *absolute continuity.* A function $f(x)$, defined and finite on an interval I, is *absolutely continuous* (henceforth abbreviated AC) if to every positive number ϵ there corresponds a $\delta > 0$ such that $\Sigma\, |f(\beta_i) - f(\alpha_i)| < \epsilon$ for all finite collections of non-overlapping subintervals of I with total length less than δ. (Intervals in R_1 are non-overlapping if they have at most end points in common.) It follows at once that if the collection consists of a single interval $I_1 = (\alpha_1, \beta_1)$ of length $\beta_1 - \alpha_1 < \delta$, then $|f(\beta_1) - f(\alpha_1)| < \epsilon$, so that an AC function is also uniformly continuous, and therefore is continuous. The converse is not true; for in a moment we shall prove that every function which is AC is of BV, while in §8 we gave an example of a continuous function which was not of BV.

Let us begin by proving

9.1. *If $f(x)$ is AC on an interval* $[a, b]$, then it is of BV.*

By hypothesis, $f(x)$ is AC on the interval $[a, b]$. So by definition, there is a δ such that $\Sigma\,|f(\beta_i) - f(\alpha_i)| < 1$ whenever $\Sigma(\beta_i - \alpha_i) < \delta$, the intervals (α_i, β_i) being non-overlapping. We insert a finite number of points $a = c_0 < c_1 < c_2 \cdots < c_n = b$ between a and b in such a way that $c_{i+1} - c_i < \delta$. We notice that if the interval $[c_i, c_{i+1}]$ be subdivided in any way by points $\alpha_0 = c_i < \alpha_1 < \cdots < \alpha_k = c_{i+1}$, then $\Sigma(\alpha_{j+1} - \alpha_j) = c_{i+1} - c_i < \delta$, so by the choice of δ we have $\Sigma\,|f(\alpha_{j+1}) - f(\alpha_j)| < 1$. Thus $T_f[c_i, c_{i+1}] \leqq 1$. Applying 8.3,

$$T_f[a, b] = T_f[c_0, c_1] + \cdots + T_f[c_{n-1}, c_n] \leqq n.$$

Hence $T_f[a, b]$ is finite, as was to be proved.

Obviously the converse of this is false; a function can have BV and not be continuous at all. But it might be suspected that a continuous function which has BV is necessarily absolutely continuous. This too is false. For we can give an example of a function which is finite, continuous and monotonic increasing (and therefore certainly of BV), and yet is not AC. This example is not quite trivially easy to construct, but it is important enough to justify some effort.

We begin with the interval $[0, 1]$, and imagine that we blacken the middle third, $(\frac{1}{3}, \frac{2}{3})$. This interval we call $I_{\frac{1}{2}}$—the reason for the strange notation will appear in a moment. If we remove $I_{\frac{1}{2}}$ from $[0, 1]$ we leave two white intervals, $[0, \frac{1}{3}]$ and $[\frac{2}{3}, 1]$. We blacken the (open) middle third of each of these and name the new black intervals $I_{\frac{1}{4}}$ and $I_{\frac{3}{4}}$ respectively. Removing these new black intervals leaves four white intervals; we blacken the middle third of each, name the new black intervals $I_{\frac{1}{8}}$, $I_{\frac{3}{8}}$, $I_{\frac{5}{8}}$, $I_{\frac{7}{8}}$ in order from left to right, and continue this process. The original interval had length 1; after removing the

* Recall the conventions as to round and square brackets adopted in §2.

first black interval, the remainder had total length $\frac{2}{3}$; after removing the next two black intervals, the remainder had total length $(\frac{2}{3})^2$, and so on; after removing the n-th set of black intervals, the remainder has total length $(\frac{2}{3})^n$.

Now we define a function $f_1(x)$ which is constant on each black interval, and has on the interval $I_r (r = m/2^n)$ the value $m/2^n$. This function is defined only on the black intervals; to extend it to all x in $[0, 1]$ we define $f(0) = 0$ and for the remaining points $f(x) = \sup \{f_1(\xi) \mid \xi \leqq x\}$. In particular, $f(1) = 1$. If x happens to be in a black interval we readily see that $f(x) = f_1(x)$. Also, $f(x)$ is monotonic increasing by 5.3; for if we increase x we increase the class of numbers $f_1(\xi)$ of which $f(x)$ is the sup. We next show that $f(x)$ is continuous. Let x_0 be any point of $[0, 1]$ and ϵ any positive number; we must exhibit an open interval (α, β) containing x_0 such that $|f(x) - f(x_0)| < \epsilon$ whenever x is in (α, β). If x_0 happens to be in a black interval this is easy, since we have only to take (α, β) to be the black interval containing x_0, and $f(x)$ will be constant on (α, β). Otherwise, choose n large enough so that $2^{-n} < \epsilon$, and consider the n-th stage of subdivision of $[0, 1]$. The point x_0 will lie in a white interval of this subdivision. The black intervals nearest to x_0 on left and right respectively will be I_p and I_q, where $p = m/2^n$ and $q = (m + 1)/2^n$. Choose a point α in I_p and a point β in I_q. Then by the definition, $f(\alpha) = f_1(\alpha) = m/2^n$ and $f(\beta) = f_1(\beta) = (m + 1)/2^n$. For all points x between α and β, and in particular for x_0 itself, we shall have $f(\alpha) \leqq f(x) \leqq f(\beta)$, since $f(x)$ is monotonic; and therefore

$$|f(x) - f(x_0)| \leqq f(\beta) - f(\alpha) = 1/2^n < \epsilon.$$

Thus $f(x)$ is proved continuous.

It remains only to show that $f(x)$ is not absolutely continuous. We do this by showing for a certain ϵ, namely $\epsilon = 1$, there is no $\delta > 0$ such that $\Sigma |f(\beta_i) - f(\alpha_i)| < \epsilon$ whenever $\Sigma(\beta_i - \alpha_i) < \delta$. Let δ be any positive number, and let the white intervals of the n-th stage of subdivision be denoted by $[\alpha_1, \beta_1], \cdots, [\alpha_p, \beta_p]$, where $\alpha_1 = 0$ and $\beta_p = 1$. We have already seen that the total length of these intervals, $\Sigma(\beta_i - \alpha_i)$, has the value $(\frac{2}{3})^n$, and we choose n large enough so that this is less than δ. Now the interval (β_i, α_{i+1}) is a black interval, and $f(x)$ is constant on black intervals; so $f(\beta_i) = f(\alpha_{i+1})$. Therefore

$$\sum_{i=1}^{p} (f(\beta_i) - f(\alpha_i)) = \sum_{i=1}^{p} (f(\beta_i) - f(\alpha_i)) + \sum_{i=1}^{p-1} (f(\alpha_{i+1}) - f(\beta_i))$$
$$= f(\beta_p) - f(\alpha_1) = f(1) - f(0) = 1.$$

Hence the sum $\Sigma \mid f(\beta_i) - f(\alpha_i) \mid$ is equal to 1, although the total length of the intervals $[\alpha_i, \beta_i]$ is equal to $(\frac{2}{3})^n < \delta$; so $f(x)$ is not absolutely continuous.

Returning now to the theory of AC functions, we wish to prove

9.2. *If $f(x)$ and $g(x)$ are both AC on the interval $[a, b]$, then so are $cf(x)$ (c a finite constant), $f(x) \pm g(x)$, and $f(x)g(x)$; and so is $1/f(x)$, if $f(x) \neq 0$.*

We shall prove only the statement about $f(x)g(x)$; the proofs of the other three are similar. Since f and g are continuous, they are bounded;* say $|f| \leqq M$ and $|g| \leqq N$. Let $[\alpha_1, \beta_1], \cdots, [\alpha_n, \beta_n]$ be non-overlapping intervals in $[a, b]$; then

$$\Sigma \mid f(\beta_i)g(\beta_i) - f(\alpha_i)g(\alpha_i) \mid = \Sigma \mid f(\beta_i)(g(\beta_i) - g(\alpha_i)) + g(\alpha_i)(f(\beta_i) - f(\alpha_i)) \mid \leqq M\Sigma \mid g(\beta_i) - g(\alpha_i) \mid + N\Sigma \mid f(\beta_i) - f(\alpha_i) \mid .$$

Since f, g are AC we can find a $\delta > 0$ such that whenever $\Sigma(\beta_i - \alpha_i) < \delta$ the first sum on the right is less than $\epsilon/2M$ and the second less than $\epsilon/2N$. Then the sum on the left is less than ϵ, proving that fg is AC.

A class of AC functions, which, among other virtues, has the property of being easily recognized, is the class of *Lipschitzian* functions. A function $f(x)$ defined on an interval† $[a, b]$ is said to be *Lipschitzian*, or to *satisfy a Lipschitz condition*, if there is a constant M such that $|f(x_1) - f(x_2)| \leqq M |x_1 - x_2|$ for all x_1 and x_2 in $[a, b]$. If f satisfies this condition, then for any set of intervals $[\alpha_1, \beta_1], \cdots, [\alpha_n, \beta_n]$ in $[a, b]$ we have

$$\Sigma \mid f(\beta_i) - f(\alpha_i) \mid \leqq \Sigma M \mid \beta_i - \alpha_i \mid = M\Sigma(\beta_i - \alpha_i);$$

so the sum on the left is less than ϵ if $\Sigma(\beta_i - \alpha_i)$ is less than $\delta = \epsilon/M$.

A still more special class is the class of functions defined and having a continuous derivative on a closed interval $[a, b]$. For if $f(x)$ is such a function, then $f'(x)$ is bounded (say $|f'(x)| \leqq M$), since it is continuous. Then by the theorem of mean value, for each pair x_1 and $x_2 \neq x_1$ of numbers in $[a, b]$ there is an $\bar{x}$ between x_1 and x_2 such that $f(x_2) - f(x_1) = (x_2 - x_1)f'(\bar{x})$. Therefore

$$|f(x_2) - f(x_1)| \leqq M |x_2 - x_1| .$$

* Cf. Courant's Differential and Integral Calculus, (English edition), p. 63, or see the first exercise in §7.

† This definition applies equally well to a function $f(x)$ defined on a set E in R_q if we replace $|x_1 - x_2|$ by $\| x_1 - x_2 \|$. The function $d(x)$ of §5, p. 26, is thus Lipschitzian.

Suppose that $g(x)$ is defined on $[a, b]$ and takes on values in $[\alpha, \beta]$, that is $\alpha \leqq g(x) \leqq \beta$; and suppose that $f(y)$ is defined on $[\alpha, \beta]$. Then $f(g(x))$ is defined on $[a, b]$. From the fact that $f(g(x))$ is continuous whenever f and g are continuous one might be tempted to surmise that $f(g(x))$ is AC whenever f and g are AC. This is not true.* However, we can prove the following theorem.

9.3. *Let $g(x)$ be defined and AC on an interval $[a, b]$, and let the values of $g(x)$ lie in an interval $[\alpha, \beta]$. Let $f(y)$ be AC on $[\alpha, \beta]$. If either one of the two conditions*

(i) *$g(x)$ monotonic,*

(ii) *$f(y)$ Lipschitzian*

is satisfied, then $f(g(x))$ is AC on the interval $[a, b]$.

Suppose first that g is monotonic. Let ϵ be an arbitrary positive number. Since $f(y)$ is AC, there is a positive number γ such that if $(\eta_1, \zeta_1), (\eta_2, \zeta_2), \cdots, (\eta_p, \zeta_p)$ is a set of non-overlapping subintervals of $[\alpha, \beta]$ with $\Sigma(\zeta_i - \eta_i) < \gamma$, the sum $\Sigma \mid f(\zeta_i) - f(\eta_i) \mid$ is less than ϵ. Since $g(x)$ is AC, there is a positive δ such that if $(\alpha_1, \beta_1), \cdots, (\alpha_p, \beta_p)$ are non-overlapping subintervals of $[a, b]$ with $\Sigma(\beta_i - \alpha_i) < \delta$, the sum $\Sigma \mid g(\beta_i) - g(\alpha_i) \mid$ is less than γ. But if $g(x)$ is monotonic increasing the intervals $(g(\alpha_i), g(\beta_i))$ are non-overlapping, and if $g(x)$ is monotonic decreasing the intervals $(g(\beta_i), g(\alpha_i))$ are non-overlapping. In either case, by the choice of γ we have $\Sigma \mid f(g(\beta_i)) - f(g(\alpha_i)) \mid < \epsilon$, so that $f(g(x))$ is AC.

Suppose next that $f(x)$ is Lipschitzian, with Lipschitz constant M. Let ϵ be a positive number. Since $g(x)$ is AC, there is a δ such that if $(\alpha_1, \beta_1), \cdots, (\alpha_p, \beta_p)$ is a set of non-overlapping subintervals of $[a, b]$ with $\Sigma(\beta_i - \alpha_i) < \delta$, the sum $\Sigma \mid g(\beta_i) - g(\alpha_i) \mid$ is less than ϵ/M. But then by the Lipschitz property

$$\sum \mid f(g(\beta_i)) - f(g(\alpha_i)) \mid < M \cdot \frac{\epsilon}{M} = \epsilon,$$

and so $f(g(x))$ is AC.

* For an example, see Carathéodory, *Vorlesungen über reelle Funktionen*, p. 554.

CHAPTER II

The Lebesgue Integral

10. The starting point in our definition of integral is the idea of the volume of an interval. If I is a closed interval $a^{(i)} \leqq x^{(i)} \leqq b^{(i)}$, $i = 1, \cdots, q$, we define the volume ΔI of the interval I to be $(b^{(1)} - a^{(1)})(b^{(2)} - a^{(2)}) \cdots (b^{(q)} - a^{(q)})$. If the interval I is partly closed or open, that is if some or all of the signs $\leqq$ in the definition of I are replaced by $<$, we still use the same definition for ΔI. From this definition it is easily seen that

10.1. (*a*) *For every interval I, ΔI is finite and non-negative, and is the same as $\Delta \bar{I}$.*

(*b*) *If a closed interval I is the union of two closed non-overlapping intervals I_1, I_2, then $\Delta I = \Delta I_1 + \Delta I_2$.*

There is a very definite reason for singling out these two properties of ΔI. In Chapter V we shall introduce a new definition of a function ΔI, not the same as the one here considered, but nevertheless satisfying 10.1(*a*) and 10.1(*b*). So every theorem on integration which uses no properties of ΔI except these two will necessarily be true for the ΔI in Chapter V, and can be used there without repetition of proofs. In preparation for Chapter V, therefore, we shall mark with an "*s*" each theorem which uses no property of ΔI other than 10.1(*a*) and 10.1(*b*), and shall omit the "*s*" from the others.

From 10.1 it follows at once that if I is a degenerate interval, $\Delta I = 0$. For by 10.1(*a*) we may as well suppose I closed; otherwise we could replace it by its closure, which is also degenerate. Then I is the union of the non-overlapping intervals I and I, so by 10.1(*b*) we have $2\Delta I = \Delta I$, and ΔI must be 0.

10.2s. *Let I be a closed interval, and let $I_1, \cdots, I_n$ be non-overlapping closed intervals whose sum is I. Then*

$$\sum_{j=1}^{n} \Delta I_j = \Delta I.$$

Before beginning the proof, we wish to remark that in it we shall not make any use of the non-negativeness of ΔI.

The intervals I, I_j are defined by inequalities

$$I: \qquad a^{(i)} \leqq x^{(i)} \leqq b^{(i)}, \qquad (i = 1, \cdots, q),$$
$$I_j: \qquad a_j^{(i)} \leqq x^{(i)} \leqq b_j^{(i)} \qquad (i = 1, \cdots, q; j = 1, \cdots, n).$$

For each value of i ($i = 1, \cdots, q$) we arrange the numbers $a^{(i)}$, $b^{(i)}$, $a_j^{(i)}$, $b_j^{(i)}$ in ascending order, rejecting repetitions. We thus obtain numbers

$$a^{(i)} = c_0^{(i)} < c_1^{(i)} < \cdots < c_{n_i-1}^{(i)} < c_{n_i}^{(i)} = b^{(i)}.$$

Consider now the aggregates of intervals

$$(*)\ J_k:\ c_k^{(i)} \leqq x^{(i)} \leqq c_{k+1}^{(i)} \qquad (i = 1, \cdots, q;\ k = 0, \cdots, n_i - 1).$$

These are closed, and their sum is I. We now show that $\Sigma\Delta J_k = \Delta I$. This is evident if the $c_h^{(i)}$ consist of the numbers $a^{(i)}$, $b^{(i)}$ alone, for then there is but one interval J_k and it is I itself. Suppose the equation satisfied whenever there are m numbers $c_j^{(i)}$; we will show it still satisfied when there are $m + 1$ of them. From the $m + 1$ numbers we discard one which is not an $a^{(i)}$ or a $b^{(i)}$, say to be specific the number $c_1^{(1)}$. Correspondingly we construct the intervals analogous to (*); these consist of the intervals (*) themselves, except that those of the intervals (*) whose defining inequalities involve $c_1^{(1)}$ are missing and in their place we have the intervals (**)

$$\begin{aligned}(**)\quad & c_0^{(1)} \leqq x^{(1)} \leqq c_2^{(1)},\\ & c_k^{(i)} \leqq x^{(i)} \leqq c_{k+1}^{(i)} \qquad (i = 2, \cdots, q;\ k = 0, \cdots, n_i - 1).\end{aligned}$$

Each of the intervals in (**) is the sum of exactly two of the intervals rejected from (*), namely

$$\begin{aligned}& c_0^{(1)} \leqq x^{(1)} \leqq c_1^{(1)},\\ & c_k^{(i)} \leqq x^{(i)} \leqq c_{k+1}^{(i)} \qquad (i = 2, \cdots, q;\ k = 0, \cdots, n_i - 1)\end{aligned}$$

and

$$\begin{aligned}& c_1^{(1)} \leqq x^{(1)} \leqq c_2^{(1)},\\ & c_k^{(1)} \leqq x^{(i)} \leqq c_{k+1}^{(i)} \qquad (i = 2, \cdots, q;\ k = 0, \cdots, n_i - 1),\end{aligned}$$

and these intervals are non-overlapping. By 10.1(b), the sum of the Δ's is the same whether or not we include $c_1^{(1)}$ among the points of division, so even with the $m + 1$ points $c_k^{(i)}$ we still have $\Sigma\Delta J_k = \Delta I$.

If I_j is any one of the intervals given in the statement of the theorem, it contains some (or none) of the J_k; these we denote by

$$(*_*\!*)\qquad J_{j,1},\ J_{j,2},\ \cdots,\ J_{j,m(j)}.$$

By the same proof just given for I itself, we have

$$\sum_{h=1}^{m(j)} \Delta J_{j,h} = \Delta I_j.$$

It is easy to see that if an interval J_k has an interior point in an interval I_j, it is entirely contained in I_j. Since the I_j are non-overlapping, no J_k is contained in two different I_j, and so the list ($*_*$*) contains no repetitions. Moreover, each J_k is contained in some I_j, for if x is an interior point of J_k it is in some I_j, hence J_k is contained in I_j. That is, the list ($*_*$*) consists of all the intervals J_k without repetitions. Hence

$$\sum_{j=1}^{n}\left[\sum_{h=1}^{m(j)} \Delta J_{j,h}\right] = \Sigma\Delta J_k = \Delta I.$$

With the preceding equation, this establishes the theorem.

A corollary is

10.3s. *If an interval I (open, closed or partly closed) is the sum of non-overlapping intervals $I_1, \cdots, I_n$, then $\Delta I = \Delta I_1 + \cdots + \Delta I_n$.*,

For by 10.1 we know that $\Delta I = \Delta\bar{I}$ and $\Delta I_j = \Delta\bar{I}_j$, where $\bar{I}$ $\bar{I}_j$ are the closures of I, I_j respectively (that is, the intervals I, I_j plus their respective boundaries). These closures are closed intervals, and no two of the $\bar{I}_j$ have any common points except boundary points, so by 10.2

$$\Delta I = \Delta\bar{I} = \Sigma\Delta\bar{I}_j = \Sigma\Delta I_j.$$

The first class of functions which we shall integrate is the class of *step-functions*. A function $s(x)$ is a step-function if it is defined on an interval I and is constant on each one of a finite set of disjoint intervals whose sum is I. Suppose then that $s(x)$ is a step-function on I, and has the constant value c_k on each of the disjoint intervals $I_k (k = 1, \cdots, n)$ whose sum is I. We define

10.4s. $\int_I' s(x)\,dx \equiv \sum_{k=1}^{n} c_k\Delta I_k.$

The reason for marking the integral sign with a prime in the definition is to avoid its confusion with other integrals later to be defined; even when applied to step-functions these later (and more widely useful) definitions will be conceptually different, and, unless we drop the "s" from certain theorems, may even be numerically different.

Our first duty is to examine the self-consistency of this definition. A step-function on I may be represented in many ways; for example, if we decompose each interval I_k into $I_{k,1}$ and $I_{k,2}$, and define $s^*(x)$ to be equal to c_k on both $I_{k,1}$ and $I_{k,2}$, then $s^*(x)$ is the same function as $s(x)$, but the intervals $I_{1,1}, I_{1,2}, I_{2,1}, I_{2,2}, \cdots, I_{n,1}, I_{n,2}$ used in representing it are not the same as $I_1, \cdots, I_n$. We must show that the integral of $s^*(x)$ is nevertheless the same as the integral of $s(x)$.

10.5s. *If $s(x)$ is a step-function on I equal to c_k on $I_k (k = 1, \cdots, n)$, and $s'(x)$ is a step-function on I equal to c'_h on $I'_h(h = 1, \cdots, m)$, and $s'(x)$ is identically equal to $s(x)$, then*

$$\int_I' s'(x)\,dx = \int_I' s(x)\,dx.$$

Each of the intersections $I_k \cap I'_h$ is either an interval or empty. Discard the empty intersections, and name the others $J_1, \cdots, J_p$. Each J_l is contained in some I_k and is also contained in some I'_h. On I_k the function $s(x)$ is constant; define $\sigma(x)$ to be constantly equal to $s(x)$ on J_l. Then $\sigma(x) = s(x) = s'(x)$. Consider any one, k, of the integers $1, \cdots, n$. The interval I_k is parted into disjoint sub-ntervals $J_{l_1}, \cdots, J_{l_s}$; so $\Delta I_k = \Sigma \Delta J_{l_i}$, by 10.3, and

$$c_k \Delta I_k = \sum_{i=1}^{s} c_k \Delta J_{l_i}.$$

Adding these equations member by member for $k = 1, \cdots, n$, the sum on the left is $\int_I' s(x)\,dx$. On the right the J_{l_i} add up to I, so by definition the sum on the right is $\int_I' \sigma(x)\,dx$. Hence

$$\int_I' s(x)\,dx = \int_I' \sigma(x)\,dx.$$

In exactly the same way

$$\int_I' s'(x)\,dx = \int_I' \sigma(x)\,dx,$$

and our theorem is proved.

We shall need just three simple theorems on the integrals of step-functions.

10.6s. *Let $s_1(x)$ and $s_2(x)$ be step-functions on an interval I and let k be a finite constant. Then*

(*a*) $$\int_I' ks_1(x)\,dx = k \int_I' s_1(x)\,dx;$$

(*b*) $$\int_I' (s_1(x) + s_2(x))\,dx = \int_I' s_1(x)\,dx + \int_I' s_2(x)\,dx;$$

(*c*) *if $s_1(x) \leqq s_2(x)$ for all x in I, then*

$$\int_I' s_1(x)\,dx \leqq \int_I' s_2(x)\,dx.$$

As we saw in proving 10.5, there is a partition $J_1, \cdots, J_l$ of I such that each of the functions $s_1(x)$, $s_2(x)$ is constant on each interval J_h. Suppose that on J_h the functions $s_1(x)$, $s_2(x)$ have the values $c_h^{(1)}$, $c_h^{(2)}$ respectively. Then $s_1(x) + s_2(x) = c_h^{(1)} + c_h^{(2)}$ on J_h, and

$$\int_{I_i}' (s_1(x) + s_2(x))\, dx = \sum_{i=1}^{l} (c_i^{(1)} + c_i^{(2)})\Delta J_i = \sum_{i=1}^{l} c_i^{(1)}\Delta J_i + \sum_{i=1}^{l} c_i^{(2)}\Delta J_i$$
$$= \int_I' s_1(x)\, dx + \int_I' s_2(x)\, dx.$$

Also $\int_I' ks_1(x)\, dx = \sum_{i=1}^{l} (kc_i^{(1)}\Delta J_i) = k \sum_{i=1}^{l} c_i^{(1)}\Delta J_i = k \int_I' s_1(x)\, dx$. If $s_1(x) \leqq s_2(x)$, then $c_i^{(1)} \leqq c_i^{(2)}$ for each i. So by this and 10.1(a),

$$\int_I' s_1(x)\, dx = \sum_{i=1}^{l} c_i^{(1)}\Delta J_i \leqq \sum_{i=1}^{l} c_i^{(2)}\Delta J_i = \int_I' s_2(x)\, dx.$$

10.7s. *Let $s(x)$ be a step-function on a closed interval I, and let $I_1, \cdots, I_n$ be non-overlapping closed intervals such that $I_1 \cup I_2 \cup \cdots \cup I_n = I$. Then*

$$\sum_{j=1}^{n} \int_{I_j}' s(x)\, dx = \int_I' s(x)\, dx.$$

As before, we may suppose that $s(x)$ is constant and equal to c_h on intervals $J_1, \cdots, J_l$, where each J_h is contained in one of the intervals I_j. We suppose that the J_h are numbered in such a way that J_h is in I_1 if $h = 1, \cdots, h_1$, J_h is in I_2 if $h = h_1 + 1, \cdots, h_2, \cdots$, J_h is in I_n if $h = h_{n-1} + 1, \cdots, h_n$ (here $h_n = l$). Then

$$\int_I' s(x)\, dx = \sum_{h=1}^{l} c_h\Delta J_h$$
$$= \sum_{h=1}^{h_1} c_h\Delta J_h + \sum_{h=h_1+1}^{h_2} c_h\Delta J_h + \cdots + \sum_{h=h_{n-1}+1}^{h_n} c_h\Delta J_h = \sum_{j=1}^{n} \int_{I_j}' s(x)\, dx.$$

In order to state the next theorem we need the concept of translation of a function.

10.8s. *Let $h = (h^{(1)}, \cdots, h^{(q)})$ be a q-tuple of real numbers. If E is a set in R_q, its translation $E^{(h)}$ is defined to be the set*

$$\{(x^{(1)}, \cdots, x^{(q)}) \mid (x^{(1)} - h^{(1)}, \cdots, x^{(q)} - h^{(q)}) \text{ in } E\}.$$

If $f(x)$ is a function defined on E, its translation $f^{(h)}(x)$ is the function defined on $E^{(h)}$ by the equation

$$f^{(h)}(x) = f(x^{(1)} - h^{(1)}, \cdots, x^{(q)} - h^{(q)}) \quad (x \text{ in } E^{(h)}).$$

Suppose then that we have some sort of integration process which assigns an integral to each one of a certain class of functions. The integral will be said to be *invariant under translation* if, whenever $f(x)$ is integrable over E, $f^{(h)}(x)$ is also integrable over $E^{(h)}$ for all sets $(h^{(1)}, \cdots, h^{(q)})$, and the two integrals are equal.

We can now state our next theorem. (Observe that it is not marked with an "s.")

10.9. *The integral, as defined in* 10.4 *for step-functions, is invariant under translation.*

Given any interval I, it is evident from the second sentence of this chapter that $\Delta I^{(h)} = \Delta I$ for all h. By definition 10.4 our theorem is seen at once to be true.

11. The next stage which we desire to reach in our development of the Lebesgue integral is the integration of continuous functions. However, it costs us little additional effort to define the Riemann integral, and we thereby show the pattern later to be used in defining the Lebesgue integral. (See also page 383.)

11.1s. *Let $f(x)$ be defined and bounded* on a closed interval I. Then the upper Darboux integral of $f(x)$ over I is the* inf *of $\int_I' s(x)\,dx$ for all step-functions $s(x) \geqq f(x)$ on I, and the lower Darboux integral of $f(x)$ over I is the* sup *of $\int_I' s(x)\,dx$ for all step-functions $s(x) \leqq f(x)$ on I:*

$$(R) \overline{\int_I} f(x)\,dx \equiv \inf \left\{ \int_I' s(x)\,dx \mid s(x) \geqq f(x) \right\},$$
$$(R) \underline{\int_I} f(x)\,dx \equiv \sup \left\{ \int_I' s(x)\,dx \mid s(x) \leqq f(x) \right\}.$$

If these upper and lower integrals are equal, $f(x)$ is Riemann integrable over I, and its Riemann integral is

$$(R) \int_I f(x)\,dx \equiv (R) \overline{\int_I} f(x)\,dx = (R) \underline{\int_I} f(x)\,dx.$$

We need only a few of the simplest properties of these integrals; in fact, it would be sufficient for our purposes to establish these properties only for continuous functions $f(x)$.

11.2s. *If $f(x)$ is defined and bounded on I, then*

$$(R) \overline{\int_I} f(x)\,dx \geqq (R) \underline{\int_I} f(x)\,dx.$$

* Although we include ∞ and $-\infty$ in our number system R^*, we retain the usual meaning for this expression; $f(x)$ is bounded on I if it has a *finite* upper bound and a *finite* lower bound on I.

Let $\sigma(x)$ be any step-function $\leqq f(x)$ on I. Then for every step-function $s(x) \geqq f(x)$ on I we have $s(x) \geqq \sigma(x)$, so by 10.6(*c*)

$$\int_I' s(x)\,dx \geqq \int_I' \sigma(x)\,dx.$$

Thus $\int_I' \sigma(x)\,dx$ is a lower bound for the integrals $\int_I' s(x)\,dx$, $s(x) \geqq f(x)$, hence it does not exceed their inf:

$$(R)\ \overline{\int_I} f(x)\,dx \geqq \int_I' \sigma(x)\,dx.$$

This holds for all $\sigma(x) \leqq f(x)$, so it holds for the sup of the $\int_I' \sigma(x)\,dx$:

$$(R)\ \overline{\int_I} f(x)\,dx \geqq (R)\ \underline{\int_I} f(x)\,dx.$$

11.3s. *If $\varphi(x)$ is continuous on a closed interval I, it is Riemann integrable over I.*

Let ϵ be an arbitrary positive number. There is a positive number δ such that $|\varphi(x_1) - \varphi(x_2)| < \epsilon$ whenever x_1 and x_2 are both in I and $\|x_1, x_2\| < \delta$ (cf. the first exercise in §7). We subdivide I into disjoint subintervals $J_1, \cdots, J_m$ of diagonal less than δ, and on each subinterval J_k we assign to $s(x)$ and $S(x)$ the respective values $\inf \varphi(J_k)$ and $\sup \varphi(J_k)$. Then

$$s(x) \leqq \varphi(x) \leqq S(x) \qquad \text{and} \qquad S(x) - s(x) \leqq \epsilon$$

for all x in I. By 11.1 and 11.2,

$$\int_I' s(x)\,dx \leqq \underline{\int_I} \varphi(x)\,dx \leqq \overline{\int_I} \varphi(x)\,dx \leqq \int_I' S(x)\,dx.$$

With 10.6 and 10.4 this yields

$$0 \leqq \overline{\int_I} \varphi(x)\,dx - \underline{\int_I} \varphi(x)\,dx \leqq \int_I' (S(x) - s(x))\,dx \leqq \int_I' \epsilon\,dx = \epsilon\Delta I\cdot$$

But the last number of this inequality is an arbitrary positive number, since ϵ is arbitrary; so the upper and lower integrals of $\varphi(x)$ are equal. This establishes the theorem.

The next theorem is not needed until much later, but we establish it because its proof is intimately related to that of 11.3.

11.4s. *If I is a closed interval, there is a denumerable collection $\mathfrak{S}$ of step-functions on I with the following property. To each positive number α and each function $\varphi(x)$ continuous on I there corresponds a function $\sigma(x)$ of the class $\mathfrak{S}$ such that $|\varphi(x) - \sigma(x)| < \alpha$ for all x in I.*

The only property of the partitions of I into subintervals J_k used in 11.3 was that for every δ there must be a partition into intervals of diagonal less than δ. Evidently a denumerable sequence of partitions is all that is needed; for instance, the first partition could consist of I alone, and each later partition could be obtained by bisecting the sides of the preceding partition. For each partition we construct the step-functions which are constant and rational-valued on the intervals of the partition; these are denumerable by 3.3 and 3.4, and the aggregate $\mathfrak{S}$ of all step-functions for all partitions used is also denumerable by 3.2. In the proof of 11.3 we suppose $\epsilon = \alpha/2$; then for the function $s(x)$ which we constructed we have $| s(x) - \varphi(x) | < \alpha/2$. On each interval J_k we can approximate the value of $s(x)$ by a rational number to within less than $\alpha/2$. This gives us a function of the class $\mathfrak{S}$ differing from φ by less than α.

11.5s. *If $f(x)$ and $g(x)$ are both Riemann integrable over a closed interval I and c is a finite constant, then*

(*a*) *c is Riemann integrable over I, and*

$$(R) \int_I c\, dx = c\Delta I;$$

(*b*) *$cf(x)$ is Riemann integrable over I, and*

$$(R) \int_I cf(x)\, dx = c \cdot (R) \int_I f(x)\, dx;$$

(*c*) *$f(x) + g(x)$ is Riemann integrable over I, and*

$$(R) \int_I (f(x) + g(x))\, dx = (R) \int_I f(x)\, dx + (R) \int_I g(x)\, dx;$$

(*d*) *if $f(x) \geqq 0$, then*

$$(R) \int_I f(x)\, dx \geqq 0;$$

(*e*) *if $f(x) \geqq g(x)$, then*

$$(R) \int_I f(x)\, dx \geqq (R) \int_I g(x)\, dx.$$

Since the constant function c is continuous, it is Riemann integrable. Also, c is at the same time a step-function $\geqq c$ and a step-function $\leqq c$, so by 11.1

$$(R) \overline{\int} c\, dx \leqq \int_I' c\, dx \leqq (R) \underline{\int_I} c\, dx.$$

This, with 11.2, completes the proof of (*a*).

Let S stand for the class of step-functions $\geqq f(x)$. If $c < 0$, the class of step-functions $\leqq cf(x)$ is the class of functions $cs(x)$ with $s(x)$ in S. So by 10.6 and 5.5

$$\begin{aligned}(R) \underline{\int}_I cf(x)\,dx &= \sup \left\{ \int_I' cs(x)\,dx \mid s(x) \text{ in } S \right\} \\ &= \sup \left\{ c \int_I' s(x)\,dx \mid s(x) \text{ in } S \right\} \\ &= c \inf \left\{ \int_I' s(x)\,dx \mid s(x) \text{ in } S \right\} \\ &= c \cdot (R) \overline{\int}_I f(x)\,dx = c \cdot (R) \int_I f(x)\,dx.\end{aligned}$$

In just the same way

$$(R) \overline{\int}_I cf(x)\,dx = c \cdot (R) \int_I f(x)\,dx.$$

This establishes (b) if $c < 0$. If $c > 0$, then by the proof just completed $(-c)f(x)$ is Riemann integrable, and so is $(-1)[(-c)f(x)]$; and

$$\begin{aligned}(R) \int_I cf(x)\,dx &= (R) \int_I (-1)(-c)f(x)\,dx = -(R) \int_I (-c)f(x)\,dx \\ &= c \cdot (R) \int_I f(x)\,dx.\end{aligned}$$

If $c = 0$, (b) is trivial (it reduces to (a) with $c = 0$).

Let ϵ be any positive number. There are step-functions $s_1(x)$, $s_2(x)$ such that $s_1(x) \leqq f(x)$, $s_2(x) \leqq g(x)$, and

$$\int_I' s_1(x)\,dx > (R) \int_I f(x)\,dx - \frac{\epsilon}{2},$$

$$\int_I' s_2(x)\,dx > (R) \int_I g(x)\,dx - \frac{\epsilon}{2}.$$

Hence by 10.6

$$\int_I' (s_1(x) + s_2(x)) > (R) \int_I f(x)\,dx + (R) \int_I g(x)\,dx - \epsilon.$$

But $s_1(x) + s_2(x) \leqq f(x) + g(x)$, so by the definition 11.1 of the lower integral of $f(x) + g(x)$

$$(R) \underline{\int}_I (f(x) + g(x))\,dx > (R) \int_I f(x)\,dx + (R) \int_I g(x)\,dx - \epsilon.$$

This holds for every $\epsilon > 0$, so by 5.2

$$(R) \underline{\int}_I (f(x) + g(x))\,dx \geqq (R) \int_I f(x)\,dx + (R) \int_I g(x)\,dx.$$

Similarly, using step-functions $s_1(x)$, $s_2(x)$ such that $s_1(x) \geqq f(x)$, $s_2(x) \geqq g(x)$, we can establish

$$(R) \overline{\int_I} (f(x) + g(x))\, dx \leqq (R) \int_I f(x)\, dx + (R) \int_I g(x)\, dx.$$

This with 11.2 establishes (*c*).

To establish (*d*), we observe that 0 is a step-function $\leqq f(x)$. So

$$(R) \int_I f(x)\, dx \geqq \int_I' 0\, dx = 0.$$

For (*e*) we have by (*b*), (*c*) and (*d*)

$$(R) \int_I f(x)\, dx - (R) \int_I g(x)\, dx = (R) \int_I (f(x) - g(x))\, dx \geqq 0.$$

Let us specialize these results to continuous functions. For such functions $\varphi(x)$ we denote the integral of $\varphi(x)$ by $\int_I \varphi(x)\, dx$, omitting the (*R*) before the integral sign; and we shall call the integral the "elementary integral" or "Cauchy integral" of $\varphi(x)$. For such functions we have already proved in 11.5 the following statements.

11.6s. *Let the functions $\varphi(x)$ and $\psi(x)$ be continuous on a closed nterval I, and let k_1 and k_2 be finite constants. Then*

(*a*) $$\int_I k_1\, dx = k_1 \Delta I,$$

(*b*) $$\int_I (k_1\varphi(x) + k_2\psi(x))\, dx = k_1 \int_I \varphi(x)\, dx + k_2 \int_I \psi(x)\, dx,$$

(*c*) *if* $\varphi(x) \geqq \psi(x)$, *then* $\int_I \varphi(x)\, dx \geqq \int_I \psi(x)\, dx.$

Two further properties of the elementary* integral will be needed.

11.7s. *Let $\varphi(x)$ be continuous on a closed interval I, and let I_1, I_2, $\cdots$, I_n be a partition of I into non-overlapping closed intervals. Then*

$$\sum_{j=1}^{n} \int_{I_j} \varphi(x)\, dx = \int_I \varphi(x)\, dx.$$

Let ϵ be any positive number. There are step-functions $s(x)$, $S(x)$ such that $s(x) \leqq \varphi(x) \leqq S(x)$ and

$$\int_I \varphi(x)\, dx + \epsilon > \int_I' S(x)\, dx \geqq \int_I \varphi(x)\, dx \geqq \int_I' s(x)\, dx > \int_I \varphi(x)\, dx - \epsilon.$$

* They are also properties of the Riemann integral.

On each I_j it is still true that $s(x) \leqq \varphi(x) \leqq S(x)$, so

$$\int'_{I_j} S(x)\,dx \geqq \int_{I_j} \varphi(x)\,dx \geqq \int'_{I_j} s(x)\,dx.$$

Adding for $j = 1, \cdots, n$, and using 10.7,

$$\int'_I S(x)\,dx \geqq \sum_{j=1}^{n} \int_{I_j} \varphi(x)\,dx \geqq \int'_I s(x)\,dx.$$

Hence the numbers $\int_I \varphi(x)\,dx$ and $\Sigma \int'_{I_j} \varphi(x)\,dx$ are both between $\int'_I S(x)\,dx$ and $\int'_I s(x)\,dx$, which differ by less than 2ϵ; so

$$\left| \int_I \varphi(x)\,dx - \sum_{j=1}^{n} \int_{I_j} \varphi(x)\,dx \right| < 2\epsilon.$$

This holds for every positive ϵ, so by 5.2 the two are equal.

11.8. *The elementary integral is invariant under translation.*

Let $\varphi(x)$ be continuous on a closed interval I, and let $h = (h^{(1)}, \cdots, h^{(q)})$ be any q-tuple. The translation $\varphi^{(h)}(x)$ is continuous on $I^{(h)}$. For each positive ϵ there are step-functions $s(x)$, $S(x)$ such that $s(x) \leqq \varphi(x) \leqq S(x)$ and

$$\int_I \varphi(x)\,dx + \epsilon > \int'_I S(x)\,dx \geqq \int'_I s(x)\,dx > \int_I \varphi(x)\,dx - \epsilon.$$

Then for the translation we have $s^{(h)}(x) \leqq \varphi^{(h)}(x) \leqq S^{(h)}(x)$, while by 10.9

$$\int'_{I^{(h)}} s^{(h)}(x)\,dx = \int'_I s(x)\,dx,$$
$$\int'_{I^{(h)}} S^{(h)}(x)\,dx = \int'_I S(x)\,dx.$$

Thus the integral of $\varphi^{(h)}$ over $I^{(h)}$ is between the integrals of s and S over I, and by the preceding inequality

$$\int_I \varphi(x)\,dx - \epsilon < \int_{I^{(h)}} \varphi^{(h)}(x)\,dx < \int \varphi(x)\,dx + \epsilon.$$

Since ϵ is arbitrary, by 5.2 the integrals are equal.

12. If we wish to generalize the concept of integral, a very natural attempt would be to define a new kind of upper and lower integral by replacing step-functions by Riemann-integrable functions in the definition 11.1. This attempt is a failure; the apparently new integral thus obtained is merely the Riemann integral back again. Another possibility is this. If a function happens to be representable in the form sup $\{f_1(x), f_2(x), \cdots\}$, where the (denumerable) set $\{f_n(x)\}$ consists of functions which we can integrate, we could define the new integral of $f(x)$ to be the sup of the integrals of all those integrable functions $g(x) \leqq f(x)$. There are some serious defects in this, the most obvious being that if $f(x)$ can be integrated thus, it is not necessarily true that $-f(x)$ can also be integrated. So this kind of integration is not a desirable end in itself. Nevertheless, we shall investigate it, not as an end in itself, but as a means to an end, just as the integration of step-functions was a means to an end. For we shall use the functions integrable by this somewhat primitive and unsymmetrical method to replace the step-functions in 11.1; and we shall find that the integral thus defined is a highly important generalization of the Riemann integral, being in fact the Lebesgue integral.

In one minor respect we shall depart from the program just suggested. We suggested studying functions which were the upper bounds of collections of Riemann-integrable functions. Instead, we shall study the functions which are the upper bounds of collections of continuous functions. In the end we lose no generality, and we gain the advantage that the continuous functions are a simpler class to handle than the Riemann integrable functions.

By 7.5, a function which is the sup of a collection of functions continuous on a closed interval I is lower semi-continuous, and obviously it is bounded below. Likewise, a function which is the inf of a collection of functions continuous on I is upper semi-continuous and is bounded above. So we now introduce the definition of the next class of functions for which the process of integration will be defined:

12.1s. *A function $u(x)$, defined on a closed interval I, will be called a U-function if (a) $u(x)$ is lower semi-continuous on I, and (b) $u(x)$ is bounded below; that is, there is a finite constant M such that $u(x) \geqq M$ for all x in I. Furthermore, a function $l(x)$, defined on a closed interval I will be called an L-function if (a) it is upper semi-continuous on I, and (b) it is bounded above.*

Before discussing the integrals of these functions, we shall stop to investigate a few of their properties.

12.2s. *Let $f(x)$ be defined on the closed interval I. Then $f(x)$ is continuous on I if and only if it is both a U-function and an L-function.*

For if $f(x)$ is continuous, it is bounded both above and below, and by 7.1 it is both upper and lower semi-continuous; so it is both a U- and an L-function. Conversely, if $f(x)$ is both a U- and an L-function, it is bounded both below and above, so it is finite. Also it is both upper and lower semi-continuous, so by 7.1 it is continuous.

12.3s. *If $u(x)$ is a U-function and $l(x)$ is an L-function, then if $0 \leqq c < \infty$ the function $c \cdot u(x)$ is a U-function and $c \cdot l(x)$ is an L-function; while if $-\infty < c \leqq 0$ the function $c \cdot u(x)$ is an L-function and $c \cdot l(x)$ is a U-function.*

There are finite numbers M, N such that $u(x) \geqq M$ and $l(x) \leqq N$. If $c < 0$, then $c \cdot u(x) \leqq cM$ and upper semi-continuous, and $c \cdot l(x) \geqq cN$ and lower semi-continuous, by 7.2; so $c \cdot u(x)$ is an L-function, and $c \cdot l(x)$ is a U-function.

If $c > 0$, then we apply the result for $c < 0$ twice; $-c \cdot u(x)$ is an L-function, so $(-1)(-c)u(x)$ is again a U-function. Similarly $c \cdot l(x)$ is an L-function.

If $c = 0$, then $c \cdot l(x)$ and $c \cdot u(x)$ are both identically 0, so they are both U-functions and both L-functions by 12.2.

12.4s. *If the functions $f_1(x), \cdots, f_n(x)$ are all U-functions (or all L-functions) on I, then the functions $f_1(x) + f_2(x) + \cdots + f_n(x)$, $\sup \{f_1(x), \cdots, f_n(x)\}$ and $\inf \{f_1(x), \cdots, f_n(x)\}$ are all U-functions (or all L-functions).*

Suppose that all the $f_i(x)$ are U-functions. Then they are lower semi-continuous, and there are finite numbers M_i such that $f_i(x) \geqq M_i$. By a simple induction, we learn from 7.7 that $f_1(x) + f_2(x) + \cdots + f_n(x)$ is lower semi-continuous. By 7.5, $\sup \{f_1(x), \cdots, f_n(x)\}$ is lower semi-continuous, and by 7.6 $\inf \{f_1(x), \cdots, f_n(x)\}$ is lower semi-continuous. Moreover,

$$\begin{aligned} f_1(x) + \cdots + f_n(x) &\geqq M_1 + \cdots + M_n, \\ \sup \{f_1(x), \cdots, f_n(x)\} &\geqq \inf \{f_1(x), \cdots, f_n(x)\} \\ &\geqq \inf \{M_1, \cdots, M_n\}. \end{aligned}$$

Hence if the f_i are U-functions, 12.4 is proved.

Next, suppose that the $f_i(x)$ are L-functions. Then the functions $-f_i(x)$ are U-functions, and by the preceding paragraph so are $(-f_1(x)) + (-f_2(x)) + \cdots + (-f_n(x))$, $\sup \{(-f_1(x)), \cdots, (-f_n(x))\}$, $\inf \{(-f_1(x)), \cdots, (-f_n(x))\}$. But these functions are the negatives of $f_1(x) + \cdots + f_n(x)$, $\inf \{f_1(x), \cdots, f_n(x)\}$ and $\sup \{f_1(x), \cdots, f_n(x)\}$ respectively; so these last three functions are all L-functions.

12.5s. *If $f(x)$ and $g(x)$ are non-negative U-functions, so is $f(x)g(x)$; and if $f(x)$ and $g(x)$ are non-negative L-functions, so is $f(x)g(x)$.*

If f and g are U-functions, they are lower semi-continuous, so by 7.8 $f(x)g(x)$ is lower semi-continuous. Since furthermore $f(x)g(x) \geqq 0$, it is a U-function.

If f and g are L-functions, then they are upper semi-continuous, and there are finite numbers M, N such that $f(x) \leqq M$ and $g(x) \leqq N$. Then by 7.8 the product $f(x)g(x)$ is uppeı semi-continuous. Moreover, $f(x)g(x) \leqq MN$, so it is an L-function.

12.6s. *If $f(x)$ is the* sup *of a collection of U-functions, it is a U-function; and if it is the* inf *of a collection of L-functions, it is an L-function.*

We prove the first statement; the second is obtained by changing signs. Since $f(x)$ is the sup of a collection of lower semi-continuous functions, it is lower semi-continuous by 7.5. If $g(x)$ is any function of the collection, it is bounded below, say $g(x) \geqq M$, because it is a U-function. So $f(x)$, which is the sup of all functions of the collection, is $\geqq g(x) \geqq M$. Therefore $f(x)$ is a U-function.

REMARK. The functions $u_n(x) \equiv n$, $n = 0, \pm 1, \pm 2, \cdots$ are all continuous, therefore by 12.2 they are U- and L-functions. Hence their sup, ∞, is a U-function and their inf, $-\infty$, is an L-function.

12.7s. Corollary. *If $f(x)$ is the limit of an increasing sequence of U-functions, it is a U-function, and if it is the limit of a decreasing sequence of L-functions it is an L-function.*

Suppose $u_1(x) \leqq u_2(x) \leqq \cdots$ and $\lim_{n\to\infty} u_n(x) = f(x)$, where the u_n are U-functions. Then $f(x) = \sup u_n(x)$, so $f(x)$ is a U-function by 12.6. The other part of the theorem follows by change of signs.

12.8s. *Let $u_1(x), u_2(x), \cdots$ be U-functions on a closed interval I, and let $l(x)$ be an L-function on I. If $u_1(x) \leqq u_2(x) \leqq \cdots$ and $\lim_{n\to\infty} u_n(x) > l(x)$ for all x in I, then there is an n_0 such that $u_n(x) > l(x)$ for all $n \geqq n_0$.*

Put $E_n = \{x \mid x \text{ in } I,\ u_n(x) \leqq l(x)\}$. Each set E_n is closed, by 7.4 and 2.7. Also $E_n \supset E_{n+1}$, for if x is in E_{n+1} then $u_n(x) \leqq u_{n+1}(x) \leqq l(x)$, so x is also in E_n. If there were a point x_0 belonging to all the sets E_n, we would have $u_n(x_0) \leqq l(x_0)$ for all n, hence $\lim_{n\to\infty} u_n(x_0) \leqq l(x_0)$. This is contrary to hypothesis, so there is no point x_0 belonging to all the sets E_n. But by 2.14, this implies that there is an n_0 such that all the sets E_n with $n \geqq n_0$ are empty. That is, if $n \geqq n_0$ there are no points at which $u_n(x) \leqq l(x)$, so for these values of n it must be true that $u_n(x) > l(x)$.

EXERCISE. If $\varphi_0(x), \varphi_1(x), \cdots$ are all continuous on a closed interval I, and $\varphi_n(x) \to \varphi_0(x)$ for each x in I, and $\varphi_1(x) \leqq \varphi_2(x) \leqq \cdots$, then $\varphi_n(x)$ tends to $\varphi_0(x)$ uniformly on I. Moreover, in this statement I can be replaced by any bounded closed set.

Let us recall that the *characteristic function* $K_E(x)$ of a set E is the function which has the value 1 if x is in E and the value 0 if x is in

CE. It is evident that the behavior of this function will depend on the structure of the set E. One relationship between the set E and its characteristic function is stated in the next theorem.

12.9s. *Let E be a set contained in a closed interval I. The characteristic function $K_E(x)$ is a U-function on I if and only if E is open relative to I, and is an L-function on I if and only if E is closed.*

If $K_E(x)$ is a U-function, it is lower semi-continuous. By 7.4, the set of x in I such that $K_E(x) > \frac{1}{2}$ is open relative to I. But this set is E itself. Conversely, suppose E open relative to I. Since $K_E(x)$ is bounded below, it will be a U-function if we can prove that it is lower semi-continuous. So let x_0 be any point of I. First, if x_0 is on CE, then $K_E(x_0) = 0$. For every $h < K_E(x_0) = 0$ there is an ϵ (for example, $\epsilon = 1$) such that $K_E(x) > h$ if x is in $N_\epsilon(x_0)I$. Thus by 7.3 $K_E(x)$ is lower semi-continuous at x_0. Second, if x_0 is in E, then $K_E(x_0) = 1$. Let h be any number less than 1. Since E is open relative to I, there is an $\epsilon > 0$ such that $N_\epsilon(x_0)I \subset E$, so that $K_E(x) = 1 > h$ for all x in $N_\epsilon(x_0)I$. Thus by 7.4 $K_E(x)$ is lower semicontinuous at x_0, which completes the proof of the first statement.

By 2.7, E is closed if and only if it is closed relative to I. This is true by 2.5 if and only if $I - E$ is open relative to I. This, as we have just proved, is true if and only if $K_{I-E}(x)$ is a U-function. This is true, by 12.3, if and only if $-K_{I-E}(x)$ is an L-function. This is true, by 12.2 and 12.4, if and only if $1 - K_{I-E}(x)$ is an L-function. But $1 - K_{I-E}(x)$ is identically equal to $K_E(x)$. So E is closed if and only if $K_E(x)$ is an L-function.

If, for example, E consists of a single point, then $K_E(x)$ is an L-function and $K_{I-E}(x)$ is a U-function. Both of these are discontinuous, which proves that the class of U-functions and the class of L-functions are definitely larger than the class of continuous functions.

13. We have now established all the facts concerning U-functions and L-functions which we need. Now we extend the definition of the integral to these functions.

13.1s. Definition. *If $u(x)$ is a U-function on a closed interval I, $\int_I u(x)\,dx$ is defined to be* sup $\int_I \varphi(x)\,dx$ *for all continuous functions $\varphi(x) \leqq u(x)$. If $l(x)$ is an L-function on I, $\int_I l(x)\,dx$ is defined to be* inf $\int_I \varphi(x)\,dx$ *for all continuous functions $\varphi(x) \geqq l(x)$.*

Remark 1. If $u(x)$ is a U-function, it is bounded below; so there really do exist continuous functions $\varphi(x) \leqq u(x)$. Likewise if $l(x)$ is an L-function there is a continuous (constant) function $\geqq l(x)$.

REMARK 2. If $u(x)$ is a *bounded* U-function, say $| u(x) | \leqq M < \infty$, then for every continuous function $\varphi(x) \leqq u(x)$ we have $\varphi(x) \leqq M$, whence

$$\int_I \varphi(x)\, dx \leqq M \cdot \Delta I.$$

Therefore the integral of u is finite, and does not exceed $M\Delta I$. Likewise every *bounded* L-function has a finite integral.

We now have definitions of the integral for three classes of functions, namely continuous functions, U-functions and L-functions. The question arises—are these definitions ever inconsistent? That is, if a function is, say, both a U- and an L-function, we have two different definitions for its integral. Are the integrals equal? By 12.2, if a function is in two of the three classes it is in all three. So our problem is to prove that if $\psi(x)$ is a continuous function, the three integrals coincide; that is,

$$(a) \qquad \int_I \psi(x)\, dx = \inf \int_I \varphi(x)\, dx$$

foı all continuous functions $\varphi(x) \geqq \psi(x)$;

$$(b) \qquad \int_I \psi(x)\, dx = \sup \int_I \varphi(x)\, dx$$

for all continuous functions $\varphi(x) \leqq \psi(x)$, where the sign $\int$ always means the elementary integral.

Since $\psi(x)$ is itself a continuous function $\geqq \psi(x)$, the right member of (a) is not greater than the left. On the other hand, for every continuous $\varphi(x) \geqq \psi(x)$ we have

$$\int_I \psi(x)\, dx \leqq \int_I \varphi(x)\, dx.$$

So the integral of ψ is a lower bound for the integrals of all such functions φ, whence $\int_I \psi(x)\, dx \leqq \inf \int_I \varphi(x)\, dx$ for all continuous $\varphi(x) \geqq \psi(x)$. We have thus established (a).

If we write (a) in the form $\int_I [-\psi(x)]\, dx = \inf \int_I [-\varphi(x)]\, dx$ for all continuous functions $-\varphi(x) \geqq -\psi(x)$, and then apply 11.6 and 5.5, we obtain (b).

Thus the integral as defined in 13.1 is a true extension of the elementary integral as defined for continuous functions, and (within the class of U- and L-functions) the symbol $\int_I f(x)\, dx$ always has exactly one meaning.

13.2s. *If $f(x)$ and $g(x)$ are both U-functions or both L-functions on a closed interval I, and $f(x) \geqq g(x)$, then*

$$\int_I f(x)\,dx \geqq \int_I g(x)\,dx.$$

Suppose them both U-functions. Then the class F of all continuous functions $\leqq f(x)$ contains the class G of all continuous functions $\leqq g(x)$, and by 5.1

$$\begin{aligned}\int f(x)\,dx &= \sup\left\{\int_I \varphi(x)\,dx \mid \varphi(x) \text{ in } F\right\}\\ &\geqq \sup\left\{\int_I \varphi(x)\,dx \mid \varphi(x) \text{ in } G\right\}\\ &= \int_I g(x)\,dx.\end{aligned}$$

The proof in case $f(x)$ and $g(x)$ are both L-functions is similar.

13.3s. *Let $u_1(x) \leqq u_2(x) \leqq \cdots$ be a sequence of U-functions on a closed interval I, and let $g(x) = \lim_{n\to\infty} u_n(x)$. Then $g(x)$ is a U-function, and*

$$\lim_{n\to\infty} \int_I u_n(x)\,dx = \int_I g(x)\,dx.$$

For each x the numbers $u_n(x)$ increase with n, and so do the numbers $\int u_n(x)\,dx$ by 13.2. Hence the limits mentioned in the theorem exist, by the exercise after 6.8. Also, $g(x)$ is a U-function, by 12.7.

Let h be any (finite) number less than $\int_I g(x)\,dx$. We can find a number k greater than h, but less than the value of the integral, and by 13.1 we can find a continuous function $\varphi_1(x) \leqq g(x)$ such that

$$\int_I \varphi_1(x)\,dx > k.$$

Define $\varphi(x) = \varphi_1(x) - (k - h)/(\Delta I + 1)$. Then $\varphi(x)$ is continuous, and $\varphi(x) < \varphi_1(x) \leqq g(x)$, and by 11.6

$$\int_I \varphi(x)\,dx = \int_I \varphi_1(x)\,dx - [(k-h)/(\Delta I + 1)]\int_I 1\,dx > k - (k-h) = h.$$

Now $\lim u_n(x) = g(x) > \varphi(x)$, so by 12.8 there is an n_0 such that $u_n(x) > \varphi(x)$ for all x if $n \geqq n_0$. Then by 13.1 or 13.2

$$\int_I u_n(x)\,dx \geqq \int_I \varphi(x)\,dx > h$$

for all $n \geqq n_0$. Hence

$$\lim_{n\to\infty} \int_I u_n(x)\,dx > h.$$

This holds for all h less than the integral of g, so by 5.2

$$\lim_{n\to\infty} \int_I u_n(x)\,dx \geqq \int_I g(x)\,dx.$$

On the other hand, for all n we have $u_n(x) \leqq g(x)$, so by 13.2

$$\int_I u_n(x)\,dx \leqq \int g(x)\,dx.$$

Hence

$$\lim_{n\to\infty} \int_I u_n(x)\,dx \leqq \int_I g(x)\,dx.$$

This completes the proof of the theorem.

13.4s. Corollary. *If* $\varphi_1(x) \leqq \varphi_2(x) \leqq \cdots$ *is a sequence of functions continuous on* I *such that* $\lim_{n\to\infty} \varphi_n(x) = g(x)$, *then*

$$\lim_{n\to\infty} \int_I \varphi_n(x)\,dx = \int_I g(x)\,dx.$$

This follows at once from 13.3 and the consistency proof just after 13.1.

13.5s. *If* $f(x)$ *is a U-function or an L-function on* I *and* c *is a finite constant, then*

$$\int_I c \cdot f(x)\,dx = c \int_I f(x)\,dx.$$

Suppose that $f(x)$ is a U-function. Let F be the class of all continuous functions $\varphi(x) \leqq f(x)$. If $c < 0$, the class of continuous functions $\geqq cf(x)$ consists of the functions $c\varphi(x)$ with $\varphi(x)$ in F. Also, the function $cf(x)$ is an L-function by 12.3. Hence, using 13.1, 11.6 and 5.5,

$$\begin{aligned}\int_I cf(x)\,dx &= \inf \left\{\int_I c\varphi(x)\,dx \mid \varphi \text{ in } F\right\}\\ &= \inf \left\{c \int_I \varphi(x)\,dx \mid \varphi \text{ in } F\right\}\\ &= c \cdot \sup \left\{\int_I \varphi(x)\,dx \mid \varphi \text{ in } F\right\}\\ &= c \cdot \int_I f(x)\,dx.\end{aligned}$$

If $f(x)$ is an L-function and $c < 0$, then $cf(x)$ is a U-function by 12.3, and by the proof just completed

$$\begin{aligned} c \int_I f(x)\,dx &= c \int_I c^{-1} \cdot [cf(x)]\,dx \\ &= c\{c^{-1} \int_I cf(x)\,dx\} \\ &= \int_I cf(x)\,dx. \end{aligned}$$

This completes the proof for $c < 0$.

If $c > 0$, by the proof just completed for negative c we have

$$\begin{aligned} \int_I cf(x)\,dx &= \int_I (-1)(-c)f(x)\,dx = (-1) \int_I (-c)f(x)\,dx \\ &= (-1)(-c) \int_I f(x)\,dx = c \int_I f(x)\,dx. \end{aligned}$$

If $c = 0$, the equation to be proved is

$$\int_I 0\,dx = 0,$$

which is obvious.

13.6s. *If $f(x)$ and $g(x)$ are both U-functions or both L-functions on I, then*

$$\int_I [f(x) + g(x)]\,dx = \int_I f(x)\,dx + \int_I g(x)\,dx.$$

Suppose them both U-functions. By 7.9, we can find sequences $\varphi_1(x) \leqq \varphi_2(x) \leqq \cdots$ and $\psi_1(x) \leqq \psi_2(x) \leqq \cdots$ of functions continuous on I such that

$$\lim_{n\to\infty} \varphi_n(x) = f(x), \qquad \lim_{n\to\infty} \psi_n(x) = g(x).$$

Then $\varphi_1(x) + \psi_1(x) \leqq \varphi_2(x) + \psi_2(x) \leqq \cdots$, and

$$\lim_{n\to\infty} [\varphi_n(x) + \psi_n(x)] = f(x) + g(x).$$

So by 13.4 and 11.6

$$\begin{aligned} \int_I [f(x) + g(x)]\,dx &= \lim_{n\to\infty} \int_I [\varphi_n(x) + \psi_n(x)]\,dx \\ &= \lim_{n\to\infty} \int_I \varphi_n(x)\,dx + \lim_{n\to\infty} \int_I \psi_n(x)\,dx \\ &= \int_I f(x)\,dx + \int_I g(x)\,dx. \end{aligned}$$

Suppose now that f and g are both L-functions. Then $f(x) + g(x)$ is an L-function, and $-f$ and $-g$ are U-functions. By the proof above and 13.5 (with $c = -1$)

$$\begin{aligned}\int_I [f(x) + g(x)]\,dx &= -\int_I [-f(x) - g(x)]\,dx \\ &= -\left[\int_I [-f(x)\,dx] + \int_I [-g(x)]\,dx\right] \\ &= \int_I f(x)\,dx + \int_I g(x)\,dx.\end{aligned}$$

13.7s. Corollary. *If $u(x)$ is a U-function and $l(x)$ is an L-function, and $u(x) \geqq l(x)$, then*

$$\int_I u(x)\,dx \geqq \int_I l(x)\,dx.$$

Since $-l(x)$ is a U-function and $u(x) + [-l(x)] \geqq 0$, by 13.5, 13.6 and 13.2

$$\int_I u(x)\,dx - \int_I l(x)\,dx = \int_I [u(x) - l(x)]\,dx \geqq \int_I 0\,dx = 0.$$

13.8s. *If $f(x)$ is a U-function or an L-function on I, and $I_1, \cdots, I_n$ are non-overlapping closed intervals such that $\bigcup I_i = I$, then*

$$\int_I f(x)\,dx = \sum_{i=1}^{n} \int_{I_i} f(x)\,dx.$$

We prove this for U-functions; the theorem for L-functions follows by change of sign. Suppose then that $f(x)$ is a U-function on I, and let $\varphi_1(x) \leqq \varphi_2(x) \leqq \cdots$ be a sequence of continuous functions tending everywhere to $f(x)$. (Such a sequence exists by 7.9.) Then the $\varphi_m(x)$ are also continuous on each I_i. By 13.4 and 11.7,

$$\begin{aligned}\int_I f(x)\,dx &= \lim_{m\to\infty} \int_I \varphi_m(x)\,dx = \lim_{m\to\infty} \left(\sum_{i=1}^{n} \int_{I_i} \varphi_m(x)\,dx\right) \\ &= \sum_{i=1}^{n} \left(\lim_{m\to\infty} \int_{I_i} \varphi_m(x)\,dx\right) = \sum_{i=1}^{n} \int_{I_i} f(x)\,dx.\end{aligned}$$

13.9. *The integrals of U-functions and of L-functions are invariant under translation.*

Let $u(x)$ be a U-function on a closed interval I, and let

$$(h^{(1)}, \cdots, h^{(q)})$$

be any q-tuple of real numbers. There is a sequence $\varphi_1(x)$, $\varphi_2(x)$, $\cdots$ of functions continuous on I such that $\varphi_1(x) \leqq \varphi_2(x) \leqq \cdots$

and lim $\varphi_n(x) = u(x)$. The translations $\varphi_i^{(h)}(x)(i = 1, 2, \cdots)$ are continuous on $I^{(h)}$, increase with increasing i, and approach $u^{(h)}(x)$. Hence by 12.7 $u^{(h)}$ is a U-function, while by 13.4 and 11.8

$$\begin{aligned}\int_{I^{(h)}} u^{(h)}(x)\,dx &= \lim_{i\to\infty} \int_{I^{(h)}} \varphi_i^{(h)}(x)\,dx \\ &= \lim_{i\to\infty} \int_I \varphi_i(x)\,dx \\ &= \int_I u(x)\,dx.\end{aligned}$$

The statement about L-functions follows by change of sign.

14. Our tools are now ready, and we proceed to use them to define the upper and lower Daniell integrals of an arbitrary function $f(x)$. These upper and lower integrals will be used in the next section to define the Lebesgue integral itself.

14.1s. Definition. *Let $f(x)$ be any function defined on the closed interval I. Then its upper Daniell integral is*

$$\overline{\int_I} f(x)\,dx \equiv \inf \int_I u(x)\,dx$$

for all U-functions $u(x) \geqq f(x)$; its lower Daniell integral is*

$$\underline{\int_I} f(x)\,dx = \sup \int_I l(x)\,dx$$

for all L-functions $l(x) \leqq f(x)$.

REMARK 1. If $f(x)$ is bounded on I, say $|f(x)| \leqq M$, it lies between the continuous functions $-M$ and M. So by 14.1 its upper and lower integrals over I are finite.

REMARK 2. In definition 14.1 the interval I is distinguished from all other intervals by being the common range of definition of all the functions mentioned in the definition. Accordingly we shall call I the "basic interval" in the definition 14.1.

The names "upper integral" and "lower integral" are justified by the following theorem.

14.2s. *For every function $f(x)$ defined on a closed interval I,*

$$\underline{\int_I} f(x)\,dx \leqq \overline{\int_I} f(x)\,dx.$$

* The class of all U-functions $\geqq f(x)$ is surely not empty, for by the remark after 12.6 the function $u(x) \equiv \infty$ is a U-function. Likewise the class of L-functions $\leqq f(x)$ is not empty, since $l(x) \equiv -\infty$ is an L-function.

Let $l(x)$ be any L-function equal to or less than $f(x)$. Then for every U-function $u(x) \geqq f(x)$ we have $u(x) \geqq l(x)$, so by 13.7

$$\int_I l(x)\,dx \leqq \int_I u(x)\,dx.$$

This holds for all U-functions $u(x) \geqq f(x)$, so we can replace the term on the right by its inf:

$$\int_I l(x)\,dx \leqq \overline{\int_I} f(x)\,dx.$$

This again holds for all L-functions $l(x) \leqq f(x)$, so we can replace the term on the left by its sup:

$$\underline{\int_I} f(x)\,dx \leqq \overline{\int_I} f(x)\,dx.$$

14.3s. *For every function $f(x)$ defined on I,*

$$\overline{\int_I} (-f(x))\,dx = -\underline{\int_I} f(x)\,dx.$$

If L is the class of all L-functions $l(x) \leqq f(x)$, their negatives $-l(x)$ form the class of U-functions $\geqq -f(x)$. Hence by 14.1, 13.5 (with $c = -1$) and 5.5

$$\begin{aligned}
\overline{\int_I} (-f(x))\,dx &= \inf \left\{ \int_I (-l(x))\,dx \mid l(x) \text{ in } L \right\} \\
&= \inf \left\{ -\int_I l(x)\,dx \mid l(x) \text{ in } L \right\} \\
&= -\sup \left\{ \int_I l(x)\,dx \mid l(x) \text{ in } L \right\} \\
&= -\underline{\int_I} f(x)\,dx.
\end{aligned}$$

14.4s. *If $f(x) \geqq g(x)$ for all x in a closed interval I, then*

$$\overline{\int_I} f(x)\,dx \geqq \overline{\int_I} g(x)\,dx \qquad \text{and} \qquad \underline{\int_I} f(x)\,dx \geqq \underline{\int_I} g(x)\,dx.$$

To prove the first statement, let F be the class of all U-functions $\geqq f(x)$, and let G be the class of U-functions $\geqq g(x)$. Since $f(x) \geqq g(x)$, the class G contains F. So by 5.1 $\inf \{\int u(x)\,dx \mid u \text{ in } F\} \geqq \inf \{\int u(x)\,dx \mid u \text{ in } G\}$. By definition 14.1, this proves the first part.

For the second part, since $-f(x) \leqq -g(x)$, we can use the preceding paragraph together with 14.3 to obtain

$$\underline{\int} f(x)\,dx = -\overline{\int} (-f(x))\,dx \geqq -\overline{\int} (-g(x)\,dx = \underline{\int} g(x)\,dx.$$

14.5s. *Let $f(x)$ and $g(x)$ be any functions defined on a closed interval I and such that $f(x) + g(x)$ is also defined on I. Then*

$$\overline{\int} (f(x) + g(x))\,dx \leqq \overline{\int} f(x)\,dx + \overline{\int} g(x)\,dx,$$
$$\underline{\int} (f(x) + g(x))\,dx \geqq \underline{\int} f(x)\,dx + \underline{\int} g(x)\,dx,$$

provided that the right members of these inequalities are defined.

If $u_1(x) \geqq f(x)$ and $u_2(x) \geqq g(x)$ are U-functions, then $u_1(x) + u_2(x)$ is a U-function by 12.4, and $u_1(x) + u_2(x) \geqq f(x) + g(x)$. So by 14.1 and 13.6

$$\overline{\int_I} [f(x) + g(x)]\,dx \leqq \int_I [u_1(x) + u_2(x)]\,dx$$
$$= \int_I u_1(x)\,dx + \int_I u_2(x)\,dx.$$

If we suppose $u_1(x)$ held fixed at any particular U-function $\geqq f(x)$, this holds for all U-functions $u_2(x) \geqq g(x)$, so it holds if we replace the last term by its lower bound:

$$\overline{\int_I} [f(x) + g(x)]\,dx \leqq \int_I u_1(x)\,dx + \overline{\int_I} g(x)\,dx.$$

Since this is true for every U-function $u_1(x) \geqq f(x)$, we can replace the first term on the right by its lower bound:

$$\overline{\int_I} [f(x) + g(x)]\,dx \leqq \overline{\int_I} f(x)\,dx + \overline{\int_I} g(x)\,dx.$$

This establishes the first inequality. The second is obtained from this by a change of sign, using 14.3:

$$\underline{\int_I} [f(x) + g(x)]\,dx = -\left\{\overline{\int_I} [-f(x) - g(x)]\,dx\right\}$$
$$\geqq -\left\{\overline{\int_I} (-f(x))\,dx + \overline{\int_I} (-g(x))\,dx\right\}$$
$$= \underline{\int_I} f(x)\,dx + \underline{\int_I} g(x)\,dx.$$

14.6s. *If $f(x)$ is defined on a closed interval I, and $c > 0$, then*

$$\overline{\int_I} cf(x)\,dx = c\,\overline{\int_I} f(x)\,dx \quad \textit{and} \quad \underline{\int_I} cf(x)\,dx = c\,\underline{\int} f(x)\,dx.$$

If U is the class of all U-functions $u(x) \geqq f(x)$, then the functions $c \cdot u(x)$, $u(x)$ in U, constitute the class of U-functions $\geqq c \cdot f(x)$ (see

12.3). Therefore by 14.1, 13.5 and 5.5,

$$\begin{aligned}\overline{\int_I} cf(x)\,dx &= \inf\left\{\int_I cu(x)\,dx \mid u(x) \text{ in } U\right\}\\ &= \inf c\cdot\left\{\int_I u(x)\,dx \mid u(x) \text{ in } U\right\}\\ &= c\cdot\inf\left\{\int_I u(x)\,dx \mid u(x) \text{ in } U\right\}\\ &= c\,\overline{\int_I} f(x)\,dx.\end{aligned}$$

To prove the second equation, we use the first equation and 14.3:

$$\underline{\int_I} cf(x)\,dx = -\overline{\int_I} c[-f(x)]\,dx = c\left[-\overline{\int_I}[-f(x)]\,dx\right] = c\,\underline{\int_I} f(x)\,dx.$$

15. As with the Riemann integral, we use the upper and lower integrals to define the (Lebesgue) integral of $f(x)$. (See page 383.)

15.1s. Definition. *If $f(x)$ is defined on the closed interval I, and its upper and lower integrals are finite and equal, $f(x)$ is said to be summable over I, and its integral is defined to be the common value of its upper and lower integrals:*

$$\int_I f(x)\,dx \equiv \overline{\int_I} f(x)\,dx \equiv \underline{\int_I} f(x)\,dx.$$

Our first duty is to show that this new integral generalizes and is consistent with the integral which we have been using in the preceding sections. As a matter of fact, in one trivial respect this is not so. The new integral must be finite, whereas the integrals of U-functions as defined in 13.1 could be $+\infty$, and the integrals of L-functions could be $-\infty$. So our first task is to show that if $f(x)$ is a U-function or an L-function whose integral as defined in 13.1 has a finite value i, then

$$\overline{\int_I} f(x)\,dx = \underline{\int_I} f(x)\,dx = i.$$

We suppose first that $f(x)$ is a U-function. On the one hand, $f(x)$ is itself a U-function $\geqq f(x)$, so by 14.1

$$(A) \qquad \overline{\int_I} f(x)\,dx \leqq i.$$

On the other hand, let h be any number less than i. By 13.1, there is a continuous function $\varphi(x) \leqq f(x)$ such that $\int_I \varphi(x)\,dx > h$. But φ

is by 12.2 an L-function $\leqq f(x)$, so by 14.1

$$\int_I \varphi(x)\,dx \leqq \underline{\int_I} f(x)\,dx.$$

Thus for every number $h < i$ the lower integral of f is greater than h, and so by 5.2

$$(B) \qquad \underline{\int_I} f(x)\,dx \geqq i.$$

From (A), (B) and 14.2,

$$\overline{\int_I} f(x)\,dx \leqq i \leqq \underline{\int_I} f(x)\,dx \leqq \overline{\int_I} f(x)\,dx,$$

so the upper and lower integrals of f are both equal to the finite number i, and the integral defined in 15.1 has the same value i as the integral defined in 13.1.

If $f(x)$ is an L-function, then $-f(x)$ is a U-function whose integral is (by 13.5) equal to $-i$. So

$$\overline{\int_I} [-f(x)]\,dx = \underline{\int_I} [-f(x)]\,dx = -i.$$

This, with 14.3, proves that the upper and lower integrals of f are both equal to i. In particular, we have shown that the Lebesgue integral defined in 15.1 is consistent with the Cauchy integral for continuous functions.

Now we begin the investigation of the more elementary properties of the integral.

15.2s. *If $f(x)$ and $g(x)$ are both summable over I, and $f(x) \geqq g(x)$ for all x in I, then*

$$\int_I f(x)\,dx \geqq \int_I g(x)\,dx.$$

This follows at once from 14.4 and the definition 15.1 of the integral.

Before proceeding with the study of our integrals we stop to establish three simple inequalities which we shall need.

15.3s. *If $a_1, a_2, \cdots, a_n$ are numbers finite or $+\infty$ and $b_1, b_2, \cdots, b_n$ are numbers finite or $-\infty$ such that $a_i \geqq b_i$ $(i = 1, \cdots, n)$ then*

(i) $$\sup \{a_1, \cdots, a_n\} \geqq \sup [b_1, \cdots, b_n],$$

(ii) $$\inf \{a_1, \cdots, a_n\} \geqq \inf \{b_1, \cdots, b_n\},$$

(iii) $$\sup \{a_1, \cdots, a_n\} - \sup \{b_1, \cdots, b_n\} \leqq \sum_{i=1}^{n} (a_i - b_i).$$

Since a_i and b_i are functions on the range $\{1, \cdots, n\}$, (i) and (ii) are corollaries of 5.4. Let a_k be the greatest of $a_1, \cdots, a_n$. Then, since each difference $a_i - b_i$ is non-negative,

$$\sup\{a_1, \cdots, a_n\} - \sup\{b_1, \cdots, b_n\} = a_k - \sup\{b_1, \cdots, b_n\}$$
$$\leqq a_k - b_k \leqq \sum_{i=1}^{n} (a_i - b_i).$$

In stating the next basic theorem it is convenient to introduce a pair of symbols for which we shall have frequent use.

15.4s. *Let $f(x)$ be defined (finite or $\pm\infty$) on a set E. The functions $f^+(x)$ and $f^-(x)$ are defined on E by the equations*

$$f^+(x) = \sup\{f(x), 0\}, \qquad f^-(x) = \sup\{-f(x), 0\}.$$

From the definitions we have at once the following trivial equations.

15.5s. *If $f(x)$ is defined on E, then for all x in E the equations*

$$f(x) = f^+(x) - f^-(x), \qquad |f(x)| = f^+(x) + f^-(x)$$

are satisfied.

In the next theorem we establish the basic rules of the arithmetic of summable functions.

15.6s. *Let $f_1(x)$ and $f_2(x)$ both be summable over a closed interval I. Then*

(*a*) *$f_1(x) + f_2(x)$ is summable over I, provided it is defined for all x in I; and in this case*

$$(\alpha) \qquad \int_I [f_1(x) + f_2(x)]\,dx = \int_I f_1(x)\,dx + \int_I f_2(x)\,dx;$$

b) for every finite constant c the function $cf_1(x)$ is summable over I, and

$$(\beta) \qquad \int_I cf_1(x)\,dx = c\int_I f_1(x)\,dx;$$

(*c*) $\sup\{f_1(x), f_2(x)\}$, $\inf\{f_1(x), f_2(x)\}$, $f_1^+(x)$, $f_1^-(x)$, $|f_1(x)|$ *are all summable over I, and*

$$(\gamma) \qquad \int_I |f_1(x)|\,dx \geqq \left|\int_I f_1(x)\,dx\right|;$$

(*d*) *if $f_1(x)$ and $f_2(x)$ are bounded,* $f_1(x)f_2(x)$ is summable over I.*

* This hypothesis is much stronger than necessary, as we shall see in 22.4 and 24.2.

By hypothesis,

$$\overline{\int_I} f_1(x)\,dx = \underline{\int_I} f_1(x)\,dx, \qquad \overline{\int_I} f_2(x)\,dx = \underline{\int_I} f_2(x)\,dx.$$

So to prove (*a*) and (α) we use 14.5 and 14.2:

$$\begin{aligned}\underline{\int_I} [f_1(x) + f_2(x)]\,dx &\geqq \underline{\int_I} f_1(x)\,dx + \underline{\int_I} f_2(x)\,dx \\ &= \int_I f_1(x)\,dx + \int_I f_2(x)\,dx = \overline{\int_I} f_1(x)\,dx \\ &\qquad + \overline{\int_I} f_2(x)\,dx \\ &\geqq \overline{\int_I} [f_1(x) + f_2(x)]\,dx \geqq \underline{\int_I} [f_1(x) + f_2(x)]\,dx.\end{aligned}$$

Hence equality holds throughout.

Statements (*b*) and (β) are clear if $c = 0$, for then $cf(x) = 0$, which has integral 0.

If $c < 0$, then by 14.6 and 14.3

$$\begin{aligned}\overline{\int_I} cf_1(x)\,dx &= \overline{\int_I} |c|(-1)f_1(x)\,dx = |c| \overline{\int_I} [-f_1(x)]\,dx \\ &= -|c| \underline{\int_I} f_1(x)\,dx = c \int_I f_1(x)\,dx = -|c| \overline{\int_I} f_1(x)\,dx \\ &= |c| \underline{\int_I} [-f_1(x)]\,dx = \underline{\int_I} |c|(-1)f_1(x)\,dx = \underline{\int_I} cf_1(x)\,dx.\end{aligned}$$

If $c > 0$, then in the proof just completed $(-c)f_1(x)$ is summable, and therefore so is $(-1)(-c)f_1(x) \equiv cf_1(x)$, and

$$\begin{aligned}\int_I cf_1(x)\,dx = \int_I (-1)(-c)f_1(x)\,dx &= -\int_I (-c)f_1(x)\,dx \\ &= c\int_I f_1(x)\,dx.\end{aligned}$$

Now let ϵ be any positive number. By definition of the integral, there are U-functions $u_1(x)$, $u_2(x)$ and L-functions $l_1(x)$, $l_2(x)$ such that $u_i(x) \geqq f_i(x) \geqq l_i(x)$ $(i = 1, 2)$ and

$$\int_I f_i(x)\,dx - \epsilon < \int_I l_i(x)\,dx, \qquad \int_I u_i(x)\,dx < \int_I f_i(x)\,dx + \epsilon \qquad (i = 1, 2).$$

Hence (using 13.6)

$$0 \leqq \int_I [u_1(x) - l_1(x)]\,dx < 2\epsilon,$$

$$0 \leqq \int_I [u_2(x) - l_2(x)]\,dx < 2\epsilon.$$

Now define $u(x) = \sup\{u_1(x), u_2(x)\}$, $l(x) = \sup\{l_1(x), l_2(x)\}$. Then by 12.4 $u(x)$ is a U-function and $l(x)$ is an L-function. Since $u_i(x) - l_i(x) \geqq 0$, we have by 13.7, 15.3iii, 13.6 and 13.5,

$$\begin{aligned} 0 &\leqq \int_I [u(x) - l(x)]\,dx \\ &\leqq \int_I u_1(x)\,dx - \int_I l_1(x)\,dx + \int_I u_2(x)\,dx - \int_I l_2(x)\,dx \\ &< 4\epsilon. \end{aligned}$$

Incidentally, this proves that the integrals of $u(x)$ and $l(x)$ are finite. For since u and l are respectively U- and L-functions the integral of $u(x)$ can not be $-\infty$, and the integral of $l(x)$ can not be $+\infty$; so if either had an infinite integral, the difference $\int u(x)\,dx - \int l(x)\,dx$ would be $+\infty$, and not $< 4\epsilon$. Again, by two applications of 15.3i,

$$u(x) \geqq \sup\{f_1(x), f_2(x)\} \geqq l(x).$$

Hence

$$\begin{aligned} \int_I l(x)\,dx &\leqq \underline{\int_I} \sup\{f_1(x), f_2(x)\}\,dx \\ &\leqq \overline{\int_I} \sup\{f_1(x), f_2(x)\}\,dx \leqq \int_I u(x)\,dx \\ &\leqq \int_I l(x)\,dx + 4\epsilon. \end{aligned}$$

This proves that the upper and lower integrals of $\sup\{f_1(x), f_2(x)\}$ are finite, and that they differ by less than 4ϵ. But ϵ is arbitrary, so the upper and lower integrals are finite and equal, and $\sup\{f_1(x), f_2(x)\}$ is proved summable.

From this and (b) (with $c = -1$) we at once see that $\inf\{f_1(x), f_2(x)\}$ is summable; for $-f_1(x)$ and $-f_2(x)$ are summable by (b), so by the part of the proof just completed and 5.5

$$-\inf\{f_1(x), f_2(x)\} = \sup\{-f_1(x), -f_2(x)\}$$

is summable, so by (b) $\inf\{f_1(x), f_2(x)\}$ is summable.

By the definition 15.4, with the part of (c) already proved, we see that $f_I^+(x)$ and $f_I^-(x)$ are summable; whence by (15.5) and (a) of the present theorem we have $|f_1(x)|$ summable. To establish (γ), we observe that $f_1(x) \leqq |f_1(x)|$ and $-f_1(x) \leqq |f_1(x)|$. Integrating and using 15.2 and (b) (with $c = -1$) we obtain

$$\int_I f_1(x)\,dx \leqq \int_I |f_1(x)|\,dx, \qquad -\int_I f_1(x)\,dx \leqq \int_I |f_1(x)|\,dx,$$

whence (γ) follows at once.

To prove (*d*), suppose first that $0 \leqq f_i(x) \leqq M$, $(i = 1, 2)$. If ϵ is any positive number, we can (as in proving (*c*)) find U-functions $u_1(x)$, $u_2(x)$ and L-functions $l_1(x)$, $l_2(x)$ such that $u_i(x) \geqq f_i(x) \geqq l_i(x)$, $(i = 1, 2)$ and

$$\int_I [u_i(x) - l_i(x)]\, dx < \epsilon, \qquad (i = 1, 2).$$

There is no harm in assuming $0 \leqq l_i(x)$ and $u_i(x) \leqq M$ $(i = 1, 2)$; for otherwise we could replace $u_i(x)$ by $\inf \{u_i(x), M\}$, which is a U-function $\geqq f_i(x)$, and we could replace $l_i(x)$ by $\sup \{l_i(x), 0\}$, which is an L-function $\leqq f_i(x)$. By 12.5, $l_1(x)l_2(x)$ is an L-function and $u_1(x)u_2(x)$ is a U-function. From $0 \leqq l_i(x) \leqq f_i(x) \leqq u_i(x) \leqq M$ $(i = 1, 2)$ we find

$$0 \leqq l_1(x)l_2(x) \leqq f_1(x)f_2(x) \leqq u_1(x)u_2(x) \leqq M^2,$$

so

$$\begin{aligned} 0 \leqq \int_I l_1(x)l_2(x)\, dx &\leqq \underline{\int_I} f_1(x)f_2(x)\, dx \leqq \overline{\int_I} f_1(x)f_2(x)\, dx \\ &\leqq \int_I u_1(x)u_2(x)\, dx \leqq \int_I M^2\, dx = M^2 \Delta I. \end{aligned}$$

This proves that the upper and lower integrals of $f_1(x)f_2(x)$ are finite. To prove them equal, we observe that $u_1u_2 - l_1l_2$ is a U-function, by 12.5, 12.3 and 12.4. For each x in the interval I the inequality

$$\begin{aligned} u_1(x)u_2(x) - l_1(x)l_2(x) &= u_1(x)[u_2(x) - l_2(x)] + l_2(x)[u_1(x) - l_1(x)] \\ &\leqq M[u_2(x) - l_2(x)] + M[u_1(x) - l_1(x)] \end{aligned}$$

is satisfied, and the two terms in the last member are U-functions by 12.3 and 12.4. Hence by 13.2 and 13.6

$$\begin{aligned} 0 &\leqq \overline{\int_I} f_1(x)f_2(x)\, dx - \underline{\int_I} f_1(x)f_2(x)\, dx \\ &\leqq \int_I u_1(x)u_2(x)\, dx - \int_I l_1(x)l_2(x)\, dx \\ &\leqq M \left\{ \int_I [u_2(x) - l_2(x)]\, dx + \int_I [u_1(x) - l_1(x)]\, dx \right\} \\ &\leqq 2M\epsilon. \end{aligned}$$

Thus the upper and lower integrals differ by less than the arbitrary positive number $2M\epsilon$, so they are equal. We have therefore established (*d*) under the supplementary hypothesis that $f_1(x)$ and $f_2(x)$ are non-negative.

This supplementary hypothesis is easily removed. For if $f_1(x)$ and $f_2(x)$ are bounded and summable, the functions f_1^+, f_1^-, f_2^+, f_2^- are all summable by (*c*), and they are bounded and non-negative. Hence all four products

$$f_1^+f_2^+,\ f_1^+f_2^-,\ f_1^-f_2^+,\ f_1^-f_2^-$$

are summable by the proof just completed, and by (*a*) and (*b*) the combination

$$\begin{aligned} f_1^+(x)f_2^+(x) - f_1^+(x)f_2^-(x) - f_1^-(x)f_2^+(x) + f_1^-(x)f_2^-(x) \\ = (f_1^+(x) - f_1^-(x))(f_2^+(x) - f_2^-(x)) \\ = f_1(x)f_2(x) \end{aligned}$$

is summable.

The next theorem is a decidedly non-trivial convergence theorem. It will form the basis for all the later theorems expressing the powerful convergence properties of the Lebesgue integral.

15.7s. *Let $f_n(x)$ ($n = 1, 2, 3 \cdots$) be a sequence of functions defined and summable on a closed interval I. Suppose that $f_1(x) \leqq f_2(x) \leqq \cdots$, and define*

$$f(x) = \lim_{n\to\infty} f_n(x).$$

Then $f(x)$ is summable over I if and only if

$$\lim_{n\to\infty} \int_I f_n(x)\,dx$$

is finite; and if this limit is finite, then

$$\int_I f(x)\,dx = \lim_{n\to\infty} \int_I f_n(x)\,dx.$$

In the statement of the theorem, we have tacitly assumed that $f_n(x)$ and $\int_I f_n(x)\,dx$ converge to limits (finite or $+\infty$) as $n \to \infty$. This is certainly true, for the numbers $f_n(x)$ are monotonic increasing functions of n by hypothesis and their integrals are monotonic increasing functions of n by 15.2. It is also clear (cf. 6.8 and the exercise after it) that $f_n(x) \leqq f(x)$ for every n and every x in I. So by 14.4

$$\int_I f_n(x)\,dx \leqq \underline{\int_I} f(x)\,dx$$

for each n, whence

$$(A) \qquad \lim_{n\to\infty} \int_I f_n(x)\,dx \leqq \underline{\int_I} f(x)\,dx.$$

This shows that if the left member of (A) is ∞, the function $f(x)$ is not summable. It remains to show that if the left member of (A) is finite $f(x)$ is summable and its integral is the left member of (A).

Let ϵ be an arbitrary positive number. For each positive integer i there is a U-function $u_i^*(x)$ and an L-function $l_i^*(x)$ such that $l_i^*(x) \leqq f_i(x) \leqq u_i^*(x)$ and

$$\int_I l_i^*(x)\,dx > \int_I f_i(x)\,dx - \frac{\epsilon}{2^{1+i}},$$
$$\int_I u_i^*(x)\,dx < \int_I f_i(x)\,dx + \frac{\epsilon}{2^{1+i}}.$$

Hence

$$\int_I (u_i^* - l_i^*)\,dx < \frac{\epsilon}{2^i}.$$

For each n we define

$$u_n(x) = \sup\,\{u_1^*(x), \cdots, u_n^*(x)\},$$
$$l_n(x) = \sup\,\{l_1^*(x), \cdots, l_n^*(x)\}.$$

These are respectively a U-function and an L-function, by 12.4. Also it is clear that

$$u_n(x) \geqq u_n^*(x) \geqq f_n(x) \geqq l_n(x).$$

By 15.3,

$$u_n(x) - l_n(x) \leqq \sum_{i=1}^{n} [u_i^*(x) - l_i^*(x)],$$

so if we integrate and use 15.6 and 15.2 we obtain

$$(B) \qquad \int_I u_n(x)\,dx - \int_I f_n(x)\,dx \leqq \int_I u_n(x)\,dx - \int_I l_n(x)\,dx$$
$$\leqq \sum_{i=1}^{n} \int_I [u_i^*(x) - l_i^*(x)]\,dx$$
$$< \sum_{i=1}^{n} \frac{\epsilon}{2^i} < \epsilon.$$

From the definition of u_n, with 5.1, we see that $u_1(x) \leqq u_2(x) \leqq \cdots$, so $u_n(x)$ tends to a limit (finite or $+\infty$) as $n \to \infty$. We call this limit $u(x)$. By 13.3, $u(x)$ is a U-function, and

$$(C) \qquad \int_I u(x)\,dx = \lim_{n\to\infty} \int_I u_n(x)\,dx.$$

We have already seen that $u_n(x) \geqq f_n(x)$ for every n, so

$$u(x) = \lim_{n\to\infty} u_n(x) \geqq \lim_{n\to\infty} f_n(x) = f(x).$$

Hence by 14.1

$$(D) \qquad \overline{\int_I} f(x)\,dx \leqq \int_I u(x)\,dx.$$

From (B) it follows readily that

$$(E) \qquad \lim_{n\to\infty} \int_I u_n(x)\,dx \leqq \epsilon + \lim_{n\to\infty} \int_I f_n(x)\,dx.$$

Inequalities (C), (D) and (E) imply that

$$\overline{\int_I} f(x)\,dx \leqq \epsilon + \lim_{n\to\infty} \int_I f_n(x)\,dx.$$

But ϵ is an arbitrary positive number, so by 5.2 we have

$$\overline{\int_I} f(x)\,dx \leqq \lim_{n\to\infty} \int_I f_n(x)\,dx.$$

By 14.2, this inequality and inequality (A) imply that the upper and lower integrals of $f(x)$ over I are equal to the limit of the integrals of the $f_n(x)$, and by definition 15.1 the conclusion of the theorem is established.

Our next theorem shows a new relationship between U- and L-functions and summable functions; every summable function $f(x)$ is caught between two functions which are limits of U- or L-functions, and differ from f to such a small extent as not to affect the value of the integral.

15.8s. *If $f(x)$ is summable over I, there exist functions $g(x)$ and $h(x)$ defined over I and having the following properties.*

(1) *There is a sequence of summable U-functions $\{u_n(x)\}$ such that $u_1(x) \geqq u_2(x) \geqq u_3(x) \geqq \cdots$ and $\lim_{n\to\infty} u_n(x) = h(x)$.*

(2) *There is a sequence of summable L-functions $\{l_n(x)\}$ such that $l_1(x) \leqq l_2(x) \leqq l_3(x) \leqq \cdots$ and $\lim_{n\to\infty} l_n(x) = g(x)$.*

(3) *The sup of each of the functions $h(x)$, $u_n(x)$ is equal to sup $f(x)$.*

(4) *The inf of each of the functions $g(x)$, $l_n(x)$ is equal to inf $f(x)$.*

(5) $g(x) \leqq f(x) \leqq h(x)$.

(6) $\int_I g(x)\,dx = \int_I f(x)\,dx = \int_I h(x)\,dx$.

It will be enough to prove the statements concerning $h(x)$, since those concerning $g(x)$ can be obtained by a change of sign.

For each positive integer n there is a U-function $u_n^*(x)$ on I such that $u_n^*(x) \geqq f(x)$ on I and

$$\int_I u_n^*(x)\,dx < \int_I f(x)\,dx + \frac{1}{n}.$$

Also, we may suppose that $\sup u_n^*(x) = \sup f(x)$ on I; otherwise, the first of these bounds would exceed the second, and we would need only to replace $u_n^*(x)$ by $\inf \{u_n^*(x), \sup f(x)\}$. This would still be a U-function $\geqq f(x)$ by 12.4, and it would have an integral less than $\int f(x)\,dx + \frac{1}{n}$ by 13.2.

Next we define $u_n(x) = \inf \{u_1^*(x), \cdots, u_n^*(x)\}$. This is a U-function, by 12.4. By 5.1 we have $u_1(x) \geqq u_2(x) \geqq \cdots$, so that the limit

$$h(x) \equiv \lim_{n \to \infty} u_n(x)$$

exists on I. Since $u_n^*(x) \geqq f(x)$ for each n, it is clear that

$$f(x) \leqq u_n(x) \leqq u_n^*(x).$$

So by the definition of $h(x)$

$$f(x) \leqq h(x) \leqq u_n(x) \leqq u_n^*(x).$$

Since $u_n^*(x)$ is a U-function $\geqq h(x)$, we have by 14.4, 14.2 and 14.1

$$\int_I f(x)\,dx \leqq \underline{\int_I} h(x)\,dx \leqq \overline{\int_I} h(x)\,dx \leqq \int_I u_n^*(x)\,dx < \int_I f(x)\,dx + \frac{1}{n}.$$

Here n is arbitrary, so the upper and lower integrals of $h(x)$ are both equal to the integral of $f(x)$, establishing (6). We saw that we could choose u_n^* so that its sup was equal to $\sup f(x)$. Hence each $u_n(x)$ is $\leqq \sup f(x)$, and $\sup h(x) \leqq \sup f(x)$. Inequality is impossible by 5.4, so (3) is established. Conclusions (5) and (1) have already been shown true, so this completes the proof.

REMARK. If $f(x)$ is non-negative and vanishes on the boundary of I, we may choose $g(x)$ and $h(x)$ so that they too vanish on the boundary of I. For we suppose $\inf g(x) = \inf f(x) = 0$, so $0 \leqq g(x) \leqq f(x)$ for all x. This forces $g(x)$ to vanish wherever $f(x)$ vanishes, in particular on the boundary of I. All the properties of the functions $u_n(x)$ mentioned in the theorem are retained if we replace $u_n(x)$ by $u_n(x) \cdot K_J(x)$, where J is the interior of I (cf. 12.9 and 12.5). So we may suppose that each $u_n(x)$ vanishes on the boundary of I. Then their limit $h(x)$ vanishes on the boundary.

15.9. *The integral defined in* 15.1 *is invariant under translation.*

Let $f(x)$ be summable over I, and let $(h^{(1)}, \cdots, h^{(q)})$ be any q-tuple of real numbers. For each positive ϵ there exist a U-function $u(x)$ and an L-function $l(x)$ such that $l(x) \leqq f(x) \leqq u(x)$ on I and

$$\int_I [u(x) - l(x)]\, dx < \epsilon.$$

The translations $u^{(h)}(x)$, $l^{(h)}(x)$ are respectively U- and L-functions on $I^{(h)}$, and satisfy the inequality

$$l^{(h)}(x) \leqq f^{(h)}(x) \leqq u^{(h)}(x) \qquad (x \text{ in } I^{(h)}).$$

This, with 13.9 and 14.2, implies that

$$\int_I l(x)\, dx \leqq \underline{\int}_{I^{(h)}} f^{(h)}(x)\, dx \leqq \overline{\int}_{I^{(h)}} f^{(h)}(x)\, dx \leqq \int_I u(x)\, dx.$$

But the integral of f over I also lies between the integrals of l and of u over I. Hence the upper and lower integrals of $f^{(h)}$ over $I^{(h)}$ both differ from the integral of f over I by less than ϵ. Since ϵ is arbitrary, this with the definition 15.1 establishes the theorem.

16. In the preceding section we showed, immediately after defining the Lebesgue integral, that it is consistent with the ordinary (Riemann or Cauchy) integral whenever the integrand is a continuous function. But in §11 we sketched the definition of the general Riemann integral, without restricting the integrand to be continuous. Although we have not used and shall not use this general Riemann integral, we shall here prove that we have actually generalized it; that every Riemann integrable function is summable, and its Lebesgue integral is equal to its Riemann integral. The first step in this proof is to establish a lemma concerning the integral of the characteristic function of an interval.

16.1. *If I_0 is an interval contained in a closed interval I and is either open or closed or contains part of its boundary, then*

$$\int_I K_{I_0}(x)\, dx = \Delta I_0.$$

First let us suppose that I_0 is a closed interval $\gamma^{(i)} \leqq x^{(i)} \leqq \delta^{(i)}$. Let I_n be the interval $\gamma^{(i)} - \frac{1}{n} < x^{(i)} < \delta^{(i)} + \frac{1}{n}$. It is possible to find a continuous function $\varphi(x)$ such that $0 \leqq \varphi(x) \leqq 1$ for all x, $\varphi(x) = 1$ if x is in I_0, and $\varphi(x) = 0$ if x is in CI_n. For example, if $d(x)$ is the distance of x from I_0 (end of §5) we can take $\varphi(x) = \sup \{0, 1 - n \cdot d(x)\}$. If we split I into a finite number of intervals one of which is $I \cap I_n$, then the sup of $\varphi(x)$ on $I \cap I_n$ is 1 and the sup of $\varphi(x)$ on each of the other intervals is 0. Hence by the definition 11.1 of the elementary integral

$$\int_I \varphi(x)\,dx \leqq 1 \cdot \Delta(I \cap I_n) \leqq \Delta I_n = \left(\delta^{(1)} - \gamma^{(1)} + \frac{2}{n}\right) \cdots \left(\delta^{(q)} - \gamma^{(q)} + \frac{2}{n}\right).$$

But $\varphi(x)$ is a continuous function nowhere less than $K_{I_0}(x)$, so by 14.1

$$\overline{\int_I} K_{I_0}(x)\,dx \leqq \int_I \varphi(x)\,dx.$$

Hence

$$\overline{\int_I} K_{I_0}(x)\,dx \leqq \prod_{i=1}^{q} \left(\delta^{(i)} - \gamma^{(i)} + \frac{2}{n}\right).$$

Here n is arbitrary, so

$$\overline{\int_I} K_{I_0}(x)\,dx \leqq \prod_{i=1}^{q} (\delta^{(i)} - \gamma^{(i)}) = \Delta I_0.$$

If it happens that $\gamma^{(i)} = \delta^{(i)}$ for some i, then $\Delta I_0 = 0$. But the characteristic function of I_0 is not negative, so

$$0 \leqq \underline{\int_I} K_{I_0}(x)\,dx \leqq \overline{\int_I} K_{I_0}(x)\,dx \leqq \Delta I_0 = 0.$$

In this case our theorem holds.

Next let I_0 be the open interval $\gamma^{(i)} < x^{(i)} < \delta^{(i)}$, where $\gamma^{(i)} < \delta^{(i)}$ for each i. For all large n we have $\gamma^{(i)} + \frac{1}{n} < \delta^{(i)} - \frac{1}{n}$. Let J_n be the interval $\gamma^{(i)} + \frac{1}{n} \leqq x^{(i)} \leqq \delta^{(i)} - \frac{1}{n}$, and let $\psi(x)$ be a continuous function such that $0 \leqq \psi(x) \leqq 1$ and $\psi(x) = 1$ on J_n, $\psi(x) = 0$ on

CI_0. Then

$$K_{I_0}(x) \geqq \psi(x), \qquad \text{so} \qquad \underline{\int_I} K_{I_0}(x)\,dx \geqq \int_I \psi(x)\,dx.$$

On the other hand, let I be split into a finite number of intervals of which J_n is one. Then the minimum of $\psi(x)$ on J_n is 1, and on the other intervals the minimum of $\psi(x)$ is $\geqq 0$. Hence by the definition of the elementary integral

$$\int_I \psi(x)\,dx \geqq 1 \cdot \Delta J_n = \prod_{i=1}^{q} \left(\delta^{(i)} - \gamma^{(i)} - \frac{2}{n}\right),$$

and

$$\underline{\int_I} K_{I_0}(x)\,dx \geqq \prod_{i=1}^{q} \left(\delta^{(i)} - \gamma^{(i)} - \frac{2}{n}\right).$$

Since this is true for all large n,

$$\underline{\int_I} K_{I_0}(x)\,dx \geqq \prod_{i=1}^{q} (\delta^{(i)} - \gamma^{(i)}) = \Delta I_0.$$

Finally, suppose that we are given a non-degenerate interval I_0 containing all, part, or none of its boundary. Then I_0 contains an interval $I_1 : \gamma^{(i)} < x^{(i)} < \delta^{(i)}$ and is contained in $I_2 : \gamma^{(i)} \leqq x^{(i)} \leqq \delta^{(i)}$. Hence

$$K_{I_1}(x) \leqq K_{I_0}(x) \leqq K_{I_2}(x).$$

Therefore, using the above proof and 14.2 and 14.4,

$$\begin{aligned}\Delta I_0 = \Delta I_1 &\leqq \underline{\int} K_{I_1}(x)\,dx \leqq \underline{\int_I} K_{I_0}(x)\,dx \leqq \overline{\int_I} K_{I_0}(x)\,dx \\ &\leqq \overline{\int_I} K_{I_2}(x)\,dx \leqq \Delta I_2 = \Delta I_0.\end{aligned}$$

By the definition 15.1 of the Lebesgue integral, this establishes 16.1.

From 16.1 we immediately conclude

16.2. *Let $I_1, \cdots, I_n$ be a decomposition of a closed interval I into disjoint intervals (each containing all, part or none of its boundary), and let $s(x)$ be a step function having a constant value c_i on each I_i. Then $s(x)$ is summable, and*

$$\int_I s(x)\,dx = \sum_{i=1}^{n} c_i \Delta I_i.$$

For

$$s(x) = \sum_{i=1}^{n} c_i K_{I_i}(x),$$

so by 15.6 $s(x)$ is summable and

$$\int_I s(x)\,dx = \sum_{i=1}^{n} c_i \int_I K_{I_i}(x)\,dx = \sum_{i=1}^{n} c_i \Delta I_i.$$

Now let $f(x)$ be any function defined and bounded on I. As in §11, we define its upper Darboux (or Riemann) integral as the inf of $\Sigma c_i \Delta I_i$ for all step-functions having on each I_i a value c_i equal to or greater than $\sup f(x)$ on I_i. Since the step-functions are summable, by 5.1 this upper Darboux integral is not less than the inf of the integrals of all summable functions $\geqq f(x)$:

$$R \bar{\int_I} f(x)\,dx \geqq \inf \left\{ \int_I g(x)\,dx \mid g(x) \text{ summable} \geqq f(x) \right\}$$
$$\geqq \bar{\int_I} f(x)\,dx,$$

by 14.4. We can treat the lower integral similarly, and thus find

$$R \underline{\int_I} f(x)\,dx \leqq \underline{\int} f(x)\,dx \leqq \bar{\int_I} f(x)\,dx \leqq R \bar{\int_I} f(x)\,dx.$$

Hence if $f(x)$ is Riemann integrable (so that its upper and lower integrals are both equal to its Riemann integral) then $f(x)$ is summable, and its Lebesgue integral is equal to its Riemann integral.

If the converse of this statement were true, the Lebesgue integral would be (at least for bounded functions) identical with the Riemann, and all our work would represent simply a new and comparatively involved way of defining the Riemann integral. But this is not so, as we shall see in §18.

In all calculus texts there is some discussion of an extension of the notion of Riemann (or Cauchy) integration to functions which have a finite number of infinite discontinuities. Thus, for example, if $f(x)$ is continuous for $a < x \leqq b$, but has an infinite discontinuity at a, it is customary to define

$$\int_a^b f(x)\,dx = \lim_{\epsilon \to 0+} \int_{a+\epsilon}^b f(x)\,dx,$$

provided that this limit exists. Now it is quite possible for a function to be integrable in this sense, even though $|f(x)|$ is not integrable.

But with Lebesgue integrals, if $f(x)$ is summable on $[a, b]$, so is $|f(x)|$. Hence the Lebesgue integral does not cover this extension of the Riemann integral. Of course, we could extend the definition of the Lebesgue integral in exactly the same way, and this extended Lebesgue integral would include the extended Riemann integral. But this is not worth while, for if we are interested in developing an integral for functions such that $|f(x)|$ is not summable there are much stronger methods available. This subject will be treated in Chapter VIII.

17. In elementary calculus we considered not only integrals over fixed intervals, but also integrals over variable intervals. That is, along with the integral over $[a, b]$ we also studied the integral over subintervals $[\alpha, \beta]$, and investigated its behavior as a function of α and β. Here we seek even more freedom. Instead of considering integrals over one fixed interval I, as has been done heretofore, we now wish to define and study the integrals of functions $f(x)$ over sets E of a high degree of arbitrariness. Since we therefore wish to disregard any contribution to the integral due to the functional values assumed on $I - E$, this suggests changing the function $f(x)$ by setting it equal to zero on $I - E$ and then integrating the resulting function over I.

17.1s. *If $f(x)$ is defined on a subset E of a closed interval I, and the function $f_0(x)$ which has the value $f(x)$ for x in E and the value 0 for x in $I - E$ is summable over I, then we say that $f(x)$ is summable over E, and we define*

$$\int_E f(x)\,dx \equiv \int_I f_0(x)\,dx.$$

It is occasionally inconvenient notationally to deal with functions defined only on a set E, especially when we are simultaneously handling functions defined over other ranges. This inconvenience can easily be avoided. We could, for example, use the $f_0(x)$ of 17.1. But for the purpose of calculating the integral of $f(x)$ over E it is really immaterial how we extend the definition of $f(x)$. For no matter what values we may choose to assign $f(x)$ on CE, in calculating its integral over E we replace all values on CE by 0. We may therefore just as well assume that all functions are defined over the whole space R_q, and then define

17.2s. *$\int_E f(x)\,dx$ is defined to be $\int_I f(x)K_E(x)\,dx$, provided that this integral exists.*

The equivalence of this definition with 17.1 is evident, for $f(x)K_E(x)$ is the same as the $f_0(x)$ of 17.1.

We can at once make one obvious statement about integrals over a set $E \subset I$. Since every statement about the integral of $f(x)$ over E is merely a statement about the integral of $f(x)K_E(x)$ over I, theorems 15.2, 15.6 and 15.7 hold without change for integrals over E. That is, in each of these theorems we have the privilege of replacing everywhere the symbol $\int_I$ by $\int_{E^*}$, where E^* is any subset of I.

All our integration so far has been performed with respect to one fixed fundamental closed interval I. But now that we are discussing integrals over subsets E of J, we must remember that a set E contained in an interval I is also contained in infinitely many other intervals. Suppose then that E is contained in I and also contained in another closed interval I_1, and suppose that $f(x)$ is defined on E. If $f(x)$ is summable over E when I is used as basic interval, is it still summable over E when I_1 is basic interval, and if so, has it the same integral? We here meet a slightly unexpected situation; we must establish two different theorems, one an "s" theorem and the other not.

For compactness we introduce a new notation, to be used in this section only. If $f(x)$ is defined on a set E contained both in I_1 and in I_2, we define

$$S_1 = \overline{\int}_{I_1} f(x)K_E(x)\,dx, \qquad S_2 = \overline{\int}_{I_2} f(x)K_E(x)\,dx,$$

$$s_1 = \underline{\int}_{I_1} f(x)K_E(x)\,dx, \qquad s_2 = \underline{\int}_{I_2} f(x)K_E(x)\,dx.$$

Then

17.3s. *Let the set E consist of points interior to each of two closed intervals I_1 and I_2, and let $f(x)$ be defined and non-negative on E. Then $S_1 = S_2$ and $s_1 = s_2$.*

Let us first suppose $I_1 \subset I_2$, and let J be the open interval consisting of the interior points of I_1. For every $\epsilon > 0$ there exists a U-function $u_1(x) \geqq f(x)K_E(x)$ and an L-function $l_1(x) \leqq f(x)K_E(x)$ such that

$$\int_{I_1} u_1(x)\,dx < S_1 + \epsilon, \qquad \int_{I_1} l_1(x)\,dx > s_1 - \epsilon.$$

By 7.10, we may suppose that $u_1(x)$ and $l_1(x)$ are defined and are respectively U- and L-functions over the whole space R_q; and we may further assume $l_1(x) \geqq 0$, since otherwise we may replace $l_1(x)$ by $\sup\{l_1(x), 0\}$. The functions $K_J(x)$ and $K_{I_1}(x)$ are respectively a U-function and an L-function, by 12.9. Hence by 12.5 the functions

$$u_2(x) \equiv u_1(x)K_J(x) \qquad \text{and} \qquad l_2(x) = l_1(x)K_{I_1}(x)$$

are respectively a U-function and an L-function, and are non-negative. We see that $l_2(x) = l_1(x)K_J(x)$; for except on the boundary of I_1 we have $K_{I_1}(x) = K_J(x)$, and on the boundary we have $0 \leqq l_1(x) \leqq f(x)K_E(x) = 0$, so that $l_1(x)K_J(x) = 0 = l_1(x)K_{I_1}(x) = l_2(x)$. It follows readily that

$$u_2(x) \geqq f(x)K_E(x) \geqq l_2(x) \text{ for all } x.$$

For if x is in J, we have $u_2(x) = u_1(x) \geqq f(x)K_E(x) \geqq l_1(x) = l_2(x)$, while if x is not in J we find

$$u_2(x) = 0 = f(x)K_E(x) = l_1(x)K_J(x) = l_2(x).$$

Now let I_2 be subdivided into a finite number of non-overlapping intervals $I_1, J_1, J_2, \cdots, J_h$, one of which is I_1 and all of which are closed. On $J_1, \cdots, J_h$ both $u_2(x)$ and $l_2(x)$ vanish identically. So by 13.8, since the factor $K_J(x)$ vanishes identically on $J_1, \cdots, J_h$,

$$\begin{aligned} S_2 &\leqq \int_{I_2} u_2(x)\,dx = \int_{I_1} u_2(x)\,dx = \int_{I_1} u_1(x)K_J(x)\,dx \\ &\leqq \int_{I_1} u_1(x)\,dx < S_1 + \epsilon, \\ s_2 &\geqq \int_{I_2} l_2(x)\,dx = \int_{I_1} l_2(x)\,dx = \int_{I_1} l_1(x)\,dx \geqq s_1 - \epsilon. \end{aligned}$$

Here ϵ is an arbitrary positive number, so

$$(A) \qquad S_2 \leqq S_1, \qquad s_2 \geqq s_1.$$

On the other hand, for every $\epsilon > 0$ there is a U-function $u(x) \geqq f(x)K_E(x)$ on I_2 and an L-function $l(x) \leqq f(x)K_E(x)$ on I_2 such that

$$\int_{I_2} u(x)\,dx < S_2 + \epsilon, \qquad \int_{I_2} l(x)\,dx > s_2 - \epsilon.$$

Using the same subdivisions of I_2 as above, on all the intervals $J_1, \cdots, J_h$ we have $l(x) \leqq f(x)K_E(x) \equiv 0 \leqq u(x)$. Hence by 13.8 and 13.2

$$\begin{aligned} S_2 + \epsilon &> \int_{I_2} u(x)\,dx \geqq \int_{I_1} u(x)\,dx \geqq S_1, \\ s_2 - \epsilon &< \int_{I_2} l(x)\,dx \leqq \int_{I_1} l(x)\,dx \leqq s_1. \end{aligned}$$

Since ϵ is an arbitrary positive number, this proves $S_2 \geqq S_1$, $s_2 \leqq s_1$. This, together with (A), establishes the theorem, provided that $I_1 \subset I_2$. The restriction $I_1 \subset I_2$ is easily removed. For we can always find a closed interval I_0 containing both I_1 and I_2, and by the above proof (with an obvious definition of S_0 and s_0)

$$S_2 = S_0 = S_1, \qquad s_2 = s_0 = s_1.$$

An immediate corollary is

17.4s. *Let E be a set whose points are interior to each of two closed intervals I_1 and I_2, and let $f(x)$ be defined on E. Then if either of the integrals*

$$\int_{I_1} f(x)K_E(x)\,dx, \qquad \int_{I_2} f(x)K_E(x)\,dx$$

is defined so is the other, and the two are then equal.

Suppose that the first of these integrals is defined. We define $f^+(x)$ and $f^-(x)$ as in 15.4. By 15.4 and 15.6, these functions are non-negative and summable over I_1. Hence by 17.3

$$\begin{aligned}\int_{I_2} f^+(x)K_E(x)\,dx &= \overline{\int}_{I_1} f^+(x)K_E(x)\,dx \\ &= \int_{I_1} f^+(x)K_E(x)\,dx \\ &= \underline{\int}_{I_1} f^+(x)K_E(x)\,dx \\ &= \underline{\int}_{I_2} f^+(x)K_E(x)\,dx,\end{aligned}$$

so that $f^+(x)K_E(x)$ is summable over I_2, and

$$\int_{I_2} f^+(x)K_E(x)\,dx = \int_{I_1} f^+(x)K_E(x)\,dx.$$

A similar statement applies to $f^-(x)$. So by 15.5 and 15.6 their difference $f(x)$ is summable over I_2, and

$$\begin{aligned}\int_{I_2} f(x)K_E(x)\,dx &= \int_{I_2} f^+(x)K_E(x)\,dx - \int_{I_2} f^-(x)K_E(x)\,dx \\ &= \int_{I_1} f^+(x)K_E(x)\,dx - \int_{I_1} f^-(x)K_E(x)\,dx \\ &= \int_{I_1} f(x)K_E(x)\,dx.\end{aligned}$$

Thus if the symbol

$$\int_E f(x)\,dx$$

has a meaning when interpreted as the integral of $f(x)K_E(x)$ over some closed interval I to which E is interior, it continues to have the same meaning with respect to every such closed interval.

So far we have considered only intervals containing the points of E as interior points. If we wish to allow boundary points of I to belong to E, we must drop the "s" from the theorem. More annoying, we must refer to two theorems (19.13 and 23.3) which we have not yet proved. But the next theorem is not needed in the proofs of theorems 19.13 and 23.3, and because of its resemblance to 17.4 we wish to have the next theorem here, not later. So we shall use these later theorems in anticipation of their proof.

17.5. *If a set E is contained in (not necessarily interior to) two closed intervals I_1 and I_2, and $f(x)$ is a function such that $f(x)K_E(x)$ is summable over I_1, then $f(x)K_E(x)$ is also summable over I_2 and has the same integral as over I_1.*

The product of I_1 and I_2 is a closed interval. Let E_i be the subset of E which is interior to I_1I_2, and let E_b be the remainder $E - E_i$ of E. Then the points of E_b will be on the boundary of I_1I_2. This boundary consists of a finite number of degenerate closed intervals, so by 16.1, 19.13 and 23.3 we have

$$\int_{I_1} f(x)K_E(x)\,dx = \int_{I_1} f(x)K_{E_i}(x)\,dx,$$
$$\int_{I_2} f(x)K_E(x)\,dx = \int_{I_2} f(x)K_{E_i}(x)\,dx.$$

But E_i is interior to I_1 and to I_2, so by 17.4 the right members of these equations are equal, and the theorem is established.

The results of this section justify our using the symbol $\int_E f(x)\,dx$ without specifying the particular interval to which E is interior. To make this specific, we adopt the following definition

17.6s. *If E is a bounded set and f is defined over E, and $f(x)K_E(x)$ is summable over some closed interval I which contains E in its interior, we define*

$$\int_E f(x)\,dx \equiv \int_I f(x)\cdot K_E(x)\,dx.$$

By 17.4 the integral thus defined is independent of I; and if we were willing to drop the "s" from the numbering of the theorem, we could even omit the words "in its interior" without altering the content of the definition. However, one slight caution is needed. If E is itself a closed interval, the symbol

$$\int_E f(x)\,dx$$

is ambiguous; it could mean either the integral as defined in 15.1, with basic interval E, or it could have the meaning stated in 17.6. With the special definition of ΔI given in the first paragraph of §10 (that is, for the Lebesgue integral) this ambiguity is harmless; the value of the integral is the same whichever way we choose to think of it. But for other interval-functions ΔI satisfying 10.1 the two kinds of integral may be different in value. As we have just agreed, the symbol will have the meaning in 17.6 unless the contrary is specifically stated; on the few occasions on which we wish to go back to the definition 15.1, we shall use some such expression as "Using I as basic interval, we compute

$$\int_I f(x)\,dx."$$

18. In the preceding section we freed ourselves of the restriction that integrals must be taken over intervals. But still there remains one trace of the restriction to intervals; for we have as yet only considered bounded sets E. Now we shall rid ourselves of this restriction too.

In elementary calculus there is a familiar way of defining integrals over an unbounded range. We define

$$\int_{-\infty}^{\infty} f(x)\,dx = \lim_{m,n\to\infty} \int_{-m}^{n} f(x)\,dx,$$

provided that the limit exists and is finite. If the limit still is finite (it necessarily must exist) when we replace $f(x)$ by $|f(x)|$ we say that the integral converges absolutely. It is this last type of extension which we shall adopt as a model. Denoting by W_n the interval $-n \leqq x^{(i)} \leqq n$ $(i = 1, \cdots, q)$ we define the integral of $f(x)$ over an unbounded set E as follows.

18.1s. *Let $f(x)$ be defined on an unbounded set E. Then $f(x)$ is summable over E if it is summable over EW_n for each n and*

$$\lim_{n\to\infty} \int_{EW_n} |f(x)|\,dx$$

is finite. In this case we define

$$\int_E f(x)\,dx = \lim_{n\to\infty} \int_{EW_n} f(x)\,dx.$$

If the limit is finite, from the inequalities

$$0 \leqq f^+(x) \leqq |f(x)|, \qquad 0 \leqq f^-(x) \leqq |f(x)|$$

we see that the integrals

$$\int_{E \cdot W_n} f^+(x)\,dx, \qquad \int_{E \cdot W_n} f^-(x)\,dx$$

also approach finite limits as $n \to \infty$. Hence so does their difference;

$$\lim_{n \to \infty} \int_{E \cdot W_n} f(x)\,dx$$

exists and is finite, so that the definition of the integral of $f(x)$ over E has a meaning.

REMARK. The word "unbounded" in 18.1 is really superfluous. If E is bounded the limiting process of 18.1 is trivial, for the set EW_n and the integral of $f(x)$ over EW_n remains constant for all n large enough so that W_n contains E in its interior. Consequently, if E is bounded and I is any interval containing E in its interior and $f(x)$ is defined on the whole space R_q, either all three of the integrals

$$\int_E f(x)\,dx, \qquad \int_I f(x)K_E(x)\,dx, \qquad \int_{R_q} f(x)K_E(x)\,dx$$

(the second being understood as in 15.1 with I as basic interval) are defined or none of them is defined, and if they are defined they are all equal. In fact, with the special definition of ΔI as the product of the edges of I we can say the same if I contains E (not necessarily in its interior).

EXERCISE. If $f(x)$ is summable over E, then

$$\lim_{n \to \infty} \int_{E - EW_n} f(x)\,dx = 0.$$

EXERCISE. Let $\{M_n\}$ be a sequence of bounded closed or open sets such that each interval W_m is contained in all sets M_n with n greater than a certain n_m. If $f(x)$ is defined on a set E, it is summable over E if and only if it is summable over EM_n for each n and the numbers

$$\int_{EM_n} |f(x)|\,dx$$

are bounded. In that case

$$\int_E f(x)\,dx = \lim_{n \to \infty} \int_{EM_n} f(x)\,dx.$$

Now we have carried the definition of the integral as far as we wish, and we shall begin the investigation of its properties. Our first theorems are extensions of 15.2 and 15.6.

18.2s. *If $f(x)$ and $g(x)$ are both summable over a set E, and $f(x) \geqq g(x)$ for all x in E, then*

$$\int_E f(x)\,dx \geqq \int_E g(x)\,dx.$$

For all x, $f(x)K_E(x) \geqq g(x)K_E(x)$. If E is contained in an interval I, then our theorem follows at once from 15.2. If E is unbounded, then by 15.2 for every n we have

$$\int_{EW_n} f(x)\,dx \geqq \int_{EW_n} g(x)\,dx.$$

As $n \to \infty$ the left and right members of this inequality tend respectively to the left and right members of the inequality in the theorem, so the theorem is established.

18.3s. *If $f(x)$ and $g(x)$ are both defined and summable over a set E, then the functions $f(x) + g(x)$, $cf(x)$ (c any finite constant),* sup $\{f(x), g(x)\}$, inf $\{f(x), g(x)\}$, *$f^+(x)$, $f^-(x)$ and $|f(x)|$ are summable over E,* provided in the case of $f(x) + g(x)$ that this sum is defined on E. *Furthermore*

$$\int_E (f(x) + g(x))\,dx = \int_E f(x)\,dx + \int_E g(x)\,dx,$$

$$\int_E cf(x)\,dx = c\int_E f(x)\,dx,$$

$$\left|\int_E f(x)\,dx\right| \leqq \int_E |f(x)|\,dx.$$

Suppose first that E is interior to an interval I. Then $f(x)K_E(x)$ and $g(x)K_E(x)$ are summable over I. So by 15.6 $(f(x) + g(x))K_E(x)$, $cf(x)K_E(x)$, sup $\{f(x)K_E(x), g(x)K_E(x)\}$, inf $\{f(x)K_E(x), g(x)K_E(x)\}$, $f^+(x)$, $f^-(x)$ and $|f(x)K_E(x)|$ are all summable over I. This establishes the summability of the functions specified in the theorem. Also the two equations in the theorem are obvious consequences of 15.6, and

$$\left|\int_E f(x)\,dx\right| = \left|\int_I f(x)K_E(x)\,dx\right| \leqq \int_I |f(x)K_E(x)|\,dx$$
$$= \int_I |f(x)|\,K_E(x)\,dx = \int_E |f(x)|\,dx.$$

Suppose now that E is unbounded. By the preceding paragraph, the functions $f(x) + g(x)$, etc., are all summable over EW_n. By the definition of summability, the integrals

$$\int_{EW_n} |f(x)|\,dx, \qquad \int_{EW_n} |g(x)|\,dx$$

approach finite limits M, N as $n \to \infty$, and therefore do not exceed M,

N respectively. For every n the set EW_n is bounded, so we may use 15.6:

$$\int_{EW_n} | f(x) + g(x) | \, dx \leqq \int_{EW_n} [\, | f(x) | + | g(x) | \,] \, dx \leqq M + N,$$

$$\int_{EW_n} | cf(x) | \, dx = | c | \int_{EW_n} | f(x) | \, dx \leqq | c | \, M,$$

$$\int_{EW_n} | \sup \{f(x), g(x)\} | \, dx \leqq \int_{EW_n} [\, | f(x) | + | g(x) | \,] \, dx \leqq M + N,$$

$$\int_{EW_n} | \inf \{f(x), g(x)\} | \, dx \leqq \int_{EW_n} [\, | f(x) | + | g(x) | \,] \, dx \leqq M + N,$$

$$\int_{EW_n} | f^+(x) | \, dx \leqq \int_{EW_n} | f(x) | \, dx \leqq M,$$

$$\int_{EW_n} | f^-(x) | \, dx \leqq \int_{EW_n} | f(x) | \, dx \leqq M.$$

Therefore these functions are all summable over E.

The formulas concluding the theorem have all been shown valid for bounded sets, in particular for EW_n. Letting $n \to \infty$ establishes these formulas for the set E.

The next theorem extends the important convergence theorem 15.7 to arbitrary sets E.

18.4s. *Let the functions $f_1(x)$, $f_2(x)$, $\cdots$ be non-negative* and summable over a set E, and let $f_1(x) \leqq f_2(x) \leqq f_3(x) \leqq \cdots$. Let $f(x) = \lim_{n \to \infty} f_n(x)$. Then $f(x)$ is summable over E if and only if*

$$\lim_{n \to \infty} \int_E f_n(x) \, dx$$

is finite; and if this limit is finite, it is equal to the integral of $f(x)$:

$$\int_E f(x) \, dx = \lim_{n \to \infty} \int_E f_n(x) \, dx.$$

Let us first suppose that E is bounded and interior to an interval I, which we shall use as our basic interval. We shall denote by $f'_n(x)$, $f'(x)$ respectively the functions $f_n(x)K_E(x)$, $f(x)K_E(x)$. (Here the symbol $'$ has nothing to do with differentiation.) Then

$$f'_1(x) \leqq f'_2(x) \leqq \cdots \leqq f'(x),$$

and $f'_n(x) \to f'(x)$. So by 15.7 $f'(x)$ is summable over I if and only if the integrals of the $f'_n(x)$ approach a finite limit, and in that case

$$\int_I f'(x) \, dx = \lim_{n \to \infty} \int_I f'_n(x) \, dx.$$

* The hypothesis that the $f_i(x)$ are non-negative is not really necessary, but it makes the proof easier. We shall remove it after establishing 23.4.

But

$$\int_I f'(x)\,dx = \int_E f(x)\,dx, \qquad \int_I f'_n(x)\,dx = \int_E f_n(x)\,dx,$$

so this concludes the case of bounded E.

Now suppose that E is not bounded. If $f(x)$ is summable, then by 18.2

$$\int_E f_i(x)\,dx \leqq \int_E f(x)\,dx$$

for every i, so

$$\text{(F)} \qquad \lim_{i\to\infty} \int_E f_i(x)\,dx \leqq \int_E f(x)\,dx < \infty.$$

Conversely, suppose that this limit is finite, say equal to M. Then for each n

$$\lim_{i\to\infty} \int_{EW_n} f_i(x)\,dx \leqq M.$$

So by the preceding proof $f(x)$ is summable over EW_n, and its integral is not greater than M. Since $f(x) \geqq 0$, this proves that

$$\lim_{n\to\infty} \int_{EW_n} |f(x)|\,dx = \lim_{n\to\infty} \int_{EW_n} f(x)\,dx \leqq M.$$

Hence $f(x)$ is summable.

Let ϵ be any positive number. There is an n such that

$$\int_{EW_n} f(x)\,dx > \int_E f(x)\,dx - \epsilon,$$

by definition 18.1. Hence, using the preceding part of the proof,

$$\lim_{i\to\infty} \int_E f_i(x)\,dx \geqq \lim_{i\to\infty} \int_{EW_n} f_i(x)\,dx = \int_{EW_n} f(x)\,dx > \int_E f(x)\,dx - \epsilon.$$

Here ϵ is an arbitrary positive number, so by 5.2

$$\lim_{i\to\infty} \int_E f_i(x)\,dx \geqq \int_E f(x)\,dx.$$

This, with (F), establishes the theorem.

18.5s. Corollary. *Let the functions $f_1(x)$, $f_2(x)$, $\cdots$ be non-positive* and summable over a set E, and let $f_1(x) \geqq f_2(x) \geqq \cdots$.*

* This hypothesis is actually superfluous, and will be removed after 23.4.

Let $f(x) = \lim_{n\to\infty} f_n(x)$. *Then* $f(x)$ *is summable if and only if*

$$\lim_{n\to\infty} \int_E f_n(x)\,dx$$

is finite; and if this limit is finite, then

$$\int_E f(x)\,dx = \lim_{n\to\infty} \int_E f_n(x)\,dx.$$

If we define $g_n(x) = -f_n(x)$, the g_n satisfy the hypotheses of 18.4. Now replacing $g_n(x)$ by $-f_n(x)$ in the conclusions of 18.4, we obtain the theorem above.

18.6s. Corollary. *If the functions* $f_n(x)$ *are non-negative and summable on* E, *and*

$$\sum_{n=1}^{\infty} \int_E f_n(x)\,dx$$

converges, then the sum of the series $\Sigma f_n(x)$ *is summable, and*

$$\int_E \left[\sum_{n=1}^{\infty} f_n(x)\right] dx = \sum_{n=1}^{\infty} \int_E f_n(x)\,dx.$$

For the partial sums $g_n(x) = f_1(x) + \cdots + f_n(x)$ satisfy the hypotheses of 18.4, and by 18.3

$$\lim_{n\to\infty} \int_E g_n(x)\,dx = \lim_{n\to\infty} \sum_{i=1}^{n} \int_E f_n(x)\,dx < \infty.$$

Let E be the aggregate of rational numbers in $[0, 1]$. The set E is denumerable, so its elements can be arranged in a sequence $r_1, r_2, \cdots$. If $f_n(x)$ vanishes except at r_n and $f_n(r_n) = 1$, by 16.1 the integral

$$\int_0^1 f_n(x)\,dx$$

exists and has the value 0. The sum of the functions $f_n(x)$ is $K_E(x)$, so by 18.6 this function also is summable over $[0, 1]$, and

$$\int_0^1 K_E(x)\,dx = 0.$$

Since the rationals are everywhere dense in $[0, 1]$, we see readily that the upper and lower Riemann integrals of $K_E(x)$ over $[0, 1]$ are respectively 1 and 0. Hence the Riemann integral does not exist. This shows that the Lebesgue integral is more general than the Riemann.

18.7. *The Lebesgue integral, defined in* 18.1, *is invariant under translation.*

Let $f(x)$ be summable and non-negative over a set E, and let $(h^{(1)}, \cdots, h^{(q)})$ be a q-tuple of real numbers. If n is a positive integer and I any closed interval containing W_n in its interior, by 15.9 we have

$$\int_{I^{(h)}} f^{(h)}(x) K_{E^{(h)}W_n^{(h)}}(x)\,dx = \int_I f(x) K_{EW_n}(x)\,dx.$$

By 17.5 and 18.1, both integrals may be taken over the entire space R_q instead of over $I^{(h)}$ and I respectively, without changing their values. Then as n increases the integrand in the right member increases and tends to $f(x)K_E(x)$; so by 18.4 its integral tends to that of $f(x)K_E(x)$. So the left member has the same finite limit, and again using 18.4 we find that the limit of the left integrand is summable, and

$$\begin{aligned}\int_{R_q} f^{(h)}(x) K_{E^{(h)}}(x)\,dx &= \lim_{n\to\infty} \int_{R_q} f^{(h)}(x) K_{E^{(h)}W_n^{(h)}}(x)\,dx \\ &= \lim_{n\to\infty} \int_{R_q} f(x) K_{EW_n}(x)\,dx = \int_{R_q} f(x) K_E(x)\,dx.\end{aligned}$$

This establishes the theorem if $f(x)$ is non-negative. For the general case, we write $f(x) = f^+(x) - f^-(x)$ as before. The integrals of f^+ and f^- are invariant under translation by the part of the proof just completed, so by 18.3 the same is true of the integral of f.

CHAPTER III

Measurable Sets and Measurable Functions

19. In 16.1 we saw that we could obtain the volume of an interval I_0 by integrating its characteristic function. Written in the notation of §17, $\Delta I_0 = \int_{I_0} 1\,dx$. This of course seems to be a complicated way of getting back what we knew to begin with. Nevertheless, it suggests the following extension. Suppose that E is an arbitrary set. Even though E is too intricately constructed to permit us to assign it a volume by elementary processes, still the number $\int_E 1\,dx$ will (if it exists) give us a measure of the "amount" of the set E, and this "amount" will agree with the elementary volume in the simpler cases, such as sets E consisting of a finite number of intervals. However, even though E has a very simple structure it may still happen that $\int_E 1\,dx$ is not defined; consider for example the set E consisting of the whole space R_q. Consequently we phrase our definition so as to bring these sets also within the fold:

19.1s. *The set E is measurable if for each of the intervals $W_n = \{x \mid -n \leqq x^{(i)} \leqq n\}$ the function 1 is summable over EW_n; and if E is measurable, its measure is defined to be*

$$mE = \lim_{n \to \infty} \int_{EW_n} 1\,dx.$$

The following theorem is almost self-evident.

19.2s. *If E is measurable, then $0 \leqq mE \leqq \infty$.*

If E is bounded and measurable, then $0 \leqq mE < \infty$.

If E is measurable and $mE < \infty$, then

$$mE = \int_E 1\,dx = \int_{R_q} K_E(x)\,dx.$$

For every n the integral of 1 over EW_n is non-negative, so letting n increase we obtain $0 \leqq mE \leqq \infty$. If furthermore E is bounded, the integral in the right member of the defining equation 19.1 remains constant for all n large enough so that W_n contains E in its interior; so mE is that constant finite value. If mE is finite, definition 19.1 is merely the definition 18.1 of $\int_E 1\,dx$.

From our theorems on integrals we obtain at once several theorems on measurable sets.

19.3s. *If E_1 and E_2 are measurable, and E_1 is contained in E_2, then $mE_1 \leqq mE_2$.*

Let n be any positive integer. Then $E_1W_n \subset E_2W_n$, so $K_{E_1W_n}(x) \leqq K_{E_2W_n}(x)$ for all x. By 18.2

$$\int_{E_1W_n} 1\,dx = \int_{R_q} K_{E_1W_n}(x)\,dx \leqq \int_{R_q} K_{E_2W_n}(x)\,dx = \int_{E_2W_n} 1\,dx.$$

Letting $n \to \infty$ yields our conclusion.

19.4s. *If E is closed, it is measurable.*

Let n be an arbitrary positive integer and I a closed interval (which we use as basic interval) containing W_n in its interior. Since EW_n is closed, its characteristic function is an L-function by 12.9. Being bounded, it is summable over I by the remark after 15.1. That is, 1 is summable over EW_n for every n.

19.5s. *The whole space R_q and the empty set Λ are measurable sets, and $m\Lambda = 0$.*

Both of these sets are closed, so they are measurable by 19.4. Also, $K_\Lambda(x) \equiv 0$, so

$$\int_{W_n} K_\Lambda(x)\,dx = 0$$

for every n; hence $m\Lambda = 0$.

19.6s. *If E_1 and E_2 are both measurable, so are $E_1 - E_2$, E_1E_2 and CE_1.*

For each n the characteristic functions of E_1W_n and E_2W_n are summable, so the functions

$$K_{[E_1-E_2]W_n}(x) = \sup\,\{K_{E_1W_n}(x) - K_{E_2W_n}(x),\,0\}$$

and

$$K_{E_1E_2W_n}(x) = K_{E_1W_n}(x)K_{E_2W_n}(x)$$

are summable by 15.6. Hence $E_1 - E_2$ and E_1E_2 are measurable by the definition 19.1. Since $CE_1 = R_q - E_1$, and R_q is measurable by 19.5 and E_1 is measurable by hypothesis, CE_1 is also measurable.

19.7s. *If E is open, it is measurable.*

For CE is closed by 2.6, hence measurable by 19.4. So by 19.6 $E = C(CE)$ is measurable.

EXERCISE. Prove that every interval is measurable.

19.8s. *If, $E_1, E_2, E_3, \cdots$ is a finite or denumerable collection of measurable sets, then $\bigcup E_i$ is measurable, and $m \bigcup E_i \leqq \Sigma mE_i$. Moreover, if the E_i are disjoint, then $m \bigcup E_i = \Sigma mE_i$; and if the E_i*

are denumerable in number and $E_1 \subset E_2 \subset E_3 \subset \cdots$, *then* $m \bigcup E_i = \lim_{n \to \infty} mE_n$.

In order to avoid considering separate cases, we suppose that the E_i are denumerable in number; if the E_i are finite in number, we add on infinitely many repetitions of the empty set Λ. Define

$$f_p(x) = \inf\{1, \sum_{i=1}^{p} K_{E_i}(x)\}, \qquad f(x) = \inf\{1, \sum_{i=1}^{\infty} K_{E_i}(x)\}.$$

Then by 1.1 $f_p(x)$ and $f(x)$ are respectively the characteristic functions of $E_1 \cup \cdots \cup E_p$ and of $\bigcup E_i$. From the definition it is clear that $0 \leqq f_1(x) \leqq f_2(x) \leqq \cdots$ and that $\lim_{p \to \infty} f_p(x) = f(x)$. For each n and each p,

$$\int_{W_n} f_p(x)\,dx \leqq \int_{W_n} 1\,dx < \infty;$$

so by 18.4 $f(x)$ is summable over W_n. That is, 1 is summable over $W_n(\bigcup E_i)$, so by definition $\bigcup E_i$ is measurable.

For each n we have by 18.4

$$mW_n(\bigcup E_i) = \int_{W_n} f(x)\,dx = \lim_{p \to \infty} \int_{W_n} f_p(x)\,dx \leqq \lim_{p \to \infty} \int_{W_n} \sum_{i=1}^{p} K_{E_i}(x)\,dx$$

$$= \lim_{p \to \infty} \sum_{i=1}^{p} \int_{W_n} K_{E_i}(x)\,dx = \lim_{p \to \infty} \sum_{i=1}^{p} mE_iW_n = \sum_{i=1}^{\infty} mE_iW_n \leqq \sum_{i=1}^{\infty} mE_i.$$

Since this holds for each n, $m \bigcup E_i = \lim_{n \to \infty} mW_n(\bigcup E_i) \leqq \sum_{i=1}^{\infty} mE_i$.

If the E_i are disjoint, for every positive integer p we have by 1.1

$$K_{E_1 \cup \cdots \cup E_p}(x) = \sum_{i=1}^{p} K_{E_i}(x).$$

Hence by 19.3 and 18.3

$$m \bigcup_{i=1}^{\infty} E_i \geqq m \bigcup_{i=1}^{p} E_i = \int_{R_q} K_{E_1 \cup \cdots \cup E_p}(x)\,dx$$

$$= \sum_{i=1}^{p} \int_{R_q} K_{E_i}(x)\,dx = \sum_{i=1}^{p} mE_i.$$

This holds for all positive integers p, so

$$m \bigcup_{i=1}^{\infty} E_i \geqq \sum_{i=1}^{\infty} mE_i.$$

Together with the reverse inequality already established, this proves

$$m \bigcup E_i = \sum_{i=1}^{\infty} mE_i.$$

Now suppose $E_1 \subset E_2 \subset \cdots$. Using again the functions $f_p(x)$, $f(x)$ of the first paragraph, we see that $f_p(x)$ is the characteristic function of $E_1 \cup \cdots \cup E_p$, which is E_p. So, as above, for each n

$$mW_n(\bigcup E_i) = \lim_{p\to\infty} \int_{W_n} f_p(x)\,dx = \lim_{p\to\infty} mW_nE_p \leqq \lim_{p\to\infty} mE_p.$$

This holds for all n, so

$$m \bigcup E_i = \lim_{n\to\infty} mW_n(\bigcup E_i) \leqq \lim_{p\to\infty} mE_p.$$

On the other hand, for every p the set E_p is contained in $\bigcup E_i$, so by 19.3 $mE_p \leqq m \bigcup E_i$, and $\lim_{p\to\infty} mE_p \leqq m \bigcup E_i$. This completes the proof of the theorem.

19.9s. *If $E_1, E_2, \cdots$ is a finite or denumerable collection of measurable sets, then $\bigcap E_i$ is measurable. Moreover, if there are denumerably many E_i and $E_1 \supset E_2 \supset E_3 \cdots$ and at least one E_i has finite measure, then $m \bigcap E_i = \lim_{i\to\infty} mE_i$.*

Since $\bigcap E_i = C(\bigcup CE_i)$, it is measurable by 19.6 and 19.8. Now suppose $E_1 \supset E_2 \supset \cdots$. If the set E_m has finite measure, neither the set $\bigcap E_i$ nor the limit $\lim mE_i$ is affected if we discard the sets $E_1, \cdots, E_{m-1}$; so we may as well suppose $mE_1 < \infty$. Define $D_i = E_1 - E_i$. Then D_i is measurable by 19.6, and $D_1 \subset D_2 \subset D \cdots$. Hence by 19.8 $\bigcup D_i$ is measurable, and $m \bigcup D_i = \lim_{i\to\infty} mD_i$. But $D_i \cup E_i = E_1$, and D_i and E_i are disjoint; so by 19.8 $mD_i = mE_1 - mE_i$. Also $(\bigcup D_i) \cup (\bigcap E_i) = E_1$ and $\bigcup D_i$ and $\bigcap E_i$ are disjoint; so $m \bigcup D_i = mE_1 - m \bigcap E_i$. Therefore

$$mE_1 - m \bigcap E_i = \lim\,(mE_1 - mE_i),$$

whence $m \bigcap E_i = \lim mE_i$.

It is interesting to observe that the hypothesis that at least one E_i has finite measure can not be removed. Suppose, for example, the E_n is the set of all (finite) numbers $x \geqq n$. Then $mE_n = \infty$ for each n, but $\bigcap E_n$ is the empty set and has measure 0.

If $E_1, E_2, \cdots$ is an infinite sequence of sets, we define the complete limit-set $\lim\sup E_i$ to be the set of all points which belong to infinitely many sets E_i, and we define the restricted limit-set lim inf E_i to be the set consisting of all points which belong to all but a finite number of the E_i. A point x belongs to infinitely many E_i if and only if it belongs to the sum $E_n \cup E_{n+1} \cup \cdots$ for every n; and this in turn is true if and only if it belongs to the intersection (product) of all such sums. Again a point x belongs to all but a finite number of the E_i if and only if it belongs to the intersection $E_n E_{n+1} E_{n+2} \cdots$ for some n; and this in turn is true if and only if it belongs to the sum of all such intersections. In symbols, we have established the following theorem.

19.10s. *If $E_1, E_2, \cdots$ is an infinite sequence of sets, then*

$$\limsup E_i = \bigcap_{n=1}^{\infty} \bigcup_{i=n}^{\infty} E_i,$$

$$\liminf E_i = \bigcup_{n=1}^{\infty} \bigcap_{i=n}^{\infty} E_i.$$

Concerning these two limit-sets we now establish two theorems.

19.11s. *If the sets $E_1, E_2, \cdots$ are all measurable, so are* lim sup E_i *and* lim inf E_i, *and*

$$(A) \qquad m(\liminf E_i) \leqq \liminf mE_i.$$

Also, if all the sets E are contained in a set of finite measure, then

$$(B) \qquad m(\limsup E_i) \geqq \limsup mE_i.$$

Consider first the set lim inf E_i. The intersection $E_n E_{n+1} \cdots$ is measurable for each n, by 19.9; so by 19.8 the sum of these intersections, which is lim inf E_i, is measurable. If lim inf mE_i is ∞, inequality (A) is obvious. Otherwise, let h be a number greater than lim inf mE_i. By 6.2, in every neighborhood of ∞ there are integers i such that mE_i is less than h. In every intersection $E_n E_{n+1} \cdots$ there appears such a set E_i, hence by 19.3

$$m \bigcap_{i=n}^{\infty} E_i < h.$$

As n increases the intersection-set $E_n E_{n+1} \cdots$ expands, so by 19.8 and the preceding inequality we have

$$m \left(\bigcup_{n=1}^{\infty} \bigcap_{i=n}^{\infty} E_i \right) = \lim_{n \to \infty} m \bigcap_{i=n}^{\infty} E_i \leqq h.$$

Since h is an arbitrary number greater than lim inf mE_i, by 5.2 our statements concerning lim inf E_i are established.

Let E be any measurable set containing all the E_i (for example, R_q itself). Then by 1.1

$$\begin{aligned}\limsup E_i &= E - [E - \limsup E_i] \\ &= E - [E - \bigcap_{n=1}^{\infty} \bigcup_{i=n}^{\infty} E_i] \\ &= E - \bigcup_{n=1}^{\infty} [E - \bigcup_{i=n}^{\infty} E_i] \\ &= E - \bigcup_{n=1}^{\infty} [\bigcap_{i=n}^{\infty} (E - E_i)] \\ &= E - \liminf (E - E_i).\end{aligned}$$

This shows that $\limsup E_i$ is measurable. Moreover, if E has finite measure, by (A) and 6.3 we have

$$\begin{aligned}m \limsup E_i &= mE - m \liminf (E - E_i) \\ &\geqq mE - \liminf (mE - mE_i) \\ &= mE - [mE - \limsup mE_i] \\ &= \limsup mE_i.\end{aligned}$$

This completes the proof.

In case $\limsup E_i$ and $\liminf E_i$ are identical, we denote them both by $\lim E_i$. From 19.11 and 6.7 it is obvious that the following is true.

19.12s. *If $E_1, E_2, \cdots$ are measurable subsets of a set E of finite measure, and* $\lim E_i$ *exists, then* $\lim mE_i$ *also exists, and*

$$m \lim E_i = \lim_{i \to \infty} mE_i.$$

19.13s. *If a set E has measure* 0, *and the set E_1 is contained in E, then $mE_1 = 0$. If $E_1, E_2, \cdots$ is a finite or denumerable collection of sets of measure* 0, *then $\bigcup E_i$ has measure* 0.

If $E_1 \subset E$, then $K_{E_1}(x) \leqq K_E(x)$. Then for each integer n and each interval I containing W_n in its interior we have

$$\begin{aligned}0 &\leqq \underline{\int_I} K_{W_nE_1}(x)\,dx \leqq \overline{\int_I} K_{W_nE_1}(x)\,dx \leqq \int_I K_{W_nE}(x)\,dx \\ &\leqq \int_{R_q} K_E(x)\,dx = 0.\end{aligned}$$

So the characteristic function of W_nE_1 is summable and its integral is 0 for every n, and by definition, $mE_1 = 0$.

The second part of the theorem follows from 19.8, for $\bigcup E_i$ is measurable, and

$$0 \leqq m \bigcup E_i \leqq \Sigma mE_i = 0.$$

The next three theorems are not "s" theorems.

19.14. *If E is a denumerable set, $mE = 0$.*

By 16.1, the measure of a set containing a single point (which can be regarded as a degenerate closed interval) is zero. If E is denumerable, it is the sum of denumerably many sets consisting of one point each, so by 19.13 it has measure zero.

The concept of a set of measure zero is so important and useful that it is frequently used in discussions in which no other concept out of the theory of Lebesgue integration is needed. Thus, for example, in the study of the Riemann integral it is shown (though we shall not prove it) that in order for a function $f(x)$ defined and bounded in an interval I to be Riemann integrable over I, it is necessary and sufficient that the set of discontinuities of $f(x)$ have measure zero. When sets of measure zero enter in such discussions, it is customary to say that a set E has measure zero if for every positive ϵ it is possible to cover E with a finite or denumerable set of open intervals $I_1, I_2, \cdots$ such that $\Sigma \Delta I_n < \epsilon$. We now proceed to show that this definition is in fact equivalent to ours. (It is not an "s" theorem.)

19.15. *In order that a set E have measure 0, it is necessary and sufficient that for every positive ϵ there exist a finite or denumerable set of open intervals $I_1, I_2, \cdots$ such that $\bigcup I_n \supset E$ and $\Sigma \Delta I_n < \epsilon$.*

The condition is sufficient. Suppose first that E is bounded, interior say to a closed interval I, which we use as basic interval. With the intervals I_n of the hypothesis,

$$K_E(x) \leqq K_{\bigcup I_n}(x),$$

$$\begin{aligned}\overline{\int_I} K_E(x)\,dx &\leqq \int_I K_{\bigcup I_n}(x)\,dx \\ &\leqq \int_I \sum K_{I_n}(x)\,dx \\ &= \sum \int_I K_{I_n}(x)\,dx \\ &= \sum \Delta I_n < \epsilon.\end{aligned}$$

Since this holds for every positive ϵ, and $K_E(x)$ is non-negative, by 5.2, 14.4 and 14.2 we have

$$0 \leqq \underline{\int_I} K_E(x)\,dx \leqq \overline{\int_I} K_E(x)\,dx \leqq 0.$$

Hence, by 19.1, $mE = 0$.

If E is unbounded, by the preceding paragraph $mEW_n = 0$ for every n, so by 19.1 we again have $mE = 0$.

The condition is necessary. Suppose first that E is bounded, interior say to a closed interval I, which we use as basic interval. Let ϵ

be a positive number. Since the integral of the characteristic function of E is zero, by 15.1 there is a U-function $u(x) \geqq K_E(x)$ such that

$$\int_I u(x)\,dx < \frac{\epsilon}{2^{q+1}}.$$

(Recall that our set E is in q-dimensional space R_q.) Let V be the subset of I on which $u(x) > \frac{1}{2}$. This is open relative to I by 7.4, and contains E. It is clear that

$$\tfrac{1}{2}K_V(x) \leqq u(x);$$

for if x is in V the left member is $\frac{1}{2}$ and the right member exceeds $\frac{1}{2}$, and otherwise the left member is zero and the right is $\geqq K_E(x)$, hence non-negative. Since $K_V(x)$ is a bounded U-function by 12.9, it is summable, and

$$\int_I K_V(x)\,dx \leqq 2\int_I u(x)\,dx < 2^{-q}\epsilon.$$

The set V is the sum of a finite or denumerable collection of closed intervals I_n^*, by 3.7. If we denote the interior of I_n^* by J_n, the sets J_n are disjoint, and their sum is contained in V. By 16.1,

$$mI_n^* = mJ_n = \Delta I_n^*.$$

Now using 19.8 and 19.3

$$\begin{aligned}\Sigma\Delta I_n^* = \Sigma mJ_n &= m \bigcup J_n \leqq mV \\ &= m\bigcup I_n^* \leqq \Sigma mI_n^* = \Sigma\Delta I_n^*,\end{aligned}$$

whence

$$\Sigma\Delta I_n^* = mV < 2^{-q}\epsilon.$$

The I_n^* are still not the intervals desired, since they are closed. Suppose that x_n is the center of I_n^*, so that I_n^* is defined by inequalities $x_n^{(i)} - \epsilon_n^{(i)} \leqq x^{(i)} \leqq x_n^{(i)} + \epsilon_n^{(i)}$. Let I_n be the interval

$$x_n^{(i)} - 2\epsilon_n^{(i)} < x^{(i)} < x_n^{(i)} + 2\epsilon_n^{(i)}.$$

Each edge of I_n is twice the corresponding edge of I_n^*, so $\Delta I_n = 2^q\Delta I_n^*$. Hence

$$\Sigma\Delta I_n = 2^q\Sigma\Delta I_n^* < \epsilon,$$

and the I_n are the intervals desired.

If E is of measure zero but unbounded, for each n the subset EW_n can be enclosed in open intervals $I_{n,1}, I_{n,2}, \cdots$ with

$$\sum_j \Delta I_{n,j} < \epsilon 2^{-n}.$$

Then the aggregate of all intervals $I_{n,j}$ $(n = 1, 2, \cdots)$ is denumerable, and

$$\sum_{n,j} \Delta I_{n,j} < \epsilon \sum 2^{-n} = \epsilon.$$

Clearly E is covered by these intervals. This completes the proof.

19.16. *Let E be a set, and $(h^{(1)}, \cdots, h^{(q)})$ a q-tuple of real numbers. If E is measurable so is its translation $E^{(h)}$, and*

$$mE^{(h)} = mE.$$

If mE is finite, this is merely 18.7 applied to $K_E(x)$. Otherwise, for each n we consider EW_n. This has finite measure, so its translation has the same finite measure:

$$mE^{(h)}W_n^{(h)} = mEW_n.$$

As n increases the sets EW_n expand, and so do their translations. So by 19.8 the sum $E^{(h)}$ of the translations $E^{(h)}W_n^{(h)}$ is measurable, and

$$mE^{(h)} = \lim_{n\to\infty} mE^{(h)}W_n^{(h)} = \lim_{n\to\infty} mEW_n = mE.$$

20. The simplest sets with which we have had to deal are the intervals. For these the measure is easily found; by 16.1, $mI = \Delta I$. (This is not an "s" theorem, but a rather more complicated device permits us to obtain the measures of intervals from the interval-function ΔI without using any properties of ΔI except those in 10.1. The proof occurs later, in 47.3.) We may regard open sets as next in simplicity, since by 3.6 each open set G is the sum of a denumerable set of disjoint intervals $I_1, I_2, \cdots$. Thus for an open set G we have $mG = \Sigma mI_n$. That is, on the basis of the theory which we have developed it is now possible to calculate the measure of an open set directly from the values of the interval-function ΔI, without explicitly computing any integrals. If F is a bounded closed set, we can enclose it in an open interval G; then mF is the difference between the measures of the open sets G and $G - F$. If F is an unbounded closed set, it is the same as $\lim FW_n$, and each set FW_n is a bounded closed set. So the measures of closed sets are also obtainable without explicitly performing an integration.

We now proceed to show how the measures of arbitrary sets can be studied through a knowledge of measures of open and closed sets. It will follow that the whole theory of measure, and with it the whole theory of integration, could be built on this basis, although we do not choose to develop it in that way. However, a word of caution may not be amiss. We have already used 19.8 in obtaining the equation $mG = \Sigma mI_n$, in the preceding paragraph. This informs us in particular that the sum ΣmI_n has the same value, no matter how we choose to decompose the set G into non-overlapping closed intervals. If we had chosen to base our theory of integration on the measures of open sets, putting the definition $mG = \Sigma mI_n$ at the beginning of the theory, we would have had to devise some other proof of this independence. This would have called for a proof somewhat like that of 10.2, but more complicated because it would involve infinite collections of intervals. And this would be only the first of several troubles, for the theorems to which we shall refer in the proofs in this section would be no longer available, and consequently the proofs would be much more difficult.

It is convenient to define first two numbers, the *exterior measure* m_eE and the *interior measure* m_iE, which are defined for all sets and which are related to the measure of E (when it exists) in much the same way that the upper and lower integrals of $f(x)$ are related to the integral of $f(x)$ (when it exists).

20.1s. *Let E be an arbitrary set in the space R_q. Then its exterior measure m_eE is the* inf *of the measures of all open sets containing E, and its interior measure m_iE is the* sup *of the measures of all closed sets contained in E.*

As an immediate consequence of the definition we have the following theorem.

20.2s. *For every set E the inequalities*

$$0 \leqq m_iE \leqq m_eE$$

are satisfied.

Let E be a set, and let F and G be respectively a closed set and an open set such that $F \subset E \subset G$. By 19.3 $0 \leqq mF \leqq mG$. For each closed set F contained in E this inequality holds for all open sets G containing E, so mF is a lower bound for mG, and cannot exceed the inf. That is,

$$mF \leqq m_eE.$$

This holds for all closed sets F contained in E, so m_eE is an upper bound for the measures of such sets, and cannot be less than their sup, which is m_iE.

Another immediate consequence of 20.1 is the following

20.3s. *If $E_1 \subset E_2$, then $m_iE_1 \leqq m_iE_2$ and $m_eE_1 \leqq m_eE_2$.*

The class of closed sets $F \subset E_1$ is a subset of the class of closed sets $F \subset E_2$, and by 5.3 this implies the first inequality. The second is proved analogously.

Still another corollary of definition 20.1 is this theorem.

20.4s. *If $E_1, E_2, \cdots$ is a finite or denumerable collection of sets in the space R_q, then*

$$m_e \bigcup E_i \leqq \Sigma m_e E_i.$$

If the right member of the inequality is infinite this is trivial Otherwise, let ϵ be an arbitrary positive number. By 20.1 we can enclose E_i in an open set G_i such that

$$mG_i < m_e E_i + \frac{\epsilon}{2^i}.$$

Then $G = \bigcup G_i$ is an open set, and by 19.8

$$mG \leqq \sum mG_i < \sum m_e E_i + \sum \left(\frac{\epsilon}{2^i}\right)$$
$$= \Sigma m_e E_i + \epsilon.$$

So by 20.1 we have

$$m_e \bigcup E_i < \Sigma m_e E_i + \epsilon,$$

and by 5.2 this implies the conclusion of the theorem.

Now we establish a connection between the exterior measure and upper integral, and between interior measure and lower integral.

20.5s. *Let E be a bounded set, interior to a closed interval I. Then*

$$m_e E = \overline{\int_I} K_E(x)\, dx \qquad \text{and} \qquad m_i E = \underline{\int_I} K_E(x)\, dx.$$

Let J be the interior of I. If G is any open set containing E, then JG is open by 2.10, contains E because J and G both contain E, and is interior to I. Hence by 19.3, 19.2, the remark after 18.1, 1.2 and 14.4

$$mG \geqq mJG = \int_{R_q} K_{JG}(x)\, dx$$
$$= \int_I K_{JG}(x)\, dx$$
$$\geqq \overline{\int_I} K_E(x)\, dx.$$

Thus by definition 20.1 we have

$$(A) \qquad m_e E \geqq \overline{\int_I} K_E(x)\, dx.$$

Likewise, if F is closed and contained in E,

$$\begin{aligned} mF &= \int_{R_q} K_F(x)\,dx \\ &= \int_I K_F(x)\,dx \\ &\leqq \underline{\int_I} K_E(x)\,dx, \end{aligned}$$

whence by 20.1

$$(B) \qquad m_iE \leqq \underline{\int_I} K_E(x)\,dx.$$

Let ϵ be an arbitrary positive number. By definition 14.1, there is a U-function $u(x) \geqq K_E(x)$ such that

$$\int_I u(x)\,dx < \overline{\int_I} K_E(x)\,dx + \frac{\epsilon}{2}.$$

Define

$$u_1(x) = u(x) + \frac{\epsilon}{2(\Delta I + 1)}.$$

This is a U-function by 12.4, and is clearly positive-valued. We define G to be the subset of I on which $u_1(x) > 1$. This set is open relative to I by 7.4. So is JG, since J is the interior of I; and by 2.7 the set JG is open. Also, JG contains E, for if x is in E it is in J by hypothesis, and

$$u_1(x) = u(x) + \frac{\epsilon}{2(\Delta I + 1)} \geqq K_E(x) + \frac{\epsilon}{2(\Delta I + 1)} > 1,$$

so that x is in G. Also the characteristic function of JG is less than $u_1(x)$; for if x is in JG the characteristic function is 1 while $u_1(x) > 1$ by definition of G, and otherwise the characteristic function is 0 while $u_1(x)$ is positive-valued. Hence by 20.1, 19.2, 18.3 and 11.6

$$\begin{aligned} m_eE \leqq mJG &= \int_I K_{JG}(x)\,dx \\ &\leqq \int_I u_1(x)\,dx \\ &= \int_I u(x)\,dx + \int_I \frac{\epsilon}{2(\Delta I + 1)}\,dx \\ &< \left[\overline{\int_I} K_E(x)\,dx + \frac{\epsilon}{2}\right] + \frac{\epsilon \Delta I}{2(\Delta I + 1)} \\ &< \overline{\int_I} K_E(x)\,dx + \epsilon. \end{aligned}$$

Since ϵ is arbitrary, by 5.2 this implies

$$m_e E \leqq \overline{\int_I} K_E(x)\, dx.$$

This, with (A), establishes the first of the two equations in the conclusion of our theorem.

Again, let ϵ be an arbitrary positive number. There is an L-function $l(x) \leqq K_E(x)$ such that

$$\int_I l(x)\, dx > \underline{\int_I} K_E(x)\, dx - \frac{\epsilon}{2}.$$

Define

$$l_1(x) = l(x) - \frac{\epsilon}{2(\Delta I + 1)},$$

and let F be the set on which $l_1(x) \geqq 0$. This is closed relative to I by 7.4, hence is closed by 2.7. It is contained in E, for at each point x of F we have

$$K_E(x) \geqq l(x) > l_1(x) \geqq 0$$

so $K_E(x)$ must be 1. Also, $K_F(x) > l_1(x)$ for all x; for if x is in F we have $K_F(x) = 1$ and $l_1(x) < l(x) \leqq K_E(x) \leqq 1$, and otherwise $K_F(x)$ is zero and $l_1(x)$ is negative.

By 20.1, 19.2, 18.3 and 11.6

$$\begin{aligned} m_i E &\geqq mF \\ &= \int_I K_F(x)\, dx \\ &\geqq \int_I l_1(x)\, dx \\ &= \int_I \left\{ l(x) - \frac{\epsilon}{2(\Delta I + 1)} \right\} dx \\ &> \left\{ \underline{\int_I} K_E(x)\, dx - \frac{\epsilon}{2} \right\} - \frac{\epsilon \Delta I}{2(\Delta I + 1)} \\ &> \underline{\int_I} K_E(x)\, dx - \epsilon. \end{aligned}$$

Since ϵ is arbitrary, by 5.2 this implies

$$m_i E \geqq \underline{\int_I} K_E(x)\, dx,$$

and this with (B) completes the proof.

From this theorem we deduce several corollaries.

20.6s. *If E is bounded, it is measurable if and only if $m_iE = m_eE$; and in that case $mE = m_eE = m_iE$.*

Let I be a closed interval containing E in its interior. Since E is bounded, by 19.2 it is measurable if and only if it has finite measure. Again by 19.2, this is true if and only if $K_E(x)$ is summable over the space R_q. By the remark after 18.1, this is true if and only if $K_E(x)$ is summable over I. By 15.1, this is true if and only if the upper and lower integrals of $K_E(x)$ are finite and equal. Since $K_E(x)$ is bounded, its upper and lower integrals over J cannot be infinite; so E is measurable if and only if the upper and lower integrals of $K_E(x)$ over I are equal. By 20.5, this is true if and only if $m_iE = m_eE$. Furthermore, if $m_iE = m_eE$, by 20.5 their common value is the integral of $K_E(x)$ over I, which by 17.4 and 18.1 is its integral over R_q, which by 19.2 is mE.

Theorem 20.6 permits us to distinguish bounded measurable sets from bounded non-measurable sets, and to compute the measures of the former. From this we could proceed as in §19 to define measurable sets to be those sets E such that EW_n is measurable for each n, and to obtain their measures as in 19.1. An alternative procedure will be mentioned after 20.10.

20.7s. *If E is a measurable set and ϵ is a positive number, there exists an open set G containing E such that $m(G - E) < \epsilon$, and there exists a closed set F contained in E such that $m(E - F) < \epsilon$.*

If E is measurable and bounded, by 20.6 we have $m_eE = mE$, so by 20.1 the set E is contained in an open set G such that $mG < mE + \epsilon$. But by 19.8 we have $mG = mE + m(G - E)$, so $m(G - E) < \epsilon$. If E is measurable but unbounded, for each positive integer n the set EW_n is measurable and bounded, and by the preceding sentence is contained in an open set G_n such that

$$m(G_n - EW_n) < \frac{\epsilon}{2^n}.$$

The set $G = \bigcup G_n$ is open by 2.10 and contains $\bigcup EW_n$, which is E. Also, $G - E$ is contained in $\bigcup (G_n - EW_n)$; for if x is in $G - E$, it is in some G_n but not in any EW_n. Hence by 19.3 and 19.8

$$\begin{aligned} m(G - E) &\leqq m \bigcup (G_n - EW_n) \\ &\leqq \sum m(G_n - EW_n) \\ &< \sum \left(\frac{\epsilon}{2^n}\right) \\ &= \epsilon. \end{aligned}$$

Thus the first conclusion is established.

If E is measurable, so is its complement CE. We enclose CE in an open set G such that $m(G - CE) < \epsilon$, and we define $F = CG$. This is closed by 2.6, and is contained in E. Moreover,

$$E - F = E \cap CF = EG = G - CE,$$

so $m(E - F) = m(G - CE) < \epsilon$. This completes the proof.

20.8s. *If E is measurable, then $m_iE = m_eE = mE$.*

If $mE = \infty$, then by 20.7 with $\epsilon = 1$ there is a closed set $F \subset E$ with $mF = \infty$. By 20.1 and 20.2 this implies $m_iE = m_eE = \infty$. If mE is finite, the sets F and G of 20.7 show that $m_iE \geqq mF > mE - \epsilon$ and $m_eE \leqq mG < mE + \epsilon$. Since ϵ is arbitrary, by 5.2 and 20.2 we have

$$m_eE \leqq mE \leqq m_iE \leqq m_eE,$$

establishing the theorem.

20.9s. *Let M be a set having finite measure, and let E be a subset of M. Then*

$$(A) \qquad m_i(M - E) = mM - m_eE$$

and

$$(B) \qquad m_e(M - E) = mM - m_iE.$$

Let ϵ be an arbitrary positive number. The numbers in equations (A) and (B) are clearly finite. By 20.7 there is an open set G containing M for which $m(G - M) < \epsilon/2$, and by 20.1 there is a closed set F contained in E for which $mF > m_iE - \frac{\epsilon}{2}$. The set $G - F = G \cap CF$ is open by 2.6 and 2.10, and contains $M - E = M \cap CE$. Hence by 20.1 and 19.8

$$\begin{aligned} m_e(M - E) &\leqq m(G - F) \\ &= mG - mF \\ &< \left[mM + \frac{\epsilon}{2}\right] - \left[m_iE - \frac{\epsilon}{2}\right] \\ &= mM - m_iE + \epsilon. \end{aligned}$$

By 5.2 this implies

$$(C) \qquad m_e(M - E) + m_iE \leqq mM.$$

On the other hand, by 20.7 there is a closed subset F_1 of M such that $m(M - F_1) < \epsilon/2$, and by 20.1 there is an open set G_1 containing $M - E$ such that $mG_1 < m_e(M - E) + \frac{\epsilon}{2}$. Then the set $F_1 - G_1 = F_1 \cap CG_1$ is closed by 2.6 and 2.10, and is contained in E. So by 19.3 and 19.8

$$\begin{aligned} mG_1 + m(F_1 - G_1) &\geqq mF_1G_1 + m(F_1 - G_1) \\ &= mF_1 \\ &> mM - \frac{\epsilon}{2}. \end{aligned}$$

But by 20.1 and the choice of G_1 this yields

$$m_e(M - E) + m_iE > mM - \epsilon,$$

and by 5.2 we find

$$m_e(M - E) + m_iE \geqq mM.$$

This and (C) together imply (B). From (B) we obtain (A) by interchanging E and $M - E$.

20.10s. *Let E be a set in the space R_q. In order that E be measurable it is necessary and sufficient that the equation*

$$(A) \qquad m_eE_1 = m_eE_1E + m_eE_1 \cap CE$$

be satisfied for every set E_1.

Suppose that E is measurable; then CE is also measurable. Let ϵ be an arbitrary positive number. By 20.7, there are open sets G_1 and G_2 such that G_1 contains E, G_2 contains CE, and

$$m(G_1 - E) < \epsilon \qquad \text{and} \qquad m(G_2 - CE) < \epsilon.$$

But

$$\begin{aligned} G_1G_2 &= G_1G_2(E \cup CE) \\ &= G_1G_2E \cup G_1G_2CE \\ &= G_2E \cup G_1CE \\ &= [G_2 - CE] \cup [G_1 - E], \end{aligned}$$

so by 19.3 and 19.8

$$mG_1G_2 \leqq m[G_2 - CE] + m[G_1 - E] < 2\epsilon.$$

We now wish to establish the inequality

$$(B) \qquad m_eE_1 \geqq m_eE_1E + m_eE_1 \cap CE$$

for all sets E_1. If $m_eE_1 = \infty$ this is trivial. Otherwise, by 20.1 there is an open set G_3 containing E_1 for which $mG_3 < m_eE_1 + \epsilon$. Then G_3G_1 and G_3G_2 are open sets by 2.10, and contain E_1E and $E_1 \cap CE$ respectively. Hence by 20.1

$$m_eE_1E \leqq mG_3G_1,$$
$$m_eE_1 \cap CE \leqq mG_3G_2.$$

The three disjoint sets $G_1 \cap CG_2$, $G_2 \cap CG_1$ and G_1G_2 fill the set $G_1 \cup G_2$, so by 19.3 and 19.8

$$\begin{aligned} mG_3 &\geqq mG_3 \cap [G_1 \cup G_2] \\ &= mG_3G_1 \cap CG_2 + mG_3G_2 \cap CG_1 + mG_3G_1G_2 \\ &= [mG_3G_1 \cap CG_2 + mG_3G_1G_2] + [mG_3G_2 \cap CG_1 + mG_3G_2G_1] \\ &\qquad - mG_3G_1G_2 \\ &\geqq mG_3G_1 + mG_3G_2 - mG_1G_2. \end{aligned}$$

It follows that

$$\begin{aligned} \epsilon + m_eE_1 &> mG_3 \\ &\geqq m_eE_1E + m_eE_1 \cap CE - 2\epsilon, \end{aligned}$$

and by 5.2 this implies inequality (B). But $E_1E \cup E_1CE = E_1$, so by 20.4 we have

$$m_eE_1 \leqq m_eE_1E + m_eE_1 \cap CE.$$

This and (B) together establish equation (A).

Conversely, suppose equation (A) satisfied for a set E. For E_1 we choose a bounded measurable set. By 20.9,

$$m_iE_1E = mE_1 - m_e(E_1 - E).$$

Since E_1 is measurable, by 20.8 equation A implies

$$m_eE_1E = mE_1 - m_e(E_1 - E).$$

Hence the exterior and interior measures of the bounded set E_1E are equal, and by 20.6 E_1E is measurable. In particular, if E_1 is the interval W_n this shows that EW_n is measurable, so by 19.1 the set E is measurable. This completes the proof of the theorem.

If we had a more complete theory of the exterior measure of sets, developed without reference to measurable sets, we could use the property expressed in 20.10 to *define* the property of measurability. That is, a set E would by definition be measurable if the equation

$$m_e E_1 = m_e E_1 E + m_e E_1 CE$$

held for every set E_1. Having selected the measurable sets by this test, we could define the measure of a measurable set to be the same as its exterior measure. From the properties of exterior measure already established we could then deduce the properties of measurable sets, and erect a theory of integration on this foundation. This is the method originated by Carathéodory. An exposition can be found in Carathéodory's *Vorlesungen über reelle Funktionen*, or Kestelman's *Modern Theories of Integration.*

EXERCISE. Show that if E is a bounded set, there are sets P and S with the following properties. P is an intersection of open sets and contains E. S is a sum of closed sets and is contained in E. The equations $mP = m_e E$ and $mS = m_i E$ hold. Furthermore, for every measurable set M we have $mPM = m_e EM$ and $mSM = m_i EM$. Extend this to unbounded sets E.

21. If we recall even a very little about the Riemann integral and the mild generalization of it mentioned in §16, we see that a function $f(x)$ which is not Riemann integrable may lack integrability for either of two very different reasons. It may have too many discontinuities, as in the example after 18.6. Or it may simply have too many large functional values; for example the function $f(x)$ which is x^{-1} for $0 < x \leqq 1$ and has the value 0 at $x = 0$. This function is not integrable, although it has only a single discontinuity. This latter type of function may clearly have many of the desirable properties of Riemann-integrable functions, even though it is not integrable.

In the case of the Lebesgue integral the situation is similar. A function which is not summable may be of so intricate a structure that its upper and lower integrals have different values. Such complicated functions we shall avoid. On the other hand, it may be non-summable merely because it has too many large functional values. In this case its structure may be as simple as that of the summable functions. In order to be able to discuss the functions of relatively simple structure we introduce the notion of a *measurable* function. The class of measurable functions is closely related to the class of summable functions, as theorems 21.4 and 22.1 will show.

21.1s. *A function $f(x)$ defined on a set E in I is measurable (on E) if for each constant a (finite or infinite) the set of points x in E at which $f(x) \geqq a$ is a measurable set.*

We use the symbol $E[f \geqq a]$ to mean the set of all x in E such that $f(x) \geqq a$ (in our previous notation: $\{x \mid f(x) \geqq a\} \cap E$), and we also use analogous symbols for sets on which $f(x)$ satisfies other conditions. The meaning will be evident in each case.

A first remark is that if $f(x)$ is measurable on E, then E is a measurable set. For $E[f \geqq -\infty]$ is then measurable, and this set is E itself.

A second simple remark is that a set E is measurable if and only if $K_E(x)$ is a measurable function on the space R_q. For if E is measurable the set $R_q[K_E \geqq a]$ is one of the measurable sets R_q, E, Λ according as $a \leqq 0$, $0 < a \leqq 1$, $a > 1$; and if K_E is a measurable function the set $E \equiv R_q[K_E \geqq 1]$ is measurable.

A third remark is that if $f(x)$ is measurable over E and $h^{(1)}, \cdots, h^{(q)}$ are arbitrary constants, then (with the notation 10.8) $f^{(h)}(x)$ is measurable over $E^{(h)}$. This follows from 19.16 if we observe that x is in $E^{(h)}$ and $f^{(h)}(x) \geqq a$ if and only if $x + h$ is in E and $f(x + h) \geqq a$. It is not an s-theorem.

A fourth remark almost as obvious is

21.2s. *If $f(x)$ is defined and measurable on a set E, and E_0 is a measurable subset of E, then $f(x)$ is measurable on E_0.*

For if a is any number, the set $E_0[f \geqq a]$ is the product of the measurable sets E_0 and $E[f \geqq a]$, so it is measurable by 19.6.

We now list five conditions each of which could serve as a definition of measurability.

21.3s. *If $f(x)$ is defined on a measurable set E, then each of the following conditions is necessary and sufficient in order that $f(x)$ be measurable on E:*

(*a*) *$E[f \leqq a]$ is measurable for all a;*

(*b*) *$E[f < a]$ is measurable for all a;*

(*c*) *$E[f > a]$ is measurable for all a;*

(*d*) *$E[b \geqq f > a]$ is measurable for all a and all b;*

(*e*) *$E[b \geqq f \geqq a]$ is measurable for all a and all b.*

(We could state two more necessary and sufficient conditions by replacing $\geqq$ by $>$ after the number b in (*d*) and (*e*).)

Suppose $f(x)$ is measurable. The set $E[f > a]$ is the sum of the sets $E[f \geqq r]$ for all rational numbers $r > a$. For if $a < \infty$ and $f(x) > a$, there is a rational r such that $f(x) \geqq r > a$, and so x is in $E[f \geqq r]$; and if x is in some one of the sets $E[f \geqq r]$, say $E[f \geqq r_0]$, then $f(x) \geqq r_0 > a$, so x is in $E[f > a]$. If $a = \infty$, the set $E[f > a]$ is empty and there are no $r > a$, so the equality of $E[f > a]$ and $\bigcup E[f \geqq r]$, $r > a$ is still valid. Each of the sets $E[f \geqq r]$ is measurable by hypothesis, and there are at most denumerably many of them because the rationals are denumerable. So their sum $E[f > a]$ is measurable by 19.8. Thus if $f(x)$ is measurable, (c) is satisfied.

Suppose that (c) is satisfied. For every a we have $E[f \leqq a] = E - E[f > a]$. But $E[f > a]$ is measurable for every a, so by 19.6 so is $E[f \leqq a]$. Therefore if (c) is satisfied, so is (a).

Suppose (a) is satisfied. If in the first paragraph of this proof we replace the signs $\geqq$, $>$ by $\leqq$, $<$ respectively and replace ∞ by $-\infty$, we have a proof that (b) is satisfied.

Suppose that (b) is satisfied. Then for every a the set $E[f < a]$ is measurable. But $E[f \geqq a] = E - E[f < a]$, so by 19.6 $E[f \geqq a]$ is measurable for every a, and so $f(x)$ is measurable. Summing up, we have shown that the measurability of $f(x)$ implies (c), which implies (a), which implies (b), which implies the measurability of $f(x)$. Hence the measurability of $f(x)$ is equivalent to (a), to (b) and to (c).

If $f(x)$ is measurable, then for every a and b the sets $E[f \leqq b]$ and $E[f > a]$ are measurable, by condition (a) and condition (c) respectively. Hence their product $E[b \geqq f > a]$ is measurable. Conversely, if $E[b \geqq f > a]$ is measurable for all a and b, then by setting $b = \infty$ we find that $E[f > a]$ is measurable for each a. This is condition (c), equivalent to measurability, so condition (d) is equivalent to measurability.

If in the preceding paragraph we replace $>$ by $\geqq$ and replace the words "condition (c)" by "the definition of measurability of $f(x)$," we have a proof that (e) is equivalent to the measurability of $f(x)$.

EXERCISE. Theorem 21.3 remains valid even if a and b are restricted to lie in a set M whose closure is R^*.

The next theorem partly exhibits the relationship between measurability and summability. This relationship will appear even closer after the demonstration of theorems 22.1 and 22.2, which are partial converses of 21.4.

21.4s. *If $f(x)$ is defined and summable over a measurable set E, it is measurable on E.*

We define $f(x)$ to be $-\infty$ on the complement of E; this does not affect its summability over E. Let a be an arbitrary real number. For each integer n we define

$$f_n(x) = \inf \{n \cdot \sup \{f(x) - a, 0\}, 1\}.$$

It is clear that $f_1(x) \leqq f_2(x) \leqq \cdots$, so the limit of $f_n(x)$ as $n \to \infty$ exists. Call this limit $g(x)$. We shall now show that $g(x)$ is the characteristic function of $E[f > a]$.

If x is in $E[f > a]$, then $f(x) - a > 0$, and $\sup \{f(x) - a, 0\} = f(x) - a > 0$. Hence for all n greater than $(f(x) - a)^{-1}$ we have $n \cdot \sup \{f(x) - a, 0\} > 1$, so that $f_n(x) = \inf \{n \cdot \sup \{f(x) - a, 0\}, 1\} = 1$. Thus $f_n(x) = 1$ for all large n, and the limit $g(x)$ of $f_n(x)$ is 1.

If x is not in $E[f > a]$, then $f(x) - a \leqq 0$, and $\sup \{f(x) - a, 0\} = 0$. Then for all n we have $n \cdot \sup \{f(x) - a, 0\} = 0$, so that $f_n(x) = \inf \{n \cdot \sup \{f(x) - a, 0\}, 1\} = \inf \{0, 1\} = 0$. Therefore $g(x) = \lim f_n(x) = 0$. Thus $g(x) = 1$ if x is in $E[f > a]$ and is 0 otherwise.

For each of the intervals W_p the intersection EW_p is measurable by 19.1, so 1 is summable over EW_p by 19.2. Also $f(x)$ is summable over EW_p by 18.1; so by 18.3 $f(x) - a$ is summable over EW_p, and so also is $\sup \{f(x) - a, 0\}$. But this vanishes on $W_p - E$, as shown above; so by the definition 17.1 it is summable over the entire interval W_p. Again by 18.3, $n \sup \{f(x) - a, 0\}$ and

$$f_n(x) = \inf \{n \sup \{f(x) - a, 0\}, 1\}$$

are summable over W_p. For all n we have (by 15.2)

$$\int_{W_p} f_n(x)\, dx \leqq \int_{W_p} 1\, dx,$$

so the integrals do not tend to ∞ as $n \to \infty$. Therefore by 18.4 the limit $g(x)$ of the $f_n(x)$ is summable over W_p. Since $g(x)$ is the characteristic function of $E[f > a]$, this implies that $W_pE[f > a]$ is measurable for all p, so for every finite number a the set $E[f > a]$ is measurable by definition 19.1.

For $a = +\infty$, $E[f > a]$ is empty, and for $a = -\infty$, $E[f > a]$ is the whole set E. In either case, it is measurable. Therefore $E[f > a]$ is measurable for all a, and by 21.3 the function $f(x)$ is measurable.

21.5s. Corollary. *If $f(x)$ is summable over a set E, then the subset E_0 of E on which $f(x) \neq 0$ is measurable.*

Let $f_0(x) = f(x)$ if x is in E, $f_0(x) = 0$ if x is in CE. Then $f_0(x)$ is summable over the whole space R_q, and by 21.4 it is measurable. By 21.3, the sets $R_q[f_0 > 0]$ and $R_q[f_0 < 0]$ are measurable. But these sets lie entirely in E, and their sum is $E[f \neq 0]$.

Theorem 21.4 would lead us to suspect that theorems analogous to those of §18 should hold for measurable functions. We now establish several such theorems.

21.6. *If $f(x)$ is measurable on E and c is any finite* constant, then $cf(x)$ is measurable on E.*

For $c = 0$ this is obvious. Suppose $c < 0$. Then for every a the set $E[cf \geqq a]$ is the same as the set $E[f \leqq ac^{-1}]$, which is measurable by 21.3a; so $cf(x)$ is measurable on E. If $c > 0$, then as we have just seen $(-c)f(x)$ is measurable, and therefore so is $(-1)(-c)f(x) = cf(x)$.

21.7s. *If $f_1(x), f_2(x), \cdots$ is a finite or denumerable collection of functions measurable on E, then* $\sup_i \{f_i(x)\}$ *and* $\inf_i \{f_i(x)\}$ *are measurable on E.*

Let $\varphi(x) = \sup_i \{f_i(x)\}$, and let a be any number. We first show $E[\varphi > a] = \bigcup E[f_i > a]$. If x is in $E[\varphi > a]$, then $\varphi(x) = \sup_i f_i(x) > a$, so for some value j of i we have $f_j(x) > a$. Thus x is in $E[f_j > a]$, and therefore is in $\bigcup E[f_i > a]$. Conversely, if x is in $\bigcup E[f_i > a]$, it is in some set $E[f_j > a]$. Hence $f_j(x) > a$, and $\varphi(x) \geqq f_j(x) > a$. Therefore x is in $E[\varphi > a]$.

Now each set $E[f_i > a]$ is measurable by 21.3c. Hence $E[\varphi > a]$ is the sum of a finite or denumerable collection of measurable sets, and is measurable by 19.8. So by 21.3c, the function $\varphi(x)$ is measurable.

The functions $-f_i(x)$ are measurable by 21.6. Hence, by the preceding part of the proof, $\sup_i [-f_i(x)]$ is measurable. But this is the same as $-\inf_i [f_i(x)]$, by 5.5; and, applying 21.6 again, $\inf_i f_i(x)$ is measurable.

21.8s. Corollary. *If $f(x)$ is measurable on E, then $|f(x)|$, $f^+(x)$ and $f^-(x)$ are also measurable on E.*

For then $-f(x)$ is measurable on E, by 21.6; so by 21.7 the functions

$$\begin{aligned} |f(x)| &= \sup \{f(x), -f(x)\}, \\ f^+(x) &= \sup \{f(x), 0\}, \\ f^-(x) &= \sup \{-f(x), 0\} \end{aligned}$$

are measurable on E.

21.9s. *If $f_1(x), f_2(x), \cdots$ is a sequence of functions all of which are measurable on E, then*

$$\limsup_{n \to \infty} f_n(x) \qquad \text{and} \qquad \liminf_{n \to \infty} f_n(x)$$

are measurable on E.

If we define $g_n(x) = \sup \{f_i(x) \mid i \geqq n\}$, then $g_n(x)$ is measurable by 21.7 and also $\inf_n \{g_n(x)\}$ is measurable by 21.7. This last is the upper limit of $f_n(x)$ as $n \to \infty$, by the paragraph following 6.1. Interchanging sup and inf proves that the lower limit is measurable.

* The theorem remains true if $c = +\infty$ or $c = -\infty$.

21.10s. *If $f(x)$ is defined on a set E, and E is the sum of a finite or denumerable collection of measurable sets $E_1, E_2, \cdots$ on each of which $f(x)$ is measurable, then $f(x)$ is measurable on E.*

For then the set $E[f \geqq a]$ is the sum of the various sets $E_i[f \geqq a]$, each of which is measurable by hypothesis. So $E[f \geqq a]$ is measurable by 19.8, that is, $f(x)$ is measurable on E.

In particular, if $f(x)$ is defined on E and is measurable on $E - E_0$, where $mE_0 = 0$, then $f(x)$ is measurable on E. For $f(x)$ is measurable on E_0, since for each a the set $E_0[f \geqq a]$ has measure 0 by 19.13.

21.11s. *If*

(α) $f_1(x), \cdots f_h(x)$ are functions defined, finite and measurable on a measurable set E_0 in q-dimensional space R_q;

*(β) S is a set of points $(z^{(1)}, \cdots, z^{(h)})$ in h-dimensional space R_h, and S is either open, or closed, or the sum of a denumerable set of closed sets;**

(γ) $\varphi(z) \equiv \varphi(z^{(1)}, \cdots, z^{(h)})$ is defined on S and is continuous or upper semi-continuous or lower semi-continuous on S;
then the subset E_1 of E_0, on which the function $\varphi(f_1(x), \cdots, f_h(x))$ is defined, is measurable and $\varphi(f_1(x), \cdots, f_h(x))$ is measurable on E_1.

Let F be the mapping of E_0 into R_h defined by

$$F(x) = (f_1(x), \cdots, f_h(x)).$$

Even if the $f_i(x)$ happen to be defined at values of x not in E_0, we shall ignore such values, so that $F(x)$ is defined only on E_0. The inverse mapping, which may be many-valued, will be denoted by $F^{-1}(z)$. Thus for each z in R_h we define $F^{-1}(z)$ to be the set of all points x in E_0 such that $F(x) = z$. Also, if S_0 is a point-set in R_h, $F^{-1}(S_0)$ is the collection of all x in E_0 such that $F(x)$ is in S_0.

Let J be any interval in R_h of the type described in 3.6, namely $\{z \mid a^{(i)} \leqq z^{(i)} < b^{(i)},\ i = 1, \cdots, h\}$. We now prove that $F^{-1}(J)$ is measurable. Clearly it consists of all x in E_0 such that $F(x)$ is in J; so it consists of all x in E_0 which satisfy all of the inequalities

$$a^{(i)} \leqq f_i(x) < b^{(i)}, \qquad i = 1, \cdots, h.$$

Otherwise stated, if $f_i^{-1}[a^{(i)}, b^{(i)})$ is the set of all x satisfying

$$a^{(i)} \leqq f_i(x) < b^{(i)},$$

the set $F^{-1}(J)$ is the intersection $\bigcap_i f_i^{-1}[a^{(i)}, b^{(i)})$. Each of these sets is measurable by 21.3, so by 19.8 their intersection is also measurable.

* Every open set is the sum of denumerably many closed sets, by 3.6.

Now suppose that S_0 is an arbitrary open set in R_h. By 3.6, S_0 is the union of intervals $J_1, J_2, \cdots$ of the above mentioned type. Since each $F^{-1}(J_k)$ is measurable the union $\bigcup F^{-1}(J_k) = F^{-1}(S_0)$ is measurable by 19.8.

If S_0 is closed, then CS_0 is open. Hence $F^{-1}(CS_0)$ is measurable, and since $F^{-1}(S_0) = E_0 - F^{-1}(CS_0)$ it too is measurable by 19.6.

If S_0 is the sum of denumerably many closed sets $T_1, T_2, \cdots$, then $F^{-1}(S_0) = \bigcup F^{-1}(T_n)$ and hence is measurable by the preceding paragraph and 19.8.

From the preceding proof, for each of the three types of set S under consideration the set E_1 on which $\varphi(f_1(x), \cdots, f_h(x))$ is defined is a measurable subset of E_0, because obviously $E_1 = F^{-1}(S)$.

Now let us suppose that $\varphi(z)$ is lower semi-continuous. Suppose first that S is open. For every number a in R^* the set $S[\varphi > a]$ is open relative to S by 7.4, and so it is open, by 2.7. The set

$$E_1[\varphi(f_1, \cdots, f_h) > a]$$

is the same as the set $F^{-1}(S[\varphi > a])$, and since $S[\varphi > a]$ is open this set is measurable. By 21.3(*c*), this implies that $\varphi(f_1(x), \cdots, f_h(x))$ is measurable on E_1. Suppose next that S is closed. The set $S[\varphi \leqq a]$ is closed relative to S for every a, by 7.4; so by 2.7 it is closed. The set $E_1[\varphi(f_1, \cdots, f_h) \leqq a]$ is the same as the set $F^{-1}(S[\varphi \leqq a])$, and by the first part of the proof this is measurable because $S[\varphi \leqq a]$ is closed. Hence by 21.3(*a*) the function $\varphi(f_1(x), \cdots, f_h(x))$ is measurable on E_1. Finally, suppose that S is the union $\bigcup T_n$ of a denumerable set of closed sets. The set $S[\varphi \leqq a]$ is the sum of the sets $T_n[\varphi \leqq a]$. Each set $T_n[\varphi \leqq a]$ is closed relative to the closed set T_n, by 7.4, so is closed by 2.7. So each set $F^{-1}(T_n[\varphi \leqq a])$ is measurable, by the first part of this proof. The sum of these sets, for $n = 1, 2, \cdots$, is $F^{-1}(S[\varphi \leqq a])$, which is the same as

$$E_1[\varphi(f_1, \cdots, f_h) \leqq a].$$

This last set is therefore measurable. By 21.3 *a*, the function

$$\varphi(f_1(x), \cdots, f_h(x))$$

is measurable.

Finally, if $\varphi(z)$ is upper semi-continuous, then $-\varphi(z)$ is lower semi-continuous. Therefore $-\varphi(f_1(x), \cdots, f_h(x))$ is measurable on E_1

by the proof above, so by 21.6 the function $\varphi(f_1(x), \cdots, f_h(x))$ is also measurable on E_1.

21.12s. Corollary. *If the functions $f_1(x), \cdots, f_h(x)$ are finite valued and measurable on a set E, then the functions $f_1(x) + \cdots + f_h(x)$ and $f_1(x)f_2(x) \cdots f_h(x)$ are also measurable. The quotient $f_1(x)/f_2(x)$ is measurable if $f_2(x) \neq 0$. If $\alpha \geqq 0$, the function $|f_1(x)|^\alpha$ is measurable on E. If $\alpha < 0$, the function $|f_1(x)|^\alpha$ is measurable if $f_1(x) \neq 0$.*

These all follow from 21.11 by special choices of $\varphi(z)$. For Σf_i we use $\varphi(z) = z^{(1)} + \cdots + z^{(h)}$, continuous for all z; for Πf_i we use $\varphi(z) = \Pi z^{(i)}$, continuous for all z. In both cases the set on which $\varphi(f_1, \cdots, f_h)$ is defined is all of E. For the quotient we use $\varphi(z^{(1)}, z^{(2)}) = z^{(1)}/z^{(2)}$, defined and continuous on the open set where $z^{(2)} \neq 0$; then if $f_2(x) \neq 0$, $\varphi(f_1(x), f_2(x)) = f_1(x)/f_2(x)$ is defined for all x in E and is measurable. If $\alpha \geqq 0$ the function $\varphi(z) = |z|^\alpha$ is continuous for all z, so $|f_1(x)|^\alpha$ is defined for all x in E and is measurable. If $\alpha < 0$, the function $|z|^\alpha$ is defined and continuous on the open set $z \neq 0$; so if $f_1(x) \neq 0$, the power $|f_1(x)|^\alpha$ is defined and measurable on E.

REMARK. It follows readily that the product of measurable functions is measurable, even though the functions are not finite-valued. Let $f(x)$ and $g(x)$ be measurable on E. Then so are

$$f_n(x) = \inf \{n, \sup \{f(x), -n\}\},$$
$$g_n(x) = \inf \{n, \sup \{g(x), -n\}\},$$

by the second remark after 21.1, 21.6 and 21.7. Also $f_n(x) \to f(x)$ and $g_n(x) \to g(x)$ for all x. By 21.12, $f_n(x)g_n(x)$ is measurable for each n; so by 21.9 the limit $f(x)g(x)$ is also measurable.

22. We saw in 21.4 that every function $f(x)$ which is summable over a measurable set E is measurable. Of course the converse is false; the function 1 is measurable over the whole space R_q, but is not summable over R_q. However, two partial converses to theorem 21.4 can be established.

22.1s. *If $f(x)$ is bounded and measurable on a set E of finite measure, it is summable over E.*

Suppose that $-M < f(x) < M$, where M is finite. We consider first the case in which E is bounded, and therefore interior to some closed interval I, which we use as basic interval. If ϵ is a positive number, we can subdivide the interval $[-M, M]$ by points $\alpha_0 = -M < \alpha_1 \cdots < \alpha_n = M$ into subintervals of length less than ϵ. Let E_i be the set $E[\alpha_{i-1} < f(x) \leqq \alpha_i]$, $i = 1, 2, \cdots, n$. Every point x of E belongs to exactly one set E_i, since each $f(x)$ lies in exactly one interval

$(\alpha_{i-1}, \alpha_i]$. Now define

$$\varphi(x) = \sum_{i=1}^{n} \alpha_{i-1} K_{E_i}(x), \qquad \psi(x) = \sum_{i=1}^{n} \alpha_i K_{E_i}(x).$$

Then $\varphi(x) < f(x)K_E(x) \leqq \psi(x)$ for all x. For if x is in CE all three are equal to 0. If x is in E it belongs to one set E_j. Then in the sums defining φ and ψ all terms except the term with $i = j$ have value 0, so

$$\varphi(x) = \alpha_{j-1} < f(x)K_E(x) = f(x) \leqq \alpha_j = \psi(x).$$

Also $0 < \psi(x) - \varphi(x) = \alpha_j - \alpha_{j-1} < \epsilon$, so $\psi(x) - \varphi(x) < \epsilon$ for all x.

By 21.3(d), each set E_i is measurable, so each characteristic function $K_{E_i}(x)$ is summable over I. By 15.6, the functions $\varphi(x)$ and $\psi(x)$ are summable over I. Therefore by 14.4 and 14.2

$$\int_I \varphi(x)\,dx \leqq \underline{\int_I} f(x)K_E(x)\,dx \leqq \overline{\int_I} f(x)K_E(x)\,dx \leqq \int_I \psi(x)\,dx.$$

This tells us, first, that the upper and lower integrals of $f(x)K_E(x)$ over I are finite, and second, that

$$0 \leqq \overline{\int_I} f(x)K_E(x)\,dx - \underline{\int_I} f(x)K_E(x)\,dx \leqq \int_I \psi(x)\,dx - \int_I \varphi(x)\,dx$$
$$\leqq \int_I \epsilon\,dx = \epsilon\Delta I.$$

Hence the difference between the upper and lower integrals of $f(x)K_E(x)$ does not exceed the arbitrarily small positive number $\epsilon\Delta I$; so the two are equal, and $f(x)K_E(x)$ is summable. That is, $f(x)$ is summable over E.

If E is not bounded, by the proof above $f(x)$ is summable over EW_n for every n. Also,

$$\int_{EW_n} |f(x)|\,dx \leqq \int_{EW_n} M\,dx \leqq MmEW_n \leqq MmE < \infty.$$

So $f(x)$ is summable over E.

EXERCISE. Let $\varphi(t)$ be continuous, strictly monotonic increasing (that is, $\varphi(t_1) < \varphi(t_2)$ if $t_1 < t_2$) and bounded for $-\infty < t < \infty$; for instance, $\varphi(t) = t/(1 + |t|)$ or $\varphi(t) = \arctan t$. A function $f(x)$ defined and finite on a measurable set E in the space R_q is measurable if and only if $\varphi(f(x))$ is summable over EW_n for every n.

We now generalize 22.1 so as to allow both the set E and the function $f(x)$ to be unbounded.

22.2s. *If $f(x)$ is measurable on E and $g(x)$ is summable over E and $|f(x)| \leqq g(x)$, then $f(x)$ is summable over E.*

Suppose first that E is bounded and that $f(x)$ is non-negative. For each positive integer n we define $f_n(x) = \inf \{f(x),n\}$. Clearly we have $0 \leqq f_1(x) \leqq f_2(x) \leqq \cdots \leqq f(x) \leqq g(x)$. Each $f_n(x)$ is measurable by 21.7, and is bounded; so by 22.1 each $f_n(x)$ is summable. Also, by 18.2

$$\int_E f_n(x)\,dx \leqq \int_E g(x)\,dx < \infty.$$

As $n \to \infty$, the functions $f_n(x)$ tend to $f(x)$. For if $f(x)$ is finite, then $f_n(x) = f(x)$ for all $n > f(x)$, so $\lim f_n(x) = f(x)$; and if $f(x) = \infty$, then $f_n(x) = n$ for all n, and $\lim f_n(x) = \infty$. Therefore by 18.4 the limit $f(x)$ of the sequence $\{f_n(x)\}$ is summable over E.

Now we remove the restriction $f(x) \geqq 0$, still supposing E bounded. The functions $f^+(x)$ and $f^-(x)$ are measurable by 21.8, and they satisfy the inequalities

$$0 \leqq f^+(x) \leqq g(x), \qquad 0 \leqq f^-(x) \leqq g(x).$$

So by the preceding paragraph they are both summable over E. But by 15.5 $f(x)$ is the difference of $f^+(x)$ and $f^-(x)$, and is therefore summable by 18.3.

Finally suppose that E is unbounded. For each p, the function $f(x)$ is summable over EW_p, by the preceding proof. Also, for all p

$$\int_{EW_p} |f(x)|\,dx \leqq \int_{EW_p} g(x)\,dx \leqq \int_E g(x)\,dx < \infty.$$

So $f(x)$ is summable.

EXERCISE. Let $\varphi(t)$ be as described in the preceding exercise. A function $f(x)$ defined and finite on a measurable set E is measurable if and only if

$$\exp\left(-\sum_{i=1}^{q} [x^{(i)}]^2\right) \varphi(f(x))$$

is summable over E.

22.3s. Corollary. *If $f(x)$ is measurable and $|f(x)|$ is summable, then $f(x)$ is summable.*

For then the hypotheses of 22.2 are satisfied, with $g(x) \equiv |f(x)|$.

From theorem 22.2 we obtain two important corollaries.

22.4s. *If $f(x)$ is summable over a measurable set E and $g(x)$ is bounded and measurable over E, then $f(x)g(x)$ is summable over E.*

Suppose $|g(x)| < M$. By 21.4 and the remark after 21.12, $f(x)g(x)$ is measurable on E. Also, $|f(x)g(x)| \leqq M|f(x)|$, which is summable by 18.3. So $f(x)g(x)$ is summable over E, by 22.2.

22.5s. *If $f(x)$ is summable over a measurable* set E_1 and E is a measurable subset of E_1, then $f(x)$ is summable over E.*

For $K_E(x)$ is bounded and measurable over the whole space R_q, so it is measurable over E_1 by 21.2. By 22.4, the product $f(x)K_E(x)$ is summable over E_1; that is, $f(x)$ is summable over E.

23. We shall shortly see that sets of measure zero are usually unimportant in the theory of Lebesgue integration. Consequently, it frequently happens that a statement about a function which is true except on a set of measure zero carries the same consequences as if it were true everywhere. This situation arises often enough to justify the introduction of a special name. A statement will be said to hold "almost everywhere" or "for almost all x" provided that it holds for all x except those belonging to a set of measure zero. (The set of measure zero may be the empty set Λ; that is, if the statement holds everywhere it holds almost everywhere.)

23.1s. *If $f(x)$ is summable over E, then $f(x)$ is finite almost everywhere in E.*

Let $f_0(x)$ be equal to $f(x)$ on E and to zero elsewhere. Then $f_0(x)$ is summable over the whole space R_q, and so is $|f_0(x)|$ by 18.3. By 19.5 and 21.4 this function is measurable, so by 21.1 the set E_0 on which it is $+\infty$ is measurable. This is the same as the set $E[|f(x)| = \infty]$. For every n we have $n \leqq |f(x)|$ for all x in E_0. Therefore for each interval W_p

$$nmE_0W_p = \int_{E_0W_p} n\,dx \leqq \int_{E_0W_p} |f(x)|\,dx \leqq \int_E |f(x)|\,dx < \infty,$$

since $|f(x)|$ is summable by 18.3. This holds for all n, which is possible only if $mE_0W_p = 0$. Since $mE_0W_p = 0$ for all p, by definition $mE_0 = \lim_{p\to\infty} mE_0W_p = 0$.

The next theorem is merely a lemma to be used in the proof of 23.3.

23.2s. *If E is a set of measure 0, and $f(x)$ is defined on E, then $f(x)$ is summable over E, and its integral is 0.*

The function $f(x)$ is measurable on E, for $E[f \geqq a]$ is contained in E and has measure 0 by 19.13. By 18.3, for each n, the integral $\int_E n\,dx$ exists and has the value $n \int_E 1\,dx = n \cdot mE = 0$. Hence by 18.4 the function ∞ is summable over E, and $\int_E \infty\,dx = 0$. Since

* By use of 21.5 it can be shown that the hypothesis that E_1 is measurable can be omitted.

$f(x)$ is measurable on E and $|f(x)| \leqq \infty$ on E, $f(x)$ is summable. Also

$$0 \leqq \left| \int_E f(x)\,dx \right| \leqq \int_E |f(x)|\,dx \leqq \int_E \infty\,dx = 0.$$

23.3s. *If $f(x)$ is summable over a set E, and $g(x)$ is defined on E and equal to $f(x)$ for almost all x in E, then $g(x)$ is also summable over E, and*

$$\int_E g(x)\,dx = \int_E f(x)\,dx.$$

Define $f_0(x)$ to be equal to $f(x)$ on E and equal to zero elsewhere, and define $g_0(x)$ analogously. Let N be the set of measure zero on which $g(x) \neq f(x)$. Since N has measure 0 both $f_0(x)$ and $g_0(x)$ are summable over N, and their integrals over N are zero. In other words

$$\int_{R_q} g_0(x)K_N(x)\,dx = \int_{R_q} f_0(x)K_N(x)\,dx,$$

both integrals existing and being equal to 0. The set CN is measurable by 19.6, so $f_0(x)$ is summable over CN by 22.5. But on CN the functions f_0 and g_0 are equal, so

$$\int_{R_q} g_0(x)K_{CN}(x)\,dx = \int_{R_q} f_0(x)K_{CN}(x)\,dx,$$

both integrals existing. Adding these equations member by member, we find that $g_0(x)$ is summable over R_q and has the same integral as $f_0(x)$, which is equivalent to the conclusion of our theorem.

Two functions $f(x)$, $g(x)$ which are both defined on the same set E are called *equivalent* if $f(x) = g(x)$ for almost all x in E. Thus 23.3 can be stated in the form that if one of a pair of equivalent functions is summable, so is the other, and the integrals of the two functions are equal.

23.4s. *Let $f(x)$ be defined on the sum of two sets E_1 and E_2, and let $m(E_1 - E_2) = m(E_2 - E_1) = 0$. Then $f(x)$ is summable over E_2 if and only if it is summable over E_1; and if it is summable, then*

$$\int_{E_1} f(x)\,dx = \int_{E_2} f(x)\,dx.$$

Suppose $f(x)$ defined in any manner on $C(E_1 \cup E_2)$. The functions $f(x)K_{E_1}(x)$ and $f(x)K_{E_2}(x)$ are equivalent on the whole space R_q. For they differ at most on the set $(E_1 - E_2) \cup (E_2 - E_1)$, which has measure 0. The first of these functions is summable over R_q if and only if $f(x)$ is summable over E_1, and the second is summable over R_q

if and only if $f(x)$ is summable over E_2. So by 23.3 $f(x)$ is summable over E_1 if and only if it is summable over E_2. Moreover, if it is summable then by 23.3

$$\int_{E_1} f(x)\,dx = \int_{R_q} f(x)K_{E_1}(x)\,dx = \int_{R_q} f(x)K_{E_2}(x)\,dx = \int_{E_2} f(x)\,dx.$$

REMARK. We can now show that the hypotheses concerning the signs of the functions $f_i(x)$ in 18.4 and 18.5 are superfluous. We consider only 18.4, since this yields 18.5 as a corollary. Let E_0 be the subset of E on which $|f_1(x)| = \infty$; by 23.1, this has measure 0. Define $D = E - E_0$, and on D define $g_n(x) = f_n(x) - f_1(x)$. These functions are summable over D, by 23.4 and 18.3, and

$$0 = g_1(x) \leqq g_2(x) \leqq \cdots.$$

They tend to $f(x) - f_1(x)$ as $n \to \infty$. So by 18.4 $f(x) - f_1(x)$ is summable over D (that is, $f(x)$ is summable over D) if and only if

$$\lim_{n\to\infty} \int_D (f_n(x) - f_1(x))\,dx < \infty;$$

and this is true if and only if

$$\lim_{n\to\infty} \int_D f_n(x)\,dx < \infty.$$

In this case, by 18.4

$$\lim_{n\to\infty} \int_D (f_n(x) - f_1(x))\,dx = \int_D (f(x) - f_1(x))\,dx;$$

that is,

$$\lim_{n\to\infty} \int_D f_n(x)\,dx = \int_D f(x)\,dx.$$

Since D and E differ only on the set E_0 of measure 0, by 23.4

$$\lim_{n\to\infty} \int_E f_n(x)\,dx = \int_E f(x)\,dx.$$

23.5s. *If $f(x) \geqq 0$ on a set E, then the integral of $f(x)$ over E exists and is zero if and only if $f(x) = 0$ for almost all x in E.*

Suppose that $f(x) = 0$ for almost all x in E. By 23.3, $f(x)$ is summable, and

$$\int_E f(x)\,dx = \int_E 0\,dx = 0.$$

Conversely, suppose that $f(x)$ is summable and that its integral is 0. Let E_1 be the set on which $f(x) > 0$; we must show $mE_1 = 0$.

As $n \to \infty$, the function $nf(x)$ tends to ∞ if x is in E_1, and is always 0 if x is not in E_1. Hence $\lim nf(x) = \infty K_{E_1}(x)$. Also, $f(x) \leqq 2f(x) \leqq 3f(x) \leqq \cdots$, and

$$\int_E nf(x)\,dx = n\int_E f(x)\,dx = 0.$$

So by 18.4 the limit function $\infty K_{E_1}(x)$ is summable. It is equal to ∞ on E_1. But by 23.1 the set of points on which a summable function has the value ∞ must have measure 0; so $mE_1 = 0$.

A simple generalization of 23.5 is

23.6s. *If $f(x)$ and $g(x)$ are both summable over a set E, and $f(x) \geqq g(x)$, and*

$$\int_E f(x)\,dx = \int_E g(x)\,dx,$$

then $f(x) = g(x)$ almost everywhere in E.

Let E_0 be the subset of E on which $|g(x)| = \infty$; by 23.1, $mE_0 = 0$. On $E - E_0$ the function $f(x) - g(x)$ is defined and non-negative, and by 23.4 and 18.3

$$\int_{E-E_0}(f(x) - g(x))\,dx = \int_{E-E_0} f(x)\,dx - \int_{E-E_0} g(x)\,dx$$
$$= \int_E f(x)\,dx - \int_E g(x)\,dx = 0.$$

So by 23.5 $f(x) = g(x)$ for all x in $E - E_0$ except those x belonging to a set E_1 of measure 0. Then $E_0 \cup E_1$ has measure 0, and $f(x) = g(x)$ on $E - (E_0 \cup E_1)$; that is $f(x) = g(x)$ almost everywhere in E.

24. The mere fact that two functions $f(x)$ and $g(x)$ are both summable is of course insufficient to assure us that their product is summable. For instance, let $f(x) = g(x) = x^{-\frac{1}{2}}$ for $0 < x \leqq 1$. Each of these is summable, but their product x^{-1} is not. In 22.4 we obtained one set of conditions under which we could be sure that $f(x)\,g(x)$ is summable. Here we shall obtain a different set of conditions for the integrability of $f(x)g(x)$, and in the process we shall establish one of the most important inequalities of analysis, the Schwarz-Hölder inequality.

Let us say that two functions $\Phi(t)$ and $\Psi(t)$ are *associated* (in the sense of W. H. Young) if (*a*) they are defined and continuous for $0 \leqq t < \infty$; (*b*) they have continuous derivatives $\varphi(t) = \Phi'(t)$ and $\psi(t) = \Psi'(t)$; (*c*) $\varphi(0) = \psi(0) = \Phi(0) = \Psi(0) = 0$; (*d*) $\varphi(t)$ is strictly monotonic increasing (that is, $\varphi(t_1) < \varphi(t_2)$ if $0 < t_1 < t_2$); (*e*) $\psi(\varphi(t)) \equiv t$. Condition (*e*) states that $\varphi(t)$ and $\psi(t)$ are inverse functions, so $\psi(t)$ is also strictly monotonic increasing. We first establish a lemma concerning associated functions.

24.1s. *Let $\Phi(t)$ and $\Psi(t)$ be associated functions, and furthermore let their second derivatives $\varphi'(t)$ and $\psi'(t)$ be continuous for* $0 < t < \infty$. Then for all non-negative numbers a and b the inequality*

$$ab \leqq \Phi(a) + \Psi(b)$$

holds.

We must show that $\Phi(a) + \Psi(b) - ab \geqq 0$. If we set $b = \varphi(a)$, this function vanishes; that is, $\Phi(a) + \Psi(\varphi(a)) - a\varphi(a)$ is identically 0. For if $a > 0$ its derivative is

$$\Phi'(a) + \Psi'(\varphi(a))\varphi'(a) - \varphi(a) - a\varphi'(a)$$
$$= \varphi(a) + \psi(\varphi(a)) \cdot \varphi'(a) - \varphi(a) - a\varphi'(a) = 0,$$

by (*e*) of the definition. So it is constant for $0 < a < \infty$. Being continuous for $a \geqq 0$, its constant value is its value at $a = 0$, which is 0 by part (*c*) of the definition.

The partial derivative of $\Phi(a) + \Psi(b) - ab$ with respect to b is $\psi(b) - a$. If $b < \varphi(a)$ then $\psi(b) < \psi(\varphi(a)) = a$, so this derivative is < 0. If $b > \varphi(a)$ then $\psi(b) > \psi(\varphi(a)) = a$, so this derivative is > 0. So for each fixed a the function $\Phi(a) + \Psi(b) - ab$ has its least value at $b = \varphi(a)$. In the first paragraph we saw that its value when $b = \varphi(a)$ is 0. So the least value of $\Phi(a) + \Psi(b) - ab$ is 0, and it is therefore non-negative.

24.2s. *If $f(x)$ and $g(x)$ are finite valued and measurable on a set E, and there are associated functions $\Phi(t)$, $\Psi(t)$ with second† derivatives defined and continuous on $0 < t < \infty$ such that $\Phi(|f(x)|)$ and $\Psi(|g(x)|)$ are summable over E, then $f(x)g(x)$ is summable over E, and*

$$\left| \int_E f(x)g(x)\,dx \right| \leqq \int_E |f(x)g(x)|\,dx \leqq \int_E \Phi(|f(x)|)\,dx + \int_E \Psi(|g(x)|)\,dx.$$

The product $f(x)g(x)$ is measurable by 21.12, while by 24.1

$$|f(x)| \cdot |g(x)| \leqq \Phi(|f(x)|) + \Psi(|g(x)|).$$

Hence, by 22.2, $f(x)g(x)$ is summable over E. By 18.2 and 18.3

$$\left| \int_E f(x)g(x)\,dx \right| \leqq \int_E |f(x)g(x)|\,dx \leqq \int_E \Phi(|f(x)|)\,dx + \int_E \Psi(|g(x)|)\,dx.$$

* We shall see in §38 that this requirement of differentiability of φ and ψ can be omitted.

† As previously remarked, the condition on the second derivatives is unnecessary.

Now we shall specialize this result to obtain the Hölder inequality. As is customary we shall say that

24.3s. *The function $f(x)$ is of class L_p on E (or is in L_p) if it is measurable on E and $|f(x)|^p$ is summable over E.*

Then

24.4s. *If $f(x)$ is of class L_p on E and $g(x)$ is of class L_q on E, where $p > 1$, $q > 1$ and $1/p + 1/q = 1$, then $f(x)g(x)$ is summable over E and*

$$\left| \int_E f(x)g(x)\,dx \right| \leqq \int_E |f(x)g(x)|\,dx$$
$$\leqq \left[\int_E |f(x)|^p\,dx \right]^{1/p} \cdot \left[\int_E |g(x)|^q\,dx \right]^{1/q}.$$

(This is the Hölder inequality.)

We may without loss of generality assume that $f(x)$ and $g(x)$ are finite-valued. For by 23.1 they are both finite except on a set E_0 of measure 0. If we define $f_0(x) = f(x)$ on $E - E_0$, $f_0(x) = 0$ on E_0, and define $g_0(x)$ similarly, then by 23.3 $f(x)g(x)$ is summable if and only if $f_0(x)g_0(x)$ is summable, and none of the integrals in the theorem is changed if we replace $f(x)$ by $f_0(x)$ and $g(x)$ by $g_0(x)$.

The functions $\Phi(t) = t^p/p$ and $\Psi(t) = t^q/q$ are associated functions. For they are continuous for $0 \leqq t < \infty$. Their derivatives are $\varphi(t) = t^{p-1}$ and $\psi(t) = t^{q-1}$ respectively, and these too are continuous and strictly monotonic increasing. The second derivatives, $(p - 1)t^{p-2}$ and $(q - 1)t^{q-2}$ are continuous for $0 < t < \infty$. Moreover, $\Phi(0) = \Psi(0) = \varphi(0) = \psi(0) = 0$, and $\psi(\varphi(t)) = (t^{p-1})^{q-1} = t$, since $(p - 1)(q - 1) = 1$. Therefore by 24.2 the product $f(x)g(x)$ is summable, and

$$\int_E |f(x)g(x)|\,dx \leqq \int_E \left(\frac{|f(x)|^p}{p} \right) dx + \int_E \left(\frac{|g(x)|^q}{q} \right) dx.$$

We still must establish the Hölder inequality. If either $f(x)$ or $g(x)$ is almost everywhere zero, then $f(x)g(x)$ is also zero almost everywhere. Hence $\int |f(x)g(x)|\,dx = 0$, by 23.5, and the inequality holds. If neither $f(x)$ nor $g(x)$ is almost everywhere zero, then the numbers

$$\alpha = \left\{ \int_I |f(x)|^p\,dx \right\}^{1/p}, \qquad \beta = \left\{ \int_I |g(x)|^q\,dx \right\}^{1/q}$$

are both positive, by 23.5. Applying the inequality of the preceding paragraph to $f(x)/\alpha$ and $g(x)/\beta$, we obtain

$$\int_I \frac{|f(x)g(x)|}{\alpha\beta}\,dx \leqq \frac{1}{p} \cdot \frac{1}{\alpha^p} \cdot \int_I |f(x)|^p\,dx + \frac{1}{q} \cdot \frac{1}{\beta^q} \cdot \int_I |g(x)|^q\,dx = 1.$$

Hence

$$\left| \int_I f(x)g(x)\,dx \right| \leqq \int_I |f(x)g(x)|\,dx$$
$$\leqq \alpha\beta = \left\{ \int_I |f(x)|^p\,dx \right\}^{1/p} \cdot \left\{ \int_I |g(x)|^q \right\}^{1/q}$$

A frequently used special case of this is the *Schwarz inequality*, obtained by setting $p = q = 2$.

24.5s. *If $f(x)$ and $g(x)$ are of class L_2 on E, then $f(x)\,g(x)$ is summable over E, and*

$$\left[\int_E f(x)\,g(x)\,dx \right]^2 \leqq \left[\int_E (f(x))^2\,dx \right] \left[\int_E (g(x))^2\,dx \right].$$

The next two theorems state some properties of the classes L_p.

24.6s. *If $f(x)$ and $g(x)$ are of class $L_p (p > 0)$ on E, and c is a finite constant, then the functions $cf(x)$, $f^+(x)$, $f^-(x)$ and $|f(x)|$ are in L_p, and so is $f(x) + g(x)$ if it is defined on E. Also*

$$\left\{ \int_E |cf(x)|^p\,dx \right\}^{1/p} = |c| \left\{ \int_E |f(x)|^p\,dx \right\}^{1/p}$$

The functions $cf(x)$, $f^+(x)$, $f^-(x)$, $|f(x)|$ are all measurable by 21.6 and 21.8. The p-th powers of their absolute values do not exceed

$$[1 + |c|^p]\,|f(x)|^p,$$

which is summable since f is of the class L_p. The equation is an obvious consequence of 18.3. It still remains to consider the sum $f(x) + g(x)$.

There is no loss of generality in supposing that $f(x)$ and $g(x)$ are finite-valued; for by 23.1 this is true except on a set of measure zero, and by 23.3 we can re-define f and g to be 0 on this set without affecting any of the integrals. The sum $f(x) + g(x)$ is measurable, and so is the p-th power of its absolute value by 21.12, Also,

$$|f(x) + g(x)|^p \leqq \{ |f(x)| + |g(x)| \}^p$$
$$\leqq [2 \sup \{ |f(x)|, |g(x)| \}]^p$$
$$= 2^p \sup \{ |f(x)|^p, |g(x)|^p \}.$$

This last is summable by 18.3; so $|f(x) + g(x)|^p$ is summable over E, and $f(x) + g(x)$ is of class L_p on E.

The next theorem states the important inequality of Minkowski.

24.7s. *If $f(x)$ and $g(x)$ are of class L_p on E, and $p \geqq 1$, then*

$$\left\{ \int_E |f(x) + g(x)|^p\,dx \right\}^{1/p} \leqq \left\{ \int_E |f(x)|^p\,dx \right\}^{1/p} + \left\{ \int_E |g(x)|^p\,dx \right\}^{1/p}.$$

If $p = 1$ this is trivial, for by 18.2

$$\int_E |f(x) + g(x)|\, dx \leqq \int_E [\,|f(x)| + |g(x)|\,]\, dx$$
$$= \int_E |f(x)|\, dx + \int_E |g(x)|\, dx.$$

If $p > 1$, define $q = p/(p - 1)$. Then $\frac{1}{p} + \frac{1}{q} = 1$. Also $(p - 1)q = p$, and $p - \frac{p}{q} = 1$. Define $s(x) = |f(x)| + |g(x)|$. Then $s(x)$ is in L_p by the preceding paragraph, and

$$\int_E (s(x))^p\, dx = \int_E s(x)^{p-1}\, s(x)\, dx = \int_E s(x)^{p-1} \{\, |f(x)| + |g(x)|\, \}\, dx$$
$$= \int_E s(x)^{p-1} |f(x)|\, dx + \int_E s(x)^{p-1} |g(x)|\, dx.$$

Since $(s(x)^{p-1})^q = s(x)^p$, which is summable, $s(x)^{p-1}$ is in L_q. So by applying 24.4 to each of the last two integrals, we obtain

$$\int_E (s(x))^p\, dx \leqq \left[\int_E s(x)^p\, dx\right]^{1/q} \left[\int_E |f(x)|^p\, dx\right]^{1/p}$$
$$+ \left[\int_E s(x)^p\, dx\right]^{1/q} \left[\int_E |g(x)|^p\, dx\right]^{1/p}.$$

Dividing both members of this equation by the common factor on the right, and recalling that $1 - \frac{1}{q} = \frac{1}{p}$,

$$\left[\int_E s(x)^p\, dx\right]^{1/p} \leqq \left[\int_E |f(x)|^p\, dx\right]^{1/p} + \left[\int_E |g(x)|^p\, dx\right]^{1/p}.$$

Since

$$\int_E |s(x)|^p\, dx = \int_E (\,|f(x)| + |g(x)|\,)^p\, dx \geqq \int_E |f(x) + g(x)|^p\, dx,$$

the inequality is established.

Exercise. Show that in the Hölder inequality the two members are equal if and only if there are numbers λ, μ not both zero such that $\lambda |f(x)|^p = \mu |g(x)|^q$ for almost all x.

Exercise. Show that in Minkowski's inequality for $p > 1$ the two members are equal if and only if there are numbers λ, μ not both zero such that $\lambda f(x) = \mu g(x)$ for almost all x.

Exercise. Let E be a set of finite measure. If $1 \leqq a < b$, and $f(x)$ is a function of class L_b on E, it is also of class L_a on E. (Define $p = b/a$. Then $|f(x)|^a$ is of class L_p on E. Apply 24.4 with f, g replaced by $|f(x)|^a$, 1 respectively.)

CHAPTER IV

The Integral as a Function of Sets; Convergence Theorems

25. Our treatment of integration has permitted us to discuss multiple integrals along with single integrals, without regard to the number of independent variables $x^{(i)}$. We shall now discuss the reduction of a multiple integral to an iterated integral. In order to avoid complexity of notation we shall suppose that all functions mentioned are defined over the whole space R_q. As we know, the integral of $f(x)$ over E can be considered as the integral of $f(x)K_E(x)$ over R_q, so this agreement does not involve any loss of generality.

Let s, t be positive integers such that $s + t = q$. The point $x \equiv (x^{(1)}, \cdots, x^{(q)})$ can be written in the form $(\sigma, \tau) = (\sigma^{(1)}, \cdots \sigma^{(s)}, \tau^{(1)}, \cdots, \tau^{(t)})$, where $(\sigma^{(1)}, \cdots, \sigma^{(s)}) = (x^{(1)}, \cdots, x^{(s)})$ and $(\tau^{(1)}, \cdots, \tau^{(t)}) = (x^{(s+1)}, \cdots, x^{(q)})$. Thus σ is a point in the s-dimensional space R_s, and τ is a point in R_t. Likewise, suppose that I is a closed interval in R_q, defined by inequalities $a^{(1)} \leqq x^{(1)} \leqq b^{(1)}$, $\cdots$, $a^{(q)} \leqq x^{(q)} \leqq b^{(q)}$. If we define S to be the closed interval $a^{(1)} \leqq \sigma^{(1)} \leqq b^{(1)}, \cdots, a^{(s)} \leqq \sigma^{(s)} \leqq b^{(s)}$, and T to be the closed interval $a^{(s+1)} \leqq \tau^{(1)} \leqq b^{(s+1)}, \cdots, a^{(q)} \leqq \tau^{(t)} \leqq b^{(q)}$, then the point $x = (\sigma, \tau)$ is in I if and only if σ is in S and τ is in T.

This relationship between I, S and T is often designated by the statement that I is the "Cartesian product of S and T," and written symbolically $I = S \times T$. More generally,

25.1s. *If M and N are sets of any kind, the collection of ordered pairs (m, n) in which the first element belongs to M and the second to N is called the Cartesian product of M and N, and denoted by $M \times N$.*

Thus if M is a point-set in R_s and N a point-set in R_t the Cartesian product $M \times N$ is in R_{s+t}. Of course not every set in R_{s+t} is representable as a Cartesian product of a set in R_s and in R_t. But by repeating the discussion in the second paragraph of this section, taking care not to alter any of the signs $<$ or $\leqq$ involved in defining I, we find that every interval I in R_q is the Cartesian product of an interval S of R_s and an interval T of R_t.

With the definition of ΔI as the product of the length of the edges of I (as in the first paragraph of §10) the following statement is evident.

25.2. *If $I = S \times T$, then $\Delta I = \Delta S \cdot \Delta T$.*

For instance, if $q = 3$, $s = 1$ and $t = 2$, theorem 25.2 states that the volume of a rectangular parallelopiped is equal to the product of the altitude and the area of the base.

The theorems of this section will not be s-theorems, for in addition to 10.1 we shall use the property stated in 25.2. However, this is the only additional property of the "volume" ΔI which we shall need; so if we encounter any function ΔI of intervals (not necessarily the product of the edges of I) which satisfies 10.1 and 25.2, all the theorems of this section will be valid for the resulting integral.

Since we are considering the three different spaces R_q, R_s and R_t simultaneously, we will have three kinds of integral and three kinds of measure present in our discussions. The notation for integrals shows clearly which kinds of space is involved. To any set E in R_q, R_s or R_t only one of the three kinds of measure can apply, according to which space contains it. Nevertheless, we shall try to introduce a theoretically superfluous clarity by writing m^qE, m^sE, m^tE for the measure of E according as E is a subset of R_q, R_s, or R_t.

The principal theorem on iterated integrals is the following.

25.3. (Fubini). *If $f(x) \equiv f(\sigma, \tau)$ is summable over the space R_q, then*

(*a*) *for all points σ of R_s except those of a set E_1 of s-dimensional measure zero the function $f(\sigma, \tau)$ is summable over R_t;*

(*b*) *the function of σ defined by the integral*

$$\int_{R_t} f(\sigma, \tau)\, d\tau$$

is summable over $R_s - E_1$;

$$\text{(c)} \quad \int_{R_s - E_1} \left[\int_{R_t} f(\sigma, \tau)\, d\tau \right] d\sigma = \int_{R_q} f(x)\, dx.$$

Let $I = S \times T$ be a closed interval in the space R_q, and let

$$f(x) = \sum_{j=1}^{n} c_j K_{I_j}(x)$$

be a step-function on I, the intervals $I_j = S_j \times T_j$ being disjoint and having I for their sum. Then by 10.4

$$\int_I' f(x)\, dx = \sum_{j=1}^{n} c_j \Delta I_j.$$

But it is easily seen that

$$K_{I_j}(\sigma, \tau) = K_{S_j}(\sigma) K_{T_j}(\tau),$$

so that by 10.4 and 10.6

$$\int_T' f(\sigma, \tau)\, d\tau = \sum_{j=1}^{n} c_j K_{S_j}(\sigma) \int_T' K_{T_j}(\tau)\, d\tau$$
$$= \sum_{j=1}^{n} c_j K_{S_j}(\sigma) \Delta T_j,$$

whence by 10.4 and 10.6

$$\int_S' \left[\int_T' f(\sigma, \tau)\, d\tau \right] d\sigma = \sum_{j=1}^{n} c_j \Delta T_j \int_S' K_{S_j}(\sigma)\, d\sigma$$
$$= \sum_{j=1}^{n} c_j \Delta T_j \Delta S_j.$$

Now from these equations and 25.2 we deduce that for step-functions $f(x)$ we have

$$(A) \qquad \int_I' f(x)\, dx = \int_S' \left[\int_T' f(\sigma, \tau)\, d\tau \right] d\sigma,$$

the integrals being defined.

Next let $f(x)$ be continuous on the closed interval I. The integral

$$(B) \qquad \int_T f(\sigma, \tau)\, d\tau$$

is defined and continuous for all σ in S. For if ϵ is any positive number there is a positive δ such that $|f(\sigma_1, \tau) - f(\sigma_2, \tau)| < \epsilon(1 + \Delta T)^{-1}$ whenever σ_1 and σ_2 are in S and τ is in T and $||\sigma_1, \sigma_2|| < \delta$. Hence for such σ_1 and σ_2 we have by 11.3 and 11.6

$$\left| \int_T f(\sigma_1, \tau)\, d\tau - \int_T f(\sigma_2, \tau)\, d\tau \right|$$
$$\leqq \int_T |f(\sigma_1, \tau) - f(\sigma_2, \tau)|\, d\tau$$
$$\leqq \int_T \epsilon(1 + \Delta T)^{-1}\, d\tau < \epsilon,$$

establishing the continuity of the integral (B).

Again, let ϵ be any positive number. By the definition 11.1 of the elementary integral there are step-functions $s(x)$, $S(x)$ on I such that $s(x) \leqq f(x) \leqq S(x)$ and

$$(C) \quad \int_I f(x)\, dx - \epsilon < \int_I' s(x)\, dx \leqq \int_I' S(x)\, dx < \int_I f(x)\, dx + \epsilon.$$

Also by 11.1 we have

$$\int_T' s(\sigma, \tau)\, d\tau \leqq \int_T f(\sigma, \tau)\, d\tau \leqq \int_T' S(\sigma, \tau)\, d\tau$$

for each σ in S. As functions of σ these three integrals are a step-function, a continuous function and a step-function respectively. Hence

$$\int_S' \left[\int_T' s(\sigma, \tau)\, d\tau \right] d\sigma \leqq \int_S \left[\int_T f(\sigma, \tau)\, d\tau \right] d\sigma \leqq \int_S' \left[\int_T' S(\sigma, \tau)\, d\tau \right] d\sigma.$$

Again, by 11.1,

$$\int_I' s(x)\, dx \leqq \int_I f(x)\, dx \leqq \int_I' S(x)\, dx.$$

But by (A) the last two sets of inequalities have equal first members and equal last members. That is, the integrals

$$\int_I f(x)\, dx \qquad \text{and} \qquad \int_S \left[\int_T f(\sigma, \tau)\, d\tau \right] d\sigma$$

are both between the integrals of s and S over I, which by (C) differ by less than 2ϵ. Since ϵ is arbitrary, equation (A) holds (the primes being omitted from the integral signs) for continuous functions $f(x)$.

Next let $f(x)$ be a U-function on the interval I. By 7.9 there is a sequence of functions $\varphi_1(x) \leqq \varphi_2(x) \leqq \varphi_3(x) \leqq \cdots$ continuous on I and converging to $f(x)$ for each x in I. By 13.4,

$$(D) \qquad \lim_{n \to \infty} \int_I \varphi_n(x)\, dx = \int_I f(x)\, dx.$$

As in the previous part of the proof, the integral

$$(E) \qquad \int_T \varphi_n(\sigma, \tau)\, d\tau$$

is a continuous function of σ on S. Since for each such fixed σ the continuous function $\varphi_n(\sigma, \tau)$ has $f(\sigma, \tau)$ as limit, this last is a U-function by 12.7, and by 13.4

$$\lim_{n \to \infty} \int_T \varphi_n(\sigma, \tau)\, d\tau = \int_T f(\sigma, \tau)\, d\tau.$$

The integrals (E) being continuous, by 12.7 the right member of this equation is a U-function on S, and by 13.4

$$\lim_{n \to \infty} \int_S \left[\int_T \varphi_n(\sigma, \tau)\, d\tau \right] d\sigma = \int_S \left[\int_T f(\sigma, \tau)\, d\tau \right] d\sigma.$$

Since equation (A) (without the primes) holds for each φ_n, this equation with (D) shows that (A) also holds for the U-function $f(x)$.

By change of sign we find that (A) (without the primes) holds also for L-functions.

Next let $f(x)$ be a summable function which vanishes at all points not interior to I, which we use as basic interval. For each positive ϵ we can find a U-function $u(x)$ and an L-function $l(x)$ such that $l(x) \leqq f(x) \leqq u(x)$ on I and

$$(F) \quad \int_I f(x)\,dx - \epsilon < \int_I l(x)\,dx \leqq \int_I u(x)\,dx < \int_I f(x)\,dx + \epsilon.$$

As in the preceding paragraph, the integrals

$$\int_T u(\sigma, \tau)\,d\tau, \qquad \int_T [-l(\sigma, \tau)]\,d\tau$$

are U-functions on S, and by 14.1 and 14.2

$$\begin{aligned}\int_T l(\sigma, \tau)\,d\tau &\leqq \underline{\int_T} f(\sigma, \tau)\,d\tau \\ &\leqq \overline{\int_T} f(\sigma, \tau)\,d\tau \\ &\leqq \int_T u(\sigma, \tau)\,d\tau\end{aligned}$$

for all σ in S. Since equation (A) holds for $u(x)$ and $l(x)$, this together with 14.2 and 14.4 implies

$$\begin{aligned}\int_I l(x)\,dx &= \int_S \left[\int_T l(\sigma, \tau)\,d\tau\right] d\sigma \\ &\leqq \underline{\int_S} \left[\underline{\int_T} f(\sigma, \tau)\,d\tau\right] d\sigma \\ &\leqq \overline{\int_S} \left[\overline{\int_T} f(\sigma, \tau)\,d\tau\right] d\sigma \\ &\leqq \int_S \left[\int_T u(\sigma, \tau)\,d\tau\right] d\sigma \\ &= \int_I u(x)\,dx.\end{aligned}$$

From this and (F) we see that the third and fourth members of the inequality differ from the integral of $f(x)$ over I by less than ϵ; and since ϵ is an arbitrary positive number,

$$\begin{aligned}\underline{\int_S} \left[\underline{\int_T} f(\sigma, \tau)\,d\tau\right] d\sigma &= \overline{\int_S} \left[\overline{\int_T} f(\sigma, \tau)\,d\tau\right] d\sigma \\ &= \int_I f(x)\,dx.\end{aligned}$$

By 14.2 and 14.4 the integrals

$$\underline{\int_S} \left[\overline{\int_T} f(\sigma, \tau)\,d\tau\right] d\sigma, \qquad \overline{\int_S} \left[\underline{\int_T} f(\sigma, \tau)\,d\tau\right] d\sigma$$

both lie between the first two members of the preceding equation, so they too are equal to the integral of $f(x)$ over I. That is, the upper and lower integrals of $f(\sigma, \tau)$ over T are both summable over S, and

$$(G) \qquad \int_S \left[\overline{\int_T} f(\sigma, \tau)\, d\tau \right] d\sigma = \int_S \left[\underline{\int_T} f(\sigma, \tau)\, d\tau \right] d\sigma = \int_I f(x)\, dx.$$

Whenever σ is not interior to S the function $f(\sigma, \tau)$ vanishes for all τ by hypothesis; so by 18.1 and the remark after it the first two of these ntegrals may be taken over R_s instead of over S without changing their values. We suppose S replaced by R_s in (G). The integrands in the first two members of (G) are finite (23.1) and equal (23.6) for all σ except those on a set E_0 of measure zero. That is, on $R_s - E_0$ the upper and lower integrals of $f(\sigma, \tau)$ over T are finite and equal, so for such σ that function is summable over T. Moreover, by 17.6 its integral over T is the same as its integral over R_t, so

$$\begin{aligned} &\int_{R_s - E_0} \left[\int_{R_t} f(\sigma, \tau)\, d\tau \right] d\sigma \\ &\quad = \int_{R_s - E_0} \left[\underline{\int_T} f(\sigma, \tau)\, d\tau \right] d\sigma && \text{(by 17.6)} \\ &\quad = \int_{R_s - E_0} \left[\overline{\int_T} f(\sigma, \tau)\, d\tau \right] d\sigma && \text{(by 15.1)} \\ &\quad = \int_{R_s} \left[\overline{\int_T} f(\sigma, \tau)\, d\tau \right] d\sigma && \text{(by 23.4)} \\ &\quad = \int_S \left[\overline{\int_T} f(\sigma, \tau)\, d\tau \right] d\sigma && \text{(by 17.6)} \\ &\quad = \int_I f(x)\, dx && \text{(by (G))} \\ &\quad = \int_{R_q} f(x)\, dx. && \text{(by 17.6)} \end{aligned}$$

So theorem 25.3 holds if $f(x)$ vanishes save on a bounded set.

Next let $f(x)$ be summable and non-negative on R_q. For each n the product $f(x)K_{W_n}(x)$ vanishes except on a bounded set; and for each x this product increases with n and approaches $f(x)$ as limit; in fact, it is equal to $f(x)$ for all large n. By 18.4,

$$\int_{R_q} f(x)\, dx = \lim_{n \to \infty} \int_{R_q} f(x) K_{W_n}(x)\, dx.$$

Since 25.3 has been proved for functions which vanish outside of a bounded set, for each n there is a set E_n in R_s having measure zero and such that

$$\int_{R_q} f(x)K_{W_n}(x)\,dx = \int_{R_s - E_n} \left\{ \int_{R_t} f(\sigma, \tau)K_{W_n}(\sigma, \tau)\,d\tau \right\} d\sigma,$$

the indicated integrations being possible. If we write $N = E_1 \cup E_2 \cup E_3 \cup \cdots$, then $m^s N = 0$ by 19.13, and by 23.4 we can replace E_n by N in the preceding equation without altering the right members. This yields

$$\int_{R_q} f(x)\,dx = \lim_{n\to\infty} \int_{R_s - N} \left[\int_{R_t} f(\sigma, \tau)K_{W_n}(\sigma, \tau)\,d\tau \right] d\sigma.$$

The limit on the right is finite, being equal to the left member, and the quantities in brackets increase with n. So by 18.4 the limit of those quantities is summable over $R_s - N$, and

$$\int_{R_q} f(x)\,dx = \int_{R_s - N} \left[\lim_{n\to\infty} \int_{R_t} f(\sigma, \tau)K_{W_n}(\sigma, \tau)\,d\tau \right] d\sigma.$$

By 23.1, the quantity in brackets is finite on all of $R_s - N$ except a set having s-dimensional measure 0. We add this set to N and call the sum E_0. By 23.4, we may replace N by E_0 in the last equation without injury. By 18.4, for all σ in $R_s - E_0$ we have

$$\begin{aligned}\lim_{n\to\infty} \int_{R_t} f(\sigma, \tau)K_{W_n}(\sigma, \tau)\,d\tau &= \int_{R_t} \left[\lim_{n\to\infty} f(\sigma, \tau)K_{W_n}(\sigma, \tau) \right] d\tau \\ &= \int_{R_t} f(\sigma, \tau)\,d\tau.\end{aligned}$$

Hence

$$\int_{R_q} f(x)\,dx = \int_{R_s - E_0} \left\{ \int_{R_t} f(\sigma, \tau)\,d\tau \right\} d\sigma,$$

and 25.3 is established for non-negative summable functions $f(x)$.

If $f(x)$ is summable, we write it in the form $f(x) = f^+(x) - f^-(x)$. These last functions being non-negative by definition 15.4 and summable by 18.3, there are sets E_1, E_2 having $m^s E_1 = m^s E_2 = 0$ and such that

$$\int_{R_q} f^+(x)\,dx = \int_{R_s - E_1} \left[\int_{R_t} f^+(\sigma, \tau)\,d\tau \right] d\sigma,$$
$$\int_{R_q} f^-(x)\,dx = \int_{R_s - E_2} \left[\int_{R_t} f^-(\sigma, \tau)\,d\tau \right] d\sigma,$$

the integrals all existing. Define $E_0 = E_1 \cup E_2$. This has measure $m^s E_0 = 0$, and by 23.4 we may replace E_1, E_2 by E_0 in the two preceding equations. On subtracting we obtain the conclusion of 25.3, and the proof is complete.

Several theorems follow as corollaries of 25.3.

25.4. *Let E be a measurable set in the space R_q. For each point σ in R_s, let $E(\sigma)$ be the set of all points τ in R_t such that* (σ, τ) is in E. Then for all points σ of R_s except those of a set E_0 of measure zero the set $E(\sigma)$ is measurable. Moreover, if measure m^qE of the set E is finite, the function $m^tE(\sigma)$ is summable over the set on which it is defined, and*

$$m^qE = \int_{R_s - E_0} m^tE(\sigma)\, d\sigma.$$

If $m^qE < \infty$, then $K_E(x)$ is summable over R_q. Hence by 25.3, except on a set E_0 with $m^sE_0 = 0$ the function $K_E(\sigma, \tau)$ is summable over R_t, and

$$m^qE = \int_{R_s - E_0} \left[\int_{R_t} K_E(\sigma, \tau)\, d\tau \right] d\sigma.$$

For each fixed σ the value of $K_E(\sigma, \tau)$ is 1 or 0 according as (σ, τ) is or is not in E; that is, $K_E(\sigma, \tau) = 1$ if τ is in $E(\sigma)$, and $K_E(\sigma, \tau) = 0$ if τ is not in $E(\sigma)$. Hence $K_E(\sigma, \tau) = K_{E(\sigma)}(\tau)$, and so

$$m^qE = \int_{R_q - E_0} \left[\int_{R_t} K_{E(\sigma)}(\tau)\, d\tau \right] d\sigma = \int_{R_q - E_0} m^tE(\sigma)\, d\sigma.$$

If E is measurable, but $mE = \infty$, then for each interval W_n the set EW_n has finite measure. Define $EW_n(\sigma)$ to be the set of points τ of R_t such that (σ, τ) is in EW_n. By the preceding paragraph, $EW_n(\sigma)$ is a measurable set in R_t except for the values of σ in a set E_n with $m^sE_n = 0$. Define $E_0 = \bigcup E_n$; then $m^sE_0 = 0$, and for all σ in $R_s - E_0$ all the sets $EW_n(\sigma)$ are measurable subsets of R_t. But $E(\sigma)$ is the sum for $n = 1, 2, \cdots$ of the sets $EW_n(\sigma)$, so if σ is in $R_s - E_0$ the set $E(\sigma)$ is a measurable set in the space R_t.

25.5. *If E is a set in R_q with q-dimensional measure $m^qE = 0$ and for each σ the set $E(\sigma)$ consists of all τ in R_t for which (σ, τ) is in E, then for almost all σ in R_s the set $E(\sigma)$ has measure $m^tE(\sigma) = 0$.*

Except on a set E_0 with $m^sE_0 = 0$, the set $E(\sigma)$ is measurable, and

$$0 = m^qE = \int_{R_s - E_0} m^tE(\sigma)\, d\sigma,$$

as was proved in 25.4. But $m^tE(\sigma) \geqq 0$, so by 23.5 $m^tE(\sigma) = 0$ for almost all σ in $R_s - E_0$, that is, for almost all σ in R_s.

* That is, $E(\sigma)$ is the projection on the $(x^{(s+1)}, \cdots, x^{(q)})$-space of the part of E lying in the "plane" $x^{(1)} = \sigma^{(1)}, \cdots, x^{(s)} = \sigma^{(s)}$.

A special consequence of 25.3 is that if $f(x, y)$ is summable over the plane, then for all fixed values of y except those of a set E of measure $m^1E = 0$ the integral

$$\int_{-\infty}^{\infty} f(x, y)\, dx$$

exists; and if we replace this integral by 0 on E_0, then

$$\int_{-\infty}^{+\infty} \left[\int_{-\infty}^{+\infty} f(x, y)\, dx\right] dy = \int\int_{R_2} f(x, y)\, d(x, y).$$

A similar statement holds if we integrate first with respect to y and then with respect to x. This might lead to false hopes that even if $f(x, y)$ is not summable, the existence of one iterated integral might imply that the other exists and has the same value; or that if both iterated integrals exist and have the same value, then $f(x, y)$ must be summable. Two simple examples lay these hopes to rest.

On the interval $-1 \leqq x \leqq 1$, $-1 \leqq y \leqq 1$ define $f(x, y) = x/y$ if $y \neq 0$, $f(x, 0) = 0$; outside of the interval we set $f(x, y) = 0$. This function is measurable; in fact it is continuous at all points (x, y) with $y \neq 0$. For each fixed y the function $f(x, y)$ is continuous in x, and its integral from $x = -1$ to $x = 1$ is 0. So

$$\int_{-1}^{1} \left[\int_{-1}^{1} f(x, y)\, dx\right] dy = \int_{-1}^{1} 0\, dy = 0.$$

But if $-1 \leqq x \leqq 1$ and $x \neq 0$, the function $f(x, y)$ is not summable from $y = -1$ to $y = 1$, for $|f(x, y)| = |x|/|y|$, which is easily seen to be non-summable as a function of y. Therefore one of the iterated integrals can exist and the other be meaningless.

Next, consider the function $f(x, y)$ which is defined on the interval $-1 \leqq x \leqq 1$, $-1 \leqq y \leqq 1$ by the formula $f(x, y) = 6xy(x^2 + y^2)^{-4}$ if $(x, y) \neq (0, 0)$, $f(0, 0) = 0$. Outside of the interval we set $f(x, y) = 0$. For each fixed y, the function is continuous in x, and is odd; that is, $f(-x, y) = -f(x, y)$. Hence its integral from* $x = -1$ to $x = 1$ is 0, and

$$\int_{-1}^{1} \left[\int_{-1}^{1} f(x, y)\, dx\right] dy = 0.$$

Likewise

$$\int_{-1}^{1} \left[\int_{-1}^{1} f(x, y)\, dy\right] dx = 0.$$

Yet $f(x, y)$ is not summable over the interval. If it were, it would also be summable over $0 \leqq x \leqq 1$, $0 \leqq y \leqq 1$, by 22.5. Then the iterated

* Or from $-\infty$ to ∞.

integral

$$\int_0^1 \left[\int_0^1 f(x, y)\, dx \right] dy$$

would exist. But the integral inside the brackets is readily calculated to be $y^{-5} - y(1 + y^2)^{-3}$, which is not summable from 0 to 1.

Having shown the falsity of some statements converse to 25.3, we establish a theorem which is in the nature of a converse.

25.6 (Tonelli). *If $f(x) = f(\sigma, \tau)$ is measurable over R_q, and for all points σ in R_s except those of a set E with $m^sE = 0$ the function $|f(\sigma, \tau)|$ is summable over R_t, and the iterated integral*

$$\int_{R_s - E} \left[\int_{R_t} |f(\sigma, \tau)|\, d\tau \right] d\sigma$$

exists, then $f(x)$ is summable over R_q.

Let $g(x)$ be a positive-valued function summable over R_q; for example, we could let $g(x)$ be the exponential of $-[(x^{(1)})^2 + \cdots + (x^{(q)})^2]$. Define $f_n(x) = \inf \{ |f(x)|, ng(x)\}$. Then $f_n(x)$ is measurable by 21.4, 21.6 and 21.7, and $|f_n(x)| \leqq ng(x)$, which is a summable function. So $f_n(x)$ is summable over R_q, by 22.2. Except for the points σ in a set E_n with $m^sE_n = 0$, the function $f_n(\sigma, \tau)$ is summable over R_t, by 25.3. Define $E_0 = E \cup E_1 \cup E_2 \cup \cdots$; then $m^sE_0 = 0$, and if σ is in $R_s - E_0$ all the functions $f_n(\sigma, \tau)$ are summable over R_t. Also, $f_n(x) \leqq |f(x)|$, so by 25.3 and 18.2

$$\int_{R_q} f_n(x)\, dx = \int_{R_s - E_0} \left[\int_{R_t} f_n(\sigma,\tau)\, d\tau \right] d\sigma \leqq \int_{R_s - E_0} \left\{ \int_{R_t} |f(\sigma, \tau)|\, d\tau \right\} d\sigma.$$

Therefore the integrals of the $f_n(x)$ are bounded. But by the definition, $f_1(x) \leqq f_2(x) \leqq \cdots$, and $\lim f_n(x) = |f(x)|$; so by 18.4 $|f(x)|$ is summable over R_q. By hypothesis $f(x)$ is measurable, so by 22.3 it is summable over R_q.

In theorem 25.6 it is annoying (though necessary) to have the measurability of $f(x)$ as a hypothesis. Likewise in 25.4 it would be desirable to conclude from the measurability of E_σ for almost all σ that E is measurable. This is not so; Sierpiński* has constructed an example of a non-measurable set contained in the plane, and yet not having more than two points in common with any line. However, we can prove

25.7. *If E^s is a measurable set in R_s and E^t is a measurable set in R_t, then the set $E_1 = E^s \times E^t = \{(\sigma, \tau) \mid \sigma \text{ in } E^s, \tau \text{ in } E^t\}$ is measurable and $m^qE = (m^sE^s)(m^tE^t)$.*

* W. Sierpiński, *Sur une problème concernant les ensembles mesurables superficiellement;* Fundamenta Mathematicae vol. 1 (1920) pp. 112–115.

Suppose that E^s and E^t are bounded. For each $\epsilon > 0$ we can by 20.1 find sets F^t, G^t in R_t such that F^t is closed, G^t is open, $F^t \subset E^t \subset G^t$, and $m^tF^t > m^tE^t - \epsilon$, $m^tG^t < m^tE^t + \epsilon$. Likewise we can find sets F^s, G^s in R_s such that F^s is closed, G^s is open, $F^s \subset E^s \subset G^s$, and $m^sF^s > m^sE^s - \epsilon$, $m^sG^s < m^sE^s + \epsilon$. Let G be $G^s \times G^t = \{(\sigma, \tau) \mid \sigma$ in G^s, τ in $G^t\}$, and let F be $F^s \times F^t$. The set G is open. For let $x = (\sigma, \tau)$ be any point of G. Since σ is in the open set G^s, there is a $\gamma > 0$ such that if $\|\bar\sigma, \sigma\| < \gamma$, then $\bar\sigma$ is in G^s. Since τ is in the open set G^t, there is a $\delta > 0$ such that if $\|\bar\tau, \tau\| < \delta$, then $\bar\tau$ is in G^t. Let α be the smaller of γ and δ. If $\bar x = (\bar\sigma, \bar\tau)$, then $\|\bar x, x\|^2 = \|\bar\sigma, \sigma\|^2 + \|\bar\tau, \tau\|^2$, by the definition of distance. So if $\|\bar x, x\| < \alpha$, then $\|\bar\sigma, \sigma\| < \alpha \leqq \gamma$ and $\|\bar\tau, \tau\| < \alpha \leqq \delta$. Therefore $\bar\sigma$ is in G^s and $\bar\tau$ is in G^t, and so $\bar x$ is in G.

The set F is closed. For let $x = (\sigma, \tau)$ be an accumulation point of F, and let $(\sigma_1, \tau_1), (\sigma_2, \tau_2), \cdots$ be a sequence of points of F tending to x. Since $\|\sigma_n, \sigma\| \leqq \|x_n, x\|$ and $\|\tau_n, \tau\| \leqq \|x_n, x\|$, it is also true that σ_n tends to σ and τ_n to τ. But σ_n is in F^s, which is closed, so σ is also in F^s. Likewise τ is in F^t, so (σ, τ) is in F.

Since G is open, it is measurable. The set $G(\sigma)$ defined like $E(\sigma)$ in 25.4 is the same as G^t for all σ in G^s and is empty for σ not in G^s. Therefore $m^tG(\sigma) = K_{G^s}(\sigma)m^tG^t$, and by 25.4

$$m^qG = \int_{R_s} m^tG(\sigma)\,d\sigma = \int_{R_s} K_{G^s}(\sigma)(m^tG^t)\,d\sigma = (m^tG^t)(m^sG^s) < (m^sE^s + \epsilon)(m^tE^t + \epsilon).$$

By 20.1, $m^q_eE < (m^sE^s + \epsilon)(m^tE^t + \epsilon)$ for all $\epsilon > 0$, so

$$m^q_eE \leqq (m^sE^s)(m^tE^t).$$

Since F is closed, it is measurable. Just as above, we find

$$m^q_iE \geqq (m^sE^s)(m^tE^t).$$

So by 20.2 $m^q_eE = m^q_iE = (m^sE^s)(m^tE^t)$, which establishes our theorem for bounded sets.

If E^s and E^t are not both bounded, we let E^s_n be the part of E^s in the cube $-n \leqq \sigma^{(i)} \leqq n$, and let E^t_n be the part of E^t in the cube $-n \leqq \tau^{(i)} \leqq n$. The set E_n is defined to be the set of (σ, τ) with σ in E^s_n and τ in E^t_n. By the preceding proof, for each n the set E_n is measurable and

$$m^qE_n = (m^sE^s_n)(m^tE^t_n).$$

As $n \to \infty$, the sets E_n, E^s_n and E^t_n swell, and their sums are respectively

E, E^s and E^t. So by 19.8 the set E is measurable, and

$$m^q E = (m^s E^s)(m^t E^t).$$

The next theorem is designed to permit us to avoid in applications the sets of measure 0 on which $f(\sigma, \tau)$ may fail to be summable over R_t. We shall prove this only for double integrals. For values of q larger than 2 the corresponding theorem is valid, but the proof is longer.

25.8. *If $f(x, y)$ is summable over the plane $-\infty < x < \infty$, $-\infty < y < \infty$, there is a function $g(x, y)$ equivalent to $f(x, y)$ such that $g(x, y)$ is summable as a function of y for all values of x and summable as a function of x for all values of y, and*

$$\iint_{R_2} g(x, y)\, d(x, y) = \int_{R_1} \left[\int_{R_1} g(x, y)\, dx \right] dy = \int_{R_1} \left[\int_{R_1} g(x, y)\, dy \right] dx;$$

or in the conventional notation of the calculus,

$$\int_{-\infty}^{+\infty} \int_{-\infty}^{+\infty} g(x, y)\, dx\, dy = \int_{-\infty}^{\infty} \left[\int_{-\infty}^{\infty} g(x, y)\, dx \right] dy$$
$$= \int_{-\infty}^{\infty} \left[\int_{-\infty}^{\infty} g(x, y)\, dy \right] dx.$$

By 25.3, except on a set X with measure $m^1 X = 0$ the function $f(x, y)$ is a summable function of y. Let E_1 be $X \times R_1$; by 25.7, this set has measure 0. Define $f_1(x, y) = f(x, y)$ if (x, y) is in CE_1, $f(x, y) = 0$ if (x, y) is in E_1. Then $f_1(x, y)$ is equivalent to $f(x, y)$. For all x it is summable as a function of y; for either x is in CX and $f_1(x, y)$ is equal to the summable function $f(x, y)$, or else x is in X and $f_1(x, y) \equiv 0$.

Again by 25.3, except for y in a set Y with $m^1 Y = 0$ the function $f_1(x, y)$ is a summable function of x. Let E_2 be $R_1 \times Y$. By 25.7, $m^2 E_2 = 0$. Define $g(x, y) = f_1(x, y)$ if (x, y) is in CE_2, $g(x, y) = 0$ if (x, y) is in E_2. As before, $g(x, y)$ is a summable function of x for all y. Consider any fixed x. For this x the functions $g(x, y)$ and $f_1(x, y)$ are equal unless y is in the set Y; that is, $g(x, y)$ is equivalent to $f_1(x, y)$ as a function of y alone. Since $f_1(x, y)$ is a summable function of y, so is $g(x, y)$. Thus $g(x, y)$ is a summable function of each variable for each fixed value of the other. The equations in the statement of this theorem are then an immediate consequence of 25.3.

In elementary calculus we have the theorem that if $f(x)$ is continuous and positive on the interval $a \leqq x \leqq b$, then $\int_a^b f(x)\, dx$ is the area under the curve $y = f(x)$; that is, it is the area of the region bounded above by $y = f(x)$, below by the x-axis, and on the left and right respectively by the lines $x = a$ and $x = b$. Using the Lebesgue integral, we can generalize this result directly. Consistently with our previous notation, we shall use x for points $(x^{(1)}, \cdots, x^{(q+1)})$, σ for points $(x^{(1)}, \cdots, x^{(q)})$ of R_q, and τ for points (real numbers) of R_1. Thus each x can be written as (σ, τ). We shall suppose that we are given an interval-function $\Delta_q I_q$ for intervals in the q-dimensional space R_q; in order that the next theorem may be an s-theorem, we do not make use of the definition of $\Delta_q I_q$ as the product of the sides of I_q, and instead utilize only the "s-properties" 10.1 and their consequences. For intervals I_{q+1} of the space R_{q+1} we define an interval function $\Delta_{q+1} I_{q+1}$ as follows. If the closure $\bar{I}_{q+1}$ of I_{q+1} is defined by inequalities $a^{(i)} \leqq x^{(i)} \leqq b^{(i)}$, $i = 1, 2, \cdots, q + 1$, and I_q is the interval $a^{(i)} \leqq x^{(i)} \leqq b^{(i)}$, $i = 1, 2, \cdots, q$, then we define

$$\Delta_{q+1} I_{q+1} = (b^{(q+1)} - a^{(q+1)}) \Delta_q I_q.$$

It will be noticed that if we take $\Delta_q I_q$ to be the product of the sides of I_q, then $\Delta_{q+1} I_{q+1}$ is merely the product of the sides of I_{q+1}. Having now a $\Delta_{q+1} I_{q+1}$ defined (and plainly satisfying 10.1) for all intervals of $(q + 1)$-space R_{q+1} we are able to define integrals and measure in R_{q+1}.

Now suppose that $f(\sigma)$ is defined and non-negative on a set E in R_q. The generalization of the "region under the curve" will be the *ordinate set* Ω. This set consists of those points $(\sigma, \tau) \equiv (x^{(1)}, \cdots, x^{(q+1)})$ such that σ is in E and $0 \leqq \tau < f(\sigma)$. The elementary theorem on the area under the curve $y = f(x)$ has the following generalization:

25.9s. *Let $f(\sigma)$ be defined and non-negative on a measurable set E in q-dimensional space, and let Ω be its ordinate set. Then*

(a) *Ω is measurable if and only if $f(\sigma)$ is a measurable function;*

(b) *Ω has finite measure $m^{q+1}\Omega$ if and only if $f(\sigma)$ is summable;*

(c) *if $m^{q+1}\Omega$ is finite, then*

$$m^{q+1}\Omega = \int_E f(\sigma)\, d\sigma.$$

To simplify notation we extend the range of definition of $f(\sigma)$ by setting it equal to zero on $R_q - E$. This leaves the integral in (c) unchanged; and also it leaves the set Ω unchanged, for at all points of $R_q - E$ the condition $0 \leqq \tau < f(\sigma)$ cannot be satisfied. The set Ω now consists of all (σ, τ) such that $0 \leqq \tau < f(\sigma)$. We observe that because of the way $\Delta_{q+1} I_{q+1}$ is defined, it satisfies the factorization requirement 25.2.

Let us first suppose that Ω is measurable. The set of all points (σ, τ) with σ in R_q and $\tau < 0$ is measurable by 25.7. We add this set to Ω and call the sum Ω^*. Then Ω^* is measurable. From its definition, it consists of all (σ, τ) with $\tau < f(\sigma)$. By 25.3, for almost all numbers τ (say for all τ in T, where $m^1(CT) = 0$), the set $\Omega^*(\tau)$ is measurable, where by $\Omega^*(\tau)$ we mean the set of all σ such that (σ, τ) is in Ω^*. Now we show that $\Omega^*(\tau)$ is the same as the set $R_q[f > \tau]$. If σ is in $\Omega^*(\tau)$, then (σ, τ) is in Ω^*, so $\tau < f(\sigma)$. Conversely, if $\tau < f(\sigma)$ then (σ, τ) is in Ω^* and σ is in $\Omega^*(\tau)$. Hence $\Omega^*(\tau) = R_q[f > \tau]$. But we have seen that $\Omega^*(\tau)$ is measurable for all τ in the set T. Hence $R_q[f > \tau]$ is measurable for all τ in the set T, which is dense in the space of real numbers; and by the exercise just after 21.3 the function $f(\sigma)$ is measurable.

Conversely, let $f(\sigma)$ be measurable. The positive rationals are a denumerable set, and can be arranged in a sequence $r_1, r_2, \cdots$. Let E_j be the set of all (σ, τ) such that $f(\sigma) > r_j$ and $0 \leqq \tau \leqq r_j$. The conditions on σ and τ being independent of each other, the set E_j is measurable by 25.7. If (σ, τ) is in E_j, then $0 \leqq \tau < f(\sigma)$, so (σ, τ) is in Ω. Therefore $\bigcup E_j$ is contained in Ω. On the other hand if (σ, τ) is in Ω, then $0 \leqq \tau < f(\sigma)$; so there is a rational number r_j such that $\tau < r_j < f(\sigma)$ and (σ, τ) is in E_j. Hence $\bigcup E_j = \Omega$. The set Ω being the sum of measurable sets is measurable by 19.8. Thus conclusion (a) is established.

For each point σ in R_q we now define $\Omega(\sigma)$ to be the set of all τ such that (σ, τ) is in Ω. This set is empty if $f(\sigma) = 0$, and is the interval $0 \leqq \tau < f(\sigma)$ if $f(\sigma) > 0$. Therefore for all σ the set $\Omega(\sigma)$ is measurable and the equation

$$m^1\Omega(\sigma) = f(\sigma)$$

holds; for if $f(\sigma) = 0$ then $\Omega(\sigma)$ is empty and has measure 0, and if $f(\sigma) > 0$ then $m^1\Omega(\sigma) = f(\sigma) - 0 = f(\sigma)$. Suppose then that $m^{q+1}\Omega < \infty$. By 25.4

$$m^{q+1}\Omega = \int_{R_q} m^1\Omega(\sigma)\, d\sigma = \int_{R_q} f(\sigma)\, d\sigma = \int_E f(\sigma)\, d\sigma,$$

since $f(\sigma) = 0$ except on E. This establishes half of (b) and also (c).

Suppose, conversely, that $f(\sigma)$ is summable. Let $K_\Omega(\sigma, \tau)$ be the characteristic function of Ω. This function is measurable, since Ω is a measurable set by (a). For each fixed σ, the function $K_\Omega(\sigma, \tau)$ is equal to 1 for $0 \leqq \tau < f(\sigma)$ and to 0 elsewhere. So

$$\int_{R_1} | K_\Omega(\sigma, \tau) |\, d\tau = f(\sigma).$$

This is summable over R_q, by hypothesis. Therefore by 25.6 the function K_Ω is summable over R_{q+1}, and $m^{q+1}\Omega$ is finite by 19.1.

An obvious analogue of 25.9 holds for non-positive functions $f(x)$. If for such functions we define the ordinate set Ω to be all (σ, τ) with σ in E and $0 \geqq \tau > f(\sigma)$, theorem 25.9 remains valid except that the left member of the equation in (c) must be multiplied by -1. The proof is unaltered except that all inequalities are reversed and $f(\sigma)$ is replaced by $-f(\sigma)$ in the last two equations in the proof.

The extension of the theorem to functions which are not necessarily non-negative is trivial. If $f(\sigma)$ is defined on a measurable set E, we define Ω_+ to be the set of points (σ, τ) such that σ is in E, $f(\sigma) > 0$, and $0 \leqq \tau < f(\sigma)$; and we define Ω_- to be the set of points (σ, τ) such that σ is in E, $f(\sigma) < 0$, and $f(\sigma) < \tau \leqq 0$. The set Ω_+ is the ordinate set for the function $f^+(\sigma) \equiv \sup \{f(\sigma), 0\}$, and the set Ω_- is the ordinate set for the function $-f^-(\sigma) \equiv -\sup \{-f(\sigma), 0\}$. Recalling 21.8, 21.12, 19.4, 19.6 and 19.8, we see that $f(\sigma)$ is a measurable function if and only if Ω_+ and Ω_- are measurable, which is true if and only if $\Omega_+ \cup \Omega_-$ is measurable. The function $f(\sigma)$ is summable if and only if $\Omega_+ \cup \Omega_-$ has finite measure, and in that case the equation

$$\int_E f(\sigma)\, d\sigma = m^{q+1}\Omega_+ - m^{q+1}\Omega_-$$

holds.

Our choice of the signs $\leqq$ or $<$ in the definition of the ordinate-set Ω was more or less arbitrary, and we chose the signs as we did for convenience in proof. For example, if $f(x)$ is non-negative we can define another ordinate-set Ω^c as the set of all points (σ, τ) of R_{q+1} with $0 \leqq \tau \leqq f(\sigma)$. This contains Ω, and $\Omega^c - \Omega$ consists of the points $(\sigma, f(\sigma))$ with $f(\sigma)$ finite. We leave it as an exercise to prove that $m^{q+1}(\Omega^c - \Omega) = 0$. From this it follows that theorem 25.9 remains valid if we replace Ω by Ω^c.

It is hardly necessary to point out that if the concept of measure in R_{q+1} has been defined by some method independent of the concept of integral, we could use the measure of the ordinate-set Ω as the definition of the integral of $f(\sigma)$.

26. Let $f(x)$ be a function defined and summable over a measurable set E_1. By 22.5, $f(x)$ is also summable over every measurable subset E of E_1. That is, if E is a measurable set contained in E_1 the number $\int_E f(x)\, dx$ is defined. Thus we have a law of correspondence by which to each set E in the class of all measurable subsets of E_1 there is assigned a (finite) real number. In other words, we have defined a function $F(E)$ in which the functional values are (finite) real numbers, but the independent variable E ranges over the class of all measurable sets contained in E_1. Such a function is called a *set-function*, or *function of sets.*

We have just had one example of a set-function. Some other examples are:

(*a*) $F(E)$ is defined for all measurable sets E, and for such sets $F(E) = mE$.

(*b*) $F(E)$ is defined for all measurable subsets E of a set E_1 of finite measure, and for these sets $F(E) = mE$.

(*c*) $F(E)$ is defined for all sets E; $F(E) = 1$ if E contains the origin 0, $F(E) = 0$ otherwise. Here $F(E)$ is given by the formula $F(E) = K_E(0)$.

(*d*) $F(E)$ is defined for all sets E, and $F(E) = \text{diam } E$; that is, $F(E) = \sup \{|| x_1, x_2 || \mid x_1, x_2 \text{ in } E\}$.

(*e*) $F(E)$ is defined for all subsets E of the interval W_1, and for such sets $F(E) = \text{diam } E$.

The first two of these examples can be expressed in the form $F(E) = \int_E 1\, dx$. But later theorems will show that there can not exist any function $f(x)$ such that $\int_E f(x)\, dx$ gives us the set functions (*c*), (*d*) or (*e*).

There are a number of properties which a set-function may have. For this section we wish to define and discuss several of these properties. First, however, we introduce a definition concerning collections of the sets themselves.

26.1s. *A class* $\{E\}$ *of sets is additive if for every finite number* E_1, $E_2, \cdots, E_n$ *of sets of* $\{E\}$ *the sum* $\bigcup E_i$ *is also a set of* $\{E\}$; *it is completely additive if for every finite or denumerable set* $E_1, E_2, \cdots$ *of sets of* $\{E\}$ *the sum* $\bigcup E_i$ *is also a set of* $\{E\}$.

Returning now to functions of sets:

26.2s. *A set-function* $F(E)$ *defined on a class* $\{E\}$ *of sets is bounded if there is a finite number* M *such that* $|F(E)| \leqq M$ *for each set* E *of* $\{E\}$.

26.3s. *A set-function* $F(E)$ *defined on a class* $\{E\}$ *of sets is of bounded variation (abbreviated "of BV") if there is a finite number* M *such that* $|F(E_1)| + \cdots + |F(E_n)| \leqq M$ *for every finite collection* $E_1, \cdots, E_n$ *of disjoint sets of* $\{E\}$.

26.4s. *A set-function* $F(E)$ *defined on an additive class* $\{E\}$ *of sets is additive if* $F(\bigcup E_i) = \Sigma F(E_i)$ *for every finite collection* $E_1, \cdots, E_n$ *of disjoint sets of* $\{E\}$. *A set-function* $F(E)$ *defined on a completely additive class* $\{E\}$ *of sets is completely additive if* $F(\bigcup E_i) = \Sigma F(E_i)$ *for every* *finite or denumerable collection* $E_1, E_2, \cdots$ *of disjoint sets of the class* $\{E\}$.

Example (*a*) is unbounded and not of BV. It is completely additive by 19.8.

Example (*b*) is completely additive, bounded, and of BV, by 19.8 and 19.3.

Example (*c*) is bounded, for $F(E) = 0$ or 1 for every set. By 1.1, if $E_1, E_2, \cdots$ is a finite or denumerable collection of disjoint sets

$$K_{\bigcup E_i}(x) = \Sigma K_{E_i}(x)$$

for every x. Substituting the origin 0 for x, we obtain

$$\Sigma \mid F(E_i) \mid = \Sigma F(E_i) = F(\bigcup E_i) = 0 \text{ or } 1,$$

which proves that $F(E)$ is completely additive and of BV.

Example (*e*) is bounded, example (*d*) is not. Neither is additive. For if E_1 consists of a single point x_1 and E_2 of a single point $x_2 \neq x_1$, then $F(E_1 \cup E_2) = \| x_1, x_2 \|$, while $F(E_1) = F(E_2) = 0$. Neither is of BV. For let S', S'' be two spheres in S_1 of radius r and distance between centers $3r$. Let $x'_1, x'_2, \cdots$ be distinct points of S' and let $x''_1, x''_2, \cdots$ be distinct points of S''. Define E_n to be the set consisting of x'_n and x''_n; these sets are disjoint. Then $F(E_n) \geqq r$, so $\mid F(E_1) \mid + \cdots + \mid F(E_n) \mid \geqq nr$, and $F(E)$ is not of BV.

It is obvious that if $F(E)$ is completely additive, it is additive. Moreover the concepts of boundedness and BV are closely related, for

26.5s. *If $F(E)$ is of BV on the class $\{E\}$ it is bounded. If $\{E\}$ is an additive class of sets and $F(E)$ is additive and bounded, it is of BV.*

If $\Sigma \mid F(E_i) \mid \leqq M$ for every finite collection $E_1, \cdots, E_n$ of sets of $\{E\}$, then in particular (taking $n = 1$) $\mid F(E) \mid \leqq M$ for each E of the class $\{E\}$.

Suppose that $\{E\}$ is additive and $F(E)$ is additive on $\{E\}$, and let M be a finite number such that $\mid F(E) \mid \leqq M$ for each E of $\{E\}$. Let $E_1, E_2, \cdots, E_n$ be a collection of disjoint sets of $\{E\}$. We subdivide the aggregate of sets $E_1, \cdots, E_n$ into two parts; the sets (which we denote by E_j^+) such that $F(E_j^+) \geqq 0$, and the other sets (which we denote by E_k^-) such that $F(E_k^-) < 0$. Then

$$\sum_{i=1}^{n} \mid F(E_i) \mid = \sum F(E_j^+) - \sum F(E_k^-)$$
$$= F(\bigcup E_j^+) - F(\bigcup E_k^-) = \mid F(\bigcup E_j^+) \mid + \mid F(\bigcup E_k^-) \mid \leqq 2M,$$

so $F(E)$ is of BV.

Next we prove a very simple theorem.

26.6s. *If $F(E)$ is finite and additive on a class of sets including the empty set Λ, then $F(\Lambda) = 0$.*

For $\Lambda = \Lambda \cup \Lambda$; therefore $F(\Lambda) = F(\Lambda) + F(\Lambda)$.

We now define continuity and absolute continuity (AC) for set-functions.

26.7s. *A set-function $F(E)$, defined on a class $\{E\}$ of sets, is continuous if to every $\epsilon > 0$ there corresponds a $\delta > 0$ such that $|F(E)| < \epsilon$ for all sets E of the class $\{E\}$ whose diameter* diam E *is less than δ.*

Examples (d) and (e) are clearly continuous. Examples (a) and (b) are also continuous.* For if diam $E < \delta$, then E can be enclosed in an interval of side 2δ and volume $(2\delta)^q$. Hence $mE \leqq (2\delta)^q$, which is less than an arbitrary ϵ if δ is small enough. But example (c) is not continuous. For let $\epsilon = \frac{1}{2}$. No matter how small a number $\delta > 0$ we choose, there is a set E (the set consisting of the origin alone) of diameter diam $E = 0 < \delta$ and such that $F(E) = 1 > \epsilon$.

26.8s. *A set-function $F(E)$ defined on a class $\{E\}$ of measurable sets E is absolutely continuous if to every $\epsilon > 0$ there corresponds a $\delta > 0$ such that $|F(E)| < \epsilon$ for all sets E of the class $\{E\}$ which have measure $mE < \delta$.*

Examples (a) and (b) are clearly absolutely continuous (we can take $\delta = \epsilon$). Examples (d) and (e), which were continuous, are not absolutely continuous. For suppose x_1 and x_2 are two (distinct) fixed points of I. If we take $\epsilon = ||x_1, x_2||$ there can be no δ satisfying 26.8. For if $\delta > 0$, the set E consisting of the two points x_1 and x_2 has $mE = 0 < \delta$ and $F(E) = \text{diam } E = ||x_1, x_2|| = \epsilon$. Example ($c$) is surely not absolutely continuous, for it is not continuous, while

26.9.† *If a set-function $F(E)$ is absolutely continuous, it is continuous.*

Let ϵ be a positive number. There is a δ such that $|F(E)| < \epsilon$ wherever $mE < \delta$. If diam $E < \delta^{1/q}/2$, then $|F(E)| < \epsilon$. For then E can be enclosed in an interval I_0 of sides less than $\delta^{1/q}$, hence of volume $\Delta I_0 < \delta$; so $mE \leqq \Delta I_0 < \delta$, and $|F(E)| < \epsilon$.

A simple consequence of absolute continuity is given in the next theorem.

26.10s. *If $F(E)$ is defined and additive on the class of all measurable subsets of E^*, it is AC if and only if for every $\epsilon > 0$ there exists a $\delta > 0$ such that $\Sigma |F(E_i)| < \epsilon$ for all finite collections $E_1, \cdots, E_n$ of disjoint measurable subsets of E^* such that $\Sigma mE_i < \delta$.*

* This is not an "s" statement.

† Observe that this is not an "s" theorem.

If the condition of 26.10 is satisfied, then $F(E)$ is AC, as we see by considering only collections consisting of one set E_1. Conversely, suppose that $F(E)$ is AC, and let ϵ be a positive number. There is a $\delta > 0$ such that $|F(E)| < \epsilon/2$ if $mE < \delta$. Let $E_1, E_2, \cdots, E_n$ be a collection of disjoint measurable subsets of I with $\Sigma mE_i < \delta$. We classify the E_i into the sets E_j^+ such that $F(E_j^+) \geqq 0$ and the remaining sets E_k^- such that $F(E_k^-) < 0$. Then

$$\Sigma |F(E_i)| = \Sigma F(E_j^+) - \Sigma F(E_k^-) = F(\bigcup E_j^+) - F(\bigcup E_k^-).$$

But $m(\bigcup E_j^+) \leqq m \bigcup E_i < \delta$ and $m(\bigcup E_k^-) \leqq m \bigcup E_i < \delta$, so each term on the right is less than $\epsilon/2$, and $\Sigma |F(E_i)| < \epsilon$.

We now establish the relationship between AC functions and functions of BV.

26.11s. *If $F(E)$ is additive and AC on the class of all measurable sets E contained in a set E^* of finite measure, then it is of BV and bounded.*

We have shown in 26.5 that $F(E)$ is of BV if and only if it is bounded. Suppose it is not bounded. Let $k = |F(E^*)| + 1$. Then there is a measurable subset E_1 of E^* such that $|F(E_1)| > 2k$. Defining $E_2 = E^* - E_1$, the sets E_1 and E_2 are disjoint and measurable, so

$$|F(E_2)| \geqq |F(E_1)| - |F(E^*)| > 2k - (k - 1) > k.$$

Either $F(E)$ is unbounded on the subsets of E_1, or it is unbounded on the subsets of E_2. For suppose $|F(E)| \leqq M_i$ for all measurable subsets of $E_i (i = 1, 2)$. Every measurable set E in I can be subdivided into the two disjoint measurable portions EE_1 and EE_2, and EE_i is a subset of $E_i (i = 1, 2)$. So

$$|F(E)| = |F(EE_1) + F(EE_2)| \leqq |F(EE_1)| + |F(EE_2)| \leqq M_1 + M_2.$$

But this contradicts the assumption that $F(E)$ is unbounded. Let us therefore suppose (a mere matter of choice of notation) that $F(E)$ is unbounded on subsets of E_1. Then there exists a measurable set $E_3 \subset E_1$ such that $|F(E_3)| > 2|F(E_1)| > k$. If we define $E_4 = E_1 - E_3$, then

$$|F(E_4)| \geqq |F(E_3)| - |F(E_1)| \geqq 2|F(E_1)| - |F(E_1)| > k.$$

Again, by the same argument as above $F(E)$ is unbounded either on the subsets of E_3 or on the subsets of E_4. Suppose E_3 is the one. We choose an E_5 in E_3 such that $|F(E_5)| > 2|F(E_3)|$, define $E_6 = E_3 - E_5$, and proceed as before. Continuing thus, we obtain a sequence of measurable sets $E_1 \supset E_3 \supset E_5 \supset \cdots$ such that $|F(E_i)| > k$ and $|F(E_i) - F(E_{i+2})| > k$ $(i = 1, 3, 5, \cdots)$. From this we shall show that $F(E)$ can not be AC. For let $\epsilon = k$, and let δ be any positive number. The sets $E_1 - E_3$, $E_3 - E_5$, $\cdots$ are all disjoint and all contained in E_1, so by 19.8

$$\sum_{i=1}^{\infty} m(E_{2i-1} - E_{2i+1}) = m\left[\bigcup_{i=1}^{\infty} (E_{2i-1} - E_{2i+1})\right] \leqq mE_1 < \infty.$$

Because the series converges, its terms tend to 0, and we can find a set $E_{2i-1} - E_{2i+1}$ of measure less than δ. But $|F(E_{2i-1} - E_{2i+1})| > k = \epsilon$, so $F(E)$ is not AC.

EXERCISE. Show that if "s" is omitted from 26.11 a simpler proof is possible along the following lines. Split E^* into a finite number of parts $E_1, \cdots, E_n$ each of measure so small that $|F(E)| < 1$ if E is contained in any one E_i. Then for all sets E, $|F(E)| \leqq |F(EE_1)| + \cdots + |F(EE_n)| < n$.

The words "absolutely continuous" have been used in this section with a meaning quite different from their meaning in §9. There is no possibility of confusion, since in the two places the words are applied to two different types of functions. Hence no proof of consistency is needed, or even possible. However, it can be shown that there is actually a relationship between AC set-functions and AC functions $f(x)$ of a single real variable x. Suppose that $F(E)$ is an AC additive function of measurable sets in an interval $I = [a, b]$ of one-dimensional space R_1. With $F(E)$ we can associate a function $f(x)$ as follows. If $a \leqq x \leqq b$ we define $f(x) = F([a, x))$; since the interval $[a, a)$ is the empty set Λ, this gives $f(a) = F([a, a)) = F(\Lambda) = 0$. The function $f(x)$ thus built out of $F(E)$ is then AC. For if $a \leqq \alpha < \beta \leqq b$ the intervals $[a, \alpha)$ and $[\alpha, \beta)$ are disjoint, so $f(\beta) = F([a, \beta)) = F([a, \alpha)) + F([\alpha, \beta)) = f(\alpha) + F([\alpha, \beta))$. That is, $F([\alpha, \beta)) = f(\beta) - f(\alpha)$. Let ϵ be any positive number. Since $F(E)$ is AC, by 26.10 there is a $\delta > 0$ such that $\Sigma\,|F(E_i)| < \epsilon$ whenever $E_1, \cdots, E_n$ are disjoint measurable subsets of I such that $\Sigma mE_i < \delta$. In particular, let $[\alpha_1\beta_1), \cdots, [\alpha_n, \beta_n)$ be a set of disjoint intervals such that $\Sigma\,|\beta_i - \alpha_i| < \delta$. Then* $m \bigcup [\alpha_i, \beta_i) = \Sigma m[\alpha_i, \beta_i) = \Sigma\,|\beta_i - \alpha_i| < \delta$, so

$$\epsilon > \Sigma\,|F([\alpha_i, \beta_i))| = \Sigma\,|f(\beta_i) - f(\alpha_i)|.$$

This proves that $f(x)$ is AC as defined in §9.

* This is not an "s" statement since we use a special property of ΔI.

EXERCISE. Show that if $F(E)$ is of BV on the class of measurable subsets of $[a, b]$, then the function $f(x)$ defined in the preceding paragraph is of BV as defined in §8.

27. Our introduction to set functions was by way of the particular set-function

$$F(E) \equiv \int_E f(x)\,dx,$$

where $f(x)$ is a function summable over a measurable set E^*. We have already remarked that this set-function is defined over the class of all measurable subsets of E^*, which is a completely additive class by 19.8. Now we proceed to study this particular set-function in more detail. In fact, we shall show that it has every one of the properties defined in §26.

27.1s. *If $f(x)$ is summable over a measurable set E^*, then the set function*

$$F(E) \equiv \int_E f(x)\,dx$$

is absolutely continuous on the class of all measurable subsets of E^*.

Let ϵ be a positive number. By 18.3, $|f(x)|$ is summable over E^*. Define $f_n(x) = \inf\{|f(x)|, n\}$. This function is measurable by 21.4 and 21.7, and $|f_n(x)| \leqq |f(x)|$, so $f_n(x)$ is summable by 22.2. Moreover,

$$\int_{E^*} f_n(x)\,dx \leqq \int_{E^*} |f(x)|\,dx.$$

Clearly $f_1(x) \leqq f_2(x) \leqq \cdots$. As $n \to \infty$, the functions $f_n(x)$ approach $|f(x)|$ as limit; for if $|f(x)| = \infty$, then $f_n(x) = n \to \infty$, and if $f(x)$ is finite, then $f_n(x) = |f(x)|$ for all $n \geqq |f(x)|$. By 18.4,

$$\lim_{n\to\infty} \int_{E^*} f_n(x)\,dx = \int_{E^*} |f(x)|\,dx.$$

So for some value p of n we have

$$\int_{E^*} f_p(x)\,dx > \int_{E^*} |f(x)|\,dx - \frac{\epsilon}{2}.$$

Now let E be any subset of E^* with $mE < \epsilon/2p$. Then

$$|F(E)| = \left|\int_E f(x)\,dx\right| \leqq \int_E |f(x)|\,dx$$

$$= \int_E [f_p(x) + (|f(x)| - f_p(x))]\,dx$$

$$\leqq \int_E p\,dx + \int_{E^*} (|f(x)| - f_p(x))\,dx < \epsilon.$$

So $F(E)$ is absolutely continuous.

The hypothesis in 27.1 that E^* is measurable is easily removed. For if we set $f(x) = 0$ on CE^* it is integrable over the whole space R_q, and its integral over subsets of E^* is unaffected.

In order to prove the complete additivity of $F(E)$ we establish a lemma which is of interest in itself. It is a partial converse to theorem 22.5.

27.2s. *If $E_1, E_2, \cdots$ is a finite or denumerable collection of disjoint sets, and $f(x)$ is defined on $\bigcup E_i$ and is summable over each E_i, and*

$$\sum \int_{E_i} |f(x)|\, dx < \infty,$$

then $f(x)$ is summable over $\bigcup E_i$ and

$$\sum \int_{E_i} f(x)\, dx = \int_{\bigcup E_i} f(x)\, dx.$$

Suppose first that $f(x) \geqq 0$. By 1.2,

$$f(x)K_{\bigcup E_i}(x) = \sum f(x)K_{E_i}(x).$$

Each of the functions on the right is summable over the whole space R_q. So this equation, together with 18.3 if the E_i are finite in number and with 18.6 if the E_i are denumerable, shows that $f(x)K_{\bigcup E_i}(x)$ is summable over R_q (that is, $f(x)$ is summable over $\bigcup E_i$) and

$$\int_{\bigcup E_i} f(x)\, dx = \int_{R_q} f(x)K_{\bigcup E_i}(x)\, dx = \sum \int_{R_q} f(x)K_{E_i}(x)\, dx$$
$$= \sum \int_{E_i} f(x)\, dx.$$

We now remove the restriction that $f(x) \geqq 0$: The functions f^+ and f^- of 15.4 are non-negative, and by 18.3, both are summable over each E_i, and

$$\int_{E_i} |f^+(x)|\, dx \leqq \int_{E_i} |f(x)|\, dx,$$
$$\int_{E_i} |f^-(x)|\, dx \leqq \int_{E_i} |f(x)|\, dx.$$

Hence by the preceding part of the proof $f^+(x)$ and $f^-(x)$ are summable over $\bigcup E_i$, and

$$\int_{\bigcup E_i} f^+(x)\, dx = \sum \int_{E_i} f^+(x)\, dx,$$
$$\int_{\bigcup E_i} f^-(x)\, dx = \sum \int_{E_i} f^-(x)\, dx.$$

Therefore by 18.3 $f^+(x) - f^-(x)$ is summable over $\bigcup E_i$, and

$$\begin{aligned}\int_{\bigcup E_i} f(x)\,dx &= \int_{\bigcup E_i} \{f^+(x) - f^-(x)\}\,dx \\ &= \sum \left\{\int_{E_i} f^+(x)\,dx - \int_{E_i} f^-(x)\,dx\right\} \\ &= \sum \int_{E_i} f(x)\,dx.\end{aligned}$$

EXERCISE. If $E_1, E_2, \cdots$ is a finite or denumerable collection of measurable sets (not necessarily disjoint) and $f(x)$ is defined on $\bigcup E_i$ and is summable over each E_i, and $\sum \int_{E_i} |f(x)|\,dx < \infty$, then $f(x)$ is summable over $\bigcup E_i$.

EXERCISE. Construct an example to show that the hypothesis "$\sum \int_{E_i} |f(x)|\,dx < \infty$" in 27.2 and the preceding exercise can not be weakened to read "$\sum \left| \int_{E_i} f(x)\,dx \right| < \infty$."

27.3s. *If $f(x)$ is summable over E^*, then the set-function*

$$F(E) = \int_E f(x)\,dx$$

is completely additive on the class of all measurable subsets of E^*.

Let $E_1, E_2, \cdots$ be a finite or denumerable collection of disjoint measurable subsets of E^*. By 1.2, $K_{\bigcup E_i}(x) = \sum K_{E_i}(x)$. If the E_i are finite in number, then by 1.2, 18.3 and 18.2

$$\begin{aligned}\sum \int_{E_i} |f(x)|\,dx = \sum \int_{R_q} |f(x)|\,K_{E_i}(x)\,dx &= \int_{R_q} |f(x)|\,K_{\bigcup E_i}(x)\,dx \\ \leqq \int_{R_q} |f(x)|\,K_{E^*}(x)\,dx &= \int_{E^*} |f(x)|\,dx < \infty.\end{aligned}$$

If the E_i are denumerable in number, this still holds for each finite subset $E_1, \cdots, E_n$ of them, so

$$\sum_1^\infty \int_{E_i} |f(x)|\,dx = \lim_{n\to\infty} \sum_1^n \int_{E_i} |f(x)|\,dx \leqq \int_{E^*} |f(x)|\,dx < \infty.$$

In either case, the sum is finite. So by 27.2

$$F(\bigcup E_i) = \int_{\bigcup E_i} f(x)\,dx = \sum \int_{E_i} f(x)\,dx = \sum F(E_i),$$

which establishes our theorem.

27.4. Corollary. *The function $F(E)$ of* 27.1 *is continuous.* This follows at once from 27.1 and 26.9.

27.5s. Corollary. *The function $F(E)$ of 27.1 is of BV and bounded.*

This follows from 27.1, 27.3 and 26.11. But a simple proof is possible without reference to 26.11. For if $E_1, \cdots, E_n$ are disjoint measurable subsets of E^*, then

$$\begin{aligned}\sum |F(E_i)| &= \sum \left| \int_{R_q} f(x) K_{E_i}(x)\,dx \right| \leqq \sum \int_{R_q} |f(x)|\, K_{E_i}(x)\,dx \\ &= \int_{R_q} |f(x)|\, K_{\bigcup E_i}(x)\,dx \leqq \int_{R_q} |f(x)|\, K_{E^*}(x)\,dx \\ &= \int_{E^*} |f(x)|\,dx < \infty.\end{aligned}$$

Hence $F(E)$ satisfies the definition 26.3 of BV, with $M = \int_{E^*} |f(x)|\,dx$. With $n = 1$, this proves that $F(E)$ is bounded.

For simple integrals over intervals it is useful to introduce the usual notation of the calculus. Let $f(x)$ be summable over an interval $[a, b]$. If $a \leqq \alpha < \beta \leqq b$ the function $f(x)$ is summable over $[\alpha, \beta)$, by 22.5. We define

$$\begin{aligned}\int_\alpha^\beta f(x)\,dx &= \int_{[\alpha,\beta)} f(x)\,dx, \\ \int_\beta^\alpha f(x)\,dx &= -\int_{[\alpha,\beta)} f(x)\,dx, \\ \int_\alpha^\alpha f(x)\,dx &= 0.\end{aligned}$$

With this notation it is easy to establish the following statement.

27.6s. *If $f(x)$ is summable over $[a, b]$, and c, d, e are all in $[a, b]$, then*

$$\int_c^e f(x)\,dx = \int_c^d f(x)\,dx + \int_d^e f(x)\,dx.$$

To prove this we re-write it by transposing all terms to the left, obtaining

$$\int_c^e f(x)\,dx + \int_e^d f(x)\,dx + \int_d^c f(x)\,dx = 0$$

The left member is unchanged if we permute the letters c, d, e cyclically, and it is only changed in sign if we interchange any two of the letters. We may therefore suppose the notation chosen so that $c \leqq d \leqq e$. Now if $c = d$ or $d = e$ the equation is trivial. Otherwise the intervals $[c, d)$ and $[d, e)$ are disjoint and their sum is $[c, e)$, so the equation is true by 27.3.

The next statement is an immediate corollary of 27.1.

27.7. *If $f(x)$ is summable over $[a, b]$, then the function*

$$F(x) = \int_a^x f(x)\,dx$$

is absolutely continuous on $[a, b]$.

If for measurable subsets E of $[a, b]$ we define $F^*(E)$ to be the integral of $f(x)$ over E, by 27.1 this is an AC function of sets. By the last part of §26, the function $F^*([a, x))$ is AC on the interval $[a, b]$. But this is exactly $F(x)$.

28. One of the outstanding virtues of the Lebesgue integral is that it permits passages to the limit under hypotheses much weaker than those needed when using Riemann integrals. We have already had an example in 18.4 of a convergence theorem valid for Lebesgue integrals, but certainly false for Riemann integrals, as the example in §18 proved. In this section we shall study several ways in which a sequence of functions can converge to a limit function, and in the next section we shall establish a number of very powerful convergence theorems.

The reader is already familiar with the idea of uniform convergence. A sequence of functions $f_n(x)$ defined on a set E converges uniformly to $f(x)$ on E if for every $\epsilon > 0$ there is an n_ϵ such that $|f_n(x) - f(x)| < \epsilon$ for all x in E whenever $n > n_\epsilon$. Moreover we have used in 18.4 the idea of convergence everywhere. If $f(x)$, $f_1(x)$, $f_2(x)$, $\cdots$ are all defined on a set E, then $f_n(x)$ converges everywhere to $f(x)$ if for each x in E the numbers $f_n(x)$ converge to the number $f(x)$. But we know that sets of measure zero are unimportant in integration, so we may suspect that a concept more suited for use with the Lebesgue integral is the following.

28.1s. *Let the functions $f(x), f_1(x), f_2(x), \cdots$ be defined on a set E. Then $f_n(x)$ converges almost everywhere to $f(x)$ if the numbers $f_n(x)$ converge to the number $f(x)$ for all x in $E - E_0$, where E_0 is a set of measure* 0.

If $f_n(x)$ tends uniformly to $f(x)$, then for every $\delta > 0$ the relation "$f_n(x)$ in $N_\delta(f(x))$" holds for all x if n is greater than a certain n_δ. If we wish to define a somewhat similar but weaker type of convergence we could require that to each positive number δ there shall correspond an integer n_δ such that $f_n(x)$ lies in $N_\delta(f(x))$ for all $n > n_\delta$ and for all x except at most those in a "small" subset of E. However, the question arises—Should this "small" subset, on which $f_n(x)$ is not in $N_\delta(f(x))$, be allowed to depend on n, or should one such subset serve for all n? We are thus led to consider two definitions:

28.2s. *Let the functions $f(x), f_1(x), \cdots$ be defined on a measurable set E. Then the functions $f_n(x)$ converge (or tend) almost uniformly to $f(x)$ if to every pair ϵ, δ of positive numbers there corresponds a set E_ϵ (depending on ϵ alone) of measure $mE_\epsilon < \epsilon$ such that $f_n(x)$ is in $N_\delta(f(x))$ for all x in $E - E_\epsilon$, provided that n is greater than a certain $n_{\epsilon,\delta}$. Here the "small" subset E_ϵ is independent of n.*

28.3s. *Let the functions* $f(x)$, $f_1(x)$, $\cdots$ *be defined on a measurable set* E. *Then the functions* $f_n(x)$ *converge in measure to* $f(x)$ *if for every pair* ϵ, δ *of positive numbers the set of* x *on which* $f_n(x)$ *is not in* $N_\delta(f(x))$ *can be enclosed in a set* $E_{n,\delta}$ *with measure* $mE_{n,\delta} < \epsilon$, *provided that* n *is greater than a certain* $n_{\epsilon,\delta}$. *Here the "small" subset* $E_{n,\delta}$ *is permitted to vary with* n.

We thus have three new kinds of convergence, each of which will turn out to be especially well designed to apply in certain cases and less useful in others. It is therefore important to establish the properties of these convergences and the interrelations between them, so that we may be able at any time to use the most appropriate formulation.

To begin with, we observe that our last two definitions can be worded rather differently. The definition of convergence in measure is merely the statement that for every $\delta > 0$ we have $\lim_{n \to \infty} mE_{n,\delta} = 0$. And if $f(x)$ be almost everywhere finite, the definition of almost uniform convergence requires merely that $f_n(x)$ converge uniformly to $f(x)$ on $E - E_\epsilon$, if we make the provision that the set of measure 0 on which $|f(x)| = \infty$ is to be included in E_ϵ.

In the next two theorems we establish some simple properties of these convergence modes.

28.4s. *Let* $f(x)$, $f_1(x)$, $f_2(x)$, $\cdots$ *be defined on a set* E. *Then if* $f_n(x)$ *converges to* $f(x)$ *in any one of our three modes, every subsequence* $\{f_\alpha(x)\}$ *also converges to* $f(x)$ *in the same mode, and* $|f_n(x)|$ *converges to* $|f(x)|$ *in the same mode.*

For convergence almost everywhere this follows from the known properties of sequences of numbers; for except on a set of measure 0 the numbers $f_n(x)$ tend to $f(x)$, so the same is true of every subsequence, and also $|f_n(x)|$ tends to $|f(x)|$ (cf. 4.3).

If $f_n(x) \to f(x)$ in measure, then for every $\delta > 0$ we can enclose the set of x for which $f_n(x)$ is not in $N_\delta(f(x))$ in a measurable set $E_{n,\delta}$ such that $mE_{n,\delta}$ tends to zero as $n \to \infty$. Then for each integer α the set of x for which $f_\alpha(x)$ is not in $N_\delta(f(x))$ is enclosed in $E_{\alpha,\delta}$; and since the numbers $mE_{\alpha,\delta}$ are a subsequence of the numbers $mE_{n,\delta}$ they too approach zero as $\alpha \to \infty$. Hence $f_\alpha(x)$ tends to $f(x)$ in measure. Also, the set $E^*_{n,\delta}$ on which $|f_n(x)|$ is not in $N_\delta(|f(x)|)$ is contained in $E_{n,\delta}$, by 4.3. Since $mE_{n,\delta} \to 0$, this proves that $|f_n(x)|$ converges in measure to $|f(x)|$.

If $f_n(x)$ converges almost uniformly to $f(x)$, then except on a set E_ϵ of measure $mE_\epsilon < \epsilon$ the functions $f_n(x)$ tend to $f(x)$ uniformly. The same is therefore true of the subsequence $f_\alpha(x)$. Also for n greater than a certain n_ϵ the numbers $f_n(x)$ are in $N_\epsilon(f(x))$ for all x in $E - E_\epsilon$, so by 4.3 $|f_n(x)|$ is in $N_\epsilon(|f(x)|)$ for all such n and x, and $|f_n(x)|$ tends uniformly to $|f(x)|$ on $E - E_\epsilon$. Hence $|f_n(x)|$ tends to $|f(x)|$ almost uniformly on E.

28.5s. *If the functions* $f(x), f_1(x), \cdots, g(x), g_1(x), \cdots$ *have finite values on a set* E, *and* $f_n(x)$ *converges to* $f(x)$ *in any one of our three modes, and* $g_n(x)$ *converges to* $g(x)$ *in the same mode, and* c *is any finite constant, then* $f_n(x) + g_n(x)$ *converges to* $f(x) + g(x)$ *and* $cf_n(x)$ *converges to* $cf(x)$ *in the same mode.*

Suppose first that $f_n(x)$ tends to $f(x)$ and $g_n(x)$ to $g(x)$ almost everywhere. Then except on a set E_0 of measure zero $f_n(x)$ tends to $f(x)$, and except on a set E_1 of measure zero $g_n(x)$ tends to $g(x)$. So except on E_0 the functional values $cf_n(x)$ tends to $cf(x)$ and except on the set $E_0 \cup E_1$ (which has measure 0) the functional values $f_n(x) + g_n(x)$ tend to $f(x) + g(x)$.

Suppose next that $f_n(x)$ converges to $f(x)$ and $g_n(x)$ to $g(x)$ almost uniformly. For every $\epsilon > 0$ there are sets $E_{1,\epsilon}$ and $E_{2,\epsilon}$ of measure less than $\epsilon/2$ such that $f_n(x)$ converges to $f(x)$ uniformly on $E - E_{1,\epsilon}$ and $g_n(x)$ converges to $g(x)$ uniformly on $E - E_{2,\epsilon}$. Therefore on $E - (E_{1,\epsilon} \cup E_{2,\epsilon})$ the functions $f_n(x) + g_n(x)$ converge uniformly to $f(x) + g(x)$ and $cf_n(x)$ converges uniformly to $cf(x)$; and $m(E_{1,\epsilon} \cup E_{2,\epsilon}) < \epsilon$. So $f_n(x) + g_n(x)$ converges to $f(x) + g(x)$ and $cf_n(x)$ to $cf(x)$ almost uniformly on E.

Suppose finally that $f_n(x)$ converges to $f(x)$ and $g_n(x)$ to $g(x)$ in measure. Let ϵ, δ be any positive numbers. There is a number n_1 such that if $n \geqq n_1$ the set of x on which $f_n(x)$ is not in $N_{\delta/2}(f(x))$ can be enclosed in a set $E_{1,n,\delta}$ of measure less than $\epsilon/2$. That is, if $n \geqq n_1$ then $|f_n(x) - f(x)| < \delta/2$ except on $E_{1,n,\delta}$. Likewise, there is a number n_2 such that if $n \geqq n_2$, then $|g_n(x) - g(x)| < \delta/2$ except on a set $E_{2,n,\delta}$ of measure less than $\epsilon/2$. Therefore, except on the set $E_{1,n,\delta} \cup E_{2,n,\delta}$, whose measure is less than ϵ, we have $|(f_n(x) + g_n(x)) - (f(x) + g(x))| < \delta$, and $f_n(x) + g_n(x)$ converges in measure to $f(x) + g(x)$. Likewise $|cf_n(x) - cf(x)| < |c|\,\delta/2$, which is an arbitrary positive number, so $cf_n(x)$ converges in measure to $cf(x)$.

We now begin to study the interrelations of the three modes of convergence.

28.6s. *Let* $f(x), f_1(x), f_2(x), \cdots$ *be defined on a measurable set* E. *If* $f_n(x)$ *tends to* $f(x)$ *almost uniformly, then* $f_n(x)$ *tends to* $f(x)$ *almost everywhere and* $f_n(x)$ *tends to* $f(x)$ *in measure.*

Let E_0 be the set on which $f_n(x)$ fails to converge to $f(x)$; we must show $mE_0 = 0$. By the definition of almost uniform convergence, for every $\epsilon > 0$ there is a set E_ϵ with $m_eE_\epsilon < \epsilon$ such that $f_n(x)$ tends to

$f(x)$ uniformly on $E - E_\epsilon$. Then no point of E_0 can be in $E - E_\epsilon$; so $E_0 \subset E_\epsilon$. Now let ϵ take on successively the values $1, \frac{1}{2}, \frac{1}{3}, \cdots$ then E_0 is contained in each set $E_{1/n}$, so it is contained in $\bigcap E_{1/n}$. This set is measurable by 19.9, and $m \bigcap E_{1/n} \leqq mE_{1/n} < 1/n$ for every n. So $m \bigcap E_{1/n} = 0$, and by 19.13 E_0 is measurable and $mE_0 = 0$. Therefore $f_n(x)$ converges to $f(x)$ almost everywhere in E.

Let ϵ, δ be any positive numbers. Except on a set E_ϵ with $mE_\epsilon < \epsilon$ we have $f_n(x)$ in $N_\delta(f(x))$ whenever $n > n_{\epsilon,\delta}$. Hence if $n > n_{\epsilon,\delta}$ the set of x on which $f_n(x)$ is *not* in $N_\delta(f(x))$ is contained in E_ϵ, and so E_ϵ will serve as the $E_{n,\delta}$ of the definition of convergence in measure. Therefore $f_n(x)$ converges in measure to $f(x)$.

Thus almost uniform convergence is the strongest of the three modes.

28.7s. *Let $f(x)$, $f_1(x)$, $\cdots$ be defined on a measurable set E. If $f_n(x)$ tends to $f(x)$ in measure, then there exists a subsequence $f_{n_1}(x)$, $f_{n_2}(x)$, $\cdots$ such that $f_{n_i}(x)$ tends to $f(x)$ almost uniformly (hence almost everywhere).*

Since $f_n(x)$ tends to $f(x)$ in measure, for each positive δ and for each n the set of points x at which $f_n(x)$ is not in $N_\delta(f(x))$ can be enclosed in a set $E_{n,\delta}$ in such a way that $mE_{n,\delta}$ tends to zero as $n \to \infty$. Let this first be done for $\delta = 1$. Since $mE_{n,1}$ tends to zero, there is a first n (call it n_1) for which $mE_{n,1} < 1$. The corresponding set $E_{n,1}$ we rename E_1. Next take $\delta = \frac{1}{2}$; there is a first $n > n_1$ (we call it n_2) for which $mE_{n,\frac{1}{2}} < \frac{1}{2}$. The corresponding set $E_{n,\frac{1}{2}}$ we rename E_2. Continuing this process, we obtain a subsequence

$$f_{n_1}(x), f_{n_2}(x), f_{n_3}(x), \cdots$$

of our original sequence which has the property that $f_{n_k}(x)$ is in $N_{2^{1-k}}(f(x))$ for all points x outside of a set E_k of measure less than 2^{1-k}.

We now prove that the functions so chosen converge almost uniformly to $f(x)$.

Let ϵ and δ be any positive numbers. Since $mE_k < 2^{-k+1}$, we can choose n_ϵ large enough so that

$$\sum_{k=n_\epsilon}^{\infty} mE_k < \sum_{n_\epsilon}^{\infty} 2^{-k+1} = 2^{-n_\epsilon+2} < \epsilon.$$

Define

$$E_\epsilon = \bigcup_{k=n_\epsilon}^{\infty} E_k;$$

then by 19.8 $mE_\epsilon < \epsilon$. Now choose a number $n_{\epsilon,\delta} > n_\epsilon$ so large that $2^{1-n_{\epsilon,\delta}} < \delta$. For any $k > n_{\epsilon,\delta}$, the set E_k is contained in E_ϵ; so on $E - E_\epsilon$ we have

$$f_{n_k}(x) \text{ in } N_{2^{-k+1}}(f(x)) \subset N_\delta(f(x)).$$

This proves our theorem.

28.8s. *Let the functions $f_1(x), f_2(x), \cdots$ be defined and measurable on a measurable set E. If $f_n(x)$ converges to $f(x)$ in any one of our three modes then $f(x)$ is measurable.*

If f_n tends almost everywhere to $f(x)$, then $f(x)$ is almost everywhere equal to $\liminf f_n(x)$, which is measurable by 21.9. If f_n tends to f almost uniformly, then f_n tends to f almost everywhere by 28.6, so f is measurable. If f_n converges in measure to $f(x)$, a subsequence converges almost everywhere to $f(x)$ by 28.7, so $f(x)$ is measurable.

Before proceeding to the proof of the next theorem we establish a lemma.

28.9s. Lemma. *If $f(x)$ and $g(x)$ are defined and measurable on a set E, then for every $\delta > 0$ the subset E_δ of E on which $g(x)$ is not in $N_\delta(f(x))$ is a measurable set.*

Let us subdivide E into the set E_1 on which $f(x)$ and $g(x)$ are both finite, the set E_2 on which $f(x)$ is finite and $|g(x)| = \infty$, the set E_3 on which $f(x) = +\infty$, and the set E_4 on which $f(x) = -\infty$. By 21.3, each of these sets is measurable. On E_1, the function $f(x) - g(x)$ is defined and is measurable by 21.12 and 21.2. So $|f(x) - g(x)|$ is measurable by 21.8, and the set $E_{1,\delta}$ on which $|f(x) - g(x)| \geqq \delta$ is measurable by 21.1. On all of E_2, $g(x)$ fails to lie in $N_\delta(f(x))$. On E_3, $g(x)$ is measurable by 21.2, and so the subset $E_{3,\delta}$ on which $g(x) \leqq 1/\delta$ is measurable. Likewise $g(x)$ is measurable on E_4, so the subset $E_{4,\delta}$ on which $g(x) \geqq -1/\delta$ is measurable. But by the definition of neighborhood, the set E_δ on which $g(x)$ is not in $N_\delta(f(x))$ consists of $E_{1,\delta} \cup E_2 \cup E_{3,\delta} \cup E_{4,\delta}$, so it is measurable.

28.10s. *Let the functions $f(x), f_1(x), f_2(x), \cdots$ be defined and measurable on a set E of finite measure. If $f_n(x)$ converges almost everywhere to $f(x)$, then $f_n(x)$ converges to $f(x)$ almost uniformly (hence in measure).*

Let ϵ and δ be positive numbers, and let $E_{n,\delta}$ be the subset of E on which $f_n(x)$ is not in $N_\delta(f(x))$. By 28.9, each $E_{n,\delta}$ is measurable. So if we define

$$E_\delta{}^k = \bigcup_{n=k}^{\infty} E_{n,\delta} \qquad \text{and} \qquad E_\delta = \bigcap_{k=1}^{\infty} E_\delta{}^k$$

we see by 19.8 and 19.9 or 19.11 that $E_\delta{}^k$ and E_δ are measurable sets.

But if x belongs to E_δ, it belongs to all E_δ^k. If it belongs to E_δ^k, it belongs to some $E_{n,\delta}$ with $n \geqq k$. So if it is in E_δ it belongs to infinitely many $E_{n,\delta}$. Therefore for infinitely many n the number $f_n(x)$ is *not* in $N_\delta f(x)$, and $f_n(x)$ does *not* tend to $f(x)$. So x must belong to the set E_0 of measure 0 on which $f_n(x)$ does not tend to $f(x)$; that is, $E_\delta \subset E_0$, and by 19.13 $mE_\delta = 0$.

By their definition, the sets E_δ^k shrink as k increases: $E_\delta^1 \supset E_\delta^2 \cdots$. Also they are all contained in the set E, which has finite measure. Hence by 19.9 $\lim_{k\to\infty} mE_\delta^k = mE_\delta = 0$, and for some k we have $mE_\delta^k < \epsilon$. Now define $E_\epsilon = E_\delta^k$, $n_{\epsilon,\delta} = k$. Then $mE_\epsilon < \epsilon$. By the definition of E_δ^k, if $n > n_{\epsilon,\delta} = k$ then $E_{n,\delta} \subset E_\delta^k = E_\epsilon$, so for every x in $E - E_\epsilon$ we know that x is not in $E_{n,\delta}$, and therefore $f_n(x)$ is in $N_\delta(f(x))$. This proves that $f_n(x)$ tends to $f(x)$ almost uniformly.

An important property of any limiting process is the uniqueness of the resulting limit. In the present case, we have no true uniqueness, for if f_n tends in any one of our three modes to $f(x)$, it tends equally well to any function $g(x)$ which differs from $f(x)$ only on a set of measure zero. But we *can* prove that no greater arbitrariness than this is allowed:

28.11s. *Let $f(x)$, $g(x)$, $f_1(x)$, $\cdots$ be defined on a set E. If $f_n(x)$ converges in any one of our three modes to $f(x)$ and also converges in any one of our three modes to $g(x)$, then $f(x) = g(x)$ for almost all x.*

If $f_n(x)$ converges in measure to $f(x)$, by 28.7 it is possible to select a subsequence $\{f_\alpha(x)\}(\alpha = n_1, n_2, \cdots)$ which converges almost everywhere to $f(x)$. If $f_n(x)$ tends to $f(x)$ almost everywhere, this is still true even if we take the whole sequence, and if $f_n(x)$ tends almost uniformly to $f(x)$ it tends almost everywhere to $f(x)$ by 28.6. So in any case we can choose a subsequence $\{f_\alpha(x)\}$ converging almost everywhere to $f(x)$.

The sequence $\{f_n(x)\}$ converged in some one of the three modes to $g(x)$, so the subsequence $\{f_\alpha(x)\}$ converges in the same mode to $g(x)$, by 28.4. As in the preceding paragraph, it is possible to select a subsequence $\{f_\beta(x)\}$ out of the sequence $\{f_\alpha(x)\}$ in such a way that $f_\beta(x)$ tends to $g(x)$ almost everywhere. The sequence $f_\beta(x)$ continues to converge almost everywhere to $f(x)$, by 28.4. Then the equations

$$\lim_{\beta\to\infty} f_\beta(x) = f(x) \qquad \text{and} \qquad \lim_{\beta\to\infty} f_\beta(x) = g(x)$$

are respectively true for all x except those belonging to two sets E_0, E_1 of measure 0. Therefore except on the set $E_0 \cup E_1$ of measure 0 we have $f(x) = g(x)$, as was to be proved.

It might be thought that 28.7 could be improved to read that if f_n converges in measure to f, then f_n converges almost everywhere to f. An example shows that this is false. Let E_1 be the interval $[0, 1]$, E_2 and E_3, the intervals $[0, \frac{1}{2}]$ and $[\frac{1}{2}, 1]$ respectively, E_4, E_5, E_6, E_7 the intervals $[0, \frac{1}{4}]$, $[\frac{1}{4}, \frac{1}{2}]$, $[\frac{1}{2}, \frac{3}{4}]$, $[\frac{3}{4}, 1]$ respectively, and so on, proceeding by successive bisections. Let $f_n = K_{E_n}$. Then f_n converges in measure to 0 on the interval $[0, 1]$. For if $\delta > 0$, then $f_n(x)$ is in $N_\delta(0)$ (in fact, is equal to 0) except at most on E_n; so we can take $E_{n,\delta} = E_n$, and $\lim_n mE_{n,\delta} = \lim_n mE_n = 0$. But at no point is $\lim f_n(x) = f(x)$; in fact, $\lim f_n(x)$ does not exist. For each x in $[0, 1]$ is contained in infinitely many E_n, so infinitely many $f_n(x)$ have value 1; while there are infinitely many E_n which do not contain x, so $f_n(x) = 0$ for infinitely many n. Hence $\lim f_n(x)$ does not exist.

29. It is at once apparent that the mere convergence, in any one of our three modes, of a sequence of functions $f_n(x)$ to a limit function $f(x)$ is not enough to guarantee that the integrals of the $f_n(x)$ will converge to the integral of $f(x)$. For example, let $f_n(x)$ be defined thus:

$$\begin{aligned} f_n(x) &= n^2x, && 0 \leqq x \leqq n^{-1}, \\ f_n(x) &= 2n - n^2x, && n^{-1} < x \leqq 2n^{-1}, \\ f_n(x) &= 0, && 2n^{-1} < x \leqq 2 \qquad (n = 1, 2, \cdots). \end{aligned}$$

These functions are all continuous and tend everywhere (hence in all three of our modes) to $f_0(x) = 0$. But

$$\int_0^2 f_n(x)\,dx = 1, \qquad \int_0^2 f_0(x)\,dx = 0,$$

so the integrals of the $f_n(x)$ do not converge to the integral of $f_0(x)$.

This example makes it clear that we must make other assumptions besides mere convergence of the $f_n(x)$ in order to obtain convergence of the integrals. In this section we set forth several such sets of assumptions. First, however, we shall make a general observation. If a sequence of summable functions $\{f_n(x)\}$ converges in any of our modes to a function $f_0(x)$, we know by 23.1 that each function $f_n(x)$ is finite except on a set E_n of measure 0. If we re-define $f_n(x)$ by assigning it the value 0 on E_n, then the new functions $f_n(x)$ have the same integrals as the old and still converge in the same mode as before to $f_0(x)$; for the change of values on the set $\bigcup E_n$ of measure 0 does not affect any of our types of convergence. Hence in the theorems of this section there is no loss of generality in assuming that the functions $f_n(x)$ have finite values. This assumption simplifies the statements of some of our conclusions.

The first theorem which we shall establish is a lemma which not only is useful for later proofs, but is of considerable importance in itself.

29.1s. (Fatou's Lemma). *If the functions $f_n(x)$ are all summable over a measurable set E, and the lower limit of their integrals is not $+\infty$, and there is a summable function $g(x)$ such that $f_n(x) \geqq g(x)$ on E, then $\liminf_{n\to\infty} f_n(x)$ is summable, and*

$$\int_E \liminf_{n\to\infty} f_n(x)\,dx \leqq \liminf_{n\to\infty} \int_E f_n(x)\,dx.$$

Define $g_n(x) \equiv \inf\{f_i(x) \mid i \geqq n\}$. Then $g(x) \leqq g_1(x) \leqq g_2(x) \leqq \cdots$; and by 6.9 the limit of $g_n(x)$ is $\liminf f_n(x)$. Moreover, $g_n(x)$ is measurable by 21.7, and $g(x) \leqq g_n(x) \leqq f_n(x)$, so $|g_n(x)| \leqq |g(x)| + |f_n(x)|$, which is summable. So by 22.2 $g_n(x)$ is summable over E, and by 18.2

$$\int_E g_n(x)\,dx \leqq \int_E f_n(x)\,dx.$$

It follows that

$$\lim_{n\to\infty} \int_E g_n(x)\,dx \leqq \liminf_{n\to\infty} \int_E f_n(x)\,dx.$$

Now the $g_n(x)$ satisfy the hypotheses of 18.4, so by that theorem

$$\lim_{n\to\infty} \int_E g_n(x)\,dx = \int_E \lim_{n\to\infty} g_n(x)\,dx = \int_E \liminf_{n\to\infty} f_n(x)\,dx.$$

Together with the preceding inequality, this establishes the theorem.

29.2s. Corollary. *If the functions $f_n(x)$ are summable over a measurable set E, and on E they converge in measure or almost everywhere to a function $f(x)$, and there is a summable function $g(x)$ such that $f_n(x) \geqq g(x)$ for all n and all x, then $f(x)$ is summable on E and*

$$\int_E f(x)\,dx \leqq \liminf_{n\to\infty} \int_E f_n(x)\,dx$$

provided that the right member of this inequality is finite.

Let us suppose that $f_n(x)$ tends to $f(x)$ almost everywhere in E. Then $\liminf f_n(x) = f(x)$ for almost all x in E. But $\liminf f_n(x)$ is summable by 29.1, so $f(x)$ is summable by 23.3. The inequality follows at once from 29.1 and 23.3.

Now let $f_n(x)$ tend to $f(x)$ in measure, and define

$$\lambda = \liminf_{n\to\infty} \int_E f_n(x)\,dx.$$

From the sequence $\{f_n(x)\}$ we can (by 6.4) select a subsequence $\{f_\alpha(x)\}$, $\alpha = n_1, n_2, \cdots$, such that

$$\lim_{\alpha \to \infty} \int_E f_\alpha(x)\,dx = \lambda.$$

The $f_\alpha(x)$ still converge in measure to $f(x)$, by 28.4. From them we select a subsequence $\{f_\beta(x)\}$ converging almost everywhere to $f(x)$, as is possible by 28.7. Then by the preceding paragraph $f(x)$ is summable, and

$$\int_E f(x)\,dx \leqq \liminf_{\beta \to \infty} \int_E f_\beta(x)\,dx = \lambda = \liminf_{n \to \infty} \int_E f_n(x)\,dx.$$

From 29.2 it is almost apparent that bounding the $f_n(x)$ both above and below by summable functions will enforce the convergence of the integrals. Because of the simplicity of the proof we give here the proof of this statement, although it is in fact a special case of theorem 29.7.

29.3s. *Let the functions $f_n(x)$ be defined, finite and measurable over a set E. If on E the functions $f_n(x)$ converge in measure or almost everywhere to a limit function $f(x)$, and there is a summable function $g(x)$ such that $|f_n(x)| \leqq g(x)$ for all n and all x in E, then $f(x)$ is summable over E, and the integrals of the $f_n(x)$ converge to a limit, and*

(a) $$\lim_{n \to \infty} \int_E f_n(x)\,dx = \int_E f(x)\,dx,$$

(b) $$\lim_{n \to \infty} \int_E |f_n(x) - f(x)|\,dx = 0.$$

By 22.2, the functions $f_n(x)$ are summable over E. The functions $f_n(x)$ are equal to or greater than the summable function $-g(x)$, and for all n

$$\int_E f_n(x)\,dx \leqq \int_E g(x)\,dx.$$

Hence the functions $f_n(x)$ satisfy the hypotheses of 29.2; and by that theorem $f(x)$ is summable, and

$$\liminf_{n \to \infty} \int_E f_n(x)\,dx \geqq \int_E f(x)\,dx.$$

On the other hand, the functions $-f_n(x)$ also satisfy the hypotheses of 29.2, and converge in measure or almost everywhere to $-f(x)$. So

$$-\limsup_{n \to \infty} \int_E f_n(x)\,dx = \liminf_{n \to \infty} \int_E (-f_n(x))\,dx \geqq -\int_E f(x)\,dx.$$

These two inequalities, together with 6.7, establish conclusion (*a*) or our theorem.

By 28.5 and 28.4, the functions $|f_n(x) - f(x)|$ converge in measure or almost everywhere to 0. Also, $|f_n(x) - f(x)| \leqq |f_n(x)| + |f(x)| \leqq g(x) + |f(x)|$, which is summable by the preceding paragraph. Hence applying the part of the proof already completed to these functions, we obtain conclusion (*b*).

Conclusion (*b*) is actually stronger than (*a*); for

$$0 \leqq \left| \int_E f_n(x)\,dx - \int_E f(x)\,dx \right| = \left| \int_E (f_n(x) - f(x)\,dx \right|$$
$$\leqq \int_E |f_n(x) - f(x)|\,dx,$$

so if the last integral tends to 0 the difference between the integrals of $f_n(x)$ and the integral of $f(x)$ must also tend to 0.

REMARK. It is easy to extend 29.3 from sequences $f_n(x)$ to functions $f(x, h)$. Let $f(x, h)$ be defined and finite for all x in a set E and all h in a set H, and let h_0 be an accumulation point of H. The statement "$f(x, h)$ converges in measure to $f(x)$ as $h \to h_0$" has a self-suggesting definition analogous to 28.3. We can now prove the following theorem.

Let $f(x, h)$ be measurable on E for each fixed h in H. If on E the function $f(x, h)$ converges almost everywhere or in measure to $f(x)$ as $h \to h_0$, and there is a summable function $g(x)$ such that $|f(x, h)| \leqq g(x)$ for all x in E and all h in H, then $f(x)$ is summable over E, and

$$\text{(a)} \qquad \lim_{h \to h_0} \int_E f(x, h)\,dx = \int_E f(x)\,dx,$$

$$\text{(b)} \qquad \lim_{h \to h_0} \int_E |f(x, h) - f(x)|\,dx = 0,$$

Let $h_1, h_2, \cdots$ be a sequence of points of H converging to h_0 and $\neq h_0$. The sequence $f(x, h_n)(n = 1, 2, \cdots)$ satisfies the hypotheses of 29.3, so $f(x)$ is summable over E and

$$\lim_{n \to \infty} \int_E f(x, h_n)\,dx = \int_E f(x)\,dx,$$
$$\lim_{n \to \infty} \int_E |f(x, h_n) - f(x)|\,dx = 0.$$

This holds for every sequence $\{h_n\}$ of points of H converging to h_0 and distinct from h_0, so by 4.5 we obtain the desired conclusions.

Analogous extensions of theorems and definitions 29.4–29.8 can be made with equal ease. We shall not state them in detail.

Let us return again to the example with which we begin this section. We notice that the convergence troubles arose because there were arbitrarily small intervals $[0, h]$ on which the integrals of the $f_n(x)$ were not arbitrarily small; in fact, these integrals were equal to 1 for all large n. This suggests that we might arrive at a convergence theorem by excluding this type of behavior. Now we know from 27.1 that for each n and for every $\epsilon > 0$ there is a $\delta > 0$ such that the integral of $f_n(x)$ over any set E of measure $mE < \delta$ has a value less than ϵ. We exclude the type of difficulty shown in our example by requiring that for each $\epsilon > 0$ there shall be a $\delta > 0$ which serves uniformly for all n. That is, in the definition of absolute continuity of $\int_E f(x)\,dx$, we ask that a single $\delta(\epsilon)$ shall serve uniformly for all n:

29.4s. *Let the functions $f_1(x)$, $f_2(x)$, $\cdots$ be all defined and all summable over a set E^*. The integrals*

$$F_n(E) = \int_E f_n(x)\,dx,$$

regarded as functions of measurable subsets E of E^*, are uniformly absolutely continuous if to each $\epsilon > 0$ there corresponds a $\delta > 0$ such that for every measurable subset E of E^* with $mE < \delta$ the inequality

$$|F_n(E)| \equiv \left|\int_E f_n(x)\,dx\,\right| < \epsilon$$

holds for all n.

A direct consequence of the definition is

29.5s. *Let the functions $f_1(x)$, $f_2(x)$, $\cdots$ be defined and summable over a set E^*. The set-functions $F_n(E)$ defined as in 29.4 are uniformly absolutely continuous if and only if to every $\epsilon > 0$ there corresponds a $\delta > 0$ such that*

$$\int_E |f_n(x)|\,dx < \epsilon$$

for every measurable subset E of E^* with $mE < \delta$.

If the condition above is satisfied, the functions $F_n(E)$ are uniformly absolutely continuous. For if E is any subset of E^* with $mE < \delta$, then

$$|F_n(E)| = \left|\int_E f_n(x)\,dx\right| \leqq \int_E |f_n(x)|\,dx < \epsilon.$$

Conversely, suppose that the $F_n(E)$ are uniformly absolutely continuous. If ϵ is a positive number, there is a $\delta > 0$ such that $F_n(E) < \epsilon/2$ for every subset E of E^* with $mE < \delta$. Let E be such a set.

It can be divided into the subset $E_{n,1}$ on which $f_n(x) \geqq 0$ and the subset $E_{n,2}$ on which $f_n(x) < 0$. These subsets are measurable, by 21.2 and 21.3. Each has measure less than δ, being contained in the set E whose measure is less than δ. So

$$\int_E |f_n(x)| \, dx = \left| \int_{E_{n,1}} f_n(x) \, dx \right| + \left| \int_{E_{n,2}} f_n(x) \, dx \right| = |F_n(E_{n,1})| + |F_n(E_{n,2})| < \epsilon.$$

This is the essential hypothesis in

29.6s. *Let the functions $f_1(x)$, $f_2(x)$, $\cdots$ be defined, finite valued and summable on a set E^* of finite measure. If*

(*a*) *on E^* the functions $f_n(x)$ converge in measure (or, more particularly, almost everywhere or almost uniformly) to a function $f(x)$;*

(*b*) *the set-functions $F_n(E) \equiv \int_E f_n(x) \, dx$ are uniformly absolutely continuous on the class of all measurable subsets of E^*; and either*

(*c*) *the integrals*

$$\int_{E^*} |f_n(x)| \, dx$$

are bounded, or

(*c'*) *the function $f(x)$ is finite for almost all x in E^*; then $f(x)$ is summable over E^*, and*

(*d*) $\lim_{n \to \infty} \int_{E^*} f_n(x) \, dx = \int_{E^*} f(x) \, dx,$

(*e*) $\lim_{n \to \infty} \int_{E^*} |f_n(x) - f(x)| \, dx = 0.$

As was remarked after 29.3, it is only necessary to prove (*e*), since (*d*) is a consequence of (*e*).

We first assume hypotheses (*a*), (*b*) and (*c*). Since the functions $|f_n(x)|$ converge in measure to $|f(x)|$, by 29.2 and hypothesis (*c*) the function $|f(x)|$ is summable. But by 28.8 $f(x)$ is measurable; so by 22.3 $f(x)$ is summable.

Let ϵ be an arbitrary positive number. Since $f(x)$ is summable, by 27.1 there is a $\delta_1 > 0$ such that

$$(A) \qquad \int_E |f(x)| \, dx < \frac{\epsilon}{3}$$

if E is a measurable subset of E^* with $mE < \delta_1$. By 29.5 and hypothesis (*b*), there is a $\delta_2 > 0$ such that

$$(B) \qquad \int_E |f_n(x)| \, dx < \frac{\epsilon}{3}$$

for all n if E is a measurable subset of E^* with $mE < \delta_2$. Let δ be the smallest of the numbers δ_1, δ_2 and $\epsilon/3mE^*$. By the definition of convergence in measure, there is an n_0 such that for all $n > n_0$ the set of all x for which $f_n(x)$ is not in $N_\delta(f(x))$ can be enclosed in a set $E_{n,\delta}$ with measure $mE_{n,\delta} < \delta$. Since $f(x)$ is summable, it is almost everywhere finite, and we can include in $E_{n,\delta}$ the points x for which $|f(x)| = \infty$ without increasing $mE_{n,\delta}$. Now we write

$$\int_{E^*} |f_n(x) - f(x)|\,dx = \int_{E_{n,\delta}} |f_n(x) - f(x)|\,dx + \int_{E^*-E_{n,\delta}} |f_n(x) - f(x)|\,dx \leqq \int_{E_{n,\delta}} |f_n(x)|\,dx + \int_{E_{n,\delta}} |f(x)|\,dx + \int_{E^*-E_{n,\delta}} |f_n(x) - f(x)|\,dx.$$

If $n > n_0$, then $mE_{n,\delta} < \delta$. So the first integral on the right is less than $\epsilon/3$ by (*B*), and the second is less than $\epsilon/3$ by (*A*). On $E^* - E_{n,\delta}$ the functional value $f_n(x)$ is in $N_\delta(f(x))$, and $f(x)$ is finite; therefore $|f_n(x) - f(x)| < \delta$. So the integrand in the third integral on the right is less than δ, and the measure of $E^* - E_{n,\delta}$ is at most equal to mE^*; therefore the third integral on the right is at most $\delta mE^* \leqq \epsilon/3$. Adding, we find that the integral on the left is less than ϵ for all $n > n_0$, which establishes conclusion (*e*). This completes the proof of our theorem under hypotheses (*a*), (*b*) and (*c*).

Suppose now that hypotheses (*a*), (*b*) and (*c'*) are satisfied. Let δ be a number such that

$$\int_E |f_n(x)|\,dx < 1$$

if E is a measurable subset of E^* with $mE < \delta$; such a δ exists by hypothesis (*b*) and 29.5. Since the $f_n(x)$ converge in measure to $f(x)$, there is an integer p such that if $n \geqq p$, the set of points x in E for which $f_n(x)$ is not in $N_1(f(x))$ can be enclosed in a set E_n with $mE_n < \delta/2$. By hypothesis the set on which $|f(x)| = \infty$ has measure 0, so we can include it in E_n without increasing mE_n. Then on $E^* - E_n$ the function $f(x)$ is finite and $f_n(x)$ is in $N_1(f(x))$; that is, $|f_n(x) - f(x)| < 1$ for x in $E^* - E_n$. Now if $n \geqq p$

$$\int_{E^*} |f_n(x) - f_p(x)|\,dx \leqq \int_{E^*-(E_n\cup E_p)} |f_n(x) - f_p(x)|\,dx + \int_{E_n\cup E_p} |f_n(x)|\,dx + \int_{E_n\cup E_p} |f_p(x)|\,dx.$$

Since $m(E_n \cup E_p) < \delta$, the last two integrals are each less than 1. On $E^* - (E_n \cup E_p)$ both $|f_n(x) - f(x)|$ and $|f_p(x) - f(x)|$ are less than 1, so $|f_n(x) - f_p(x)| < 2$. So the first integral on the right is at most $2mE^*$. We therefore have for $n \geqq p$

$$\int_{E^*} |f_n(x)|\,dx \leqq \int_{E^*} |f_p(x)|\,dx + \int_{E^*} |f_n(x) - f_p(x)|\,dx < \int_{E^*} |f_p(x)|\,dx + 2 + 2mE^*.$$

Consequently hypothesis (c) is satisfied, the integrals of the $|f_n(x)|$ being not greater than the largest of the numbers

$$\int_{E^*} |f_1(x)|\,dx, \cdots, \int_{E^*} |f_{p-1}(x)|\,dx, \int_{E^*} |f_p(x)|\,dx + 2 + 2mE^*.$$

Therefore if the hypotheses (a), (b) and (c') are satisfied, so are (a), (b) and (c); and the proof of our theorem is complete.

EXERCISE. Let $f(x)$ be defined on a set E of finite measure, and let $f_1(x), f_2(x), \cdots$ be summable over E. In order that the conclusions of 29.6 shall hold it is necessary that hypotheses (a), (b), (c) and (c') be satisfied.

Neither of the theorems 29.3 and 29.6 contains the other as a special case. For in 29.3 the set E^* could be of infinite measure, while we can show by a simple example that the finiteness of the measure of E^* is essential in 29.6. Let $f(x) \equiv 0$, and let $f_n(x) = 1/n$ if $n \leqq x \leqq 2n$, $f_n(x) = 0$ otherwise. Then $f_n(x)$ converges uniformly (hence almost uniformly, in measure and everywhere) to $f(x)$. But for the integrals over the whole space R_1 we have

$$\int_{R_1} f_n(x)\,dx = 1, \qquad \int_{R_1} 0\,dx = 0.$$

On the other hand, the hypotheses of 29.6 can be satisfied without the existence of the summable function $g(x)$ such that $|f_n(x)| \leqq g(x)$. For example, let $E^* = [0, 1]$, and let $f(x) \equiv 0$. For each positive integer n let $f_n(x) = x^{-\frac{3}{2}}$ for $(n+1)^{-1} \leqq x \leqq n^{-1}$, and let $f_n(x) = 0$ elsewhere. Then

$$\int_{[0,1]} |f_n(x)|\,dx = \int_{(n+1)^{-1}}^{n^{-1}} x^{-\frac{3}{2}}\,dx = -2x^{-\frac{1}{2}}\Big|_{(n+1)^{-1}}^{n^{-1}} = 2(\sqrt{n+1} - \sqrt{n}) \to 0,$$

and the hypotheses of 29.6 are satisfied. But the intervals $[(n+1)^{-1}, n^{-1}]$ cover $(0, 1]$, and any function greater than all the $f_n(x)$ would have to exceed $x^{-\frac{3}{2}}$ on $(0, 1]$, and could not be summable. However, it is not difficult now to establish a theorem which is more general than either 29.3 or 29.6.

29.7s. *Let the functions* $f_1(x)$, $f_2(x)$, $\cdots$ *be defined, finite-valued and summable over a set* E^*. *If*

(*a*) *on* E^* *the functions* $f_n(x)$ *converge in measure, or almost everywhere, or almost uniformly, to a function* $f(x)$;

(*b*) *for every* $\epsilon > 0$ *there is a set* E_ϵ *of finite measure contained in* E^* *and such that* (i) *the set-functions*

$$\int_E f_n(x)\,dx$$

are uniformly absolutely continuous on the class of all measurable subsets of E_ϵ; (ii)

$$\int_{E^*-E_\epsilon} |f_n(x)|\,dx < \epsilon$$

for all n;

(*c*) *the integrals*

$$\int_{E^*} |f_n(x)|\,dx$$

are bounded; or

(*c'*) $f(x)$ *is finite for almost all* x; *then* $f(x)$ *is summable over* E^*, *and*

(*d*) $\lim_{n\to\infty} \int_{E^*} f_n(x)\,dx = \int_{E^*} f(x)\,dx,$

(*e*) $\lim_{n\to\infty} \int_{E^*} |f_n(x) - f(x)|\,dx = 0.$

We first assume that (*a*), (*b*) and (*c*) hold. As in the proof of 29.6, $f(x)$ is summable over E^*. Let ϵ be any positive number, and let E_ϵ be the set described in (*b*). On E_ϵ, the hypotheses of 29.6 are satisfied, so

$$\lim_{n\to\infty} \int_{E_\epsilon} |f_n(x) - f(x)|\,dx = 0.$$

Since $|f_n(x)|$ tends in measure, or almost everywhere, or almost uniformly to $|f(x)|$ on $E^* - E_\epsilon$, by 29.2 we have

$$\int_{E^*-E_\epsilon} |f(x)|\,dx \leqq \liminf_{n\to\infty} \int_{E^*-E_\epsilon} |f_n(x)|\,dx \leqq \epsilon.$$

Therefore, using 6.5, 6.12 and 6.14,

$$\begin{aligned} 0 &\leqq \liminf_{n\to\infty} \int_{E^*} |f_n(x) - f(x)|\,dx \leqq \limsup_{n\to\infty} \int_{E^*} |f_n(x) - f(x)|\,dx \\ &\leqq \limsup_{n\to\infty} \int_{E_\epsilon} |f_n(x) - f(x)|\,dx + \limsup_{n\to\infty} \int_{E^*-E_\epsilon} |f_n(x)|\,dx \\ &\qquad\qquad + \int_{E^*-E_\epsilon} |f(x)|\,dx \\ &< 2\epsilon. \end{aligned}$$

This holds for all $\epsilon > 0$, so the upper and lower limits of the integrals of the $|f_n(x) - f(x)|$ are both 0. So conclusion (*e*) is established. As before, (*d*) follows from (*e*).

If hypotheses (*a*), (*b*) and (*c'*) hold, take $\epsilon = 1$. By hypothesis (*b*) there is a set E_1 of finite measure such that

$$\int_{E^*-E_1} |f_n(x)|\, dx < 1$$

for all n. But since E_1 is of finite measure, we can show exactly as in the last part of the proof of 29.6 that the integrals

$$\int_{E_1} |f_n(x)|\, dx$$

are bounded. Hence, adding, the integrals of the $|f_n(x)|$ over E^* are bounded, and hypothesis (*c*) is satisfied.

It is clear that 29.7 is more general than 29.6; for if the hypotheses of 29.6 hold we can take $E_\epsilon = E^*$ for all $\epsilon > 0$, and the hypotheses of 29.7 are satisfied. Furthermore, 29.7 includes 29.3. For suppose that the hypotheses of 29.3 hold. Then hypotheses (*a*) and (*c*) of 29.7 clearly are satisfied. Since $g(x)$ is summable over E^*, for every $\epsilon > 0$ there is an interval W_n: $-n \leqq x^{(i)} \leqq n$ such that

$$\left| \int_{E^*} g(x)\, dx - \int_{E^*W_n} g(x)\, dx \right| < \epsilon;$$

that is,

$$\int_{E^*-E^*W_n} g(x)\, dx < \epsilon.$$

Since $|f_n(x)| \leqq g(x)$ for all n, part (*ii*) of hypothesis (*b*) is satisfied if we take $E_\epsilon = E^*W_n$. By 27.1, for every number $\gamma > 0$ there is a $\delta > 0$ such that

$$\int_E g(x)\, dx < \gamma$$

if E is a measurable subset of E_ϵ with $mE < \delta$. Hence for all such E

$$\int_E |f_n(x)|\, dx < \gamma,$$

and part (*i*) of (*b*) is also satisfied.

If we are willing to drop the "*s*" from theorem 29.6 we can omit hypotheses (*c*) and (*c'*) completely:

29.8. *Let the functions $f_1(x), f_2(x), \cdots$ be defined and summable on a set E^* of finite measure. If*

(a) *on E^* the functions $f_n(x)$ converge in measure (or almost everywhere, or almost uniformly) to a function $f(x)$; and*

(b) *the set-functions $F_n(E) = \int_E f_n(x)\,dx$ are uniformly absolutely continuous on the class of all measurable subsets of E^*; then the conclusions of theorem* 29.6 *hold.*

By theorem 29.5, there is a $\delta > 0$ such that $\int_E |f_n(x)|\,dx < 1$ for all measurable subsets E of E^* with $mE < \delta$. For some interval W_p we have $mW_pE^* > mE^* - \delta$; therefore $m(E^* - W_p) < \delta$. Let t be a positive number less than $\delta^{1/q}$; then every interval I whose sides are all equal to t has measure less than δ. We can cover the interval W_p with a finite number of intervals $I_1, \cdots, I_h$ of this type. Then E^* is contained in $(E^* - W_p) \cup E^*I_1 \cup \cdots \cup E^*I_h$, and each of these sets has measure less than δ. So

$$\int_{E^*} |f_n(x)|\,dx \leqq \int_{E^*-W_p} |f_n(x)|\,dx + \int_{E^*I_1} |f_n(x)|\,dx + \cdots + \int_{E^*I_h} |f_n(x)|\,dx$$

$$\leqq h + 1.$$

Therefore hypothesis (c) of 29.6 is satisfied. The other hypotheses of 29.6 have been assumed as hypotheses here also, so the conclusions of 29.6 hold.

Since hypotheses (b) and (c) are vital in theorem 29.6, it is interesting to have criteria which will guarantee their satisfaction. We have already seen (after 29.7) that these conditions are satisfied if $|f_n(x)| \leqq g(x)$, where $g(x)$ is summable. But this merely brings us back to a special case of theorem 29.3. A criterion of an essentially different nature is

29.9s. (Nagumo). *Let $\Phi(t)$ be a function defined and non-negative on $0 \leqq t < \infty$ and such that $\lim_{t\to\infty} \Phi(t)/t = \infty$. If $\{f(x)\}$ is an aggregate of functions finite valued and measurable on a set E^* of finite measure, and there is a number H such that for every $f(x)$ in the class $\{f(x)\}$ the integral*

$$\int_{E^*} \Phi(|f(x)|)\,dx$$

exists and is less than H, then the integrals

$$\int_E f(x)\,dx$$

are uniformly absolutely continuous and the integrals

$$\int_{E^*} | f(x) | \, dx$$

are uniformly bounded for all functions $f(x)$ in the class $\{f(x)\}$.

Let ϵ be an arbitrary positive number and E an arbitrary measurable subset of E^*. By hypothesis, there exists a positive number t_1 such that $\Phi(t)/t \geqq 2H/\epsilon$ if $t \geqq t_1$. If $f(x)$ is a function belonging to the class $\{f(x)\}$, we subdivide E into the subset E_1 on which $| f(x) | < t_1$ and the subset E_2 on which $| f(x) | \geqq t_1$. These sets are measurable by 21.8 and 21.3. If x is in E_2, then $| f(x) | \geqq t_1$ and so

$$\Phi(| f(x) |) \cdot \epsilon/2H \geqq | f(x) | . \quad \text{Hence}$$

$$\begin{aligned} \left| \int_E f(x) \, dx \right| &\leqq \int_E | f(x) | \, dx \\ &= \int_{E_1} | f(x) | \, dx + \int_{E_2} | f(x) | \, dx \\ &\leqq \int_{E_1} t_1 \, dx + \int_{E_2} \Phi(| f(x) |) \frac{\epsilon}{2H} \, dx \\ &\leqq t_1 m E_1 + \frac{\epsilon}{2H} \int_E \Phi(| f(x) |) \, dx \\ &\leqq t_1 m E_1 + \frac{\epsilon}{2}. \end{aligned}$$

If we first take $\epsilon = 1$ and $E = E^*$, this proves that the integrals of the functions $| f(x) |$ over E^* are uniformly bounded. Second, if for an arbitrary ϵ we restrict the measure of E to be less than $\delta = \epsilon/2t_1$, we find

$$\left| \int_E f(x) \, dx \right| < \epsilon,$$

establishing the uniform absolute continuity of the integrals.

30. In §1 we mentioned certain fundamental properties of our definition of distance in R_q. In order to stress this importance of the four properties, we introduce a definition.

30.1s. *Let D be a collection of things (we call them "points") and $\rho(p, q)$ a real-valued function of pairs of points of D. The set D is a metric space with distance-function $\rho(p, q)$ if the following four conditions are satisfied.*

(1) *For all p, q in D, $\rho(p, q) \geqq 0$.*

(2) *$\rho(p, q) = 0$ if and only if $p = q$.*

(3) $\rho(p, q) = \rho(q, p)$.
(4) *For all p, q, r in D,*

$$\rho(p, r) \leqq \rho(p, q) + \rho(q, r).$$

As we saw in §1, the spaces R_q are metric spaces with the distance-function $\| x, y \|$ which we have been using. Again, let Y be an arbitrary set and D the collection of all bounded functions on Y. For any two such functions $p(y)$, $q(y)$ we define $\rho(p, q) = \sup | p(y) - q(y) |$ on Y. Properties (1), (2) and (3) of 30.1 are evidently satisfied, and by 5.4 and 5.6 and the inequality

$$| p(y) - r(y) | \leqq | p(y) - q(y) | + | q(y) - r(y) |$$

property (4) is also satisfied. So this too is a metric space. It is not difficult to show that with this distance-function the relation $p_n \to p_0$ (by which we of course mean $\rho(p_n, p_0) \to 0$) is equivalent to

$$(A) \qquad \lim_{n \to \infty} p_n(y) = p_0(y) \text{ uniformly on } Y.$$

For let ϵ be a positive number. If $p_n \to p_0$, then for all n greater than a certain n_ϵ we have $\rho(p_n, p_0) < \epsilon$, whence

$$\begin{aligned} | p_n(y) - p_0(y) | &\leqq \sup | p_n(y) - p_0(y) | \\ &= \rho(p_n, p_0) < \epsilon. \end{aligned}$$

This is the definition 6.16 of uniform convergence. Conversely, if (A) holds, there is an n_ϵ such that if $n > n_\epsilon$ then

$$| p_n(y) - p_0(y) | < \frac{\epsilon}{2},$$

so that $\rho(p_n, p_0) \leqq \epsilon/2 < \epsilon$.

Again, let S be the class of all functions $f(x)$ defined, finite and measurable on a measurable set E. For any two such functions $f(x)$, $g(x)$ let $\rho(p, q)$ be the inf of all numbers α such that $| f(x) - g(x) | < \alpha$ except on a set of measure less than α. Properties (1) and (3) are obvious. For property (4), let f, g, h belong to S, and let ϵ be an arbitrary positive number. Then

$$| f(x) - g(x) | < \rho(f, g) + \epsilon$$

except on a set E_1 of measure less than $\rho(f, g) + \epsilon$, and

$$| g(x) - h(x) | < \rho(g, h) + \epsilon$$

except on a set E_2 of measure less than $\rho(g, h) + \epsilon$. So except on the set $E_1 \cup E_2$, whose measure is less than $\rho(f, g) + \rho(g, h) + 2\epsilon$, we have

$$|f(x) - h(x)| < \rho(f, g) + \rho(g, h) + 2\epsilon.$$

That is,

$$\rho(f, h) \leqq \rho(f, g) + \rho(g, h) + 2\epsilon.$$

Since ϵ, is an arbitrary positive number, by 5.2 this implies property (4) of 30.1.

But with property (2) it is different; for if $f(x)$ and $g(x)$ differ on a set of measure zero we have $\rho(f, g) = 0$ without having $f = g$. This difficulty can be removed in either of two ways. We can alter the concept of equality by regarding two functions as identical if they are equivalent. This however has certain disadvantages; for example, everywhere else the meaning of "=" has been identity, not a conventional relationship. The alternative is to lump together in a single class all functions equivalent to each other and use these equivalence-classes as the points of our space S. Properties (1), (3) and (4) are undisturbed, for if they hold for f, g, h they hold for all functions respectively equivalent to f, g, h. If two points (equivalence-classes) of S are coincident, then functions f, g representing these points are equivalent, and $\rho(f, g) = 0$; that is, the distance from a point to itself is 0. Conversely, if $\rho(f, g) = 0$, let n be any positive integer. If $\alpha < 1/n$, the set $E[\,|f - g| \geqq 1/n]$ is contained in the set $E[\,|f - g| \geqq \alpha]$, whose measure is less than α. This holds for all α less than $1/n$, so

$$mE\left[\,|f - g| \geqq \frac{1}{n}\right] = 0.$$

Adding these sets for $n = 1, 2, \cdots$, we find by 19.3

$$mE[\,|f - g| > 0] = 0,$$

so $f(x)$ and $g(x)$ are equivalent. That is, if $\rho(f, g) = 0$ the functions $f(x)$ and $g(x)$ represent the same point of S.

Once this is understood, there is little danger of misunderstanding if we speak of functions $f(x)$ as belonging to S, instead of using the longer and more accurate statement that $f(x)$ is a member of an equivalence class which in turn is a point of S.

With the distance defined above, convergence of f_n to f_0 in S is equivalent to convergence of $f_n(x)$ to $f_0(x)$ in measure on E. For let $\rho(f_n, f_0)$ tend to zero. If ϵ and δ are positive numbers, there is an n_0 such that $\rho(f_n, f_0) < \inf\{\epsilon, \delta\}$ when n exceeds n_0. By definition of ρ, this implies that $|f_n(x) - f_0(x)| < \epsilon$ except on a set of measure less than δ, so $f(x)$ converges in measure to $f_0(x)$. Conversely, let $f_n(x)$ converge in measure to $f_0(x)$, and let γ be a positive number. If in the definition 28.3 we choose $\epsilon = \delta = \gamma$, we find that for all n greater than a certain n_0 we have $|f_n(x) - f_0(x)| < \gamma$ except on a set of measure less than γ, so $\rho(f_n, f_0) < \gamma$. That is, $\rho(f_n, f_0)$ tends to zero, and f_n tends to f_0 in S.

For our final example, let E be a measurable set and p a number not less than 1. The points of our space will be the finite-valued functions of class L_p on E; for two such functions $f(x)$, $g(x)$ we define the distance by the equation

$$\rho(f, g) = \left\{ \int_E |f(x) - g(x)|^p \, dx \right\}^{1/p}$$

Properties (1) and (3) of 30.1 are clearly satisfied; property (4) holds by 24.7. But as in the preceding example property (3) is not satisfied, since $\rho(f, g) = 0$ whenever $f(x)$ and $g(x)$ differ only on a set of measure zero. As before, we circumvent this trouble by defining the points of our space $L_p(E)$ to be equivalence-classes of functions of class L_p on E. And as before we shall say that a function $f(x)$ belongs to $L_p(E)$ when we really mean that it is one of an equivalence-class of functions all of which are of class L_p on E.

Theorems 2.1 to 2.10 and Theorem 2.13 depended only on the properties of distance listed in 30.1, so they hold for every metric space. This is not true of the other three theorems of §2; they utilized a special property (Dedekind continuity) of the real numbers.

Of the many properties which metric spaces may possess, there is one which is of outstanding importance whenever limiting processes occur. This is the property of *completeness*. Briefly, a space is complete if the Cauchy criterion for convergence (6.15) is valid in it. It is convenient to define the concept in two stages.

30.2s. *A sequence $\{p_n\}$ of points of a metric space D is a regular, or Cauchy, sequence if to every positive number ϵ there corresponds an n_ϵ such that $\rho(p_m, p_n) < \epsilon$ whenever $m \geqq n_\epsilon$ and $n \geqq n_\epsilon$.*

30.3s. *A metric space D is complete if every regular sequence of points of D converges to a point of D.*

Theorem 6.15 informs us that one-dimensional space R_1 is complete (the set E of 6.15 being here the set of positive integers). If $\{x_n\}$ is a regular sequence of points of R_q, the inequalities

$$|x_n^{(i)} - x_m^{(i)}| \leqq ||x_n, x_m|| \qquad (i = 1, \cdots, q)$$

show that for each i the sequence $\{x_n^{(i)}\}$ is a regular sequence in R_1.

Hence by 6.15 the $x_n^{(i)}$ approach a limit $x_0^{(i)}$, and we readily verify that $x_n \to x_0$. So R_q is complete.

If Y is an arbitrary set, the space of bounded functions on Y with the metric $\rho(f, g) = \sup | f(y) - g(y) |$ is complete; this is merely a re-wording of 6.17.

Our next two proofs both use a device which we state in the next lemma.

30.4s. *If* $\{p_n\}$ *is a regular sequence in a metric space* D, *and a subsequence* $\{p_\alpha\}(\alpha = n_1, n_2, \cdots)$ *converges to a limit* p_0, *then the whole sequence converges to* p_0.

Let ϵ be a positive number. For all m and n greater than a certain n_ϵ we have

$$\rho(p_m, p_n) < \frac{\epsilon}{2},$$

since the sequence is regular. For all α greater than a certain α_0 we have

$$\rho(p_\alpha, p_0) < \frac{\epsilon}{2},$$

since $p_\alpha \to p_0$. Choose an $\alpha = n_i$ larger than the greater of n_ϵ and α_0. The two inequalities above both hold with $m = \alpha$, and by (4) of 30.1

$$\rho(p_n, p_0) \leqq \rho(p_n, p_\alpha) + \rho(p_\alpha, p_0) < \epsilon$$

for all n greater than n_ϵ. Therefore $p_n \to p_0$.

We now proceed to show the completeness of the other spaces mentioned above.

30.5s. *Let* E *be a measurable set. The space* S *of functions finite and measurable on* E, *with the metric defined above, is complete.*

Let $\{f_n\}$ be a regular sequence. We define a sequence of integers as follows. The integer n_1 is the least one such that $\rho(f_m, f_n) < 2^{-1}$ whenever m and n are at least equal to n_1. The integer n_2 is the least integer greater than n_1 such that $\rho(f_m, f_n) < 2^{-2}$ whenever m and n are at least equal to n_2; and so on. For compactness we write

$$g_i(x) = f_{n_i}(x) \qquad (i = 1, 2, 3, \cdots);$$

then by the choice of the n_i we have

$$(A) \qquad \rho(g_i, g_j) < 2^{-i} \quad \text{if} \quad j \geqq i.$$

We need only show that the g_i form a convergent sequence, since by 30.4 the whole sequence will then converge.

Let E_i be the subset of E on which $| g_i(x) - g_{i+1}(x) | \geqq 2^{-i}$. By (A) and the definition of distance, this set has measure less than 2^{-i}. Define

$$M_i = E_i \cup E_{i+1} \cup E_{i+2} \cup \cdots .$$

Then by 19.8 we find that

$$(B) \qquad mM_i < \sum_{n=i}^{\infty} 2^{-n} = 2^{1-i}.$$

On $E - M_i$ all the inequalities

$$| g_j(x) - g_{j+1}(x) | < 2^{-j} \qquad (j = i, i + 1, \cdots)$$

are satisfied. So if j and k are both at least equal to i (we choose the notation so that $j \leqq k$) we find

$$\begin{aligned} | g_j(x) - g_k(x) | &\leqq \sum_{n=j}^{k-1} | g_n(x) - g_{n+1}(x) | \\ &< \sum_{n=j}^{k-1} 2^{-n} \\ &< 2^{1-j}. \end{aligned}$$

That is, for each x in $E - M_i$ the numbers $g_j(x)$ form a regular sequence, and by 6.15 they converge to a limit. The set N of points of E at which the $g_j(x)$ fail to converge is therefore contained in M_i. This is true for each i, so N is contained in $\bigcap M_i$. By 19.9 and (B) this has measure zero, so by 19.13 the set N has measure zero.

Let $f(x)$ be defined in E by the equations

$$\begin{aligned} f(x) &= \lim_{n \to \infty} g_n(x) \qquad && (x \text{ in } E - N) \\ f(x) &= 0 && (x \text{ in } N). \end{aligned}$$

Then $g_n(x)$ converges to $f(x)$ almost everywhere in E. By 28.10 $g_n(x)$ converges to $f(x)$ in measure in E. By our discussion of the distance function in S, this implies that $\rho(g_n, f)$ tends to zero, and by 30.4 the proof is complete.

30.6s. *If E is a measurable set and $p \geqq 1$, the space $L_p(E)$ is complete.*

Every function in $L_p(E)$ is measurable on E, and may be assumed without loss of generality to be finite-valued on E. So it belongs to the space S of functions finite and measurable on E. In order to prevent confusion, if f and g belong to $L_p(E)$ we denote their distance in $L_p(E)$ by $\rho_p(f, g)$; their distance using the distance-function of the space S will be denoted by $\rho(f, g)$ as before. We first wish to establish a relation between these distances.

Let f and g be any functions belonging to $L_p(E)$, and let α be any number such that

$$(A) \qquad \alpha > [\rho_p(f, g)]^{p/(p+1)}.$$

Let E_α be the subset of E on which $| f(x) - g(x) | \geqq \alpha$. Then

$$\begin{aligned}\rho_p(f, g) &= \left\{ \int_E | f(x) - g(x) |^p \, dx \right\}^{1/p} \\ &\geqq \left\{ \int_{E_\alpha} \alpha^p \, dx \right\}^{1/p} \\ &= \alpha \{ mE_\alpha \}^{1/p}.\end{aligned}$$

With (A), this yields

$$\alpha^{(p+1)/p} > \alpha \{ mE_\alpha \}^{1/p}$$

or

$$mE_\alpha < \alpha.$$

By definition of $\rho(f, g)$, this implies $\rho(f, g) \leqq \alpha$. But this is true for every α satisfying (A), so by 5.2 we have

$$(B) \qquad \rho(f, g) \leqq [\rho_p(f, g)]^{p/(p+1)}.$$

Now let $\{f_n\}$ be a regular sequence in $L_p(E)$. If ϵ is an arbitrary positive number, there is then an n_ϵ such that

$$\rho_p(f_m, f_n) < \epsilon^{(p+1)/p}$$

whenever m and n are at least equal to n_ϵ. With (B), this implies $\rho(f_m, f_n) < \epsilon$ for all such m and n; so our sequence $\{f_n\}$ is also a regular sequence in S. By 30.5, it converges to some limit f in S. That is, there is a function $f(x)$ finite and measurable on E such that $f_n(x)$ converges in measure to $f(x)$ on E.

By 28.7, there is a subsequence of the sequence $\{f_n(x)\}$ which converges almost everywhere in E to $f(x)$. This subsequence we denote by $\{g_i(x)\}$, $i = 1, 2, \cdots$. According to 30.4 it will be sufficient to prove that f is in $L_p(E)$ and that g_i tends to f in $L_p(E)$, for then the whole sequence f_n will have the limit f in $L_p(E)$.

For almost all x in E we have

$$\lim_{i \to \infty} g_i(x) = f(x).$$

It follows at once that for each fixed j and for almost all x

$$\lim_{i \to \infty} | g_j(x) - g_i(x) | = | g_j(x) - f(x) |,$$

and so

$$(C) \qquad \lim_{i\to\infty} | g_j(x) - g_i(x) |^p = | g_j(x) - f(x) |^p.$$

Let ϵ be a positive number. There is an n_ϵ such that

$$\rho_p(g_j, g_i) < \frac{\epsilon}{2}$$

whenever i and j are at least equal to n_ϵ. That is,

$$\int_E | g_j(x) - g_i(x) |^p \, dx < \left(\frac{\epsilon}{2}\right)^p$$

for all such i and j. Fixing j at any value $\geqq n_\epsilon$, by 29.2 and (C) we find that $| g_j(x) - f(x) |^p$ is summable, and

$$(D) \qquad \int_E | g_j(x) - f(x) |^p \, dx \leqq \liminf_{i\to\infty} \int_E | g_j(x) - g_i(x) |^p \, dx \leqq \left(\frac{\epsilon}{2}\right)^p < \epsilon^p.$$

Since $g_j(x) - f(x)$ is measurable by 21.12, this proves that it is in $L_p(E)$. But $g_j(x)$ is itself in $L_p(E)$, so by 24.6 the difference $f(x) = g_j(x) - [g_j(x) - f(x)]$ is in $L_p(E)$. Furthermore, from (D) we find

$$\rho_p(g_j, f) < \epsilon$$

whenever $j \geqq n_\epsilon$. Hence g_j tends to f in $L_p(E)$, and the proof is complete.

As an application of this theorem we shall prove the Riesz-Fischer theorem on Fourier series. First, however, we recall some elementary definitions and theorems concerning Fourier series. If $f(x)$ is summable over the interval $[-\pi, \pi]$ the integrals

$$(E) \qquad a_n = \frac{1}{\pi}\int_{-\pi}^{\pi} f(x) \cos nx \, dx \qquad (n = 0, 1, 2, \cdots)$$

$$b_n = \frac{1}{\pi}\int_{-\pi}^{\pi} f(x) \sin nx \, dx \qquad (n = 1, 2, 3, \cdots)$$

exist, by 22.4. They are called the Fourier coefficients of $f(x)$, and the series

$$(F) \qquad \frac{a_0}{2} + (a_1 \cos x + b_1 \sin x) + \cdots + (a_n \cos nx + b_n \sin nx) + \cdots$$

is the Fourier series corresponding to $f(x)$. We say nothing here about convergence; for the time being, the expansion is purely formal.

By a straightforward integration we find that if m and n are positive integers the following equations hold.

$$\int_{-\pi}^{\pi} \sin mx \cos nx \, dx = 0, \qquad (m, n = 0, 1, 2, \cdots),$$

$$(J) \qquad \int_{-\pi}^{\pi} \cos mx \cos nx = 0 \qquad (m \neq n),$$
$$= \pi \qquad (m = n),$$
$$\int_{-\pi}^{\pi} \sin mx \sin nx = 0 \qquad (m \neq n)$$
$$= \pi \qquad (m = n).$$

Given a summable function $f(x)$ we can always find its Fourier coefficients by equations (E). Given an infinite sequence a_0, a_1, b_1, a_2, b_2, $\cdots$ it is not always possible to find a function $f(x)$ of which they are the Fourier coefficients; with arbitrary a_n and b_n we cannot always solve the infinite system of equations (E) for the unknown function $f(x)$. The Riesz-Fischer theorem, which we now prove, states one set of conditions on the numbers a_n and b_n which is sufficient to guarantee the existence of a function $f(x)$ with which equations (E) hold.

30.7. *If a_0, a_1, b_1, a_2, b_2, $\cdots$ are real numbers such that $\Sigma(a_n^2 + b_n^2)$ converges, there exists a function $f(x)$ of class L_2 on $[-\pi, \pi]$ whose coefficients are the given numbers a_n and b_n.*

Let us define

$$s_n(x) = \tfrac{1}{2}a_0 + (a_1 \cos x + b_1 \sin x) + \cdots + (a_n \cos nx + b_n \sin nx)$$

If $m \geqq n$, by the formulas (J) above we find

$$\int_{-\pi}^{\pi} [s_m(x) - s_n(x)]^2 \, dx = \pi \sum_{i=n+1}^{m} [a_i^2 + b_i^2].$$

Since the squares of the a_n and the b_n form a convergent series, to each positive ϵ there corresponds an n_ϵ such that

$$\sum_{i=n+1}^{m} [a_i^2 + b_i^2] < \frac{\epsilon^2}{\pi}$$

whenever $m \geqq n \geqq n_\epsilon$. With the preceding equation, this shows that for all such m and n we have

$$\rho_2(s_m, s_n) < \epsilon.$$

Thus the sequence $\{s_n\}$ is a regular sequence in $L_2[-\pi, \pi]$, and there is a function $f(x)$ in $L_2[-\pi, \pi]$ to which the s_n converge in the metric of that space. It remains to show that the a_n and b_n are the Fourier coefficients of $f(x)$.

Let n be a fixed positive integer. The formulas (J) show that if $m \geqq n$ the equation

$$\frac{1}{\pi}\int_{-\pi}^{\pi} s_m(x)\cos nx\,dx = a_n$$

is satisfied. Hence by 24.5

$$\begin{aligned}&\left|\frac{1}{\pi}\int_{-\pi}^{\pi} f(x)\cos nx\,dx - a_n\right|\\ &= \left|\frac{1}{\pi}\int_{-\pi}^{\pi}[f(x) - s_m(x)]\cos nx\,dx\right|\\ &\leqq \frac{1}{\pi}\left\{\int_{-\pi}^{\pi}[f(x) - s_m(x)]^2\,dx\right\}^{\frac{1}{2}}\left\{\int_{-\pi}^{\pi}\cos^2 nx\,dx\right\}^{\frac{1}{2}}\\ &= \sqrt{\frac{1}{\pi}}\,\rho_2[f, s_m].\end{aligned}$$

But the quantity on the left is independent of m, and that on the right approaches zero as $m \to \infty$. Therefore the left member is 0, and the first set of equations (E) is satisfied. Repeating the argument with $\sin nx$ in place of $\cos nx$ proves that the second set of equations (E) is also satisfied, and the theorem is established.

REMARK. In proving the statement below it is convenient to observe that the Hölder inequality (conclusion of 24.4) can be written in the form

$$\left|\int_E f(x)g(x)\,dx\right| \leqq \rho_p(f, 0)\rho_q(g, 0).$$

EXERCISE. Let E be a measurable set, and let p, q be numbers greater than 1 such that $1/p + 1/q = 1$. If $\{f_n\}$ is a sequence of functions converging to f in $L_p(E)$, and $\{g_n\}$ is a sequence of functions converging to g in $L_q(E)$, then

$$\lim_{n\to\infty}\int_E f_n(x)g_n(x)\,dx = \int_E f(x)g(x)\,dx.$$

(By the Hölder inequality

$$\int_E |f_ng_n - fg|\,dx \leqq \rho_p(f_n, f)\rho_q(g_n, 0) + \rho_p(f, 0)\rho_q(g_n, g).$$

The factor $\rho_q(g_n, 0)$ is less than $\rho_q(g, 0) + 1$ if n is large.)

EXERCISE. If $\{f_n\}$ is a regular sequence in $L_p(E)$ $(p \geqq 1)$ it is a regular sequence in S; if it converges to f in $L_p(E)$, it converges to f in S.

EXERCISE. Let E have finite measure, and let $p \geqq p' \geqq 1$. If $\{f_n\}$ is a regular sequence on $L_p(E)$, it is regular on $L_{p'}(E)$; if it converges to f on $L_p(E)$, it converges to f on $L_{p'}(E)$.

CHAPTER V

Differentiation

31. So far we have dealt exclusively with integrals, making no mention of derivatives. But the interrelations between the processes of integration and differentiation are of fundamental importance in analysis. In this chapter we investigate these relations, restricting ourselves for the sake of simplicity to functions of a single real variable.

We shall of course define the derivative $f'(x_0)$ as the limit of the difference-quotient $[f(x) - f(x_0)]/(x - x_0)$ as x approaches x_0; but since this limit may fail to exist, it is desirable to have related expressions which may serve us where there is no derivative. These expressions are obtained by using the upper and lower one-sided limits (see 6.10) of the difference-quotient, and are called the "Dini derivates." Their definitions are as follows.

31.1. *Let $f(x)$ be defined and finite on an interval $[\alpha, \beta]$, and let x_0 be a point in $[\alpha, \beta]$. Then*

(*a*) *the upper derivate of $f(x)$ at x_0 is*

$$\bar{D}f(x_0) = \limsup_{x \to x_0} \frac{f(x) - f(x_0)}{x - x_0};$$

(*b*) *the lower derivate of $f(x)$ at x_0 is*

$$\underline{D}f(x_0) = \liminf_{x \to x_0} \frac{f(x) - f(x_0)}{x - x_0}.$$

If $x_0 < \beta$, then

(*c*) *the upper right derivate of $f(x)$ at x_0 is*

$$D^+f(x_0) = \limsup_{x \to x_0+} \frac{f(x) - f(x_0)}{x - x_0};$$

(*d*) *the lower right derivate of $f(x)$ at x_0 is*

$$D_+f(x_0) = \liminf_{x \to x_0+} \frac{f(x) - f(x_0)}{x - x_0}.$$

If $\alpha < x_0$, then

(*e*) *the upper left derivate of $f(x)$ at x_0 is*

$$D^-f(x_0) = \limsup_{x \to x_0-} \frac{f(x) - f(x_0)}{x - x_0};$$

(*f*) *the lower left derivate of* $f(x)$ *at* x_0 *is*

$$D_-f(x_0) = \liminf_{x \to x_0 -} \frac{f(x) - f(x_0)}{x - x_0}.$$

EXERCISE. If $f(x)$ is finite in $[a, b]$ and x_0 is in $[a, b]$, then $\bar{D}f(x_0)$ is the greatest number which is the limit of a sequence of difference-quotients $[f(\beta_n) - f(\alpha_n)]/(\beta_n - \alpha_n)$ where α_n and β_n tend to x_0 subject to the condition $\alpha_n < \beta_n$, $\alpha_n \leqq x_0 \leqq \beta_r$. An analogous statement holds for the lower derivate.

31.2. *Let* $f(x)$ *be defined and finite on the interval* $[\alpha, \beta]$, *and let* x_0 *be a point in* $[a, b]$. *Then*

(*a*) *if* $x_0 < \beta$, *and* $D^+f(x_0) = D_+f(x_0)$, *we call their common value the right derivative of* $f(x)$ *at* x_0, *and denote it by* $f'(x_0+)$;

(*b*) *if* $\alpha < x_0$, *and* $D^-f(x_0) = D_-f(x_0)$, *we call their common value the left derivative of* $f(x)$ *at* x_0, *and denote it by* $f'(x_0-)$;

(*c*) *if* $\bar{D}f(x_0) = \underline{D}f(x_0)$, *we call their common value the derivative of* $f(x)$ *at* x_0, *and denote it by* $Df(x_0)$, *or* $f'(x_0)$, *or* $\frac{d}{dx}f(x)|_{x=x_0}$.

In the rest of this section we shall investigate the simpler properties of these derivates.

From the definition it is obvious that

$$\bar{D}f(\alpha) = D^+f(\alpha) \qquad \text{and} \qquad \underline{D}f(\alpha) = D_+f(\alpha).$$

For at $x = \alpha$ it makes no difference whether we write $x \to \alpha$ or $x \to \alpha+$ under the symbol lim sup or lim inf; the condition $x \geqq \alpha$ is forcibly satisfied in either case. Likewise

$$\bar{D}f(\beta) = D^-f(\beta) \qquad \text{and} \qquad \underline{D}f(\beta) = D_-f(\beta).$$

Slightly less trivial is

31.3. *If* $\alpha < x_0 < \beta$, *then* $\bar{D}f(x_0) = \sup \{D^+f(x_0), D^-f(x_0)\}$ *and* $\underline{D}f(x_0) = \inf \{D_+f(x_0), D_-f(x_0)\}$.

By 6.4, there is a sequence of numbers $x_n > x_0$ tending to x_0 such that

$$\lim_{n \to \infty} \frac{f(x_n) - f(x_0)}{x_n - x_0} = D^+f(x_0).$$

But by the second part of 6.4, this proves that

$$\bar{D}f(x_0) \geqq D^+f(x_0).$$

Likewise, $\bar{D}f(x_0) \geqq D^-f(x_0)$, so we have

$$(A) \qquad \bar{D}f(x_0) \geqq \sup \{D^+f(x_0), D^-f(x_0)\}.$$

On the other hand, by 6.4 there is a sequence x_n of numbers different from x_0 and tending to x_0 for which

$$\lim_{n\to\infty} \frac{f(x_n) - f(x_0)}{x_n - x_0} = \bar{D}f(x_0).$$

Either there are infinitely many $x_n > x_0$, or there are infinitely many $x_n < x_0$. In the first case, we select the $x_n > x_0$ and denote them by x_n'. Then $x_n' > x_0$, and

$$\lim_{n\to\infty} \frac{f(x_n') - f(x_0)}{x_n' - x_0} = \bar{D}f(x_0),$$

By the second part of 6.4, this yields

$$D^+f(x_0) \geqq \bar{D}f(x_0),$$

hence

$$(B) \qquad \sup\{D^+f(x_0), D^-f(x_0)\} \geqq \bar{D}f(x_0).$$

In the second case, we prove similarly that $D^-f(x_0) \geqq \bar{D}f(x_0)$, so that (B) holds. From (A) and (B) we obtain the first part of our conclusion. The second part is established similarly, or can be obtained from the first after theorem 31.7 is proved.

The derivates being upper and lower limits, all the theorems of §6 are immediately available for use. If in any of the theorems of §6 we replace the pair of symbols lim sup, lim inf respectively by $\bar{D}$, $\underline{D}$, or by D^+, D_+, or by D^-, D_-, we obtain a theorem on derivates. Particularly useful are the consequences of 6.12 and 6.13:

31.4. *Let $f_1(x)$ and $f_2(x)$ be defined and finite on the interval $[\alpha, \beta]$. Then for each x in $[\alpha, \beta]$ the inequalities*

$$\begin{array}{ll}
(\alpha) & \bar{D}[f_1(x) + f_2(x)] \leqq \bar{D}f_1(x) + \bar{D}f_2(x), \\
(\beta) & \underline{D}[f_1(x) + f_2(x)] \geqq \underline{D}f_1(x) + \underline{D}f_2(x), \\
(\gamma) & \bar{D}[f_1(x) + f_2(x)] \geqq \bar{D}f_1(x) + \underline{D}f_2(x), \\
(\delta) & \underline{D}[f_1(x) + f_2(x)] \leqq \underline{D}f_1(x) + \bar{D}f_2(x), \\
(\epsilon) & \bar{D}[f_1(x) - f_2(x)] \leqq \bar{D}f_1(x) - \underline{D}f_2(x), \\
(\zeta) & \bar{D}[f_1(x) - f_2(x)] \geqq \bar{D}f_1(x) - \bar{D}f_2(x) \geqq \underline{D}[f_1(x) - f_2(x)], \\
(\eta) & \underline{D}[f_1(x) - f_2(x)] \geqq \underline{D}f_1(x) - \bar{D}f_2(x), \\
(\vartheta) & \bar{D}[f_1(x) - f_2(x)] \geqq \underline{D}f_1(x) - \underline{D}f_2(x) \geqq \underline{D}[f_1(x) - f_2(x)]
\end{array}$$

hold, provided that the additions are possible. Moreover, if $x < \beta$, we may replace $\bar{D}$, $\underline{D}$ by D^+, D_+ throughout; if $\alpha < x$, we may replace $\bar{D}$, $\underline{D}$ by D^-, D_- throughout.

Each of these twenty four relationships follows at once from the corresponding part of 6.13.

The next theorem is an immediate corollary of 31.4.

31.5. *If $f(x)$ and $g(x)$ are defined and finite on the interval $[\alpha, \beta]$, and $f(x) - g(x)$ is monotonic increasing, then*

$$\bar{D}f(x) \geqq \bar{D}g(x) \qquad \text{and} \qquad \underline{D}f(x) \geqq \underline{D}g(x)$$

for all x in $[\alpha, \beta]$. If $\alpha < x \leqq \beta$ we can replace $\bar{D}$, $\underline{D}$ by D^-, D_- respectively in these inequalities, and if $\alpha \leqq x < \beta$ we can replace $\bar{D}$, $\underline{D}$ by D^+, D_+ respectively.

Since $f(x) - g(x)$ is monotonic increasing, its difference-quotient is non-negative, so its lower derivate is non-negative. By 31.4γ,

$$\begin{aligned}\bar{D}f(x) &= \bar{D}[g(x) + (f(x) - g(x))]\\ &\geqq \bar{D}g(x) + \underline{D}(f(x) - g(x))\\ &\geqq \bar{D}g(x).\end{aligned}$$

The second inequality follows by a similar argument, with use of 31.4β. By 31.4, the statements about D^+, etc., are provable in exactly the same way.

31.6. *If $f(x)$ and $g(x)$ are defined and finite on the interval $[\alpha, \beta]$, then:*

(a) if $g(x)$ has a right derivative at x,

$$\begin{aligned}D^+[f(x) + g(x)] &= D^+f(x) + g'(x+),\\ D_+[f(x) + g(x)] &= D_+f(x) + g'(x+);\end{aligned}$$

(b) if $g(x)$ has a left derivative at x,

$$\begin{aligned}D^-[f(x) + g(x)] &= D^-f(x) + g'(x-),\\ D_-[f(x) + g(x)] &= D_-f(x) + g'(x-);\end{aligned}$$

(c) if $g(x)$ has a derivative at x, all four of the preceding equations hold, where $g'(x+) = g'(x-) = g'(x)$.

All these conclusions follow from 6.14.

Given a function $f(x)$ defined on $[\alpha, \beta]$, the function $f(-x)$ is defined on $[-\beta, -\alpha]$; that is, if we write $\xi = -x$, $g(\xi) \equiv f(x)$, then $g(\xi)$ is defined for $-\beta \leqq \xi \leqq -\alpha$. We then have

31.7. *Let $f(x)$ be defined and finite on $[\alpha, \beta]$. Between the derivates of $f(x)$, $-f(x)$, $f(-x)$ and $-f(-x)$ the following relationships hold:*

$$\begin{array}{lll}(a) & D^+(-f(x)) = -D_+f(x), & D_+(-f(x)) = -D^+f(x);\\ (b) & D^+(f(-x)) = -D_-f(x), & D_+(f(-x)) = -D^-f(x);\\ (c) & D^+(-f(-x)) = D^-f(x), & D_+(-f(-x)) = D_-f(x).\end{array}$$

In (a), (b), (c) we may everywhere interchange the affixes + and − on the letter D.

Also,

$$(d)\qquad \bar{D}(-f(x)) = \bar{D}(f(-x)) = -\underline{D}f(x),$$
$$\underline{D}(-f(x)) = \underline{D}(f(-x)) = -\bar{D}f(x),$$
$$(e)\qquad \bar{D}(-f(-x)) = \bar{D}f(x),\ \underline{D}(-f(-x)) = \underline{D}f(x).$$

Here we understand, for example, that $D^+(f(-x))$ denotes the upper right derivate of $g(\xi) \equiv f(-\xi)$ with respect to ξ at the place $\xi = -x$.

To prove (a), we have by 6.3

$$D^+(-f(x)) = \limsup_{x\to x_0+} \frac{-f(x) - (-f(x_0))}{x - x_0} = -\liminf_{x\to x_0+} \frac{f(x) - f(x_0)}{x - x_0} = -D_+f(x).$$

This gives the first part of (a). Replacing f by $-f$ gives the second. The analogues for the upper left and lower left derivates are similarly obtained.

For (b), we use $g(\xi) \equiv f(x)$, where $x = -\xi$, and note that $\xi > \xi_0$ if and only if $x < x_0 = -\xi_0$. Then

$$D^+(f(-x_0)) \equiv D^+g(\xi_0) = \limsup_{\xi\to\xi_0+} \frac{g(\xi) - g(\xi_0)}{\xi - \xi_0}$$
$$= \limsup_{x\to x_0-} \left[-\frac{f(x) - f(x_0)}{x - x_0} \right] = -\liminf_{x\to x_0-} \frac{f(x) - f(x_0)}{x - x_0} = -D_-f(x).$$

The second part of (b) can be similarly established; or it can be derived from the first and (a) by replacing f by $-f$. If in (b) we replace x by $-x$ we obtain the formulas with the affixes $+$ and $-$ interchanged.

Formulas (c) follow from (a) and (b), for

$$D^+(-f(-x)) = -D_+(f(-x)) = D^-f(x),$$
$$D_+(-f(-x)) = -D^+(f(-x)) = D_-f(x).$$

The interchange of $f(x)$ and $-f(-x)$ gives the formulas with the affixes $+$ and $-$ interchanged.

Formulas (d) and (e) follow readily from (a), (b) and (c). For example, by (a) and (5.5)

$$\bar{D}(-f(x)) = \sup\,\{D^+(-f(x)),\ D^-(-f(x))\}$$
$$= \sup\,\{-D_+f(x),\ -D_-f(x)\} = -\inf\,\{D_+f(x),\ D_-f(x)\}$$
$$= -\underline{D}f(x).$$

It is not our intention to make a detailed study of the differential calculus. But there are a few simple theorems which we shall need in the study of integrals, and these we shall now establish.

31.8. *Let $f(x)$ be defined and finite on $[a, b]$, and let x_0 be a point of $[a, b]$. If $\bar{D}f(x_0)$ and $\underline{D}f(x_0)$ are both finite, then $f(x)$ is continuous at x_0.*

Let $M - 1$ be the greater of $|\bar{D}f(x_0)|$ and $|\underline{D}f(x_0)|$. By the definition of these derivates, there is a $\delta > 0$ such that

$$-M \leqq \frac{f(x) - f(x_0)}{x - x_0} \leqq M$$

if $x \neq x_0$ is in $[a, b]$ and in $N_\delta(x_0)$. Therefore

$$|f(x) - f(x_0)| \leqq M |x - x_0|$$

for such x. If ϵ is positive, and γ is the smaller of δ and ϵ/M, then

$$|f(x) - f(x_0)| < \epsilon$$

if x is in $[a, b]$ and $|x - x_0| < \gamma$. This completes the proof.

31.9. *Let $f(x)$ and $g(x)$ be defined and finite on $[a, b]$, and let x_0 be a point of $[a, b]$. If the derivatives $f'(x_0)$ and $g'(x_0)$ both exist and are finite, then the derivative $\frac{d}{dx}(f(x)g(x))$ also exists at x_0, and*

$$\frac{d}{dx}[f(x)g(x)]\bigg|_{x=x_0} = f'(x_0)g(x_0) + f(x_0)g'(x_0).$$

By 31.8, both $f(x)$ and $g(x)$ are continuous at $x = x_0$.

We now write

$$\frac{f(x)g(x) - f(x_0)g(x_0)}{x - x_0} = \frac{f(x) - f(x_0)}{x - x_0}\, g(x) + f(x_0)\,\frac{g(x) - g(x_0)}{x - x_0}.$$

As $x \to x_0$, the factors in the first term approach $f'(x_0)$ and $g(x_0)$ respectively, those in the second term approach $f(x_0)$ and $g'(x_0)$ respectively. Hence

$$\lim_{x \to x_0} \frac{f(x)g(x) - f(x_0)g(x_0)}{x - x_0} = f'(x_0)g(x_0) + f(x_0)g'(x_0),$$

establishing the theorem.

31.10. *Let $f(y)$ be defined and finite on the interval $\alpha \leqq y \leqq \beta$, and let y_0 be in $[\alpha, \beta]$. Let $g(x)$ be defined on the interval $a \leqq x \leqq b$ and have its values in the interval $[\alpha, \beta]$, and let x_0 be in $[a, b]$. If $g(x_0) = y_0$, and the derivatives $f'(y_0)$, $g'(x_0)$ exist and are finite, then the function $f(g(x))$ has a derivative at x_0, and*

$$Df(g(x_0)) = f'(g(x_0))g'(x_0).$$

For all $y \neq y_0$ in $[\alpha, \beta]$ we define

$$(A) \qquad \mu(y) = \frac{f(y) - f(y_0)}{y - y_0} - f'(y_0),$$

and we set $\mu(y_0) = 0$. By the definition of $f'(y_0)$ we have

$$\lim_{y \to y_0} \mu(y) = 0.$$

Let ϵ be an arbitrary positive number. There is a $\gamma > 0$ such that $|\mu(y)| < \epsilon$ if $|y - y_0| < \gamma$ and y is in $[\alpha, \beta]$. But $g(x)$ is continuous at x_0, by 31.8. Hence there is a $\delta > 0$ such that $|g(x) - g(x_0)| < \gamma$ if x is in $[a, b]$ and $|x - x_0| < \delta$. That is, if x is in $N_\delta(x_0) \cap [a, b] - (x_0)$, then $\mu(g(x))$ is in $N_\epsilon(0)$; so

$$\lim_{x \to x_0} \mu(g(x)) = 0.$$

Now by (A) we can write

$$\frac{f(g(x)) - f(g(x_0))}{x - x_0} = \frac{[f'(g(x_0)) + \mu(g(x))][g(x) - g(x_0)]}{x - x_0}.$$

As x tends to x_0, the factor $f'(g(x_0)) + \mu(g(x))$ tends to $f'(g(x_0))$, the other factor on the right tends to $g'(x_0)$. This establishes our theorem.

32. Now we investigate some properties of derivates which are not so purely formal as those in the preceding section.

32.1. *Let $f(x)$ be defined and finite on the interval $[a, b]$. If $f(x)$ is monotonic increasing, or monotonic decreasing, or continuous, then D^+f, D^-f, D_+f, D_-f, $\bar{D}f$, $\underline{D}f$ are all measurable on $[a, b]$.*

If we can prove that the first four of these are measurable, then by 21.7 the last two also are measurable, for by 31.3 we know that $\bar{D}f = \sup\{D^+f, D^-f\}$ and $\underline{D}f = \inf\{D_+f, D_-f\}$. If $f(x)$ satisfies our hypotheses, so does $-f(x)$; so by 31.7 and 21.6 it is enough to prove $D^+f(x)$ and $D^-(f(x))$ measurable. We discuss only the first of these; the proof of the measurability of the upper left derivate is the same except for trivial alterations. If we define

$$\varphi(x, \alpha) = \sup \frac{f(x + h) - f(x)}{h},$$

$$\psi(x, \alpha) = \sup \frac{f(x + \rho) - f(x)}{\rho}.$$

where h ranges over all *real* numbers such that $0 < h < \alpha$ and ρ over all *rational* numbers such that $0 < \rho < \alpha$, then by 31.1 and 6.9

we have $D^+f(x) = \lim_{\alpha\to 0} \varphi(x, \alpha)$. We now show that it is also true that $D^+f = \lim_{\alpha\to 0} \psi(x, \alpha)$; in fact, that $\psi(x, \alpha) = \varphi(x, \alpha)$. First, for every α it is obvious (5.3) that $\psi(x, \alpha) \leqq \varphi(x, \alpha)$. Second, let μ be any number less than $\varphi(x, \alpha)$. Then for some positive $h < \alpha$ we have

$$\frac{f(x + h) - f(x)}{h} > \mu.$$

If $f(x)$ is monotonic increasing, we let the rational number ρ $(0 < \rho < \alpha)$ approach h from the right. Then $f(x + \rho) - f(x) \geqq f(x + h) - f(x)$, so that $[f(x + \rho) - f(x)]/\rho \geqq [f(x + h) - f(x)]/\rho$, and if ρ is near enough to h this last quotient is greater than μ. If $f(x)$ is monotonic decreasing we let ρ tend to h from the left, and obtain the same inequality. If $f(x)$ is continuous, then $\lim f(x + \rho) = f(x + h)$ and $\lim \rho = h$ as $\rho \to h$ from either side; so for ρ near enough to h the same inequality holds. In any case, then, we have

$$\psi(x, \alpha) = \sup \left\{\frac{f(x + \rho) - f(x)}{\rho}\right\} > \mu.$$

This holds for all $\mu < \varphi(x, \alpha)$, so by 5.2 $\psi(x, \alpha) \geqq \varphi(x, \alpha)$. As we already know that $\psi \leqq \varphi$, the two are equal. So

$$D^+f = \lim_{\alpha\to 0} \psi(x, \alpha).$$

For each fixed ρ, the quotient $[f(x + \rho) - f(x)]/\rho$ is measurable by 21.12 and 21.6 and the third remark after 21.1. For each fixed α, $\psi(x, \alpha)$ is the sup of a denumerable set of measurable functions, so is measurable by 21.7. Now let α take on successively values $\alpha_1, \alpha_2, \cdots$ tending to 0. Then D^+f is the limit of the sequence of measurable functions $\psi(x, \alpha_i)$, and is measurable by 21.9.

32.2. *If $\varphi(x)$ is defined and continuous on the interval $[a, b]$, and k is a constant, and E is a set of positive measure* contained in $[a, b]$ and such that $D^+\varphi(x) \geqq k$ for all x in E (or $D^-\varphi(x) \geqq k$ for all x in E), then for every $\epsilon > 0$ it is possible to find a finite set of disjoint intervals $(\alpha_1, \beta_1), \cdots, (\alpha_p, \beta_p)$ in $[a, b]$ such that $\Sigma(\beta_i - \alpha_i) > mE - \epsilon$ and*

$$\varphi(\beta_i) - \varphi(\alpha_i) \geqq (k - \epsilon)(\beta_i - \alpha_i), \qquad i = 1, \cdots, p.$$

We consider the case $D^+\varphi(x) \geqq k$ for all x in E; the changes needed to treat the case $D^-\varphi(x) \geqq k$ are obvious.

* It would be enough to assume $m_eE > 0$.

For each positive integer n we define E_n to be the set of all x in $[a, b]$ such that $[\varphi(x + h) - \varphi(x)]/h \geqq k - \epsilon$ for some $h \geqq 1/n$ (wherein $x + h$ must also be in $[a, b]$). We establish these properties of the sets E_n.

(1) *The sets* E_n *expand:* $E_1 \subset E_2 \subset E_3 \subset \cdots$. This is clear from the definition.

(2) *The set* E *is contained in* $\bigcup E_n$. For let x be a point of E. Then there exists an $h > 0$ such that $[\varphi(x + h) - \varphi(x)]/h \geqq D^+\varphi(x) - \epsilon \geqq k - \epsilon$; and if $1/n$ is less than this h, we see that x is in E_n.

(3) *The sets* E_n *are closed.* For let x_0 be the limit of a sequence $\{x_i\}$ of points of E_n. To each x_i corresponds an $h_i \geqq 1/n$ such that $[\varphi(x_i + h_i) - \varphi(x_i)]/h_i \geqq k - \epsilon$. By 2.12 we can select a subsequence of the h_i tending to a limit h_0; to save notation, we suppose that h_i is already such a sequence, $h_i \to h_0$. Since $h_i \geqq 1/n$, it is clear that $h_0 \geqq 1/n$. The points x_i and $x_i + h_i$ are in $[a, b]$, hence so are their limits x_0, $x_0 + h_0$. Then by the continuity of φ, we have $[\varphi(x_0 + h_0) - \varphi(x_0)]/h_0 = \lim_{i \to \infty} [\varphi(x_i + h_i) - \varphi(x_i)]/h_i \geqq k - \epsilon$, and x_0 is in E_n.

From (1), (2) and (3) we learn by 19.4, 19.8 and 19.3 that $\lim mE_n = m \bigcup E_n \geqq mE$; so for some n we have $mE_n > mE - \epsilon$. We now choose and fix such a value of n. Now let α_1 be the first point x in E_n; that there is such a point is certain, since E_n is closed. For some $h_1 \geqq 1/n$ we have $[\varphi(\alpha_1 + h_1) - \varphi(\alpha_1)]/h_1 \geqq k - \epsilon$; we choose such an h_1 and set $\beta_1 = \alpha_1 + h_1$. We now define α_2 to be the first point x of E_n for which $x \geqq \beta_1$, choose an $h_2 \geqq 1/n$ for which $[\varphi(\alpha_2 + h_2) - \varphi(\alpha_2)]/h_2 \geqq k - \epsilon$, and set $\beta_2 = \alpha_2 + h_2$; and we continue in this manner. Since each interval (α_i, β_i) has length $h_i \geqq 1/n$, this process must terminate; there will be a last interval (α_p, β_p). The intervals (α_i, β_i) will then enclose all of E_n, so that

$$\Sigma(\beta_i - \alpha_i) \geqq mE_n > mE - \epsilon.$$

And by the choice of the β_i, for each i we have

$$\varphi(\beta_i) - \varphi(\alpha_i) \geqq (k - \epsilon)(\beta_i - \alpha_i),$$

which completes the proof.

32.3. Corollary. *Let* $\varphi(x)$ *be continuous and monotonic increasing on the interval* $[a, b]$, *and let* E_k *be the set on which* $\bar{D}\varphi(x) \geqq k$. *Then* E_k *is measurable, and* $\varphi(b) - \varphi(a) \geqq \frac{1}{2}kmE_k$.

The derivative $\bar{D}\varphi(x)$ is measurable by 32.1, so the set E_k is measurable by definition 21.1. If $k \leqq 0$, the inequality is trivial; so we suppose $k > 0$. The set E_k is the sum of the (measurable) sets E_k^+ and E_k^- on which the respective inequalities $D^+\varphi(x) \geqq k$ and $D^-\varphi(x) \geqq k$ hold, by 31.3; so one of these (say D_k^+) has measure $\geqq \frac{1}{2}mE_k$. Let ϵ be a positive number less than k. By 32.2 there are intervals $(\alpha_1, \beta_1), \cdots,$ (α_p, β_p) such that $a \leqq \alpha_1 < \beta_1 \leqq \alpha_2 < \cdots \leqq \alpha_p < \beta_p \leqq b$ and

$$\Sigma(\beta_i - \alpha_i) > mE_k^+ - \epsilon \qquad \text{and} \qquad \varphi(\beta_i) - \varphi(\alpha_i) \geqq (k - \epsilon)(\beta_i - \alpha_i).$$

Therefore

$$\Sigma[\varphi(\beta_i) - \varphi(\alpha_i)] \geqq (k - \epsilon)\Sigma(\beta_i - \alpha_i) > (k - \epsilon)(mE_k^+ - \epsilon).$$

By the monotoneity of φ we have

$$\begin{aligned} \varphi(\alpha_1) - \varphi(a) &\geqq 0, \\ \varphi(\alpha_{i+1}) - \varphi(\beta_i) &\geqq 0, \qquad i = 1, \cdots, p - 1, \\ \varphi(b) - \varphi(\beta_p) &\geqq 0. \end{aligned}$$

Adding these inequalities,

$$\varphi(b) - \varphi(a) > (k - \epsilon)(mE_k^+ - \epsilon).$$

Since ϵ is arbitrary, this establishes the inequalities

$$\varphi(b) - \varphi(a) \geqq kmE_k^+ \geqq (\tfrac{1}{2}) \cdot kmE_k.$$

REMARK. It would not be difficult to improve the conclusion by establishing the inequality

$$\varphi(b) - \varphi(a) \geqq kmE_k.$$

But 32.3 is adequate for our needs.

33. We are now ready to study the derivatives of indefinite integrals. The first two theorems are merely lemmas, the principal theorem being 33.3.

33.1. *If $u(x)$ is a summable U-function on the interval $I = [a, b]$, and*

$$U(x) = \int_a^x u(x)\,dx \qquad (a \leqq x \leqq b),$$

then $\underline{D}U(x) \geqq u(x)$ for all x in I. Likewise, if $l(x)$ is a summable L-function on I, and

$$L(x) = \int_a^x l(x)\,dx \qquad (a \leqq x \leqq b),$$

then $\bar{D}L(x) \leqq l(x)$ for all x in I.

We prove the statement about U-functions; the statement about L-functions then follows by a change of signs.

Let x_0 be any number in $[a, b]$, and let μ be any number less than $u(x_0)$. By 7.4 the set $I[u > \mu]$ is open relative to I. Since x_0 is contained in this set, so is a neighborhood of x_0 relative to I; that is, there is a positive number δ such that if x is in I and $|x - x_0| < \delta$ then $u(x) > \mu$. If x satisfies these conditions and exceeds x_0, then by 18.2

$$\frac{U(x) - U(x_0)}{x - x_0} = \frac{1}{x - x_0}\int_{x_0}^{x} u(x)\,dx \geqq \frac{1}{x - x_0}\int_{x_0}^{x} \mu\,dx = \mu.$$

If x satisfies the conditions but is less than x_0, we can establish the inequality

$$\frac{U(x) - U(x_0)}{x - x_0} \geqq \mu$$

in the same way by interchanging x and x_0. Hence this last inequality holds whenever x is in $N_\delta(x_0) \cap I - (x_0)$. By 6.1,

$$\underline{D}U(x_0) = \liminf_{x \to x_0} \frac{U(x) - U(x_0)}{x - x_0} \geqq \mu;$$

and since this holds for all $\mu < u(x_0)$ we have by 5.2

$$\underline{D}U(x_0) \geqq u(x_0).$$

The next theorem is a corollary of 33.1.

33.2. *If $\varphi(x)$ is continuous on $[a, b]$ and*

$$\Phi(x) = \int_a^x \varphi(x)\,dx,$$

then for every x on $[a, b]$ the derivative $\Phi'(x)$ exists and is equal to $\varphi(x)$.

For $\varphi(x)$ is an L-function and is also a U-function, so by 33.1

$$\bar{D}\Phi(x) \leqq \varphi(x) \leqq \underline{D}\Phi(x) \leqq \bar{D}\Phi(x),$$

and therefore the upper and lower derivates of $\Phi(x)$ are both equal to $\varphi(x)$.

We are now ready to prove our principal theorem.

33.3. *Let $f(x)$ be summable on the interval $[a, b]$, and let*

$$F(x) = \int_a^x f(x)\,dx.$$

Then for almost all x in $[a, b]$ the function $F(x)$ has a derivative which is finite and equal to $f(x)$.

Let k and ϵ be arbitrary positive numbers. Since $f(x)$ is summable, there is a U-function $u(x) \geqq f(x)$ such that

$$(A) \qquad \int_a^b u(x)\,dx < \int_a^b f(x)\,dx + k\frac{\epsilon}{2}.$$

Define

$$U(x) = \int_a^x u(x)\,dx.$$

The difference $U(x) - F(x)$ is monotonic increasing, since if $a \leqq \alpha < \beta \leqq b$ we have

$$[U(\beta) - F(\beta)] - [U(\alpha) - F(\alpha)] = \int_\alpha^\beta u(x)\,dx - \int_\alpha^\beta f(x)\,dx \geqq 0.$$

Let E_k be the set on which $\bar{D}[U - F] \geqq k$. This set is measurable by 32.1, and by 32.3 and inequality (A)

$$k\frac{\epsilon}{2} > [U(b) - F(b)] - [U(a) - F(a)] \geqq k\frac{mE_k}{2}.$$

Hence $mE_k < \epsilon$. Except on E_k we have by 31.4η and 33.1

$$\begin{aligned} \underline{D}F(x) &\geqq \underline{D}U(x) - \bar{D}[U(x) - F(x)] \\ &> u(x) - k \\ &\geqq f(x) - k. \end{aligned}$$

Therefore the measure of the set on which the inequality $\underline{D}F(x) > f(x) - k$ fails is less than ϵ, and this set does not depend on ϵ. Its measure (being non-negative) is therefore zero, by 5.2.

In particular, if k is the reciprocal of a positive integer, the set on which the inequality

$$(B) \qquad \underline{D}F(x) > f(x) - \frac{1}{n}$$

fails is a set N_n of measure 0. The sum $\bigcup N_n$ also has measure zero by 19.13, and on the rest of $[a, b]$ inequality (B) holds for every positive integer n. By 5.2, except on $\bigcup N_n$ we must have

$$(C) \qquad \underline{D}F(x) \geqq f(x).$$

The function $-f(x)$ is summable by 18.3, and its integral from a to x is $-F(x)$. By the above proof, the inequality

$$\underline{D}[-F(x)] \geqq -f(x)$$

holds except on a set of measure zero. By 31.7, this implies that

$$(D) \qquad \bar{D}F(x) \leqq f(x)$$

for almost all x in $[a, b]$. Since by 6.5 we have $\underline{D}F(x) \leqq \bar{D}F(x)$, inequalities (C) and (D) imply that except on a set N of measure zero we have

$$\bar{D}F(x) = \underline{D}F(x) = f(x);$$

that is, $F'(x)$ exists and is equal to $f(x)$ except on N. The set M on which $f(x)$ is $\pm\infty$ is of measure zero, by 23.1. So except on the set $M \cup N$ of measure zero the derivative $F'(x)$ exists and is equal to the finite number $f(x)$, which completes the proof.

A corollary to this theorem is

33.4. *If $f(x)$ and $g(x)$ are both summable on $[a, b]$, and*

$$\int_a^x f(x)\,dx = \int_a^x g(x)\,dx$$

for all x in $[a, b]$, then $f(x)$ and $g(x)$ differ only on a set of measure 0.

For let $F(x)$ be the common value of the two integrals. Then $F'(x)$ exists almost everywhere. If we consider $\int f(x)\,dx$, we see that $F'(x) = f(x)$ for almost all x; and if we consider $\int g(x)\,dx$, we see that $F'(x) = g(x)$ for almost all x. Hence $f(x) = g(x)$ for almost all x.

34. We have just shown that the derivative of an indefinite integra- is almost everywhere equal to the integrand. The converse question suggests itself. When is a function equal to the integral of its derival tive? Before investigating this, it is reasonable to look for conditions under which we can be sure that a function *has* almost everywhere a derivative which it is possible to integrate; we shall find that all functions of bounded variation (among others) have this property. Moreover, there is a condition which a function clearly must satisfy before it can be the integral of anything—it must be absolutely continuous, for by 27.7 all indefinite integrals are absolutely continuous. We shall see this condition is also sufficient.

In preparation, we establish a lemma.

34.1. *If $f(x)$ is defined and finite on $[a, b]$, and*

(*a*) $\limsup_{x \to x_0-} f(x) \leqq f(x_0) \qquad (a < x_0 \leqq b),$

(*b*) $\liminf_{x \to x_0+} f(x) \geqq f(x_0) \qquad (a \leqq x_0 < b),$

and one of the two following conditions is satisfied:

(*c*) $D^+f(x) \geqq 0$ *for all points x in $[a, b]$ except at most those of a denumerable set E_0,*

(c′) *$D^-f(x) \geqq 0$ for all points x in $[a, b]$ except at most those of a denumerable set E_0,*
then $f(x)$ is monotonic increasing on $[a, b]$.

A remark about hypotheses (c) and (c′) may not be amiss. We assume it known that $f(x)$ has a non-negative upper right (or upper left) derivate on $[a, b] - E_0$. Concerning derivates at the points of E_0 we do not require any information to begin with. From the conclusion of the lemma we can deduce that all derivates of $f(x)$ are non-negative at all points of $[a, b]$, including those of E_0.

First let us suppose that $D^+f(x)$ is positive at all points of $[a, b] - E_0$. If the conclusion of the lemma is false, there are numbers α, β such that $a \leqq \alpha < \beta \leqq b$ and $f(\alpha) > f(\beta)$. The values of $f(x)$ corresponding to points x in E_0 form at most a denumerable set, so they cannot fill the interval $(f(\beta), f(\alpha))$. Hence there is a number μ such that

$$f(\beta) < \mu < f(\alpha),$$

and *for all x in E_0*

$$\mu \neq f(x).$$

There are points x in $[\alpha, \beta]$ such that $f(x) \geqq \mu$; for instance, α itself is such a point. Let ξ be the sup of all such x. We shall now prove that $f(\xi) = \mu$. First suppose $f(\xi) > \mu$. Then $\xi \neq \beta$, for $f(\beta) < \mu$. On the interval $(\xi, \beta]$ we have $f(x) < \mu$, by definition of ξ. Hence from the definitions 6.10 and 6.1 we have

$$\liminf_{x \to \xi+} f(x) \leqq \mu < f(\xi),$$

contrary to hypothesis (b). Second, suppose $f(\xi) < \mu$. Then $\xi \neq \alpha$, for $f(\alpha) > \mu$. By the definition of ξ, for every positive δ there is an x which is in $[\alpha, \beta]$ and in $(\xi - \delta, \xi]$ such that $f(x) \geqq \mu$. This x cannot be ξ itself, since $f(\xi) < \mu$; hence it is in $[\alpha, \xi) \cap N_\delta(\xi)$. Thus the sup of $f(x)$ on $N_\delta(\xi) \cap [\alpha, \xi)$ is at least μ, and by definition 6.1 we have

$$\limsup_{x \to \xi-} f(x) \geqq \mu > f(\xi).$$

This contradicts hypothesis (a), so $f(\xi)$ must be equal to μ.

But now

$$\frac{f(x) - f(\xi)}{x - \xi} < 0$$

for all x such that $\xi < x \leqq \beta$, so

$$D^+f(\xi) \leqq 0.$$

Since $f(\xi) = \mu$, while $f(x) \neq \mu$ for all x in E_0, the point ξ is not in E_0. Hence we have arrived at a contradiction, and the lemma is established provided that $D^+f(x)$ is positive except on E_0.

Suppose next that hypotheses (a), (b) and (c) hold. For each positive number ϵ the function $f(x) + \epsilon x$ satisfies hypotheses (a) and (b) and also has by 31.6 an upper derivate

$$D^+[f(x) + \epsilon x] = D^+f(x) + \epsilon \geqq \epsilon > 0$$

except perhaps on E_0. Hence $f(x) + \epsilon x$ is monotonic increasing by the proof just completed. That is, if $a \leqq \alpha < \beta \leqq b$ then

$$f(\beta) - f(\alpha) \geqq -\epsilon(\beta - \alpha).$$

Since ϵ is an arbitrary positive number, the right number of this inequality is an arbitrary negative number, and by 5.2 the left number is not negative. That is, $f(x)$ is monotonic increasing on $[a, b]$.

To prove the lemma under hypotheses (a), (b) and (c'), we define $g(x) = -f(-x)$, $-b \leqq x \leqq -a$, and let E_1 be the denumerable set consisting of all x such that $-x$ is in E_0. Then $g(x)$ satisfies (a) and (b), and if x is not in E_1 then $D^+g(x) = D^-f(x) \geqq 0$. That is, $g(x)$ satisfies (a), (b) and (c), and by the preceding proof is monotonic increasing. It follows at once that $f(x) = -g(-x)$ is also monotonic increasing.

With the help of this lemma, we can prove

34.2. *Let $f(x)$ be monotonic increasing and finite for $a \leqq x \leqq b$. Then for almost all x the function $f(x)$ has a finite derivative $f'(x)$; and if we set $\dot{f}(x) = f'(x)$ where $f'(x)$ is defined and finite and $\dot{f}(x) = 0$ elsewhere, then $\dot{f}(x)$ is summable and*

$$f(x) = \int_a^x \dot{f}(x)\,dx + h(x),$$

$h(x)$ being a monotonic increasing function with $h'(x) = 0$ almost everywhere.

By 32.1, D^+f is measurable. Hence if we define $g_n(x) = \inf\{n, D^+f\}$, we know by 22.1 that $g_n(x)$ is summable. It is possible to find an L-function $l_n(x) \leqq g_n(x)$ such that

$$\int_a^b g_n(x)\,dx < \int_a^b l_n(x)\,dx + \frac{1}{n}.$$

Now define

$$L_n(x) = \int_a^x l_n(x)\,dx.$$

The function $f(x) - L_n(x)$ then satisfies the hypotheses of 34.1

For $L_n(x)$ is continuous and $f(x)$ is monotonic increasing, so by 6.11 and 6.14

$$\lim_{x \to x_0 -} [f(x) - L_n(x)] \leqq f(x_0) - L_n(x_0)$$

if $a < x_0 \leqq b$, while

$$f(x_0) - L_n(x_0) \leqq \lim_{x \to x_0 +} [f(x) - L_n(x)]$$

if $a \leqq x_0 < b$. And by 31.4ς and 33.1, for each x

$$\begin{aligned} D^+(f(x) - L_n(x)) &\geqq D^+f(x) - D^+L_n(x) \\ &\geqq D^+f(x) - l_n(x) \\ &\geqq D^+f(x) - g_n(x) = D^+f(x) - \inf \{n, D^+f(x)\} \geqq 0. \end{aligned}$$

So from 34.1, $f(x) - L_n(x)$ increases monotonically. Therefore, for all numbers α, β such that $a \leqq \alpha < \beta \leqq b$ we know that $f(\beta) - L_n(\beta) \geqq f(\alpha) - L_n(\alpha)$; that is, $L_n(\beta) - L_n(\alpha) \leqq f(\beta) - f(\alpha)$. Recalling the meaning of $L_n(x)$,

$$\begin{aligned} \int_\alpha^\beta g_n(x)\, dx < \int_\alpha^\beta l_n(x)\, dx + \frac{1}{n} = L_n(\beta) - L_n(\alpha) + \frac{1}{n} \\ \leqq f(\beta) - f(\alpha) + \frac{1}{n}. \end{aligned}$$

The functions $g_n(x)$ form a monotonic increasing sequence of non-negative functions and tend everywhere to D^+f. By the last inequality, the integrals from α to β of the g_n are bounded. Hence by 29.2 D^+f is summable over (α, β), and

$$\begin{aligned} \int_\alpha^\beta D^+f(x)\, dx &= \int_\alpha^\beta \lim_{n \to \infty} g_n(x)\, dx \\ &\leqq \liminf_{n \to \infty} \int_\alpha^\beta g_n(x)\, dx \\ &\leqq \liminf_{n \to \infty} \left[f(\beta) - f(\alpha) + \frac{1}{n} \right] \\ &= f(\beta) - f(\alpha). \end{aligned}$$

In particular, if we take $\alpha = a$, $\beta = b$, we see that D^+f is summable over (a, b). Also, the function

$$g(x) \equiv f(x) - \int_a^x D^+f(x)\, dx$$

is monotonic increasing; for if we find its values at α and at $\beta > \alpha$

we have

$$g(\beta) - g(\alpha) = \left(f(\beta) - \int_a^\beta D^+f(x)\,dx\right) - \left(f(\alpha) - \int_a^\alpha D^+f(x)\,dx\right)$$
$$= f(\beta) - f(\alpha) - \int_\alpha^\beta D^+f(x)\,dx \geqq 0.$$

We now repeat the whole argument, using $g(x)$ instead of $f(x)$ and D^-g instead of D^+f. We find that D^-g is summable over (a, b), and that the function

$$h(x) = g(x) - \int_a^x D^-g(x)\,dx$$

is monotonic increasing.

By 33.3, the last term has almost everywhere a finite derivative equal to its integrand D^-g. For all such x we have by 31.6

$$(A) \qquad D^-h(x) = D^-g(x) - \frac{d}{dx}\int_a^x D^-g(x)\,dx = 0.$$

In the equation for $h(x)$ we substitute the definition of $g(x)$; we thus find

$$(B) \qquad h(x) = f(x) - \int_a^x D^+f(x)\,dx - \int_a^x D^-g(x)\,dx.$$

By 33.3, for almost all x both these integrals have finite derivatives equal to their integrands. At all such x we have by 31.6

$$D^+h(x) = D^+f(x) - D^+f(x) - D^-g(x) = -D^-g(x).$$

But since $h(x)$ and $g(x)$ are both monotonic, $D^+h(x)$ and $D^-g(x)$ are both non-negative, and the last equation is possible only if both are zero. Thus

$$(C) \qquad D^+h(x) = D^-g(x) = 0 \text{ for almost all } x \text{ in } [a, b].$$

A first consequence of (C) is that the integral of $D^-g(x)$ vanishes, and therefore (B) takes the form

$$h(x) = f(x) - \int_a^x D^+f(x)\,dx.$$

Second, $D^+h(x)$ is almost everywhere 0, by (C). But then $0 \leqq D_+h(x) \leqq D^+h(x) = 0$; and by (A) we have $D^-h(x) = 0$ for almost all x in $[a, b]$, and the inequality

$$0 \leqq D_-h(x) \leqq D^-h(x) \leqq 0$$

shows that $D_-h(x)$ is also zero almost everywhere. So almost everywhere the four derivates of $h(x)$ are all 0; that is, $h'(x)$ exists and is zero.

Except on a set E_1 of measure zero, $h(x)$ has a derivative and $h'(x) = 0$. Except on a set E_2 of measure zero the indefinite integral of $D^+f(x)$ has a derivative which is finite and equal to $D^+f(x)$. Hence, from the equation

$$(D) \qquad f(x) = h(x) + \int_a^x D^+f(x)\,dx,$$

we see that except on $E_1 \cup E_2$ the function $f(x)$ has a derivative which is finite and equal to $D^+f(x)$. That is, except on the set $E_1 \cup E_2$ the equations

$$\dot{f}(x) = f'(x) = D^+f(x)$$

hold. Equation (D) can then be written in the form

$$f(x) = h(x) + \int_a^x \dot{f}(x)\,dx,$$

and the proof is complete.

An immediate consequence is

34.3. *Let $f(x)$ be a function of bounded variation on the interval $a \leqq x \leqq b$. Then for almost all x the function $f(x)$ has a unique finite derivative $f'(x)$; and if we set $\dot{f}(x)$ equal to $f'(x)$ where $f'(x)$ is defined and finite and $\dot{f}(x) = 0$ elsewhere, then*

$$f(x) = \int_a^x \dot{f}(x)\,dx + h(x),$$

where $h(x)$ is of bounded variation and $h'(x) = 0$ almost everywhere.

By 8.6, we can represent $f(x)$ in the form $f(x) = P(x) - N(x)$, where $P(x)$ and $N(x)$ are bounded monotonic increasing functions. Applying 34.2, P and N have finite derivatives almost everywhere, and

$$(A) \qquad P(x) = \int_a^x \dot{P}(x)\,dx + h_1(x),$$

$$(B) \qquad N(x) = \int_a^x \dot{N}(x)\,dx + h_2(x),$$

where $h_1(x)$ and $h_2(x)$ are monotonic and $h_1'(x) = h_2'(x) = 0$ almost everywhere. For those x (almost all of them) at which $P'(x)$ and $N'(x)$ exist and are finite we know by 31.6 that $f'(x)$ exists, and that $f'(x) = P'(x) - N'(x)$. Therefore $\dot{f}(x) = \dot{P}(x) - \dot{N}(x)$ almost everywhere. From equations (A) and (B) we obtain

$$\begin{aligned} f(x) = P(x) - N(x) &= \int_a^x (\dot{P}(x) - \dot{N}(x))\,dx + h_1(x) - h_2(x) \\ &= \int_a^x \dot{f}(x)\,dx + h(x), \end{aligned}$$

where $h(x) \equiv h_1(x) - h_2(x)$ has derivative 0 almost everywhere, by 31.6, and is of BV, by the last sentence of §8.

With the help of lemma 34.1 we can establish a rather interesting property of the derivates.

34.4. *Let $f(x)$ be defined and continuous on the interval $[a, b]$.* The sup *of $[f(x_1) - f(x_2)]/(x_1 - x_2)$ for x_1 and x_2 in $[a, b]$ is also the* sup *of each of the functions $D^+f(x)$, $D^-f(x)$, $D_+f(x)$, $D_-f(x)$ for x in $[a, b]$; and likewise for the* inf.

Let M be the sup of $[f(x_1) - f(x_2)]/[x_1 - x_2]$ for x_1, x_2 in $[a, b]$. For each x in $[a, b]$ all four derivates at x are upper or lower limits of the values of this quotient under restrictions on x_1 and x_2; hence at each x each of the derivates is $\leqq M$, and M is *an* upper bound for all four functions $D^+f(x)$, etc.

Let m be the sup of $D_+f(x)$ on $[a, b]$; we have seen that $m \leqq M$. Suppose $m < M$; then there is a finite number k such that $m < k < M$. The function

$$g(x) = kx - f(x)$$

is continuous on $[a, b]$, and its upper right derivate is

$$D^+g(x) = k - D_+f(x) > 0.$$

So by 34.1 $g(x)$ is monotonic increasing, and for all x_1, x_2 in $[a, b]$ with $x_2 > x_1$ we have

$$0 \leqq g(x_2) - g(x_1) = k(x_2 - x_1) - [f(x_2) - f(x_1)],$$

or

$$\frac{f(x_2) - f(x_1)}{x_2 - x_1} \leqq k.$$

This also holds if $x_2 < x_1$; hence $k < M$ is an upper bound for the quotient, contradicting the definition of M.

In the same way, the sup of $D_-f(x)$ is also M. Since $D^+f(x) \geqq D_+f(x)$, it is clear that

$$M \geqq \sup D^+f(x) \geqq \sup D_+f(x) = M;$$

so $D^+f(x)$ also has its sup equal to M. Likewise the sup of $D^-f(x)$ is M.

Replacing $f(x)$ by $-f(x)$ and recalling 31.7 gives the conclusion about the greatest lower bounds.

As a corollary,

34.5. *If $f(x)$ is defined and finite on $[a, b]$, and its four derived numbers are bounded on $[a, b]$, then $f(x)$ satisfies a Lipschitz condition* on $[a, b]$.*

Let M be an upper bound for $|D^+f(x)|$, $|D^-f(x)|$, $|D_+f(x)|$ and $|D_-f(x)|$, for all x on $[a, b]$. Then by 34.4 M is an upper bound and $-M$ a lower bound for $[f(x_1) - f(x_2)]/[x_1 - x_2]$ for x_1 and x_2 in $[a, b]$; that is

$$\left|\frac{f(x_1) - f(x_2)}{x_1 - x_2}\right| \leqq M,$$

and

$$|f(x_1) - f(x_2)| \leqq M|x_1 - x_2|.$$

35. We have so far found functions for which the derivative exists and is summable, but have yet to find functions for which the integral of the derivative is the function itself. If, in 34.3, the function $h(x)$ were a constant, we would have the desired result. So we prove a lemma.

35.1. *If $\varphi(x)$ is absolutely continuous on the interval $a \leqq x \leqq b$, and $D^+\varphi(x) \geqq 0$ for almost all x in $[a, b]$, then $\varphi(x)$ is monotonic increasing.*

Let ϵ be any positive number, and let α and $\beta > \alpha$ be numbers in the interval $[a, b]$. Since φ is absolutely continuous, there is a $\gamma > 0$ (which we may assume less than ϵ) such that whenever (a_1, b_1), (a_2, b_2), $\cdots$, (a_n, b_n) are non-overlapping subintervals of $[a, b]$ with total length less than γ, the inequality $\Sigma|\varphi(b_i) - \varphi(a_i)| < \epsilon$ holds.

Let E_1 be the subset of $[\alpha, \beta]$ on which $D^+\varphi \geqq 0$; since this equation holds for almost all x in $[a, b]$, the measure of E_1 is $\beta - \alpha$.

We now apply 32.2, with k replaced by 0 and ϵ by γ. We can therefore find a set (α_1, β_1), $\cdots$ (α_n, β_n) of disjoint subintervals of $[\alpha, \beta]$ such that

$$\Sigma(\beta_i - \alpha_i) > mE_1 - \gamma = (\beta - \alpha) - \gamma$$

and

$$(A) \qquad \varphi(\beta_i) - \varphi(\alpha_i) \geqq -\gamma(\beta_i - \alpha_i) \geqq -\epsilon(\beta_i - \alpha_i).$$

The remainder of $[\alpha, \beta]$ consists of the intervals $[\alpha, \alpha_1]$, $[\beta_1, \alpha_2]$, $\cdots$, $[\beta_{n-1}, \alpha_n]$, $[\beta_n, \beta]$, and the total length of these must be less than γ. Therefore, by the definition of γ,

$$[\varphi(\alpha_1) - \varphi(\alpha)] + [\varphi(\alpha_2) - \varphi(\beta_1)] + \cdots + [\varphi(\alpha_n) - \varphi(\beta_{n-1})] + [\varphi(\beta) - \varphi(\beta_n)] \geqq -\epsilon.$$

* Cf. §9.

Adding the inequalities (A) to this last, we obtain

$$\varphi(\beta) - \varphi(\alpha) \geqq -\epsilon(1 + \Sigma(\beta_i - \alpha_i)) \geqq -\epsilon(1 + \beta - \alpha).$$

This being true for all $\epsilon > 0$, it follows that $\varphi(\beta) - \varphi(\alpha) \geqq 0$, and so $\varphi(x)$ is monotonic increasing.

As an easy consequence, we find

35.2. *If $h(x)$ is absolutely continuous on the interval $a \leqq x \leqq b$, and $h'(x) = 0$ for almost all x in $[a, b]$, then $h(x)$ is a constant.*

Let α and $\beta > \alpha$ be any numbers in the interval (a, b). The function $h(x)$ satisfies the hypotheses of 35.1, hence $h(\beta) \geqq h(\alpha)$. But the function $-h(x)$ also satisfies the same hypotheses, hence $-h(\beta) \geqq -h(\alpha)$. These inequalities are both possible only if $h(\beta) = h(\alpha)$, so that $h(x)$ is a constant.

This essentially completes the proof of our principal theorem, which is

35.3. *If $f(x)$ is absolutely continuous on the interval $a \leqq x \leqq b$, then the derivative $f'(x)$ exists and is finite for almost all x in $[a, b]$, and (defining $\dot{f}(x)$ as in* 34.2)

$$f(x) = \int_a^x \dot{f}(x)\, dx + f(a).$$

Since $f(x)$ is absolutely continuous, it is of bounded variation; and by 34.3 $f'(x)$ exists and is finite for almost all x in $[a, b]$, and

$$f(x) = \int_a^x \dot{f}(x)\, dx + h(x),$$

where $h(x)$ has derivative 0 almost everywhere. But $f(x)$ is absolutely continuous by hypothesis, and the integral of $\dot{f}(x)$ is absolutely continuous by 27.7; hence their difference $h(x)$ is absolutely continuous (9.2). By 35.2, $h(x)$ can only be a constant. We evaluate this constant by setting $x = a$; we find $f(a) = 0 + h(a)$, so $h(x) = f(a)$, and the theorem is proved.

Summing up the results thus far obtained in this chapter, we have the following. Every function of bounded variation has a derivative almost everywhere, and this derivative (considered on the set where it exists) is summable. Every absolutely continuous function $\varphi(x)$ has a derivative almost everywhere, and $\varphi(x) - \varphi(a)$ is the integral of this derivative. The integral from a to x of any summable function $f(x)$ is an absolutely continuous function $F(x)$, and $F(x)$ has a finite derivative almost everywhere, and this derivative $F'(x)$ is almost everywhere equal to $f(x)$.

36. In the remainder of this chapter we shall derive some of the useful theorems of the calculus—the formula for integration by parts, the mean value theorems and the theorem on integration by substitution.

Let $\varphi(x)$ and $\psi(x)$ be two functions absolutely continuous on the intervals $a \leqq x \leqq b$. Then by 9.2 their product $u(x) = \varphi(x)\psi(x)$ is also absolutely continuous. According to 35.3, for almost all x both derivatives $\varphi'(x)$, $\psi'(x)$ exist and are finite; and by 31.9, for all such x the product $u(x)$ is also differentiable, and $u'(x) = \varphi'(x)\psi(x) + \varphi(x)\psi'(x)$. Consequently, if we define $\dot{u}$, $\dot{\varphi}$, $\dot{\psi}$ as in 34.2, the equation

$$\dot{u} = \dot{\varphi}\psi + \varphi\dot{\psi}$$

holds for almost all x. By 34.3 $\dot{\varphi}$ and $\dot{\psi}$ are summable, and therefore by 22.4 so are $\dot{\varphi}\psi$ and $\varphi\dot{\psi}$. Now by 35.3

$$u(b) - u(a) = \int_a^b \dot{u}\, dx = \int_a^b \dot{\varphi}\psi\, dx + \int_a^b \varphi\dot{\psi}\, dx.$$

Substituting $\varphi\psi$ for u, this gives us the following theorem:

36.1. *If $\varphi(x)$ and $\psi(x)$ are absolutely continuous on the interval $a \leqq x \leqq b$, then*

$$\int_a^b \varphi\dot{\psi}\, dx = \varphi(b)\psi(b) - \varphi(a)\psi(a) - \int_a^b \psi\dot{\varphi}\, dx.$$

37. The first mean value theorem of the integral calculus is quite trivial.

37.1. *If on the interval $a \leqq x \leqq b$ the functions $f(x)$ and $p(x)$ are summable, and if moreover $p(x) \geqq 0$ and $m \leqq f(x) \leqq M$, then*

$$(A) \qquad m \int_a^b p(x)\, dx \leqq \int_a^b f(x)p(x)\, dx \leqq M \int_a^b p(x)\, dx.$$

In particular, if $f(x)$ is continuous there is a number ξ $(a \leqq \xi \leqq b)$ such that

$$\int_a^b f(x)p(x)\, dx = f(\xi) \int_a^b p(x)\, dx.$$

From the hypotheses it is obvious that

$$mp(x) \leqq f(x)p(x) \leqq Mp(x)$$

for all x in $[a, b]$. Integrating and recalling 15.2 gives the inequality (A). Assume now that f is continuous. By what we have just proved, there is a number k, $m \leqq k \leqq M$, such that

$$\int_a^b f(x)p(x)\, dx = k \int_a^b p(x)\, dx.$$

Since $f(x)$ is continuous and k lies between its maximum and minimum values, there* is a ξ in $[a, b]$ such that $f(\xi) = k$, completing the proof.

The theorem which is known under the name of the "second mean-value theorem," and which is of considerable importance in analysis, is the following:

37.2. *If $f(x)$ is summable over the interval $a \leqq x \leqq b$, and $g(x)$ is bounded and monotonic increasing on this interval, and α, β are two finite numbers such that $\alpha \leqq g(a+)$ and $\beta \geqq g(b-)$, then there is a number ξ such that $a \leqq \xi \leqq b$ and*

$$\int_a^b f(x)g(x)\,dx = \alpha \int_a^\xi f(x)\,dx + \beta \int_\xi^b f(x)\,dx.$$

If we add the same finite number to $g(x)$, α and β the hypotheses remain satisfied and both members of the equation in the conclusion are changed by the same amount. Therefore there is no loss of generality in assuming that $\beta = 0$.

Consider first the case in which the graph of $g(x)$ is a polygon, with $g(a) = \alpha$ and $g(b) = \beta = 0$. Define $F(x) = \int_a^x f(x)\,dx$. The derivates D^+g, etc., of $g(x)$ assume only a finite number of values; so these derivates are bounded, and by 34.5 $g(x)$ is AC. By 33.3, $\dot{F}(x) = f(x)$ for almost all x in $[a, b]$. So by 36.1

$$\int_a^b f(x)g(x)\,dx = F(b)g(b) - F(a)g(a) - \int_a^b F(x)\dot{g}(x)\,dx.$$

The integrated terms vanish, since $F(a) = g(b) = 0$. Also $\dot{g}(x) \geqq 0$, so by 37.1 the last integral has the value $-F(\xi)\int_a^b \dot{g}(x)\,dx$, where $a \leqq \xi \leqq b$. But

$$F(\xi) = \int_\alpha^\xi f(x)\,dx \qquad \text{and} \qquad \int_a^b \dot{g}(x)\,dx = g(b) - g(a) = -g(a) = -\alpha.$$

Making these substitutions, the theorem is proved.

Now let us return to the general case.

For each n we subdivide the interval (a, b) into n equal parts by points $x_0 = a, x_1, x_2, \cdots, x_n = b$, and we construct the function $g_n(x)$ whose graph is a polygon with the successive vertices (a, α), $(x_1, g(x_1)), \cdots (x_{n-1}, g(x_{n-1}))$, (b, β). As n increases, $g_n(x)$ tends almost everywhere to $g(x)$. For $g(x)$, being monotonic, is continuous

* The demonstration of this property of continuous functions will be found in almost any text on real variables.

at almost all* points x in $[a, b]$. Let x be a point of continuity of $g(x)$ such that $a < x < b$, and let x belong to the interval (x_i, x_{i+1}) of the nth subdivision of (a, b). As n increases both x_i and x_{i+1} approach x; hence $g(x_i)$ and $g(x_{i+1})$ both approach $g(x)$. Since $g_n(x_i) = g(x_i)$ and $g_n(x_{i+1}) = g(x_{i+1})$, and $g_n(x)$ lies between these numbers, $g_n(x)$ also approaches $g(x)$. Thus $\lim g_n(x) = g(x)$ except perhaps at the points of discontinuity of $g(x)$, which form a set of measure zero.

By what has already been proved, for each n there is a ξ_n such that

$$(B) \qquad \int_a^b f(x)g_n(x)\,dx = \alpha \int_a^{\xi_n} f(x)\,dx + \beta \int_{\xi_n}^b f(x)\,dx.$$

By 2.12, there is a subsequence $\{\xi_\alpha\}$ of the $\{\xi_n\}$ $(\alpha = n_1, n_2, \cdots)$ which converges to a limit ξ. To save notation, we suppose that $\{\xi_n\}$ is itself this subsequence. Since an integral is a continuous function of its limits, it follows that the two integrals on the right of (B) tend respectively to the integrals of $f(x)$ over (a, ξ) and (ξ, b). The functions $g_n(x)$ are all between α and β in numerical value; hence $|f(x)g_n(x)|$ is not greater than the summable function $|f(x)|(|\alpha| + |\beta|)$, $n = 1, 2, \cdots$. Also $f(x)g_n(x)$ tends almost everywhere to $f(x)g(x)$. Hence, by 29.3,

$$\int_a^b f(x)g(x)\,dx = \lim \int_a^b f(x)g_n(x)\,dx$$
$$= \lim \left[\alpha \int_a^{\xi_n} f(x)\,dx + \beta \int_{\xi_n}^b f(x)\,dx\right] = \alpha \int_a^{\xi} f(x)\,dx + \beta \int_{\xi}^b f(x)\,dx.$$

38. The first theorem which we wish to prove concerning change of variable is the following.

38.1. *Let $f(x)$ be bounded and measurable on the interval $[a, b]$. Let $\varphi(t)$ be absolutely continuous on the interval $[\alpha, \beta]$, and let the inequality $a \leqq \varphi(t) \leqq b$ hold for all t in $[\alpha, \beta]$. Then $f(\varphi(t))\dot\varphi(t)$ is summable, and*

$$(A) \qquad \int_{\varphi(\alpha)}^{\varphi(\beta)} f(x)\,dx = \int_\alpha^\beta f(\varphi(t))\dot\varphi(t)\,dt,$$

where $\dot\varphi$ is defined as in 34.2.

If we define $F(x) = \int_a^x f(x)\,dx$, the equation to be proved is

$$(B) \qquad F(\varphi(\beta)) - F(\varphi(\alpha)) = \int_\alpha^\beta f(\varphi(t))\dot\varphi(t)\,dt.$$

* This follows from 34.2 and 31.8. However, more than this is true; $g(x)$ is continuous for all x in $[a, b]$ except those of a denumerable set. Cf. exercise after 6.11.

The function $F(x)$ satisfies a Lipschitz condition, for if M is an upper bound for $|f(x)|$, then by 37.1 (with $p \equiv 1$) the inequality

$$-M(x_2 - x_1) \leqq \int_{x_1}^{x_2} f(x)\,dx = F(x_2) - F(x_1) \leqq M(x_2 - x_1)$$

holds whenever $a \leqq x_1 \leqq x_2 \leqq b$. Hence by 9.3 the function $g(t) \equiv F(\varphi(t))$ is absolutely continuous.

As a first step in our proof, we suppose that $f(x)$ is continuous. Then by 33.2 the equation $F'(x) = f(x)$ holds for all x in $[a, b]$, and so for almost all t in $[\alpha, \beta]$ (namely, all t for which $\varphi'(t)$ exists and is finite) the derivative $g'(t)$ exists and is finite, and

$$\dot{g}(t) = g'(t) = F'(\varphi(t))\varphi'(t) = f(\varphi(t))\dot{\varphi}(t),$$

by 31.10. By 35.3,

$$F(\varphi(\beta)) - F(\varphi(\alpha)) = g(\beta) - g(\alpha) = \int_{\alpha}^{\beta} \dot{g}(t)\,dt$$
$$= \int_{\alpha}^{\beta} f(\varphi(t))\dot{\varphi}(t)\,dt.$$

Thus if $f(x)$ is continuous our theorem holds.

Now suppose that (A) has been established for a class of functions, and that $\{\psi_n(x)\}$ is a sequence of uniformly bounded functions of the class (say $|\psi_n| \leqq M$) which tend everywhere to a limit function $f(x)$. Then by hypothesis the equation

$$\int_{\varphi(\alpha)}^{\varphi(\beta)} \psi_n(x)\,dx = \int_{\alpha}^{\beta} \psi_n(\varphi(t))\dot{\varphi}(t)\,dt$$

holds for each n. The functions $\psi_n(\varphi(t))\dot{\varphi}(t)$ do not exceed the summable function $M\,|\dot{\varphi}(t)|$ in absolute value, while $|\psi_n(x)| \leqq M$. And for all x we have $\lim \psi_n(x) = f(x)$, while for all t we have $\lim \psi_n(\varphi(t))\dot{\varphi}(t) = f(\varphi(t))\dot{\varphi}(t)$. So by 29.3 this last function is summable, and equation (A) holds.

In particular, (A) holds for continuous functions; hence it holds for bounded U- or L-functions, since these are limits of sequences of uniformly bounded continuous functions (7.9). If $f(x)$ is bounded and measurable, by 15.8 there are functions $g(x)$ and $h(x)$ which are respectively the limits of sequences of uniformly bounded L- and U-functions, and which satisfy the conditions

$$g(x) \leqq f(x) \leqq h(x) \text{ for all } x \text{ in } [a, b],$$
$$\int_a^b g(x)\,dx = \int_a^b f(x)\,dx = \int_a^b h(x)\,dx.$$

These three integrals may be regarded as integrals from $-\infty$ to ∞

if we set $g = f = h = 0$ outside of $[a, b]$, so by 23.6 we must have

$$g(x) = f(x) = h(x) \text{ for almost all } x \text{ in } [a, b].$$

By the preceding paragraph equation (A) holds for both $g(x)$ and $h(x)$, and therefore for every number τ in $[\alpha, \beta]$ the equation

$$\int_\alpha^\tau g(\varphi(t))\dot{\varphi}(t)\,dt = \int_{\varphi(\alpha)}^{\varphi(\tau)} g(x)\,dx = \int_{\varphi(\alpha)}^{\varphi(\tau)} h(x)\,dx$$
$$= \int_\alpha^\tau h(\varphi(t))\dot{\varphi}(t)\,dt$$

is valid. So by 33.4 the equation

$$g(\varphi(t))\dot{\varphi}(t) = h(\varphi(t))\dot{\varphi}(t)$$

holds for almost all t in $[\alpha, \beta]$. But $g(x) \leqq f(x) \leqq h(x)$ for all x in $[a, b]$; so for almost all t in $[\alpha, \beta]$ we have

$$g(\varphi(t))\dot{\varphi}(t) = f(\varphi(t))\dot{\varphi}(t) = h(\varphi(t))\dot{\varphi}(t).$$

Therefore $f(\varphi(t))\dot{\varphi}(t)$ is almost everywhere equal to a summable function, and is summable; and

$$\int_{\varphi(\alpha)}^{\varphi(\beta)} f(x)\,dx = \int_{\varphi(\alpha)}^{\varphi(\beta)} g(x)\,dx = \int_\alpha^\beta g(\varphi(t))\dot{\varphi}(t)\,dt$$
$$= \int_\alpha^\beta f(\varphi(t))\dot{\varphi}(t)\,dt.$$

A corollary of 38.1 which is sometimes useful is the following.

38.2. *Let $\varphi(t)$ be AC on the interval $[\alpha, \beta]$. If X is a point-set of measure 0, and T is a set of values of t in $[\alpha, \beta]$ such that $\varphi(t)$ is in X whenever t is in T, then $\varphi'(t) = 0$ for almost all t in the set T.*

Let $f(x)$ be the characteristic function of the set X. Since $mX = 0$, for all numbers τ in $[\alpha, \beta]$ we have by 38.1

$$0 = \int_{\varphi(\alpha)}^{\varphi(\tau)} f(x)\,dx = \int_\alpha^\tau f(\varphi(t))\dot{\varphi}(t)\,dt.$$

Therefore by 33.4 the equation

$$(B) \qquad f(\varphi(t))\dot{\varphi}(t) = 0$$

holds for almost all t in $[\alpha, \beta]$. In particular, on almost all of the set T equation (B) holds, and $\varphi'(t)$ exists and is equal to $\dot{\varphi}(t)$; and for all t in T we have $f(\varphi(t)) = 1$. So, for almost all t in T we have

$$1 \cdot \varphi'(t) = 0,$$

which was to be proved.

This yields a ready extension of 38.1:

38.3. *Let $f(x)$ be defined on $[a, b]$ and equal to a bounded measurable function $g(x)$ for almost all x in $[a, b]$. Let $\varphi(t)$ be AC on $[\alpha, \beta]$ and such that $a \leqq \varphi(t) \leqq b$ for all t in $[\alpha, \beta]$. Then $f(\varphi(t))\dot{\varphi}(t)$ is summable over $[\alpha, \beta]$, and*

$$\int_{\varphi(\alpha)}^{\varphi(\beta)} f(x)\,dx = \int_{\alpha}^{\beta} f(\varphi(t))\dot{\varphi}(t)\,dt.$$

This equation holds if $f(x)$ is replaced by $g(x)$ by 38.1. Since $f(x) = g(x)$ except on a set X of measure 0,

$$\int_{\varphi(\alpha)}^{\varphi(\beta)} f(x)\,dx = \int_{\varphi(\alpha)}^{\varphi(\beta)} g(x)\,dx.$$

If $\varphi(t)$ is not in X, then $f(\varphi(t)) = g(\varphi(t))$, and so

$$(C) \qquad f(\varphi(t))\dot{\varphi}(t) = g(\varphi(t))\dot{\varphi}(t).$$

For almost all t such that $\varphi(t)$ is in X we have $\dot{\varphi}(t) = 0$, by 38.2; so equation (C) is still valid. That is, (C) holds for almost all t in the interval $[\alpha, \beta]$; so

$$\int_{\alpha}^{\beta} f(\varphi(t))\dot{\varphi}(t)\,dt = \int_{\alpha}^{\beta} g(\varphi(t))\dot{\varphi}(t)\,dt.$$

This completes the proof.

Theorems 38.1 and 38.3 are diminished in usefulness by the fact that they apply only to functions which are equivalent to bounded functions. This restriction cannot be simply omitted, as the following example shows. Let $f(x) = x^{-\frac{2}{3}}/3$, $-8 \leqq x \leqq 8$, and let $\varphi(t) = t^3 \cos^3 (\pi/t)$, $-2 \leqq t \leqq 2$. Then if we set

$$F(x) = \int_0^x \tfrac{1}{3}x^{-\frac{2}{3}}\,dx = x^{\frac{1}{3}},$$

equation (A) takes the form (replacing $[\alpha, \beta]$ by $[0, \tau]$)

$$F(\varphi(\tau)) = \int_0^{\tau} f(\varphi(t))\dot{\varphi}(t)\,dt.$$

This is impossible, for

$$F(\varphi(\tau)) = \tau \cos\left(\frac{\pi}{\tau}\right),$$

which is not AC.

However, we can remove the hypothesis of boundedness of $f(x)$ if we assume that $\varphi(t)$ is monotonic, as the following theorem shows.

38.4. *Let $f(x)$ be summable over the interval $a \leqq x \leqq b$. Let $\varphi(t)$ be absolutely continuous and monotonic on the interval $\alpha \leqq t \leqq \beta$, and* *suppose* $a \leqq \varphi(t) \leqq b$. *Then* $f(\varphi(t))\dot{\varphi}(t)$ *is summable, and*

$$\int_{\varphi(\alpha)}^{\varphi(\beta)} f(x)\,dx = \int_{\alpha}^{\beta} f(\varphi(t))\dot{\varphi}(t)\,dt.$$

We consider only the case in which $\varphi(t)$ is monotonic increasing, if $\varphi(t)$ is monotonic decreasing, the following proof holds with only trivial changes.

Suppose to begin with that $f(x) \geqq 0$. Define $f_n(x) = \inf\{n, f(x)\}$. For each n we know by 38.1 that

$$(D) \qquad \int_{\varphi(\alpha)}^{\varphi(\beta)} f_n(x)\,dx = \int_{\alpha}^{\beta} f_n(\varphi(t))\dot{\varphi}(t)\,dt.$$

Now let n increase without bound. The functions $f_n(x)$ increase and tend everywhere to $f(x)$; and since $0 \leqq f_n(x) \leqq f(x)$ and $f(x)$ is summable, by 29.3 the left member of (D) tends to

$$(E) \qquad \int_{\varphi(\alpha)}^{\varphi(\beta)} f(x)\,dx.$$

Consider now the right member of (D). Since $\dot{\varphi}(t) \geqq 0$, as n increases these (non-negative) integrands are non-decreasing and tend everywhere to $f(\varphi(t))\dot{\varphi}(t)$. But the integrals on the right of (D) are bounded, because the left members of (D) are bounded, being not greater than (E). Hence by 18.4 the limit function $f(\varphi(t))\dot{\varphi}(t)$ is summable, and the right members of (D) tend to

$$\int_{\alpha}^{\beta} f(\varphi(t))\dot{\varphi}(t)\,dt.$$

This completes the proof for the case $f \geqq 0$.

Returning to the general case, we can represent the summable function $f(x)$ as the difference of two non-negative summable functions:

$$f(x) = f^+(x) - f^-(x),$$

as in 15.4. We apply the preceding result to f^+ and f^-, and obtain

$$\int_{\varphi(t_0)}^{\varphi(t_1)} f^+(x)\,dx = \int_{t_0}^{t_1} f^+(\varphi(t))\dot{\varphi}(t)\,dt,$$
$$\int_{\varphi(t_0)}^{\varphi(t_1)} f^-(x)\,dx = \int_{t_0}^{t_1} f^-(\varphi(t))\dot{\varphi}(t)\,dt.$$

If we subtract the second of these equations from the first, we obtain the equation in the conclusion of the theorem. The proof is thus complete.

Incidentally, this gives us another proof of the first part of 9.3. For if $F(y)$ is AC on an interval $[a, b]$ and $\varphi(x)$ is AC and monotonic on

$[\alpha, \beta]$ and assumes value in $[a, b]$, then

$$F(\varphi(x)) - F(\varphi(a)) = \int_{\varphi(a)}^{\varphi(x)} \dot{F}(y)\, dy$$
$$= \int_a^x \dot{F}(\varphi(x))\dot{\varphi}(x)\, dx,$$

so $F(\varphi(x))$ is AC.

EXERCISE. Let $\varphi(x)$ be a function which is defined and strictly monotonic increasing on $[0, \infty)$, vanishes at $x = 0$ and is AC on every interval $[0, a]$. Let $\psi(x)$ be its inverse function, and define

$$\Phi(t) = \int_0^t \varphi(x)\, dx, \quad \Psi(t) = \int_0^t \psi(x)\, dx.$$

Then the conclusion of 24.1 remains valid. (In the definition of $\Psi(\varphi(a))$, substitute $x = \varphi(y)$ and use 38.4 and 36.1 to establish $\Phi(a) + \Psi(\varphi(a)) - a\varphi(a) = 0$. An alternative proof can be based on 27.7, 9.3, 31.3, 33.2 and 35.2.)

39. We complete this chapter with some theorems on differentiation under the integral sign. If a function $f(x, t)$ is defined for all t in an interval (α, β) and all x in a set E, and for each fixed t in (α, β) the function $f(x, t)$ is summable over E when regarded as a function of x alone, then its integral is a function $\varphi(t)$ of t alone. We wish to compute the derivative of $\varphi(t)$ in terms of the partial derivative of f with respect to t.

39.1s. *Let $f(x, t)$ be defined for all x in a set E and all t in an interval (α, β). Let $f(x, t)$ be summable over E for each fixed t in (α, β); define*

$$\varphi(t) = \int_E f(x, t)\, dx.$$

Let t_0 be a value of t in (α, β). If there exists a summable function $g(x)$ such that

$$\left| \frac{f(x, t) - f(x, t_0)}{t - t_0} \right| \leqq g(x)$$

for all t in (α, β), and the partial derivative $f_t(x, t_0)$ exists for all x in E except those of a set E_0 of measure 0, then $\varphi'(t_0)$ exists, and

$$\varphi'(t_0) = \int_{E - E_0} f_t(x, t_0)\, dx.$$

Since $mE_0 = 0$, from the definition of $\varphi(t)$ and 23.4 we have

$$\frac{\varphi(t) - \varphi(t_0)}{t - t_0} = \int_{E - E_0} \frac{f(x, t) - f(x, t_0)}{t - t_0}\, dx$$

if t is in (α, β) and $t \neq t_0$. The integrand on the right does not exceed $g(x)$ in absolute value, and tends to $f_t(x, t_0)$ as t tends to t_0. So by the remark after 29.3 the limit of the integral on the right exists and is equal to the integral of $f_t(x, t_0)$:

$$\lim_{t \to t_0} \frac{\varphi(t) - \varphi(t_0)}{t - t_0} = \int_{E - E_0} f_t(x, t_0)\, dx.$$

39.2. Corollary. *Let $f(x, t)$ be defined for all x in a set E and all t in an interval (α, β) and summable over E for each t in (α, β). Let*

$$\varphi(t) = \int_E f(x, t)\, dt.$$

For all t in (α, β) and for all x in $E - E_0$, where $mE_0 = 0$, let $f_t(x, t)$ be defined and smaller in absolute value than a summable function $g(x)$. Then for all t in (α, β) the derivative $\varphi'(t)$ exists, and

$$\varphi'(t) = \int_{E - E_0} f_t(x, t)\, dx.$$

The hypotheses of 39.1 are all obviously satisfied except for the requirement that

$$\left| \frac{f(x, t) - f(x, t_0)}{t - t_0} \right| \leqq g(x).$$

But by the theorem of mean value, for all x in $E - E_0$

$$\left| \frac{f(x, t) - f(x, t_0)}{t - t_0} \right| = | f_t(x, \bar{t}) | \leqq g(x)$$

where $\bar{t}$ is between t and t_0. (Alternatively, we could obtain this inequality from 34.3 or 37.1.) So this hypothesis, too, is satisfied.

CHAPTER VI

Continuity Properties of Measurable Functions

40. The class of measurable functions is of course much larger than the class of continuous functions; a measurable function may be discontinuous at every point, as e.g. the characteristic function of the set of irrational numbers. Nevertheless, there are several important relationships between these classes.

The first of these relationships will be touched on only briefly. Suppose that a function $f(x)$ is defined and continuous in the generalized sense on a set E; that is, let the relationship

$$\lim_{x \to x_0} f(x) = f(x_0)$$

hold for all x_0 in EE', whether $f(x_0)$ is finite or $+\infty$ or $-\infty$. It may or may not be possible to extend the range of definition of $f(x)$ to the closure $\bar{E}$ of E in such a way that $f(x)$ is still continuous in the generalized sense on all of $\bar{E}$. For instance, if E is $(0, 1)$ and $f(x) = \sin(1/x)$, this extension is impossible, because $f(x)$ does not tend to any limit as $x \to 0$. If the extension is possible, we say that $f(x)$ is of Baire class 0 on E. Obviously if E is already closed and $f(x)$ continuous on E, then $f(x)$ is of class 0 on E. Suppose now that $f(x)$ is not of class 0 on E, but that there is a sequence of functions $f_1(x), f_2(x), \cdots$, each of class 0 on E, such that $\lim_{n\to\infty} f_n(x) = f(x)$ for every x in E. We then say that $f(x)$ is of class 1 on E. We continue thus by induction. If $f(x)$ is not of class $\leqq k$ on E, but there is a sequence of functions $f_1(x), f_2(x), \cdots$, each of class $\leqq k$ on E, such that $\lim_{n\to\infty} f_n(x) = f(x)$ for each x in E, then we say that $f(x)$ is of class $k + 1$ on E.

Thus, for example, if $f(x)$ is a U-function or an L-function on a closed interval I, then $f(x)$ is of Baire class 0 or 1, by definition 12.1 and theorem 7.9. Since U-functions and L-functions need not be continuous (cf. for example, 12.9) there actually are functions of class 1. It is indeed true that there are functions of class k for each positive integer k, but we shall not prove this.

The next theorem serves as a lemma for a later proof.

40.1s. *If $\varphi(z)$ is defined and continuous (in the generalized sense) for $-\infty \leqq z \leqq \infty$ and $f(x)$ is of Baire class $\leqq k$ on a set E, then $\varphi(f(x))$ is of Baire class $\leqq k$ on E.*

If $f(x)$ is of Baire class 0 on E, it can be extended to be continuous in the generalized sense on $\bar{E}$. Then $\varphi(f(x))$ is also continuous in the generalized sense on $\bar{E}$, and is of class 0 on E. We now proceed by induction. Suppose the theorem true for $k = p$. Let $f(x)$ be of class $\leqq p + 1$ on E. Then $f(x)$ is the limit on E of a sequence of functions $f_1(x)$, $f_2(x)$, $\cdots$, each of class $\leqq p$. By the induction hypothesis, $\varphi(f_n(x))$ is of class $\leqq p$ for each n, and by the continuity of φ we find that $\lim \varphi(f_n(x)) = \varphi(f(x))$ for all x in E. Hence $\varphi(f(x))$ is of class $\leqq p + 1$, and the theorem is established.

We now investigate the connection between Baire functions and measurable functions.

40.2s. *If E is a measurable set, and $f(x)$ is of Baire class p $(p \geqq 0)$ on E, then $f(x)$ is measurable on E.*

Suppose first $p = 0$. Then $f(x)$ can be extended to be continuous in the generalized sense on $\bar{E}$. It follows that for every number a the set $\bar{E}[f \geqq a]$ is a closed set, hence measurable (19.4). So $f(x)$ is measurable on $\bar{E}$ by definition 21.1, and by 21.2 it is measurable on E.

Now suppose the theorem true for $p = k$, and let $f(x)$ be of class $\leqq k + 1$ on E. It is then the limit of functions $f_n(x)$ of class $\leqq k$ on E, and by the induction hypothesis these are measurable. So by 21.9 $f(x)$ is measurable on E, and the theorem holds for $p = k + 1$. This completes the induction.

The converse of this theorem is not true, as will be shown by an example in §44. However, it is possible to prove the following theorem.

40.3s. *If $f(x)$ is measurable on a bounded* measurable set E, there are functions $g(x)$, $h(x)$ of Baire class $\leqq 2$ on E (and in fact on a closed interval containing E) such that $g(x) \leqq f(x) \leqq h(x)$ for all x in E and $g(x) = h(x)$ for almost all x in E.*

Suppose first that $f(x)$ is bounded and non-negative on E. If I is a closed interval containing E in its interior, and we set $f(x) = 0$ on the complement of E, then $f(x)$ remains summable on I. Now theorem 15.8 assures us of the existence of functions $g(x)$, $h(x)$ such that $g(x) \leqq f(x) \leqq h(x)$ for all x in I (hence for all x in E) and

$$\int_I g(x)\,dx = \int_I f(x)\,dx = \int_I h(x)\,dx.$$

Moreover, by the remark after 15.8 we may assume that g and h

* The hypothesis that E is bounded can be discarded, but the proof would be less simple.

vanish on the boundary of I. By 17.4 the integrals of f, g, and h over the whole space R_q are equal, if we define $g(x) = h(x) = 0$ outside of I. By 23.6 this implies that $g(x) = f(x) = h(x)$ for almost all x in R_q, and in particular for almost all x in E. On I the functions $g(x)$ and $h(x)$ are limits of sequences of L-functions and of U-functions respectively. Since U-functions and L-functions are of Baire class $\leqq 1$, the functions $g(x)$ and $h(x)$ are of Baire class $\leqq 2$ on I. Hence they are of Baire class $\leqq 2$ on E.

Next suppose only that $f(x)$ is measurable. Define $\mu(z) = z/(1 + |z|)$ for z finite, $\mu(-\infty) = -1$, $\mu(+\infty) = +1$. Then $\mu(z)$ is continuous and strictly monotonic increasing on $-\infty \leqq z \leqq \infty$, and it has an inverse

$$\mu^{-1}(y) = \frac{y}{1 - |y|}, \qquad -1 < y < 1,$$
$$\mu^{-1}(-1) = -\infty, \qquad \mu^{-1}(1) = +\infty.$$

The function $f^*(x) = 1 + \mu(f(x))$ then satisfies the inequality $0 \leqq f^*(x) \leqq 2$. Moreover it is measurable on E. For if $0 \leqq a \leqq 2$ the set $E[f^* \geqq a]$ is the same as $E[f \geqq \mu^{-1}(a - 1)]$, which is measurable because f is a measurable function, while if $a < 0$ the set $E[f^* \geqq a]$ is the set E itself and if $a > 2$ the set $E[f^* \geqq a]$ is empty. By the part of the proof already completed, there exist functions $g^*(x)$, $h^*(x)$ which are of Baire class $\leqq 2$ on an interval containing E in its interior and are such that $g^*(x) \leqq f^*(x) \leqq h^*(x)$. By the remark after 15.8, we may assume $0 \leqq g^*(x) \leqq h^*(x) \leqq 2$.

The inverse function $\mu^{-1}(y)$ defined above is monotonic increasing and continuous in the generalized sense on the interval $-1 \leqq y \leqq 1$. So if we define

$$\begin{aligned} \psi(y) &= \mu^{-1}(y), && -1 \leqq y \leqq +1, \\ \psi(y) &= +\infty, && y > 1, \\ \psi(y) &= -\infty, && y < 1, \end{aligned}$$

the function $\psi(y)$ is monotonic increasing and is continuous in the generalized sense, and

$$(*) \qquad \psi(\mu(z)) \equiv z, \qquad -\infty \leqq z \leqq \infty.$$

By theorem 40.1, the functions

$$g(x) \equiv \psi(g^*(x) - 1), \qquad h(x) \equiv \psi(h^*(x) - 1)$$

are of Baire class $\leqq 2$, while from (*) we deduce

$$\psi(f^*(x) - 1) = \psi(\mu(f(x))) = f(x).$$

Since ψ is monotonic increasing, from the inequality $g^* \leqq f^* \leqq h^*$ we deduce $g(x) \leqq f(x) \leqq h(x)$ for all x in E, and since $g^*(x) = f^*(x) = h^*(x)$ for almost all x in E, it follows that $g(x) = f(x) = h(x)$ for almost all x in E.

REMARK. If m and M are respectively inf f and sup f, we may assume that $g(x)$ and $h(x)$ satisfy the inequalities

$$m \leqq g(x) \leqq h(x) \leqq M.$$

For let $\varphi(t) = t$ if $m \leqq t \leqq M$, $\varphi(t) = M$ if $t > M$, and $\varphi(t) = m$ if $t < m$. By 40.1, the functions $\varphi(g(x))$ and $\varphi(h(x))$ have the desired bounds and still satisfy the requirements in 40.3.

By use of the concept of Baire functions we can obtain a considerable generalization of Theorem 21.11. This generalization will be established in theorem 40.5, in preparation for which we prove a lemma.

40.4s. *If $f(x)$ is of Baire class $\leqq p$ $(p > 0)$ on E, it is the limit on E of a sequence of functions $f_n(x)$ each of which is bounded and of Baire class $\leqq p - 1$ on E.*

If $f(x)$ is of Baire class $\leqq p$ on E, it is the limit of a sequence of functions $\{g_n(x)\}$ each of Baire class $\leqq p - 1$ on E. Now define

$$\begin{aligned} f_n(x) &= g_n(x) && \text{if} \quad -n \leqq g_n(x) \leqq n, \\ f_n(x) &= n && \text{if} \quad g_n(x) > n, \\ f_n(x) &= -n && \text{if} \quad g_n(x) < -n. \end{aligned}$$

By 40.1, these $f_n(x)$ are of Baire class $\leqq p - 1$. Each is bounded. Also, $f_n(x) \to f(x)$. For if $f(x)$ is finite, then $f_n(x) = g_n(x)$ whenever $n > |f(x)| + 1$ and $|g_n(x) - f(x)| < 1$. This holds for all but a finite number of values of n, so $\lim f_n(x) = \lim g_n(x) = f(x)$. If $f(x) = \infty$, then for every $\epsilon > 0$ the numbers $f_n(x)$ exceed $1/\epsilon$ whenever $n > 1/\epsilon$ and $g_n(x) > 1/\epsilon$; that is, for all large n. So again $\lim f_n(x) = \infty = f(x)$. A similar proof holds if $f(x) = -\infty$, and the lemma is established.

40.5s. *Let $f_1(x), \cdots, f_h(x)$ be measurable and finite valued on a set E in q-dimensional space R_q. Let S be a set in h-dimensional space which is the sum of a finite or denumerable collection of closed sets $S_1, S_2, \cdots$. Let $\varphi(z^1, \cdots, z^h)$ be defined on S, and of some finite Baire class p_j on each set S_j (here p_j may vary with j). Then the subset E^* of E on which $\varphi(f_1(x), \cdots, f_h(x))$ is defined is measurable, and $\varphi(f_1(x), \cdots, f_h(x))$ is measurable on E^*.*

Consider first the case in which S is a single closed set S_1 and $\varphi(z^1, \cdots, z^h)$ is continuous in the ordinary sense—that is, finite-valued and of class 0—on S_1. By Theorem 21.11 the subset E^* of E on which $(f_1(x), \cdots, f_h(x))$ is in S_1 is a measurable set, and $\varphi(f_1(x), \cdots, f_h(x))$ is measurable on E^*.

If $p_1 \leqq 1$, by Lemma 40.4 there is a sequence of functions defined, bounded and of class 0 on S_1 such that $\lim_n \varphi_n = \varphi$ for all z in S_1. By the preceding paragraph, each function $\varphi_n(f_1(x), \cdots, f_h(x))$ is measurable on E^*, so the limit function $\varphi(f_1(x), \cdots, f_h(x))$ is measurable on E^*. Proceeding by an obvious induction, we show that if φ is of finite Baire class p_1 on S_1 the function $\varphi(f_1(x), \cdots, f_h(x))$ is measurable on E^*.

Now we remove the restriction that S is closed. Under the hypotheses of the theorem, the preceding part of the proof shows that for each set S_j the subset E_j of E on which $(f_1(x), \cdots, f_h(x))$ lies in S_j is a measurable subset, and on it the function $\varphi(f_1(x), \cdots, f_h(x))$ is measurable. The set E^* of the theorem is then $\bigcup_j E_j$, and so is measurable. (Incidentally, this would follow from Theorem 21.11.) For every number a, the set $E_j[\varphi(f_1(x), \cdots, f_h(x)) \geqq a]$ is a measurable set, since $\varphi(f_1, \cdots, f_h)$ is measurable on E_j. Hence the set

$$E^*[\varphi(f_1(x), \cdots, f_h(x)) \geqq a] \equiv \bigcup_j E_j[\varphi(f_1(x), \cdots, f_h(x)) \geqq a]$$

is measurable, and $\varphi(f_1(x), \cdots, f_h(x))$ is therefore measurable on E^*.

41. If E is an open set in one-dimensional space R_1, and x_0 is in E, then for all sufficiently small intervals $[\alpha, \beta]$ containing x_0 the set E completely fills $[\alpha, \beta]$. We shall now show that if E is measurable, then for almost all x_0 in E the set E fills all but an arbitrarily small percentage of every small interval $[\alpha, \beta]$ containing x_0. We make this notion precise by defining metric density.

41.1. *Let E be a measurable set in R_1, and let x_0 be in R_1. If the quotient*

$$\frac{m(E \cap [\alpha, \beta])}{\beta - \alpha}$$

approaches a limit as $\beta - \alpha$ tends to zero subject to the conditions $\alpha \leqq x_0 \leqq \beta$, this limit is called the metric density of the set E at x_0.

It is thus apparent that if E is open and x_0 is in E, the metric density of E at x_0 exists and has the value 1. For then the set $E \cap [\alpha, \beta]$ is the entire interval $[\alpha, \beta]$ whenever $\beta - \alpha$ is small, so $m(E \cap [\alpha, \beta])/(\beta - \alpha) = 1$ for all such α and β. The theorem which we wish to prove, and which we stated crudely in the preceding paragraph, is

41.2. *If E is a measurable set in one-dimensional* space R_1, then for almost all x in E the metric density of E at x exists and is 1, while the metric density of E has the value 0 at almost all points of the complement of E.*

Since E is measurable, its characteristic function $K_E(x)$ is summable over every finite interval, by definition or by 22.1. So the integral

$$\varphi(x) = \int_0^x K_E(x)\,dx$$

is defined for all x. By 33.3, for all x except those of a set E_0 of measure 0 the derivative $\varphi'(x)$ exists and is equal to $K_E(x)$.

We now stop to prove the following statement. If $\psi(x)$ is defined on an interval $[a, b]$ and has a finite derivative at a point x_0 interior to $[a, b]$, then the limit of the function

$$\frac{\psi(\beta) - \psi(\alpha)}{\beta - \alpha}$$

exists as β and α tend to x_0 subject to the conditions $\alpha \leqq x_0 \leqq \beta$, $\alpha < \beta$, and the limit is $\psi'(x_0)$. For if $\epsilon > 0$ there is a $\delta > 0$ such that

$$(A) \qquad \left| \frac{\psi(x) - \psi(x_0)}{x - x_0} - \psi'(x_0) \right| < \epsilon$$

if x is in $[a, b]$ and $0 < |x - x_0| < \delta$. Now let α and β both be in $(x_0 - \delta, x_0 + \delta)$ where $\alpha \leqq x_0 \leqq \beta$ and $a \leqq \alpha < \beta \leqq b$. We shall show

$$(B) \qquad \left| \frac{\psi(\beta) - \psi(\alpha)}{\beta - \alpha} - \psi'(x_0) \right| < \epsilon.$$

If either α or β is equal to x_0, this is the same as (A). Otherwise $\alpha < x_0 < \beta$, and then

$$\left| \frac{\psi(\beta) - \psi(\alpha)}{\beta - \alpha} - \psi'(x_0) \right| = \left| \frac{\beta - x_0}{\beta - \alpha} \left\{ \frac{\psi(\beta) - \psi(x_0)}{\beta - x_0} - \psi'(x_0) \right\} + \frac{x_0 - \alpha}{\beta - \alpha} \left\{ \frac{\psi(x_0) - \psi(\alpha)}{x_0 - \alpha} - \psi'(x_0) \right\} \right|$$

$$\leqq \frac{\beta - x_0}{\beta - \alpha} \cdot \epsilon + \frac{x_0 - \alpha}{\beta - \alpha} \cdot \epsilon = \epsilon,$$

which establishes inequality (B). Our statement is therefore proved.

* The theorem is true in q-dimensional space, but the proof is more difficult. It follows readily from 72.4.

Applying this statement to $\varphi(x)$, we find that for all x in $R_1 - E_0$ the equation

$$(C) \qquad \lim \frac{\varphi(\beta) - \varphi(\alpha)}{\beta - \alpha} = \varphi'(x_0) = K_E(x_0)$$

holds, when $\beta - \alpha \to 0$ subject to the restriction that $[\alpha, \beta]$ contains x_0 (i.e., $\alpha \leqq x_0 \leqq \beta$), so that $\alpha \to x_0$ and $\beta \to x_0$. By the definition of $\varphi(x)$,

$$\varphi(\beta) - \varphi(\alpha) = \int_\alpha^\beta K_E(x)\, dx,$$

and by the definition of measure this is the measure of the set of points of E in $[\alpha, \beta]$, that is,

$$\varphi(\beta) - \varphi(\alpha) = m(E \cap [\alpha, \beta]).$$

Substituting this in equation (C), we find that for all x except those in the set E_0 of measure 0 the set E has metric density $K_E(x)$ at x; that is it has metric density 1 at almost all points in E and metric density 0 at almost all points of its complement.

Turning now to functions, we notice that if $f(x)$ is continuous on an interval (a, b), and x_0 is in (a, b), then for every $\epsilon > 0$ there is a $\delta > 0$ such that $|f(x) - f(x_0)| < \epsilon$ if $|x - x_0| < \delta$. That is, the set of x such that $|f(x) - f(x_0)| < \epsilon$ contains a neighborhood of x_0, and a fortiori has metric density 1 at x_0. Let us say that a function $f(x)$, defined and measurable on a measurable set E in R_1, is *approximately continuous* at a point x_0 of E if for every $\epsilon > 0$ the set $E[f(x_0) - \epsilon < f < f(x_0) + \epsilon]$ has metric density 1 at x_0. Then the preceding remark shows that if $f(x)$ is continuous at an interior point x_0 of E it is approximately continuous at x_0. Of course the converse is false. For example, if $f(x)$ is 1 for all rational x and 0 for all irrational x, then the set $R_1[-\epsilon < f(x) < \epsilon]$ contains all irrationals and has metric density 1 at all points. Hence $f(x)$ is approximately continuous for all irrational x, although it is everywhere discontinuous. We shall now prove the following theorem.

41.3s. *If $f(x)$ is finite and measurable on a measurable set E in R_1, then $f(x)$ is approximately continuous at almost all points of E.*

The rational numbers are denumerable, and so can be arranged in a sequence $r_1, r_2, \cdots$. Let E_{ij} be the set $E[r_i < f \leqq r_j]$; this is measurable by 21.3*d*. By 41.2, the set E_{ij} has metric density 1 at all its points except those of a set E_{ij}^0 of measure 0. These sets E_{ij}^0 are denumerable in number, so their sum

$$E_0 = \bigcup_{i,j=1}^{\infty} E_{i,j}^0$$

also has measure 0, by 19.13. We shall now show that $f(x)$ is approximately continuous at all points of $E - E_0$.

Let x_0 be in $E - E_0$, and let ϵ be a positive number. Choose rational numbers r_i, r_j such that $f(x_0) - \epsilon < r_i < f(x_0) < r_j < f(x_0) + \epsilon$. Then the set E_{ij}, which is $E[r_i < f \leqq r_j]$, is contained in the set $E_1 \equiv E[f(x_0) - \epsilon < f < f(x_0) + \epsilon]$. Since x_0 is not in E_0, it is not in E_{ij}^0, so E_{ij} has metric density 1 at x_0. But E_1 contains E_{ij}, so for all numbers α, β with $\alpha < x_0 < \beta$ and $\alpha < \beta$ we have $m(E_1 \cap [\alpha, \beta]) \geqq m(E_{ij} \cap [\alpha, \beta])$. If we let α and β tend to x_0, then

$$1 \geqq \limsup \frac{m(E_1 \cap [\alpha, \beta])}{\beta - \alpha} \geqq \liminf \frac{m(E_1 \cap [\alpha, \beta])}{\beta - \alpha}$$
$$\geqq \lim \frac{m(E_{ij} \cap [\alpha, \beta])}{\beta - \alpha} = 1.$$

Hence

$$\lim \frac{m(E_1 \cap [\alpha, \beta])}{\beta - \alpha} = 1.$$

That is, for every positive ϵ the set $E_1 \equiv E[f(x_0) - \epsilon < f < f(x_0) + \epsilon]$ has metric density 1 at x_0, and so $f(x)$ is approximately continuous at x_0.

42. In the two preceding sections we have seen that a measurable function is almost everywhere equal to a function which is of Baire class $\leqq 2$, i.e., which is the limit of functions which are themselves limits of sequences of continuous functions; and we have seen that a measurable function of one variable has a property which is a weakened form of the property of continuity. Now we shall see that summable functions can be approximated, in a certain average sense, by continuous functions. That is, if $f(x)$ is an arbitrary summable function we cannot expect to find continuous functions $\varphi(x)$ which approximate $f(x)$ uniformly; but we can make the *integral* of the difference $|f - \varphi|$ arbitrarily small.

42.1s. *If $f(x)$ is summable over a measurable set E, for every positive number ϵ there is a function $\varphi(x)$ continuous on the entire space R_q and such that*

$$\int_E |f(x) - \varphi(x)| \, dx < \epsilon.$$

We suppose $f(x)$ set equal to zero on CE. There is then no loss of generality in supposing that E is R_q, for if the inequality of the theorem holds for R_q it holds for every measurable set E.

First we suppose that $f(x)$ vanishes except on a bounded set, which we enclose in a closed interval J. Let I be a closed interval containing J in its interior; we use I as basic interval. As in §16, we construct a function $\psi(x)$ continuous on R_q, having the value 1 on J and the value 0 on CI, and such that $0 \leqq \psi(x) \leqq 1$ for all x. By 17.4, the integral of f over R_q is the same as its integral over I. By 14.1, given any positive number ϵ there is an L-function $l(x) \leqq f(x)$ such that

$$(A) \qquad \int_I l(x)\,dx > \int_I f(x)\,dx - \frac{\epsilon}{2}.$$

By 13.1 there is a function $\varphi_0(x)$ continuous on I such that $\varphi_0(x) \geqq l(x)$ and

$$(B) \qquad \int_I \varphi_0(x)\,dx < \int_I l(x)\,dx + \frac{\epsilon}{2}.$$

Then by (A) and (B) we have

$$\begin{aligned} \int_I |f(x) - \varphi_0(x)|\,dx &\leqq \int_I \{|f(x) - l(x)| + |l(x) - \varphi_0(x)|\}\,dx \\ &= \int_I [f(x) - l(x)]\,dx + \int_I [\varphi_0(x) - l(x)]\,dx \\ &< \epsilon. \end{aligned}$$

The equation $f(x) = f(x)\psi(x)$ holds for all x; for if x is in J the factor $\psi(x)$ is 1, and otherwise $f(x) = 0$. Define

$$\begin{aligned} \varphi(x) &= \varphi_0(x)\psi(x) \textit{ on } I, \\ \varphi(x) &= 0 \textit{ elsewhere.} \end{aligned}$$

Then on I we have

$$\begin{aligned} |f(x) - \varphi_0(x)| &\geqq \psi(x)\,|f(x) - \varphi_0(x)| \\ &= |f(x)\psi(x) - \varphi_0(x)\psi(x)| \\ &= |f(x) - \varphi(x)|, \end{aligned}$$

so

$$\begin{aligned} \int_{R_q} |f(x) - \varphi(x)|\,dx &= \int_I |f(x) - \varphi(x)|\,dx \\ &\leqq \int_I |f(x) - \varphi_0(x)|\,dx \\ &< \epsilon. \end{aligned}$$

Next let $f(x)$ be any summable function. We may suppose that $f(x)$ is finite-valued, since the set on which $|f| = \infty$ has measure zero by 23.1 and re-defining $f(x)$ to be zero on this set leaves the integral of $| f(x) - \varphi(x) |$ unchanged. Define

$$f_n(x) = f(x)K_{W_n}(x).$$

As n increases, $f_n(x)$ converges everywhere to $f(x)$ and never exceeds the summable function $| f(x) |$, so by 29.3 the integral of $| f(x) - f_n(x) |$ approaches zero. We choose and fix n so large that

$$\int_{R_q} | f(x) - f_n(x) | \, dx < \frac{\epsilon}{2}.$$

But $f_n(x)$ vanishes except inside of W_n, so by the part of the proof already completed there is a function $\varphi(x)$ continuous on the whole space such that

$$\int_{R_q} | f_n(x) - \varphi(x) | \, dx < \frac{\epsilon}{2}.$$

The last two inequalities establish the conclusion of the theorem.

REMARK. As was shown in our definition of $\varphi(x)$, in 42.1 we may require that $\varphi(x)$ vanish except on a bounded set.

From this we can derive another continuity property of summable functions. For convenience we suppose that $f(x)$ is summable over the whole space R_q. This involves no loss of generality, for if $f(x)$ is summable over E we can define $f(x)$ to be zero on CE, and the function thus extended is summable over R_q.

42.2s. *If $f(x)$ is summable over R_q, and h is a vector $(h^{(1)}, \cdots h^{(q)})$, then*

$$\lim_{||h|| \to 0} \int_{R_q} | f(x + h) - f(x) | \, dx = 0.$$

Let ϵ be a positive number. By 42.1, there is a function $\varphi(x)$ continuous on the space R_q, vanishing outside of some interval W_n, and such that

$$\int_{R_q} | f(x) - \varphi(x) | \, dx < \frac{\epsilon}{3}.$$

If $| | h | | < 1$, the function $\varphi(x + h)$ vanishes except on W_{n+1}. Denoting by M the supremum of $| \varphi(x) |$ on W_n, which is finite, we see that both $\varphi(x)$ and $\varphi(x + h)$ have absolute values not greater than M on W_{n+1} and vanish outside of W_{n+1}, whence

$$| \varphi(x + h) - \varphi(x) | \leqq 2MK_{W_{n+1}}(x).$$

The right member of this inequality is summable, and the left member tends to zero with $| | h | |$, so by the remark after 29.3 the integral of the left member tends to zero. Therefore there is a positive number δ such that

$$\int_{R_q} |\varphi(x + h) - \varphi(x)| \, dx < \frac{\epsilon}{3}$$

if $||h|| < \delta$.

Now in the inequality

$$\int_{R_q} |f(x + h) - f(x)| \, dx \leqq \int_{R_q} |f(x + h) - \varphi(x + h)| \, dx + \int_{R_q} |\varphi(x + h) - \varphi(x)| \, dx + \int_{R_q} |\varphi(x) - f(x)| \, dx$$

the third term on the right is less than $\epsilon/3$, the first term on the right has the same value as the third term by 18.7, and the second term is less than $\epsilon/3$ if $||h|| < \delta$. So if $||h|| < \delta$ the left member of the inequality is less than ϵ, which is the conclusion of our theorem.

The next theorem is a generalization of 33.3.

42.3. *Let $f(x)$ be summable over an interval $[a, b]$. For almost all x_0 in $[a, b]$ the limit*

$$\lim_{h \to 0} \frac{1}{h} \int_{x_0}^{x_0 + h} |f(x) - f(x_0)| \, dx$$

exists and is equal to zero, h being restricted to be such that $x_0 + h$ is also in $[a, b]$.

Let $r_1, r_2, \cdots$ be an ordering of all the rationals. By 33.3, for each r_i the function

$$(A) \qquad F_i(x) \equiv \int_a^x |f(x) - r_i| \, dx$$

has a derivative which is finite and equal to its integrand for all x in $[a, b]$ except those of a set N_i of measure zero. The set $N = \bigcup N_i$ also has measure zero, by 19.13. Now let x_0 be any point in $[a, b] - N$, and let ϵ be a positive number. There is a rational number r_i such that $|f(x_0) - r_i| < \epsilon/2$. Since x_0 is not in N, it is not in N_i, so the function $F_i(x)$ has at x_0 a derivative equal to $|f(x_0) - r_i|$. Whether h is positive or negative,

$$\begin{aligned} 0 &\leqq \frac{1}{h} \int_{x_0}^{x_0 + h} |f(x) - f(x_0)| \, dx \\ &\leqq \frac{1}{h} \int_{x_0}^{x_0 + h} |f(x) - r_i| \, dx + \frac{1}{h} \int_{x_0}^{x_0 + h} |r_i - f(x_0)| \, dx \\ &= \frac{[F_i(x_0 + h) - F_i(x_0)]}{h} + |r_i - f(x_0)| . \end{aligned}$$

The last term is less than $\epsilon/2$, and as $h \to 0$ the first has a limit which is

$$F_i'(x_0) = |f(x_0) - r_i| < \frac{\epsilon}{2}.$$

So for all h near zero it is less than $\epsilon/2$. That is, if h is in a neighborhood of zero

$$\left| \frac{1}{h} \int_{x_0}^{x_0+h} |f(x) - f(x_0)| \, dx \right| < \epsilon,$$

which completes the proof.

Theorem 42.1 is a special case of the following theorem.

42.4s. *Let E be a measurable set. If $f(x)$ belongs to the class L_p on E $(p \geqq 1)$, that is, if $f(x)$ is measurable on E and $|f(x)|^p$ is summable over E, then for every positive number ϵ there is a function $\varphi(x)$ continuous on the whole space R_q and such that*

$$(A) \qquad \left[\int_E |f(x) - \varphi(x)|^p \, dx \right]^{1/p} < \epsilon.$$

As before, there is no loss of generality in supposing that E is the entire space R_q and that $f(x)$ is finite-valued on E. For each positive integer n and each x we define

$$\begin{aligned} f_n(x) &= f(x) && \text{if} \quad -n \leqq f(x) \leqq n, \\ f_n(x) &= n && \text{if} \quad f(x) > n, \\ f_n(x) &= -n && \text{if} \quad f(x) < -n. \end{aligned}$$

Then for each x the functions $f_n(x)$ approach $f(x)$ as $n \to \infty$. Therefore the function

$$|f_n(x) - f(x)|^p$$

approaches zero as $n \to \infty$, remaining less than the summable function $|f(x)|^p$. By 29.3, its integral approaches zero, and we can choose an n for which the integral is less than $(\epsilon/2)^p$; that is,

$$(B) \qquad \left[\int_{R_q} |f_n(x) - f(x)|^p \, dx \right]^{1/p} < \frac{\epsilon}{2}.$$

On the interval $[0, 2n]$ the function t^{p-1} is increasing, and has its greatest value $(2n)^{p-1}$ at $t = 2n$. That is,

$$(C) \qquad t^p \leqq (2n)^{p-1} t$$

if $0 \leqq t \leqq 2n$. By 42.1, there is a function $\varphi(x)$ continuous on R_q for which

$$(D) \qquad \int_{R_q} |f_n(x) - \varphi(x)| \, dx < \left(\frac{\epsilon}{2}\right)^p (2n)^{1-p}.$$

We may assume $|\varphi(x)| \leqq n$, since otherwise we may replace $\varphi(x)$ by n wherever $\varphi(x) > n$ and by $-n$ wherever $\varphi(x) < -n$ without disturbing the continuity of $\varphi(x)$ or the validity of the preceding inequality. Then $|f_n(x) - \varphi(x)| \leqq 2n$, so by (C)

$$|f_n(x) - \varphi(x)|^p \leqq (2n)^{p-1}|f_n(x) - \varphi(x)|$$

for all x. This and (D) together yield

$$\int_{R_q} |f_n(x) - \varphi(x)|^p\,dx < \left(\frac{\epsilon}{2}\right)^p,$$

or

$$(E) \qquad \left\{\int_{R_q} |f_n(x) - \varphi(x)|^p\,dx\right\}^{1/p} < \frac{\epsilon}{2}.$$

By Minkowski's inequality (24.7), inequalities (B) and (E) imply inequality (A), and the proof of the theorem is complete.

Remark. Since our $\varphi(x)$ was obtained from 42.1, we are again privileged to suppose that it vanishes except on a bounded set.

Exercise. If $f(x)$ is of class L_p on R_q $(p \geqq 1)$, and h is a vector $(h^{(1)}, \cdots, h^{(q)})$, then

$$\lim_{\|h\| \to 0} \left\{\int_{R_q} |f(x+h) - f(x)|^p\,dx\right\}^{1/p} = 0.$$

42.5s. *There exists a denumerable collection Σ of functions $s(x)$, each of which is a step-function on some interval W_n and vanishes identically outside of W_n, having the following property. For every measurable set E, every number $p \geqq 1$, every function of class L_p on E and every positive number ϵ there is a function $s(x)$ in the class Σ such that*

$$(A) \qquad \left\{\int_E |f(x) - s(x)|^p\,dx\right\}^{1/p} < \epsilon.$$

By 11.4, to each interval W_n: $-n \leqq x^i \leqq n$ there corresponds a denumerable collection Σ_n of step-functions on W_n such that each function $\varphi(x)$ continuous on W_n can be uniformly approximated by functions belonging to Σ_n. We extend the range of definition of each function $s(x)$ of Σ_n by setting it equal to 0 outside of W_n; and we now proceed to prove that the collection

$$\Sigma \equiv \Sigma_1 \cup \Sigma_2 \cup \Sigma_3 \cup \cdots$$

has the required properties.

As in 42.1, there is no loss of generality in supposing that E is the entire space R_q. Let $f(x)$ be defined and of class L_p $(p \geqq 1)$ on R_q, and let ϵ be a positive number. By 42.4 and the remark following it, there is a function $\varphi(x)$ continuous on R_q, vanishing except on the interior of some interval W_n, and such that

$$(B) \qquad \left\{ \int_{R_q} | f(x) - \varphi(x) |^p \right\}^{1/p} < \frac{\epsilon}{2}.$$

Let δ be a positive number so small that

$$\delta(mW_n)^{1/p} < \frac{\epsilon}{2}.$$

By the definition of the sets Σ_n, there is a function $s(x)$ in Σ_n which vanishes except on W_n and differs from $\varphi(x)$ by less than δ on W_n. Since $\varphi(x)$ also vanishes except on W_n, this implies

$$| s(x) - \varphi(x) | < \delta K_{W_n}(x)$$

for all x. Hence

$$\begin{aligned}(C) \quad \left\{ \int_{R_q} | s(x) - \varphi(x) |^p \, dx \right\}^{1/p} &= \left\{ \int_{W_n} | s(x) - \varphi(x) |^p \, dx \right\}^{1/p} \\ &\leqq \left\{ \int_{W_n} \delta^p \, dx \right\}^{1/p} \\ &= \left\{ \delta^p m W_n \right\}^{1/p} \\ &= \delta(mW_n)^{1/p} < \frac{\epsilon}{2}.\end{aligned}$$

By Minkowski's inequality (24.7), inequalities (B) and (C) imply inequality (A), which completes the proof of the theorem.

Theorem 42.5 has a corollary of considerable importance in the theory of Fourier series and other orthogonal expansions. This is the following theorem.

42.6s. (*Riemann-Lebesgue*): *If I is a closed interval, and $\{g_n(x)\}$ is a sequence of functions such that*

(i) *all the $g_n(x)$ have the same bound, $| g_n(x) | \leqq M$, on I;*

(ii) *each $g_n(x)$ is measurable on I;*

(iii) *on each subinterval I_0 of I the relation*

$$\lim_{n \to \infty} \int_{I_0} g_n(x) \, dx = 0$$

holds; then for every function $f(x)$ summable over I and for every subinterval I_0 of I it is true that

$$\lim_{n \to \infty} \int_{I_0} f(x) g_n(x) \, dx = 0.$$

The existence of the integrals mentioned in this theorem is assured by 22.1 and 22.4. Let ϵ be a positive number. By 42.5, with $p = 1$, there is a step-function $s(x)$ such that

$$\int_I | f(x) - s(x) | \, dx < \frac{\epsilon}{2M}.$$

Then

$$(A) \qquad \left| \int_{I_0} [f(x) - s(x)] g_n(x) \, dx \right| \leqq \int_I | f(x) - s(x) | \, | g_n(x) | \, dx$$
$$\leqq M \int_I | f(x) - s(x) | \, dx < \frac{\epsilon}{2}.$$

Since $s(x)$ is a step function, the interval I_0 can be subdivided into intervals $I_1, \cdots, I_k$ on each of which $s(x)$ is constant, say $s(x) = r_j$ for all x in I_j, $j = 1, \cdots, k$. Then

$$\left| \int_{I_0} s(x) g_n(x) \, dx \right| = \left| \sum_{j=1}^{k} \int_{I_j} r_j g_n(x) \, dx \right|$$
$$\leqq \sum_{j=1}^{k} | r_j | \left| \int_{I_j} g_n(x) \, dx \right|.$$

By hypothesis (iii), each of the last-mentioned integrals tends to zero as $n \to \infty$. So we can find an n_ϵ large enough so that

$$(B) \qquad \left| \int_{I_0} s(x) g_n(x) \, dx \right| < \frac{\epsilon}{2}$$

whenever $n > n_\epsilon$. Combining (A) and (B), we have

$$\left| \int_{I_0} f(x) g_n(x) \, dx \right| < \epsilon$$

whenever $n > n_\epsilon$, establishing the theorem.

We recall that if $f(x)$ is a function of a single real variable x and is defined and summable over the interval $[-\pi, \pi]$, its Fourier coefficients are defined to be

$$a_n = \frac{1}{\pi} \int_{-\pi}^{\pi} f(x) \cos nx \, dx, \qquad b_n = \frac{1}{\pi} \int_{-\pi}^{\pi} f(x) \sin nx \, dx.$$

The Riemann-Lebesgue theorem on the Fourier coefficients is

42.7. *If $f(x)$ is summable on $[-\pi, \pi]$ and a_n, b_n are its Fourier coefficients, then*

$$\lim_{n \to \infty} a_n = \lim_{n \to \infty} b_n = 0.$$

In theorem 42.6 we take $g_n(x) = \cos nx$. Then each $g_n(x)$ is measurable, being continuous; the inequality $|g_n(x)| \leqq 1$ holds for all n and all x; and if I_0 is an interval $[\alpha, \beta]$ contained in $[-\pi, \pi]$, then

$$\left| \int_{I_0} g_n(x)\,dx \right| = \left| \int_\alpha^\beta \cos nx\,dx \right| = \left| \left(\frac{1}{n}\right) [\sin n\beta - \sin n\alpha] \right| \leqq \frac{2}{n}.$$

Hence hypothesis (iii) of 42.6 is also satisfied. Therefore by 42.6 we have $\lim_{n\to\infty} a_n = 0$. Repeating the argument with $g_n(x) = \sin nx$ yields $\lim_{n\to\infty} b_n = 0$.

EXERCISE. If $f(x)$ is absolutely continuous on $[-\pi, \pi]$ and $f(-\pi) = f(\pi)$, then

$$\lim_{n\to\infty} na_n = \lim_{n\to\infty} nb_n = 0.$$

[Use 35.3, 36.1 and 42.7.]

EXERCISE. If $f(x)$ is bounded and monotonic, or more generally if $f(x)$ is of BV, on $[-\pi, \pi]$, then the numbers na_n and nb_n are bounded.

[Use 8.6 and 37.2.]

In 27.7 we showed that the indefinite integral of a summable function is absolutely continuous, and therefore of BV. By the use of 42.5 we can obtain the value of its total variation, as follows.

42.8. *If $f(x)$ is summable over $[a, b]$, and $F(x)$ is its integral from a to x, then*

$$(A) \qquad T_F[a, b] = \int_a^b |f(x)|\,dx.$$

If $a = \alpha_1 < \alpha_2 < \cdots < \alpha_n = b$ is a partition of $[a, b]$, then

$$\begin{aligned}\sum_{i=1}^{n-1} |F(\alpha_{i+1}) - F(\alpha_i)| &= \sum_{i=1}^{n-1} \left| \int_{\alpha_i}^{\alpha_{i+1}} f(x)\,dx \right| \\ &\leqq \sum_{i=1}^{n-1} \int_{\alpha_i}^{\alpha_{i+1}} |f(x)|\,dx \\ &= \int_a^b |f(x)|\,dx,\end{aligned}$$

so that

$$(B) \qquad T_F[a, b] \leqq \int_a^b |f(x)|\,dx.$$

If $f(x)$ is a constant, equation (A) is trivial. If $f(x)$ is a step-function, the total variation of $F(x)$ over each interval of constancy of $f(x)$ is the integral of $|f(x)|$ over that interval, so by addition equation (A) holds for the step-functions $f(x)$. Now let $f(x)$ be summable and let ϵ be a positive number. By 42.5, there is a step-function $s(x)$ on $[a, b]$ such that

$$(C) \qquad \int_a^b |f(x) - s(x)|\, dx < \epsilon.$$

If we denote by $S(x)$ the integral of $s(x)$ over $[a, x]$, and denote by $M(x)$ the integral of $s(x) - f(x)$ over $[a, x]$, then by 8.1

$$T_F[a, b] \geqq T_S[a, b] - T_M[a, b].$$

But equation (A) holds for the first of these and inequality (B) for the second, so by (C)

$$\begin{aligned} T_F[a, b] &\geqq \int_a^b |s(x)|\, dx - \int_a^b |s(x) - f(x)|\, dx \\ &\geqq \int_a^b |f(x)|\, dx - 2\epsilon. \end{aligned}$$

Since ϵ is arbitrary, this implies the reverse of inequality (B), so equation (A) holds.

The next theorem may also be regarded as a corollary of 42.5.

42.9s. *Let $f(x)$ be defined on the space R_q and equal to zero on the complement of a bounded set E. Let I be a closed interval containing E in its interior. In order that $f(x)$ be summable over E, it is necessary and sufficient that there exist a sequence $\{s_n(x)\}$ of step-functions on I such that*

$$(A) \qquad \lim_{n\to\infty} s_n(x) = f(x)$$

for almost all x in I, and

$$(B) \qquad \lim_{m,n\to\infty} \int_I |s_m(x) - s_n(x)|\, dx = 0.$$

In this case the integrals of the functions $s_n(x)$ converge, and

$$(C) \qquad \lim_{n\to\infty} \int_I s_n(x)\, dx = \int_E f(x)\, dx.$$

Suppose first that $f(x)$ is summable over E; it is then summable over I and in fact over R_q, since it vanishes except on E. By 42.5, for each positive integer n there is a function $s_n(x)$ of the type described in 42.5 for which

$$\int_{R_q} |f(x) - s_n(x)|\, dx < \frac{1}{n}.$$

On I the function $s_n(x)$ is a step-function, and the preceding inequality implies

$$(D) \qquad \int_I |f(x) - s_n(x)|\, dx < \frac{1}{n}.$$

If ϵ is a positive number and $E_{n,\epsilon}$ the set on which the integrand is at least ϵ, the integral is at least equal to $\epsilon m E_{n,\epsilon}$. So

$$\epsilon m E_{n,\epsilon} < \frac{1}{n},$$

and $mE_{n,\epsilon}$ tends to zero as n increases. That is, $s_n(x)$ converges to $f(x)$ in measure. By 28.7, a subsequence converges almost everywhere to $f(x)$; without loss of generality we may suppose that $\{s_n(x)\}$ is already such a sequence. From (D) we immediately deduce (B) and (C).

Conversely, suppose that there exists a sequence such as was described in the statement of the theorem. The measurable functions $s_n(x)$ converge almost everywhere to $f(x)$, so $f(x)$ is measurable by 28.8. Let ϵ be an arbitrary positive number. There is an integer p such that if m and n are at least equal to p we have

$$\int_I |s_n(x) - s_m(x)|\, dx < \frac{\epsilon}{2}.$$

Fix m at any value $\geqq p$ and let n increase. The integrand tends almost everywhere to $|f(x) - s_m(x)|$, so by 29.2 this is summable and

$$(E) \qquad \int_I |f(x) - s_m(x)|\, dx \leqq \frac{\epsilon}{2} < \epsilon \qquad (m \geqq p).$$

Now $f(x)$ is measurable and does not exceed the summable function $s_p(x) + |f(x) - s_p(x)|$, so it is summable by 22.2. And from (E) we have

$$\lim_{m\to\infty} \int_I |f(x) - s_m(x)|\, dx = 0,$$

which implies (C).

If we use the special definition of ΔI as the product of the edges of I, and define sets of measure zero as in 19.15, this gives us a means of passing directly from the integrals 10.4 of step functions to the Lebesgue integral. F. Riesz has developed the theory of the Lebesgue integral on this basis.*

*F. Riesz, *Sur l'intégrale de Lebesgue*, Acta Mathematica, vol. 42 (1919–1920), pp. 191–205.

43. Having found several methods in which the continuous functions and the measurable functions are related, we shall now show that every finite measurable function can be made into a continuous function by expunging an arbitrarily small portion of its range of definition.

43.1s. (Lusin.) *If $f(x)$ is finite and measurable on a measurable set E, then for every positive number ϵ there is a subset E_ϵ of E such that $m(E - E_\epsilon)$ is less than ϵ, and such that the function $f(x)$, when considered as defined on E_ϵ only, is continuous. Moreover, we can take E_ϵ to be a closed set.*

Let us first suppose that $f(x)$ is summable over E. Then, by 42.1, for every n there is a function $\varphi_n(x)$ continuous on I such that

$$\int_E | f(x) - \varphi_n(x) | \, dx < n^{-1}.$$

For every positive number δ let $E_{n,\delta}$ be the subset of E on which $| f(x) - \varphi_n(x) | \geqq \delta$. Then

$$n^{-1} > \int_E | f(x) - \varphi_n(x) | \, dx \geqq \int_{E_{n,\delta}} | f(x) - \varphi_n(x) | \, dx$$
$$\geqq \int_{E_{n,\delta}} \delta \, dx = \delta m E_{n,\delta},$$

whence $mE_{n,\delta} < 1/n\delta$. So $mE_{n,\delta}$ tends to 0 as $n \to \infty$, and by definition 28.3 we see that $\varphi_n(x)$ tends in measure to $f(x)$. By 28.7, there is a subsequence $\{\varphi_\alpha(x)\}$ of the sequence $\{\varphi_n(x)\}$ which tends to $f(x)$ almost uniformly. Hence (28.2) if ϵ is a positive number, there is a subset E_0 of E such that $m(E - E_0) < \epsilon/2$ and such also that $\lim \varphi_\alpha(x) = f(x)$ uniformly on E_0. Now on E_0 the function $f(x)$ is the limit of a uniformly convergent sequence of continuous functions $\varphi_\alpha(x)$, so it is itself continuous. (This does *not* mean that the function $f(x)$ defined on E is continuous at each point of E_0; it means that $f(x)$, considered as defined only on E_0, is continuous on E_0. In symbols, for each x_0 in E_0 and for every $\epsilon > 0$ there is a $\delta > 0$ such that $| f(x) - f(x_0) | < \delta$ if x is in E_0 and the distance $\| x, x_0 \|$ is less than δ.)

By 20.7, there is a closed subset E_ϵ of E_0 such that $m(E_0 - E_\epsilon) < \epsilon/2$. Since $f(x)$ is continuous on E_0, it is continuous on E_ϵ; and $m(E - E_\epsilon) = m(E - E_0) + m(E_0 - E_\epsilon) < \epsilon$. The theorem is then established when E is bounded and $f(x)$ is summable over E.

Suppose now that $f(x)$ is finite and measurable, but not summable, over the measurable set E. Define $\mu(z) = z/(1 + | z |)$; then $| \mu(z) | < 1$ for $- \infty < z < \infty$. This function has an inverse $\mu^{-1}(y) = y/(1 - | y |)$ defined and continuous on the interval

$-1 < y < 1$. Let $\varphi(x)$ be a positive-valued continuous function which is summable over R_q; for example, we could choose

$$\varphi(x) = \exp\left\{-\sum_{i=1}^{q} (x^{(i)})^2\right\}.$$

By 21.11, the function $\mu(f(x))$ is measurable over E, and its absolute value is always less than 1. By 22.5, $\varphi(x)$ is summable over E. So by 22.4 the product $\varphi(x)\mu(f(x))$ is summable over E. Then by the part of the proof just completed there is a closed subset E_ϵ of E such that $m(E - E_\epsilon) < \epsilon$ and $\varphi(x)\ \mu(f(x))$ is continuous on E_ϵ. But $\varphi(x)$ is continuous and positive, so $\mu(f(x))$ is continuous on E_ϵ. Its value is in the open interval $(-1, 1)$, so at $\mu(f(x))$ the inverse μ^{-1} is continuous. That is,

$$f(x) = \mu^{-1}(\mu(f(x)))$$

is continuous on E_ϵ, which completes the proof.

An interesting formulation of the preceding theorem can be obtained with the help of a known theorem on extension of range of functions: *If $f(x)$ is continuous on a closed set E in q-dimensional space R_q, there is a function $\varphi(x)$ continuous on the whole space R_q and identical with $f(x)$ on E.* Since this will be used only in proving 43.2, which we shall not need in any later proofs, we shall omit its proof. If the reader is interested, he will be able to find proofs in many places, for instance in a foot-note by Hassler Whitney (Transactions of the Amer. Math. Soc., vol. 36 (1934), p. 63) or in a note by the present author (Bulletin of the Amer. Math. Soc., vol. 40 (1934), p. 837.)

From this and 43.1 we now deduce the following theorem.

43.2s. *If $f(x)$ is finite and measurable on a measurable set E in q-dimensional space R_q, then for every positive number ϵ there is a function $\varphi(x)$ continuous on all of R_q such that $\varphi(x) = f(x)$ for all points of E except those of a set E_0 of measure less than ϵ.*

By 43.1, there is a closed set E_ϵ contained in E such that $m(E - E_\epsilon) < \epsilon$ and $f(x)$ is continuous on E_ϵ. As just remarked, there is a function $\varphi(x)$ continuous on all of R_q such that $\varphi(x) = f(x)$ on E_ϵ. If we define $E_0 = E - E_\epsilon$ the proof of the theorem is complete.

44. So far we have given no proof that non-measurable sets or functions can exist, or that measurable functions exist which are not contained in the Baire classes. This gap we now proceed to fill.

Consider the class of all real-valued functions $f(x)$ defined for all real numbers x and not identically zero. We classify these functions into pairs, each function being paired with its negative. Thus we obtain an infinitude of pairs $[f(x), \ -f(x)]$. These are to be understood to be non-ordered sets, so that the pair $[f(x), \ -f(x)]$ is the same as $[-f(x), \ f(x)]$. Thus each function (except $f(x) \equiv 0$) appears in exactly one pair, namely the pair composed of itself and its negative. Now from each pair we choose just one function. This gives us a class N of functions such that for each function $f(x)$ not identically zero either $f(x)$ is in N, or $-f(x)$ is in N, but not both.

For each irrational number a we define a function $f_a(x)$ which has the value $+1$ if $x = r + a$ where r is rational, has the value -1 if $x = r - a$ where r is rational, and is zero if x has neither of these forms. There is no ambiguity in these definitions; for if a number x had a representation $r_1 + a$ and another representation $r_2 - a$ we would have $r_1 + a = r_2 - a$, hence $a = \frac{1}{2}(r_2 - r_1)$. This is impossible, since a is irrational.

The functions $f_a(x)$ allow us to classify the irrationals into two classes. Let H be the class of all irrationals a such that $f_a(x)$ belongs to the class N, and let K be the class of all irrationals such that $f_a(x)$ is not in the class N. Clearly $H \cup K$ is the class of all irrationals, while no number a belongs both to H and to K.

A property of these sets which will be needed is the following.

(*A*) *Let r be any rational number. For every irrational number a, the number $b = 2r - a$ belongs to H if a belongs to K and belongs to K if a belongs to H.*

We first show that $f_b(x) = -f_a(x)$ for all x. If $f_a(x) = 1$, then $x = r_1 + a$, where r_1 is rational. But $a = 2r - b$, so $x = r_1 + 2r - b$. Here $r_1 + 2r$ is rational, so by definition of $f_b(x)$ we find $f_b(x) = -1$. If $f_a(x) = -1$, then $x = r_2 - a$, where r_2 is rational. Then $x = r_2 - (2r - b) = (r_2 - 2r) + b$. Here $r_2 - 2r$ is rational, so $f_b(x) = +1$. Since a and b enter symmetrically in this discussion, we may interchange them in the preceding statements. This proves that $f_b(x) = 1$ if and only if $f_a(x) = -1$ and $f_b(x) = -1$ if and only if $f_a(x) = +1$. The only other value allowed to $f_a(x)$ and $f_b(x)$ is zero, so this implies that $f_a(x) = 0$ if and only if $f_b(x) = 0$. Therefore $f_b(x) = -f_a(x)$. By the definition of the class N, exactly one of the two functions $f_a(x)$ and $f_b(x) = -f_a(x)$ is in N. Therefore exactly one of the two numbers a and b is in H. Since each of the numbers a, b is in either H or K, we see that one of them is in H and the other is in K. This establishes our statement.

Now we shall show that the sets H, K, CH and CK all have interior measure zero. Since H and K are disjoint, H is contained in CK and K is contained in CH, so by 20.3 it is enough to prove that CH and CK

have interior measure zero. We give the proof for CK; the proof for CH requires only the interchange of the letters H and K.

Suppose that CK has positive interior measure. By 20.1, there is a closed set F contained in CK and having positive measure. By 41.2 there is a point x_0 at which F has metric density 1. Referring to the definition (41.1) of metric density, this implies that there are *rational* numbers α, β such that $\alpha < x_0 < \beta$ and

$$\frac{m\,(F \cap [\alpha, \beta])}{(\beta - \alpha)} > \frac{1}{2}.$$

Now let G be the set obtained by removing all the rationals from the set $F \cap [\alpha, \beta]$. Since the set of all rationals has measure zero by 19.14, this set is measurable, and

(*B*) *G is contained in $[\alpha, \beta]$ and $mG = m(F \cap [\alpha, \beta]) > \frac{1}{2}(\beta - \alpha)$.*

The set F is contained in CK, which consists of H plus the rationals, so G is contained in H. Let G^* be the set of all numbers $\alpha + \beta - x$ with x in G. The number $r = (\alpha + \beta)/2$ is rational, and if x is in G it is in H; so by statement (*A*) all the points of G^* are in K, therefore not in H. So

(*C*) *G and G^* are disjoint.*

If x is in G, then $\alpha \leqq x \leqq \beta$, so $\alpha \leqq \alpha + \beta - x \leqq \beta$, and

(*D*) G^* is contained in $[\alpha, \beta]$.

From the definition of G^* we have

$$K_{G^*}(\alpha + \beta - x) \equiv K_G(x).$$

The function $\alpha + \beta - x$ is AC and monotonic, so by 38.1 or 38.4

$$(E) \quad mG^* = \int_\alpha^\beta K_{G^*}(x)\,dx = \int_\beta^\alpha K_{G^*}(\alpha + \beta - x)(-1)\,dx$$
$$= \int_\alpha^\beta K_G(x)\,dx = mG.$$

By (*B*), (*C*) and (*E*), together with 19.8,

$$m(G \cup G^*) = mG + mG^* = 2mG > \beta - \alpha.$$

But $G \cup G^*$ is contained in $[\alpha, \beta]$, and by 19.3 we must have $m(G \cup G^*) \leqq \beta - \alpha$. This contradiction proves that CK has interior measure 0.

This enables us to show that the sets H and K are not merely non-measurable, but that for every set E of positive measure the sets EH and EK are non-measurable. Consider H, to be specific. Let E be a set of positive measure. For some interval W_n the set EW_n has finite positive measure. The sets $H \cap EW_n$ and $CH \cap EW_n$ have interior measure zero by 20.3. So by 20.9 they each have exterior measure equal to mEW_n, which is positive. By 20.8, neither is measurable. But if HE were measurable the intersection $H \cap EW_n$ would also be measurable; so HE is non-measurable. Likewise $E \cap CH$ is non-measurable. The same proof applies to EK and $E \cap CK$.

It is easy to see that the characteristic function $K_H(x)$ is non-measurable, and so is $K_{HE}(x)$ for every set E of positive measure. For the set of x such that $K_H(x) > \frac{1}{2}$ (or $K_{HE}(x) > \frac{1}{2}$) consists of the set H (or HE) itself, and this is non-measurable.

In §9 we constructed a set P contained in the interval [0, 1] and a function $f(x)$, defined, continuous and monotonic increasing on [0, 1], such that

(i) *the set P is closed, and its complement is a set of open intervals* $I_1, I_2, \cdots$ *such that* $\Sigma mI_j = 1$;

(ii) $f(x)$ *is constant on each interval* I_j;

(iii) $f(0) = 0$ *and* $f(1) = 1$.

Define now $g(x) = f(x) + x$. This function is continuous, and $g(0) = 0$, $g(1) = 2$. It is strictly increasing, for if $x_2 > x_1$ then $g(x_2) - g(x_1) = f(x_2) - f(x_1) + x_2 - x_1 \geqq x_2 - x_1 > 0$. Hence it has a continuous inverse function $g^{-1}(t)$, $0 \leqq t \leqq 2$.

Consider now the image of an interval $I_j \equiv (\alpha_j, \beta_j)$ under the mapping $t = g(x)$. On I_j the function $f(x)$ has a constant value c_j, by (ii). So on this interval $g(x) \equiv c_j + x$, and the interval (α_j, β_j) is mapped on $(c_j + \alpha_j, c_j + \beta_j)$. That is, the image of each I_j is an open interval T_j of the same length as I_j. The intervals T_j do not overlap, since $g(x)$ is monotonic. Hence

$$m \bigcup T_j = \Sigma m T_j = \Sigma m I_j = 1.$$

But the interval [0, 2] has measure 2, and all points not in $\bigcup T_j$ are necessarily in the image Q of the set P. This set is closed, being the complement of the open set $\bigcup T_j$; so it is measurable, and

$$mQ = m[0, 2] - m \bigcup T_j = 1.$$

With the set H of the preceding example, we now define the function $f(t)$ to be the characteristic function of the set HQ. This we have already seen to be non-measurable. Define $\varphi(x) = f(g(x))$, $0 \leqq x \leqq 1$. Then $\varphi(x) = 0$ unless $g(x)$ is in HQ, which is true only if x is in P. Hence $\varphi(x)$ vanishes except on a subset of the set P of measure zero, and it is a measurable function. If it were of a finite

Baire class, by 40.5 the function $\varphi(g^{-1}(t))$ would be measurable on $[0, 2]$. But $\varphi(g^{-1}(t)) = f(t)$, which is non-measurable. So $\varphi(x)$ is a measurable function which is not of Baire class p for any integer p.

If it is desired to exhibit a non-measurable set in q-dimensional space R_q, we need only to define E to be the set of all points $(x^1, \cdots, x^q)$ with x^1 in the set H previously constructed. This is non-measurable. For if it were measurable, then for almost all points $(x^2, \cdots x^q)$ of $(q - 1)$ space the set of all x^1 with $(x^1, \cdots, x^q)$ in E would be measurable (25.4). But this set of x^1 is always the non-measurable set H.

The example constructed above also serves to show that a measurable function of a continuous function is not necessarily measurable. For $\varphi(x)$ is a measurable function of x and $g^{-1}(t)$ is a continuous function of t, but $\varphi(g^{-1}(t))$ is the non-measurable function $f(t)$.

CHAPTER VII

The Lebesgue-Stieltjes Integral

45. In elementary text-books on physics the reader will frequently encounter such statements as "The moment of inertia of a body about the z-axis is $\int (x^2 + y^2)dm$." This is usually purely heuristic, and the integral is reduced to the form

$$\int (x^2 + y^2)\rho(x, y, z)\, dx\, dy\, dz$$

before any use is made of it (ρ being the density). Nevertheless, the fact remains that a new kind of integral has introduced itself in a natural way, and it seems reasonable to assume that such an integral, if rigorously defined and discussed, would be useful in physics. As a matter of fact, this new kind of integral is of great importance in several branches of pure mathematics. However, we must prepare for its study by discussing functions of several variables which have bounded variation.

The definition of total variation which we are about to give is by no means the only possible generalization of the notion in §8. It is, however, the one which is useful to us here, and other possible definitions will not be discussed.

In studying functions $f(x)$ of one variable we frequently made use of the difference $f(\beta) - f(\alpha)$, to which we could have assigned a symbol,

$$f(x) \Big|_{\alpha}^{\beta} = f(\beta) - f(\alpha).$$

This symbol we now generalize. Let $\alpha = (\alpha^{(1)}, \cdots, \alpha^{(q)})$ and $\beta = (\beta^{(1)}, \cdots, \beta^{(q)})$ be two points of an interval on which $g(x)$ is defined. There are 2^q points $v_j = (v_j^{(1)}, \cdots, v_j^{(q)})$ $(j = 1, \cdots, 2^q)$ such that each $v_j^{(i)}$ is either $\alpha^{(i)}$ or $\beta^{(i)}$. Let $n(v_j)$ be the number of lower symbols $\alpha^{(i)}$ among the coordinates of v_j. Then

$$g(x) \Big|_{\alpha}^{\beta} \equiv \sum_{v=1}^{2^q} (-1)^{n(v_j)} g(v_j).$$

For example, if $q = 1$, then

$$g(x) \Big|_{\alpha}^{\beta} = g(\beta^{(1)}) - g(\alpha^{(1)}).$$

If $q = 2$, then

$$g(x) \Big|_{\alpha}^{\beta} = g(\beta^{(1)}, \beta^{(2)}) - g(\alpha^{(1)}, \beta^{(2)}) - g(\beta^{(1)}, \alpha^{(2)}) + g(\alpha^{(1)}, \alpha^{(2)}).$$

If $q = 3$, then

$$g(x)\,\Big|_\alpha^\beta = g(\beta^{(1)}, \beta^{(2)}, \beta^{(3)}) - g(\alpha^{(1)}, \beta^{(2)}, \beta^{(3)}) - g(\beta^{(1)}, \alpha^{(2)}, \beta^{(3)})$$
$$- g(\beta^{(1)}, \beta^{(2)}, \alpha^{(3)}) + g(\alpha^{(1)}, \alpha^{(2)}, \beta^{(3)}) + g(\alpha^{(1)}, \beta^{(2)}, \alpha^{(3)})$$
$$+ g(\beta^{(1)}, \alpha^{(2)}, \alpha^{(3)}) - g(\alpha^{(1)}, \alpha^{(2)}, \alpha^{(3)});$$

and so on.

EXERCISE. Show that the operator $f(x)\,\Big|_\alpha^\beta$ in q-dimensional space can be generated from the preceding operator $f(x)\,\Big|_\alpha^\beta = f(\beta) - f(\alpha)$ by first applying $\Big|_{\alpha^{(1)}}^{\beta^{(1)}}$ to the variable $x^{(1)}$ in $f(x)$, then applying $\Big|_{\alpha^{(2)}}^{\beta^{(2)}}$ to the variable $x^{(2)}$ in the result of the first operator, and so on.

For functions $g(x)$, $h(x)$ of a single real variable the formulas

(a) $g(x)\,\Big|_\alpha^\gamma + g(x)\,\Big|_\gamma^\beta = g(x)\,\Big|_\alpha^\beta$,

(b) $g(x)\,\Big|_\alpha^\alpha = 0$,

(c) $g(x)\,\Big|_\beta^\alpha = -g(x)\,\Big|_\alpha^\beta$,

(d) if $g(x)$ is constant, then $g(x)\,\Big|_\alpha^\beta \equiv 0$,

(e) $[ag(x) + bh(x)]\,\Big|_\alpha^\beta = a(g(x)\,\Big|_\alpha^\beta) + b(h(x)\,\Big|_\alpha^\beta)$.

are very easily verified. They have their analogues for functions of several variables, also, but these are naturally a little more complicated.

45.1s. *Let $g(x)$ and $h(x)$ be defined and finite on an interval I^*, and let the points*

$$\alpha = (\alpha^{(1)}, \cdots, \alpha^{(q)}), \qquad \beta = (\beta^{(1)}, \cdots, \beta^{(q)}),$$
$$\alpha' = (\alpha^{(1)}, \cdots, \alpha^{(j-1)}, \beta^{(j)}, \alpha^{(j+1)}, \cdots, \alpha^{(q)}),$$
$$\beta' = (\beta^{(1)}, \cdots, \beta^{(j-1)}, \alpha^{(j)}, \beta^{(j+1)}, \cdots, \beta^{(q)}),$$
$$\gamma_1 = (\alpha^{(1)}, \cdots, \alpha^{(j-1)}, \gamma^{(j)}, \alpha^{(j+1)}, \cdots, \alpha^{(q)}),$$
$$\gamma_2 = (\beta^{(1)}, \cdots, \beta^{(j-1)}, \gamma^{(j)}, \beta^{(j+1)}, \cdots, \beta^{(q)})$$

all lie in I^*. Then

(a) $g(x)\,\Big|_\alpha^{\gamma_2} + g(x)\,\Big|_{\gamma_1}^\beta = g(x)\,\Big|_\alpha^\beta$,

(b) $g(x)\,\Big|_\alpha^{\beta'} = 0$,

(c) $g(x)\,\Big|_{\alpha'}^{\beta'} = -g(x)\,\Big|_\alpha^\beta$,

(d) *if $g(x)$ is independent of any one of the variables $x^{(1)}, \cdots, x^{(q)}$, then $g(x)\,\Big|_\alpha^\beta = 0$,*

(e) *for all finite numbers a and b,*

$$[ag(x) + bh(x)]\,\Big|_\alpha^\beta = a[g(x)\,\Big|_\alpha^\beta] + b[h(x)\,\Big|_\alpha^\beta].$$

In forming the differences in (a) we need the values of $g(x)$ at those points x for which $x^{(k)} = \alpha^{(k)}$ or $x^{(k)} = \beta^{(k)}$, $k \neq j$, while $x^{(j)}$ is either $\alpha^{(j)}$, $\beta^{(j)}$ or $\gamma^{(j)}$. There are 2^{q-1} of these points with the j-th coordinate equal to $\alpha^{(j)}$, and likewise for $\beta^{(j)}$ and for $\gamma^{(j)}$. We denote the points of these three aggregates by u_i, v_i, w_i $(i = 1, \cdots, 2^{q-1})$

respectively; thus $u_i^{(k)}$, $v_i^{(k)}$ and $w_i^{(k)}$ are each equal either to $\alpha^{(k)}$ or to $\beta^{(k)}$, $k \neq j$, while $u_i^{(j)} = \alpha^{(j)}$, $v_i^{(j)} = \beta^{(j)}$, $w_i^{(j)} = \gamma^{(j)}$. If x is one of the points u_i, v_i or w_i we define $n(x)$ to be the number of symbols $\alpha^{(1)}$, $\cdots$, $\alpha^{(q)}$ among the coordinates of x, and we define $n'(x)$ to be the number of symbols $\alpha^{(1)}, \cdots, \alpha^{(j-1)}, \gamma^{(j)}, \alpha^{(j+1)}, \cdots, \alpha^{(q)}$ among the coordinates of x. In calculating $g(x)\,|_\alpha^\beta$ and $g(x)\,|_\alpha^{\gamma_2}$ the lower symbols are the $\alpha^{(i)}$, so by definition

$$(A)\qquad g(x)\,|_\alpha^\beta = \sum_{i=1}^{2^{q-1}} (-1)^{n(u_i)} g(u_i) + \sum_{i=1}^{2^{q-1}} (-1)^{n(v_i)} g(v_i),$$

$$(B)\qquad g(x)\,|_\alpha^{\gamma_2} = \sum_{i=1}^{2^{q-1}} (-1)^{n(u_i)} g(u_i) + \sum_{i=1}^{2^{q-1}} (-1)^{n(w_i)} g(w_i).$$

In calculating $g(x)\,|_{\gamma_1}^\beta$ the lower symbols are $\alpha^{(1)}, \cdots \alpha^{(j-1)}, \gamma^{(j)}, \alpha^{(j+1)}$, $\cdots$, $\alpha^{(q)}$, so

$$(C)\qquad g(x)\,|_{\gamma_1}^\beta = \sum_{i=1}^{2^{q-1}} (-1)^{n'(w_i)} g(w_i) + \sum_{i=1}^{2^{q-1}} (-1)^{n'(v_i)} g(v_i).$$

Each w_i has $\gamma^{(j)}$ for its j-th coordinate, and so $\alpha^{(i)}$ is not its j-th coordinate. By the meaning of $n(w_i)$ and $n'(w_i)$ this shows that $n'(w_i) = n(w_i) + 1$. Each v_i has $\beta^{(j)}$ for its j-th coordinate, so in calculating $n(v_i)$ and $n'(v_i)$ the j-th coordinate is not counted, and $n(v_i) = n'(v_i)$. So adding equations (B) and (C) member by member yields

$$g(x)\,|_\alpha^{\gamma_2} + g(x)\,|_{\gamma_1}^\beta = \sum_{i=1}^{2^{q-1}} (-1)^{n(u_i)} g(u_i) + \sum_{i=1}^{2^{q-1}} (-1)^{n(v_i)} g(v_i)$$
$$= g(x)\,|_\alpha^\beta,$$

establishing (a).

If in (a) we set $\gamma^{(j)} = \alpha^{(j)}$, then $\gamma_1 = \alpha$ and $\gamma_2 = \beta'$, and (a) becomes

$$g(x)\,|_\alpha^{\beta'} + g(x)\,|_\alpha^\beta = g(x)\,|_\alpha^\beta,$$

establishing (b). Further, let us write (a) with β' in place of β, and then choose $\gamma^{(j)} = \beta^{(j)}$. This gives $\gamma_1 = \alpha'$, $\gamma_2 = \beta$, and by (a) and (b)

$$g(x)\,|_\alpha^\beta + g(x)\,|_{\alpha'}^{\beta'} = g(x)\,|_\alpha^{\beta'} = 0.$$

Thus (c) is established.

To prove (d), let $g(x)$ be independent of $x^{(j)}$. We write out the expression for $g(x)\,|_\alpha^\beta$ and in it replace $\beta^{(j)}$ by $\alpha^{(j)}$ whenever $\beta^{(j)}$ occurs. This alters no term, because $g(x)$ is independent of $x^{(j)}$. Hence after the substitution the sum is still equal to $g(x)\,|_\alpha^\beta$. But if β^j is replaced by $\alpha^{(j)}$, the sum has the form $g(x)\,|_{\alpha}^{\beta'}$, which is zero. Therefore $g(x)\,|_{\alpha}^{\beta} = 0$.

Statement (*e*) is evident, for with the notation of the definition (*not* that of the preceding paragraphs of this proof!)

$$\begin{aligned}[ag(x) + bh(x)]\,|_{\alpha}^{\beta} &= \sum_{j=1}^{2^q} (-1)^{n(v_j)}[ag(v_j) + bh(v_j)] \\ &= a \sum_{j=1}^{2^q} (-1)^{n(v_j)} g(v_j) + b \sum_{j=1}^{2^q} (-1)^{n(v_j)} h(v_j) \\ &= a(g(x)\,|_{\alpha}^{\beta}) + b(h(x)\,|_{\alpha}^{\beta}).\end{aligned}$$

EXERCISE. Show that if γ is a fixed point, then

$$(g(x)\,|_{\gamma}^{x})\,|_{\alpha}^{\beta} = g(x)\,|_{\alpha}^{\beta}.$$

Next we define the symbol $\Delta_g I$.

45.2s. *If $g(x)$ is defined at the vertices of an interval I whose closure is defined by the inequalities $\alpha^{(i)} \leqq x^{(i)} \leqq \beta^{(i)}$, $i = 1, \cdots, q$, then*

$$\Delta_g I \equiv g(x)\,|_{\alpha}^{\beta}.$$

Thus if $g(x)$ is defined on an interval I^*, the expression $\Delta_g I$ assigns a number to each subinterval I of I^*, and is therefore called a function of intervals. This number $\Delta_g I$ does not depend on the topological character (openness or closedness) of I, but only on its vertices. It has the property of being additive; that is:

45.3s. *If $g(x)$ is defined and finite on I^*, and I is a subinterval of I^* which is the sum of two non-overlapping intervals I_1 and I_2, then*

$$\Delta_g I = \Delta_g I_1 + \Delta_g I_2.$$

(Recall that two intervals are non-overlapping if no point of either is interior to the other.)

Suppose that the closures of I, I_1, I_2 are defined respectively by the inequalities

$$\alpha^{(i)} \leqq x^{(i)} \leqq \beta^{(i)}, \qquad i = 1, \cdots, q;$$
$$\alpha^{(i)} \leqq x^{(i)} \leqq \beta^{(i)},\ i = 1, \cdots, j-1, j+1, \cdots, q;\ \alpha^{(j)} \leqq x^{(j)} \leqq \gamma^{(j)};$$
$$\alpha^{(i)} \leqq x^{(i)} \leqq \beta^{(i)},\ i = 1, \cdots, j-1, j+1, \cdots, q;\ \gamma^{(j)} \leqq x^{(j)} \leqq \beta^{(j)}.$$

With the notation of 45.1, we have

$$\Delta_g I = g(x)\,|_{\alpha}^{\beta}, \qquad \Delta_g I_1 = g(x)\,|_{\alpha}^{\gamma_2}, \qquad \Delta_g I_2 = g(x)\,|_{\gamma_1}^{\beta}.$$

Our theorem is then a re-statement of 45.1(*a*).

Our interval-function $\Delta_g I$ has the properties 10.1 except that it may have negative values. Since the proofs of 10.2 and 10.3 made no use of the non-negativeness of ΔI, they continue to hold for $\Delta_g I$ also. That is, we have the following theorem.

45.4s. *If $g(x)$ is defined and finite on I^*, and the subinterval I of I^* is the sum of non-overlapping intervals $I_1, \cdots, I_n$, then*

$$\Delta_g I = \sum_{i=1}^{n} \Delta_g I_i.$$

We have seen that a function of intervals of the form $\Delta_g I$ is always additive. However, we may encounter additive functions of intervals defined in other ways. For instance, in the plane ($q = 2$) the area of I is an additive function of intervals I; so is the function $\varphi[I]$ which is the length of the segment of the line $x^{(2)} = x^{(1)}$ contained in I. The question suggests itself: Can all such functions be represented in the form $\Delta_g I$ for some properly chosen function $f(x)$? This is certainly true for both of our examples; for the first example we can choose $g(x^{(1)}, x^{(2)}) = x^{(1)} x^{(2)}$, and for the second we can take $g(x^{(1)}, x^{(2)}) = -\sqrt{2} \sup (x^{(1)}, x^{(2)})$. We now prove that it is true in general.

45.5s. *If $\varphi[I]$ is an additive function of intervals in q-dimensional space R_q, there is a function $g(x)$, defined and finite for all x in R_q, such that $\Delta_g I = \varphi[I]$ for all intervals I in R_q.*

If α and β are any two points, we denote by $\nu(\alpha, \beta)$ the number of superscripts i such that $\alpha^{(i)} > \beta^{(i)}$, and we denote by $I(\alpha, \beta)$ the interval $\inf \{\alpha^{(i)}, \beta^{(i)}\} \leqq x^{(i)} \leqq \sup \{\alpha^{(i)}, \beta^{(i)}\}$, $i = 1, \cdots, q$. Then we define

$$\psi\binom{\beta}{\alpha} \equiv (-1)^{\nu(\alpha,\beta)} \varphi[I(\alpha, \beta)].$$

Thus, in particular, if $\alpha^{(i)} \leqq \beta^{(i)}$ for every i, then

$$\psi\binom{\beta}{\alpha} = \varphi[I],$$

where I is the interval $\alpha^{(i)} \leqq x^{(i)} \leqq \beta^{(i)}$, $i = 1, \cdots, q$

With the notation of 45.1 we see that

$$\psi\binom{\beta'}{\alpha'} = -\psi\binom{\beta}{\alpha}.$$

For if $\alpha^{(j)} \neq \beta^{(j)}$, in counting the number $\nu(\alpha', \beta')$ of coordinates of α' in the left member which exceed the corresponding coordinates of β' and in counting the corresponding number $\nu(\alpha, \beta)$ on the right there is a difference of 1, because of the interchange of $\alpha^{(j)}$ and $\beta^{(j)}$; while if $\alpha^{(j)} = \beta^{(j)}$ both members of the equation vanish. Next we prove that the function ψ satisfies the identity

$$\psi\begin{pmatrix}\beta^{(1)}, & \cdots, & \beta^{(j-1)}, \beta^{(j)}, \beta^{(j+1)}, & \cdots, & \beta^{(q)}\\ \alpha^{(1)}, & \cdots, & \alpha^{(j-1)}, \alpha^{(j)}, \alpha^{(j+1)}, & \cdots, & \alpha^{(q)}\end{pmatrix} + \psi\begin{pmatrix}\beta^{(1)}, & \cdots, & \beta^{(j-1)}, \gamma^{(j)}, \beta^{(j+1)}, & \cdots, & \beta^{(q)}\\ \alpha^{(1)}, & \cdots, & \alpha^{(j-1)}, \beta^{(j)}, \alpha^{(j+1)}, & \cdots, & \alpha^{(q)}\end{pmatrix} + \psi\begin{pmatrix}\beta^{(1)}, & \cdots, & \beta^{(j-1)}, \alpha^{(j)}, \beta^{(j+1)}, & \cdots, & \beta^{(q)}\\ \alpha^{(1)}, & \cdots, & \alpha^{(j-1)}, \gamma^{(j)}, \alpha^{(j+1)}, & \cdots, & \alpha^{(q)}\end{pmatrix} = 0.$$

We may suppose that $\alpha^{(k)} \leqq \beta^{(k)}$ for $k \neq j$; otherwise we would interchange $\alpha^{(k)}$ and $\beta^{(k)}$ in each term, thereby changing the sign of each term. Also, we may suppose that $\alpha^{(j)} \leqq \gamma^{(j)} \leqq \beta^{(j)}$; otherwise we would need only to re-name the numbers so as to bring this about. Then the intervals

$$I_1 : \alpha^{(k)} \leqq x^{(k)} \leqq \beta^{(k)}, \qquad k \neq j; \qquad \alpha^{(j)} \leqq x^{(j)} \leqq \gamma^{(j)}$$

and

$$I_2 : \alpha^{(k)} \leqq x^{(k)} \leqq \beta^{(k)}, \qquad k \neq j; \qquad \gamma^{(j)} \leqq x^{(j)} \leqq \beta^{(j)}$$

are non-overlapping, and their sum is

$$I : \alpha^{(i)} \leqq x^{(i)} \leqq \beta^{(i)}, \qquad i = 1, \cdots, q.$$

Hence $\varphi[I] = \varphi[I_1] + \varphi[I_2]$; that is

$$\psi\binom{\beta}{\alpha} = \psi\begin{pmatrix}\beta^{(1)}, & \cdots, & \beta^{(j-1)}, \gamma^{(j)}, \alpha^{(j+1)}, & \cdots, & \alpha^{(q)}\\ \alpha^{(1)}, & \cdots, & \alpha^{(j-1)}, \alpha^{(j)}, \alpha^{(j+1)}, & \cdots, & \alpha^{(q)}\end{pmatrix} + \psi\begin{pmatrix}\beta^{(1)}, & \cdots, & \beta^{(j-1)}, \beta^{(j)}, \beta^{(j+1)}, & \cdots, & \beta^{(q)}\\ \alpha^{(1)}, & \cdots, & \alpha^{(j-1)}, \gamma^{(j)}, \alpha^{(j+1)}, & \cdots, & \alpha^{(q)}\end{pmatrix}.$$

If we interchange $\alpha^{(j)}$ and $\gamma^{(j)}$ in the first term on the right and interchange $\gamma^{(j)}$ and $\beta^{(j)}$ in the second, and then transpose, the identity is established.

Now let I be an interval $\alpha^{(i)} \leqq x^{(i)} \leqq \beta^{(i)}$. Then, by the definition of ψ and the identity just proved,

$$\varphi[I] = \psi\binom{\beta}{\alpha} = \psi\begin{pmatrix}0, \beta^{(2)}, & \cdots, & \beta^{(q)}\\ \alpha^{(1)}, \alpha^{(2)}, & \cdots, & \alpha^{(q)}\end{pmatrix} + \psi\begin{pmatrix}\beta^{(1)}, \beta^{(2)}, & \cdots, & \beta^{(q)}\\ 0, \alpha^{(2)}, & \cdots, & \alpha^{(q)}\end{pmatrix}.$$

If we insert a 0 in the second column of arguments of each term on the right, we obtain

$$\varphi[I] = \psi\begin{pmatrix}0, 0, \beta^{(3)}, & \cdots, & \beta^{(q)}\\ \alpha^{(1)}, \alpha^{(2)}, \alpha^{(3)}, & \cdots, & \alpha^{(q)}\end{pmatrix} + \psi\begin{pmatrix}0, \beta^{(2)}, \beta^{(3)}, & \cdots, & \beta^{(q)}\\ \alpha^{(1)}, 0, \alpha^{(3)}, & \cdots, & \alpha^{(q)}\end{pmatrix} + \psi\begin{pmatrix}\beta^{(1)}, 0, \beta^{(3)}, & \cdots, & \beta^{(q)}\\ 0, \alpha^{(2)}, \alpha^{(3)}, & \cdots, & \alpha^{(q)}\end{pmatrix} + \psi\begin{pmatrix}\beta^{(1)}, \beta^{(2)}, \beta^{(3)}, & \cdots, & \beta^{(q)}\\ 0, 0, \alpha^{(3)}, & \cdots, & \alpha^{(q)}\end{pmatrix}.$$

Proceeding thus, we express $\varphi[I]$ as the sum of 2^q values of ψ; in each term, the j-th column is either $\binom{0}{\alpha^{(j)}}$ or $\binom{\beta^{(j)}}{0}$. If we interchange $\alpha^{(j)}$ and 0 in each column of the form $\binom{0}{\alpha^{(j)}}$, the lower row of arguments becomes $(0, \cdots, 0)$, while the upper is of the form $v_j = (v_j^{(1)}, \cdots, v_j^{(q)})$, each $v_j^{(i)}$ being either $\alpha^{(i)}$ or $\beta^{(i)}$. If $n(v_j)$ is the number of $\alpha^{(i)}$'s present among the numbers $v_j^{(1)}, \cdots, v_j^{(q)}$, then $n(v_j)$ is also the number of inversions of columns, since each $\alpha^{(i)}$ reaches the upper row only by inverting a column; so the sign of $\psi\binom{v_j}{0}$ is the same as that of the term from which it arose if $n(v_j)$ is even, and is different if $n(v_j)$ is odd.

Therefore, if we denote $\psi\binom{x}{0}$ by $g(x)$, then $\varphi[I]$ is the sum of 2^q terms, each term being of the form $(-1)^{n(v_j)}g(v_j)$. But then by definition

$$\varphi[I] = \Delta_g I,$$

which was to be proved.

One minor consequence of this theorem is useful in that it allows us, if we wish, to extend functions $g(x)$ defined on an interval I^* in such a way that the extension is defined on the whole space, and $\Delta_g I$ is not altered for intervals I in I^*.

45.6s. *If $g(x)$ is defined and finite on an interval I^*, there is a function $h(x)$, defined and finite on the whole space, such that $\Delta_h I = \Delta_g(I \cap I^*)$ for all intervals I. In particular, $\Delta_h I = \Delta_g I$ if I is contained in I^*.*

Let $\varphi[I]$ be the set function $\Delta_g(I \cap I^*)$. This is additive; for if I is the sum of non-overlapping intervals I_1 and I_2 then $I_1 \cap I^*$ and $I_2 \cap I^*$ are non-overlapping, and their sum is $I \cap I^*$. So, by the additivity of $\Delta_g I$,

$$\varphi[I] = \Delta_g(I \cap I^*) = \Delta_g(I_1 \cap I^*) + \Delta_g(I_2 \cap I^*) = \varphi[I_1] + \varphi[I_2].$$

Therefore by 45.5 there is a function $h(x)$, defined and finite on the whole space, such that $\Delta_h I = \varphi[I] = \Delta_g(I \cap I^*)$ for all intervals I.

46. By use of the interval function $\Delta_g I$ we are able to extend the definitions and theorems of §8 without difficulty.

46.1s. *Let $g(x)$ be defined and finite on a closed interval I^* of q-dimensional space R_q. Then $g(x)$ is called positively monotonic on I^* if $\Delta_g I \geqq 0$ for every interval I contained in I^*; it is called negatively monotonic on I^* if $\Delta_g I \leqq 0$ for every interval I contained in I^*.*

Here we can substitute R_q itself for I^*.

46.2s. *If $g(x)$ is defined and finite on I^*, the total variation $T_g[I^*]$ is defined to be the* sup *of the sum $|\Delta_g I| + \cdots + |\Delta_g I_n|$ for all subdivisions of I^* into non-overlapping intervals $I_1, \cdots, I_n$. The function $g(x)$ is of bounded variation (BV) on I^* if $T_g[I^*]$ is finite.*

We now state the generalizations of some theorems of §8. Except for 46.4 and 46.9, the proofs of these theorems are obtained from the corresponding proofs in §8 by replacing $(f(\beta) - f(\alpha))$, etc., by $\Delta_f I$.

46.3s. *If $f(x)$ and $g(x)$ are of BV on I^*, and μ and ν are finite numbers, then $h(x) \equiv \mu f(x) + \nu g(x)$ is of BV on I^*, and*

$$T_h[I^*] \leqq |\mu| T_f[I^*] + |\nu| T_g[I^*].$$

The proof is like that of 8.1. It is interesting to observe here a phenomenon without any striking analogue for functions of one variable.

46.4s. *If $g(x) = g_1(x) + \cdots + g_q(x)$, where each*

$$g_j(x) \ (j = 1, \cdots, q)$$

is defined, finite and independent of the variable $x^{(j)}$ on I^*, then $\Delta_g I = 0$ for all intervals I in I^*, and $T_g[I^*] = 0$.

Since $g_j(x)$ is independent of $x^{(j)}$, for every interval I: $\alpha^{(j)} \leqq x^{(j)} \leqq \beta^{(j)}$ we have

$$\Delta_{g_j} I = g_j(x) \Big|_\alpha^\beta = 0$$

by 45.1 (*d*). By 45.1 (*e*),

$$\Delta_g I = \sum_{j=1}^{q} \Delta_{g_j} I = 0.$$

It follows at once that for every partition of I^* into non-overlapping intervals $I_1, \cdots, I_n$ the equality

$$\sum_{j=1}^{n} |\Delta_g I_j| = 0$$

holds; hence, by the definition of total variation, $T_g[I^*] = 0$.

46.5s. *If $g(x)$ is finite and positively monotonic on I^*, it is of BV on I^*, and*

$$T_g[I^*] = \Delta_g I^*.$$

For then if $I_1 \cdots I_n$ is any partition of I^* into non-overlapping intervals, we have by 45.4

$$\sum_{i=1}^{n} |\Delta_g I_i| = \sum_{i=1}^{n} \Delta_g I_i = \Delta_g I^*.$$

46.6s. *If $g(x)$ is defined and finite on I^*, and $I_1, \cdots, I_n$ are non-overlapping intervals whose sum is I^*, and $I'_1, \cdots, I'_m$ are non-over-*

lapping intervals whose sum is I^* and which are each contained in one of the I_j, $j = 1, \cdots, n$, then

$$\sum_{i=1}^{m} | \Delta_g I'_i | \geqq \sum_{j=1}^{n} | \Delta_g I_j | .$$

The proof is like that of 8.2.

46.7s. *If $g(x)$ is of BV on an interval I^*, it is of BV on every interval contained in I^*, and $T_g[I]$ is an additive function of subintervals of I^*.*

The proof is similar to that of theorems 8.3 and 8.4.

46.8s. *If $g(x)$ is of BV on I^*, then $T_g[I^*] \geqq | \Delta_g I^* |$.*

The proof is trivial, like that of 8.5.

We now come to the analogue of theorem 8.6. However, besides extending this to q-dimensional space we wish to strengthen it in other ways.

46.9s. *Let $g(x)$ be defined on the space R_q and of BV on every interval I^*. Then there exist functions $P(x)$, $N(x)$ with the following properties.*

(a) $P(x)$ and $N(x)$ are defined, finite and positively monotonic on the whole space;

(b) $P(x) - N(x) = g(x)$;

(c) $T_g[I] = \Delta_P I + \Delta_N I$ for every interval I;

(d) if $\pi(x)$, $\nu(x)$ have properties (a) and (b), then $\Delta_\pi I \geqq \Delta_P I$ and $\Delta_\nu I \geqq \Delta_N I$ for every interval I.

A pair of functions $P(x)$, $N(x)$ with properties (a), (b), (c) and (d) is called a *minimum decomposition* of $g(x)$.

If an interval I is the sum of non-overlapping intervals $I_1, \cdots, I_n$, let I^* be any interval containing I. By 46.7, $T_g[I] = T_g[I_1] + \cdots + T_g[I_n]$, and $T_g[I]$ is an additive function of intervals. By 45.5, there is a function $\tau(x)$, defined and finite on the whole space, such that

$$\Delta_\tau I = T_g[I]$$

for every interval I. We now define

$$P(x) = \tfrac{1}{2}(\tau(x) + g(x)), \qquad N(x) = \tfrac{1}{2}(\tau(x) - g(x)).$$

From the definition it is obvious that $P(x) - N(x) = g(x)$. Both functions are defined and finite for all x. To prove $P(x)$ positively monotonic, we write (using 45.1e)

$$\Delta_P I = \tfrac{1}{2}(\Delta_\tau I + \Delta_g I) \geqq \tfrac{1}{2}(T_g[I] - | \Delta_g I |),$$

which is non-negative by 46.8. Likewise

$$\Delta_N I = \tfrac{1}{2}(\Delta_\tau I - \Delta_g I) \geqq \tfrac{1}{2}(T_g[I] - | \Delta_g I |) \geqq 0.$$

So (*a*) and (*b*) are established. Statement (*c*) is easy to prove, for $P(x) + N(x) = \tau(x)$, so that

$$\Delta_P I + \Delta_N I = \Delta_\tau I = T_g[I].$$

To establish (*d*), we first observe that

$$\Delta_P I - \Delta_N I = \Delta_g I = \Delta_\pi I - \Delta_\nu I,$$

so that

$$\Delta_P I - \Delta_\pi I = \Delta_N I - \Delta_\nu I.$$

If there exists an interval I such that $\Delta_P I > \Delta_\pi I$, then for the same interval it is also true that $\Delta_N I > \Delta_\nu I$. Now $g(x) = \pi(x) - \nu(x)$, and $\pi(x)$ and $\nu(x)$ are positively monotonic; so by 46.3, 46.5 and (*c*) of this theorem

$$\begin{aligned} T_g[I] &\leqq T_\pi[I] + T_\nu[I] = \Delta_\pi I + \Delta_\nu I \\ &< \Delta_P I + \Delta_N I = T_g[I]. \end{aligned}$$

This contradiction establishes statement (*d*).

The requirement that $g(x)$ shall be of BV on every interval I^* may seem rather stringent. However, suppose that $g(x)$ is defined and of BV on an interval I^*. By 45.6, there is a function $h(x)$, defined and finite on the whole space, such that $\Delta_h I = \Delta_g(I \cap I^*)$ for every interval I. Thus for every interval I' and every partition of I' into non-overlapping intervals $I, \cdots, I_n$ we have

$$\sum_{j=1}^{n} |\Delta_h I_j| = \sum_{j=1}^{n} |\Delta_g(I_j \cap I^*)| \leqq T_g[I^*],$$

because the intervals $I_j \cap I^*$ are non-overlapping subintervals of I^*. The function $h(x)$ satisfies the hypotheses of theorem 46.9, and $\Delta_h I = \Delta_g I$ for all intervals I in I^*, so for our purposes $h(x)$ serves as well as $g(x)$.

47. If $g(x)$ is a function which is defined, finite and positively monotonic on all of q-dimensional space R_q, the function of intervals $\Delta_g I$ is non-negative by definition, and is additive by 45.4. It therefore satisfies conditions (*a*) and (*b*) of 10.1. Remembering that no other properties of ΔI were used in the proofs of the "*s*"-theorems and statements of the "*s*"-definitions, we see at once that

47.1. *If $g(x)$ is defined, finite and positively monotonic on the space R_q, and in* §§10–43 *the symbol ΔI is understood to mean $\Delta_g I$, all theorems and definitions marked with "s" remain valid.*

However, it is desirable to change notation somewhat, in order to exhibit the dependence of integrals and measures on the functions $g(x)$.

If $f(x)$ is defined on a set E and its integral (when ΔI is understood to mean $\Delta_g I$) exists, we say that $f(x)$ is g-summable over E, and we denote its integral by

$$\int_E f(x)dg(x).$$

This integral is a special case of the *Lebesgue-Stieltjes* integral, whose general definition will be given in 51.1. Correspondingly, if E is a set which is measurable when ΔI is understood to mean $\Delta_g I$, we say that E is g-measurable, and we denote its measure by $m_g E$.

If $g(x)$ is defined, finite and positively monotonic on a closed interval I^*, the function of intervals $\Delta_g I$ no longer satisfies 10.1, for it is not defined for all intervals. But this difficulty is trivial. By 45.6, there is a function $h(x)$, defined on the entire space, such that

$$\Delta_h I = \Delta_g(I \cap I^*) \geqq 0$$

for all intervals I. Hence we can use this $h(x)$ in place of $g(x)$, obtaining the interval function $\Delta_h I$ which satisfies 10.1 and which coincides with $\Delta_g I$ whenever I is contained in I^*.

The theorems not marked "*s*" are not valid for integrals with respect to general positively monotonic functions $g(x)$. However, we can establish some theorems which serve as useful substitutes for 16.1.

47.2. *If $g(x)$ is defined, finite and positively monotonic on the space R_q, and I_1 is a closed interval contained in an open interval J_2, and J_1 is the interior of I_1, then*

$$m_g J_1 \leqq m_g I_1 \leqq \Delta_g J_2 = \Delta_g \bar{J}_2$$

and

$$\Delta_g J_1 \leqq m_g J_2 \leqq m_g \bar{J}_2.$$

The equality of $\Delta_g J_2$ and $\Delta_g \bar{J}_2$ follows from 10.1. The inequalities $m_g J_1 \leqq m_g I_1$ and $m_g J_2 \leqq m_g \bar{J}_2$ follow from 19.3.

Let I_1, J_2 be defined respectively by the inequalities

$$\alpha_1^{(i)} \leqq x^{(i)} \leqq \beta_1^{(i)}, \qquad \alpha_2^{(i)} < x^{(i)} < \beta_2^{(i)}.$$

Then by hypothesis the inequalities $\alpha_2^{(i)} < \alpha_1^{(i)} \leqq \beta_1^{(i)} < \beta_2^{(i)}$ hold. Define δ to be the least of the positive numbers $\alpha_1^{(i)} - \alpha_2^{(i)}$, $\beta_2^{(i)} - \beta_1^{(i)}$. Let $d(x)$ be the distance of x from I_1. The function

$$\varphi(x) \equiv \sup\left\{0, 1 - \frac{d(x)}{\delta}\right\}$$

is clearly continuous, and $0 \leqq \varphi(x) \leqq 1$ for all x. Also, if x is in I_1, then $d(x) = 0$ and $\varphi(x) = 1$; while if x is not in J_2 then $d(x) \geqq \delta$ and $\varphi(x) = 0$.

Let W be a closed interval containing the closure of J_2 in its interior. The characteristic functions of I_1 and of J_2 are step-functions on W, and from the properties of $\varphi(x)$ just established we have

$$(A) \qquad K_{I_1}(x) \leqq \varphi(x) \leqq K_{J_2}(x)$$

for all x. Hence the integral of $\varphi(x)$ over W lies between the integrals (in the sense of 10.4) of these two characteristic functions. These two integrals are easily computed; we find

$$(B) \qquad \Delta_g I_1 \leqq \int_W \varphi(x)\, dg(x) \leqq \Delta_g J_2.$$

On the other hand, from (A) and 18.2 we have

$$(C) \quad m_g I_1 = \int_{R_q} K_{I_1}(x)\, dg(x) \leqq \int_{R_q} \varphi(x)\, dg(x) \leqq \int_{R_q} K_{J_2}(x)\, dg(x) = m_g J_2.$$

But $\varphi(x)$ vanishes save at points interior to W, so its integral over W is the same as its integral over R_q. With this in mind, we immediately deduce the remaining inequalities of our theorem from (B) and (C).

47.3. *If I is the closed interval $\alpha^{(i)} \leqq x^{(i)} \leqq \beta^{(i)}$, $i = 1, \cdots, q$, and $\{I_n\}$ is a sequence of intervals $\alpha_n^{(i)} \leqq x^{(i)} \leqq \beta_n^{(i)}$ such that $\alpha_n^{(i)} < \alpha^{(i)} \leqq \beta^{(i)} < \beta_n^{(i)}$ and $\lim_{n\to\infty} \alpha_n^{(i)} = \alpha^{(i)}$, $\lim_{n\to\infty} \beta_n^{(i)} = \beta^{(i)}$, then $\lim_{n\to\infty} \Delta_g I_n = m_g I$.*

If I is the open interval $\alpha^{(i)} < x^{(i)} < \beta^{(i)}$, $i = 1, \cdots, q$, and $\{I_n\}$ is a sequence of intervals $\alpha_n^{(i)} \leqq x^{(i)} \leqq \beta_n^{(i)}$ such that $\alpha^{(i)} < \alpha_n^{(i)} \leqq \beta_n^{(i)} < \beta^{(i)}$ and $\lim_{n\to\infty} \alpha_n^{(i)} = \alpha^{(i)}$, $\lim_{n\to\infty} \beta_n^{(i)} = \beta^{(i)}$, then $\lim_{n\to\infty} \Delta_g I_n = m_g I$.

We prove the first statement; the proof of the second is quite similar.

Let J_n be the interval $\alpha^{(i)} - n^{-1} < x^{(i)} < \beta^{(i)} + n^{-1}$, $i = 1, \cdots, q$, $n = 1, 2, \cdots$. Then $J_1 \supset J_2 \supset J_3 \supset \cdots$, and $\bigcap J_n = I$. By 19.9, the g-measure of J_n tends to that of I. So if ϵ is an arbitrary positive number, there is an m such that

$$m_g I \leqq m_g J_m < m_g I + \epsilon.$$

For all large n, the closed interval I_n is interior to J_m, while I is interior to I_n. Hence by 47.2, for all large n we have

$$m_g I \leqq \Delta_g I_n \leqq m_g J_m < m_g I + \epsilon.$$

This establishes the theorem.

EXERCISE. If $g(x)$ is continuous, then $m_g I = \Delta_g I$ for every interval I.

We now study a particular class of functions $g(x)$ for which there is a more immediate relation between $\Delta_g I$ and $m_g I$ than that expressed in 47.3. This is the class of *left continuous* functions.

47.4. *A function $g(x)$ defined on R_q is left continuous on R_q if for every x the relation*

$$\lim_{\epsilon_1 \to 0, \cdots, \epsilon_q \to 0} g(x^{(1)} - \epsilon_1, \cdots, x^{(q)} - \epsilon_q) = g(x^{(1)}, \cdots, x^{(q)})$$

holds, subject to the conditions $\epsilon_1 > 0, \cdots, \epsilon_q > 0$.

For these functions we can establish the following theorem.

47.5. *If $g(x)$ is defined, finite, positively monotonic and left continuous on R_q, then for every right-open interval $I = [\alpha, \beta)$ the equation $m_g I = \Delta_g I$ holds.*

Let I be the interval $[\alpha, \beta)$; that is, I is defined by the inequalities $\alpha^{(i)} \leqq x^{(i)} < \beta^{(i)}$, $i = 1, \cdots, q$, where $\alpha^{(i)} < \beta^{(i)}$, $i = 1, \cdots, q$. Choose a positive integer n_0 large enough so that $2/n_0$ is less than the least of the numbers $\beta^{(i)} - \alpha^{(i)}$, $i = 1, \cdots, q$. Then for all positive integers n greater than n_0 the intervals

$$I_n\colon \alpha^{(i)} - \frac{2}{n} \leqq x^{(i)} \leqq \beta^{(i)} - \frac{2}{n}, \qquad i = 1, \cdots, q,$$

$$I'_n\colon \alpha^{(i)} - \frac{3}{n} \leqq x^{(i)} \leqq \beta^{(i)} - \frac{1}{n}, \qquad i = 1, \cdots, q,$$

$$I''_n\colon \alpha^{(i)} - \frac{1}{n} \leqq x^{(i)} \leqq \beta^{(i)} - \frac{3}{n}, \qquad i = 1, \cdots, q$$

are well-defined, since then $\alpha^{(i)} - \frac{1}{n}$ is less than $\beta^{(i)} - \frac{3}{n}$.

The set I is the same as both lim sup I'_n and lim inf I'_n, as defined in 19.10. Hence by 19.12

$$\lim_{n \to \infty} m_g I'_n = m_g I.$$

In the same way

$$\lim_{n \to \infty} m_g I''_n = m_g I.$$

By 47.2, for each n the inequalities

$$m_g I'_n \geqq \Delta_g I_n \geqq m_g I''_n$$

hold. This, with the preceding limiting relationships, implies

$$(A) \qquad \lim_{n \to \infty} \Delta_g I_n = m_g I.$$

But from the definitions of I_n and Δ_g, the left continuity of $g(x)$ yields the equation

$$\lim_{n\to\infty} \Delta_g I_n = \Delta_g I.$$

These two equations establish the theorem.

REMARK. In this proof the left continuity was not used until the second-last sentence. Hence, with the notation in 47.5, equation (A) holds for every positively monotonic function $g(x)$.

48. Let us now consider some special instances of integrals with respect to $g(x)$. Of course the whole theory in which $\Delta_g I$ is the product of the edges of I is an example; for this, we can take $g(x^{(1)}, \cdots, x^{(q)}) \equiv x^{(1)}x^{(2)} \cdots x^{(q)}$.

Examples of a different nature may be drawn from physics. Suppose, for instance, that we are given a mass-distribution in three-dimensional space. To be precise, we suppose that we are given an interval-function $M(I)$ which is additive and non-negative; physically, if I is the interval $\alpha^{(i)} \leqq x^{(i)} \leqq \beta^{(i)}$ the function $M(I)$ is the mass contained in the half-open interval $\alpha^{(i)} \leqq x^{(i)} < \beta^{(i)}$. For the sake of simplicity, we suppose that the distribution is bounded; there is an interval I^* of the type $[\alpha, \beta)$ such that for every interval I containing no interior point of I^* we have $M(I) = 0$. We seek to assign to each such mass-distribution a number called its moment of inertia about the $x^{(3)}$-axis. This number is to have the following properties. If a distribution is the sum of several distributions (for a physical example, if a body is regarded as the sum of several portions), its moment of inertia about the $x^{(3)}$-axis is to be the sum of the moments of inertia of the several distributions. If a mass m is distributed entirely in an interval I, so that $M(I) = m$, and I lies in the set $\{x \mid a^2 \leqq (x^{(1)})^2 + (x^{(2)})^2 \leqq b^2\}$, then the moment of inertia of the distribution is not less than ma^2 and not greater than mb^2.

By 45.5, there is a function $m(x)$ such that $\Delta_m I = M(I)$ for every interval I. Let $S(x)$, $s(x)$ be step-functions on I^* such that

$$s(x) \leqq (x^{(1)})^2 + (x^{(2)})^2 \leqq S(x)$$

for all x in I. If I_j is one of the intervals on which $s(x)$ is a constant c_j, by hypothesis $c_j\Delta_m I_j$ is an underestimate for the moment of inertia of the part of the mass in I_j. Adding, the integral $\int'_{I^*} s(x)dm(x)$ in the sense of 10.4 is an estimate from below for the moment of inertia. This holds for every step function $s(x)$ satisfying the inequality above, hence the moment of inertia μ is at least equal to the sup of such integrals. But $(x^{(1)})^2 + (x^{(2)})^2$ is continuous, so this sup is its integral on I^*:

$$\mu \geqq \int_{I^*} [(x^{(1)})^2 + (x^{(2)})^2] dm(x).$$

Using the step-functions $S(x)$, we find that their integrals in the sense of 10.4 are overestimates for μ, so

$$\mu \leqq \int_{I^*} [(x^{(1)})^2 + (x^{(2)})^2] dm(x).$$

Comparing this with the preceding inequality, we find that the moment of inertia is given by the integral

$$\int_{I^*} [(x^{(1)})^2 + (x^{(2)})^2] dm(x).$$

The extension to unbounded distributions is obvious.

Even in this example we recognize an advantage of the Lebesgue-Stieltjes integral. The single formula above gives the moment of inertia of the matter irrespective of the distribution. The mass may all be concentrated in a finite or denumerable set of points or may be continuously distributed, and if continuously distributed, may lack the derivative called density; in any of these cases or in any combination of them the moment of inertia is still represented by the same Lebesgue-Stieltjes integral. Thus in recent work in analysis it has sometimes been found advantageous to use the Lebesgue-Stieltjes integral in order that a single proof may cover both the ordinary Lebesgue integral and the case of infinite sums.

For another example, let P be a finite or denumerable set of points $p_1, p_2, \cdots$ in R_q with no accumulation point. We could for example let P consist of all the points $(x^{(1)}, \cdots, x^{(q)})$ in which each $x^{(i)}$ is a positive integer. If I is an interval whose closure is

$$\{x \mid \alpha^{(i)} \leqq x^{(i)} \leqq \beta^{(i)}, i = 1, \cdots, q\},$$

we define $\varphi[I]$ to be the number of points of P belonging to the interval $\{x \mid \alpha^{(i)} \leqq x^{(i)} < \beta^{(i)}, i = 1, \cdots, q\}$. This function of intervals is obviously additive, so there is a function $g(x)$ such that $\Delta_g I = \varphi[I]$ for all intervals I. In the special case in which P consists of all points p with each $p^{(i)}$ a positive integer, the process used in 45.5 yields for each point x of R_q the equation $g(x) = n^{(1)}n^{(2)} \cdots n^{(q)}$, where $n^{(i)}$ is the greatest integer less than $x^{(i)}$ if $x^{(i)} > 0$ and $n^{(i)} = 0$ if $x^{(i)} \leqq 0$. Let Q be the complement of P. Each point p_j of P can be enclosed in intervals

$$I_n: p_j^{(i)} - \frac{1}{n} \leqq x^{(i)} \leqq p_j^{(i)} + \frac{1}{n}, \qquad i = 1, \cdots, q.$$

If n is large enough, I_n contains no point of P other than p_j, and $\Delta_g I_n = 1$. The point p_j can be regarded as a degenerate closed interval, and by 47.3 the set $[p_j]$ consisting of the single point p_j has g-measure $m_g[p_j] = 1$.

If I is a closed interval contained in Q, it is interior to a closed interval I_1 which is contained in Q and so contains no point of P. Then by 47.2, $0 \leqq m_g I \leqq \Delta_g I_1 = 0$. The set Q can be represented as the sum of a denumerable set of closed intervals, hence by 19.13 we have $m_g Q = 0$. It follows that an arbitrary set E is the sum of the set EQ, of g-measure zero, and the set EP, which is a subset of the set P and has g-measure equal to the number of points p_j which it contains.

Every function $f(x)$ defined on a set $[p_j]$ consisting of a single point p_j of P is evidently constant on $[p_j]$. Hence

$$\int_{[p_j]} f(x)dg(x) = f(p_j)m_g[p_j] = f(p_j).$$

If E is bounded, it contains a finite number of the points p_j, say p_{i_k}, $k = 1, \cdots, m$. Then

$$\int_E f(x)dg(x) = \sum_{k=1}^{m} \int_{[p_{i_k}]} f(x)dg(x) + \int_{EQ} f(x)dg(x) = \sum_{k=1}^{m} f(p_{i_k}).$$

If E is unbounded, we evaluate the integral of $f(x)$ over E as in §18. This integral exists if and only if $\Sigma f(p_j)$ converges absolutely, and it is then equal to the sum $\Sigma f(p_j)$. Thus the absolutely convergent series Σf_n or $\Sigma\Sigma \cdots \Sigma f_{n_1 n_2 \cdots n_q}$ are included as special cases of the integral

$$\int_{R_q} f(x)dg(x)$$

by proper choice of the integrator $g(x)$.

Merely as an exercise, we shall give another proof of the statements about this last example. Let I^* be a closed interval, $a^{(i)} \leqq x^{(i)} \leqq b^{(i)}$. We suppose that the notation is so chosen that the points of P in the interval $a^{(i)} \leqq x^{(i)} < b^{(i)}$ are $p_1, \cdots, p_m$. If $s(x)$ is a step-function constant on each of a set of intervals

$$I_j: \alpha_j^{(i)} \leqq x^{(i)} < \beta_j^{(i)}, \qquad j = 1, \cdots, m$$

whose closures cover I^*, we may suppose that the I_j are small enough so that each contains at most one point of P. Then in the sum

$$\int_{I^*}' s(x)dg(x) = \sum_{j=1}^{m} c_j\Delta_g I_j,$$

where c_j is the value of $s(x)$ on I_j, the factor $\Delta_g I_j$ vanishes unless I_j contains a point of P and is 1 if I_j contains such a point. Hence

$$\int_{I^*}' s(x)dg(x) = \sum_{j=1}^{m} s(p_j).$$

If $\varphi(x)$ is continuous on I^*, it is the uniform limit of a sequence $s_1(x) \leqq s_2(x) \leqq \cdots$ of step-functions of the type just considered. So

$$\int_{I^*} \varphi(x)dg(x) = \lim_{n\to\infty} \int_{I^*}' s_n(x)dg(x) = \lim_{n\to\infty} \sum_{j=1}^{m} s_n(p_j) = \sum_{j=1}^{m} \varphi(p_j).$$

A similar argument shows that

$$\int_{I^*} f(x)dg(x) = \sum_{j=1}^{m} f(p_j)$$

if $f(x)$ is a U-function or an L-function. If $f(p_j)$ is finite,

$$j = 1, \cdots, m,$$

we set $u(x) = f(x)$ for $x = p_1, \cdots, p_m$, $u(x) = \infty$ elsewhere, and we set $l(x) = f(x)$ for $x = p_1, \cdots, p_m$ and $l(x) = -\infty$ elsewhere Then $u(x)$ is a U-function and $l(x)$ is an L-function, and $l(x) \leqq f(x) \leqq u(x)$. Therefore

$$\sum_{j=1}^{m} f(p_j) = \int_{I^*} u(x)dg(x) \geqq \overline{\int_{I^*}} f(x)dg(x) \geqq \underline{\int_{I^*}} f(x)dg(x)$$

$$\geqq \int_{I^*} l(x)dg(x) = \sum_{j=1}^{m} f(p_j).$$

This shows that every function finite on PI^* is g-summable. If $f(x)$ is defined on the whole space, we extend the integral to the whole space by 18.1; we find that $f(x)$ is g-summable over the space if and only if $\Sigma f(p_j)$ is absolutely convergent, in which case the integral has the value $\Sigma f(p_j)$. In particular, by 19.1, every set E is g-measurable, and its g-measure is the number of points of P which it contains.

In §§10 to 43 several of the theorems were not marked with an "s." The preceding example serves to show that they state properties which do not hold for all integrals $\int f(x)dg(x)$. Thus, for instance, it is not necessarily true (19.14) that a finite or denumerable set should have g-measure zero—a set consisting of a single point of P has g-measure 1.

At the end of §17 we commented that the symbol

$$\int_I f(x)dg(x)$$

(which we then were calling $\int_I f(x)\,dx$) could have two interpretations. It could mean

$$\int_{R_q} f(x)K_I(x)dg(x),$$

and this was the meaning we chose to give it. But it could also be interpreted as in 15.1, with I as basic interval. When we choose ΔI to be the product of the edges of I this distinction was seen in §17 to be only conceptual; the two interpretations led to one and the same value for the integral. For other interval-functions $\Delta_g I$ the two interpretations can lead to different values of the integral. We use the preceding example to show this. Let I^* be an interval $\alpha^{(i)} \leqq x^{(i)} \leqq \beta^{(i)}$ $(i = 1, \cdots, q)$ such that β is in the set P but no other point of I^* is in P. Then $\Delta_g I = 0$ for every interval I contained in I^*, and if we use I^* as basic interval we find

$$\int_{I^*} f(x)dg(x) = 0$$

for every function $f(x)$. But the point β has g-measure 1, so if we give the integral over I^* the meaning agreed upon we find that if $f(\beta)$ is finite

$$\int_{I^*} f(x)dg(x) = \int_{R_q} f(x)K_I(x)dg(x)$$
$$= f(\beta).$$

The two interpretations thus lead to different values for the integral, unless $f(\beta)$ happens to be 0.

Incidentally, this same example shows that 16.1, 16.2 and 17.5 do not hold for Lebesgue-Stieltjes integrals.

The convention adopted may seem to be rather a nuisance at times. If we are interested in a function or functions defined on an interval I^*, and have a $g(x)$ defined on I^*, it would appear pointless to extend the range of $g(x)$ to the whole space and then treat our subsets of I^*

just as though they were unbounded. It is in fact unnecessary. We could use the integral as defined in §15 and ignore all extensions to unbounded sets; no trouble would arise. Or, if we wish to preserve the same formalism for bounded and unbounded sets, we observe that by 45.6 there is a function $h(x)$ defined on the whole space and such that $\Delta_h I = \Delta_g(I \cap I^*)$ for every interval I. (Of course where $g(x)$ is already defined outside of I^* it will not usually coincide with this $h(x)$). So if $s(x)$ is a step-function defined on an interval I' containing I^* and having constant value c_i on I_i, $i = 1, \cdots, n$, where the I_i are disjoint intervals whose sum is I', then

$$\int_{I'}' s(x)dh(x) = \sum_{i=1}^{n} c_i \Delta_h I_i = \sum_{i=1}^{n} c_i \Delta_g(I_i \cap I^*).$$

The intervals $I_i \cap I^*$ which are not empty form a non-overlapping set covering I^*, so we have

$$\int_{I'}' s(x)dh(x) = \int_{I^*}' s(x)dg(x).$$

In the usual manner, we can extend this equality to continuous functions and to U- and L-functions, and finally find that $f(x)$ is h-summable over I' if and only if it is g-summable over I^*, in which case the integrals are equal. Letting I' expand as in §18, we find that the integrals

$$\int_{R_q} f(x)dh(x), \qquad \int_{I^*} f(x)dg(x)$$

(the latter in the sense of §15) either both exist or both fail to exist, and if they exist they are equal.

Again, theorems 26.9 and 27.4 cannot be generalized. For let P consist of a single point, say the origin. The set function $F(E) = \int_E 1dg(x)$ is AC, by 27.1. But if E is any set containing P then $F(E) = 1$, no matter how small the diameter of E may be, so $F(E)$ is not continuous.

It may seem strange that 27.4 fails while 27.1 remains valid. But if E is any set of g-measure less than 1, then E contains no point of P, and $m_g E = 0$. For all such sets,

$$F(E) = \int_E f(x)dg(x) = 0.$$

Thus $|F(E)|$ can be made smaller than any positive ϵ by choosing $m_g E < \delta = 1$.

Theorem 15.9 also fails to generalize. For suppose P contains just one point, say the point $(2, 2, \cdots 2)$. Then the interval $I: -1 \leqq x^{(i)} \leqq 1$ has g-measure 0, but if it is translated to $I_h: 1 \leqq x^{(i)} \leqq 3$ we find $m_g I_h = 1$.

Although §25 is not an "s"-section, it needs only the special property of $g(x)$ that if I has the projections I^s and I^t on the spaces R_s and R_t respectively, there shall be interval functions $\Delta_s I^s$ and $\Delta_t I^t$ whose product is $\Delta_g I$. The reader will be able to verify that this holds true whenever $g(x)$ is of the form $\varphi(\sigma)\psi(\tau)$, where $\varphi(\sigma)$ and $\psi(\tau)$ are positively monotonic on the spaces R_s and R_t respectively.

Theorem 29.8 is a conspicuous theorem without an "s." Again let P consist of a single point p, and let $f_n(x)$ be constantly equal to n on the whole space, $n = 1, 2, \cdots$. If $f(x) \equiv \infty$, then $f_n(x)$ tends everywhere to $f(x)$. For the set E_ϵ we can take P itself. The functions $\int_E f_n(x)dg(x)$ are uniformly absolutely continuous over P, as we see if we take $\delta = 1$ in the definition 29.4. However, $f(x)$ is not g-summable and the other two conclusions of 29.8 are meaningless.

Sections 31 to 38 and §41 we abandon in toto.

49. The preceding examples have shown us that the family of g-measurable sets depends on the function $g(x)$. More definitely, if $g(x) = x^{(1)}x^{(2)} \cdots x^{(q)}$ the g-measure is Lebesgue measure, and there are non-measurable sets. If $g(x) \equiv 0$ every set is g-measurable. This suggests a question. Is there any large family of sets which we can guarantee to be g-measurable for every positively monotonic function $g(x)$?

We shall now show that there is such a family, the family of *Borel sets.* This family is defined thus.

49.1. *The family of Borel sets (in R_q) is the smallest family $\mathfrak{F}$ of sets in R_q which satisfies the conditions*

(*a*) *every interval of the form $[\alpha, \beta)$ belongs to $\mathfrak{F}$;*

(*b*) *if E_1 and E_2 are in $\mathfrak{F}$, and $E_2 \subset E_1$, then $E_1 - E_2$ belongs to $\mathfrak{F}$;*

(*c*) *if $E_1, E_2, \cdots$, are in $\mathfrak{F}$, so is $\bigcup E_n$.*

This definition is meaningless unless we can show that such a smallest family exists. There are families $\mathfrak{F}$ which satisfy (*a*), (*b*) and (*c*); for example, the family of all sets in R_q is such a family. We now define $\mathfrak{B}$ to be the family consisting of all sets E which belong to *every* family $\mathfrak{F}$ which satisfies (*a*), (*b*) and (*c*). Every interval of the form $[\alpha, \beta)$ belongs to every such $\mathfrak{F}$, by (*a*); hence every such interval belongs to $\mathfrak{B}$, and the family $\mathfrak{B}$ satisfies (*a*). If E_1 and $E_2 \subset E_1$ are in $\mathfrak{B}$, and $\mathfrak{F}$ is any family satisfying (*a*), (*b*) and (*c*), then E_1 and E_2 are in $\mathfrak{F}$, and by (*b*) so is $E_1 - E_2$. Therefore $E_1 - E_2$ is in $\mathfrak{B}$. Likewise, if $E_1, E_2, \cdots$ all belong to $\mathfrak{B}$, and $\mathfrak{F}$ is any family satisfying (*a*), (*b*) and (*c*), then the E_n all belong to $\mathfrak{F}$. By (*c*), so does $\bigcup E_n$. Since $\mathfrak{F}$ is any family satisfying (*a*), (*b*) and (*c*), the set $\bigcup E_n$ belongs to $\mathfrak{B}$. Therefore $\mathfrak{B}$ satisfies the three conditions. By its definitions it is the smallest such family; for if $\mathfrak{F}$ satisfies the three conditions, then every set E of $\mathfrak{B}$ belongs to $\mathfrak{F}$.

From the definition of Borel sets we conclude readily that every open set is a Borel set, every closed set is a Borel set, and every product $\bigcap E_n$ of Borel sets is a Borel set. For every open set is a sum $\bigcup I_n$ of intervals $I_n = [\alpha_n, \beta_n)$, by 3.6; so by 49.1(*a*) and (*c*) it is a Borel set. In particular, the whole space R_q is a Borel set. By this and (*b*) of 49.1, the complement of every Borel set is a Borel set. Every closed set is the complement of an open set, and therefore is a Borel set. If the sets $E_1, E_2, \cdots$ are Borel sets, so are $CE_1, CE_2, \cdots$. By (*c*) of 49.1, so is $\bigcup CE_n$, and hence so is the complement $C(\bigcup CE_n)$ of this sum. But $C(\bigcup CE_n) = \bigcap E_n$, by the formulas preceding 1.1, so $\bigcap E_n$ is a Borel set.

It is almost immediate that

49.2. *If $g(x)$ is any positively monotonic function and E is a Borel set, then E is g-measurable.*

If I is an interval $\alpha^{(i)} \leqq x^{(i)} < \beta^{(i)}$, $i = 1, \cdots, q$, for each i we define $A^{(i)}$ to be the half-space $x^{(i)} \geqq \alpha$ and $B^{(i)}$ to be the half-space $x^{(i)} < \beta^{(i)}$. The $A^{(i)}$ are closed, hence g-measurable (19.4), and the $B^{(i)}$ are open, hence g-measurable (19.7). The interval I is the product set $A^{(1)} \cap B^{(1)} \cap A^{(2)} \cap \cdots \cap A^{(q)} \cap B^{(q)}$, hence g-measurable (19.9). (Clearly this device could be used to show that every interval is g-measurable, however the signs $\leqq$ and $<$ are distributed in the inequalities defining I.) Therefore the family of g-measurable sets satisfies (*a*) of 49.1. By 19.6 it satisfies (*b*), and by 19.8 it satisfies (*c*). Hence the family of Borel sets is contained in the family of g-measurable sets.

An analogous question arises concerning sets having finite g-measure. It follows immediately from 49.2 and the remark after 18.1 that every bounded Borel set has finite g-measure. However, it is interesting to have another proof based more directly on the structural properties of the sets. We shall in fact prove the following.

49.3. *If $\mathfrak{F}$ is a family of sets such that*

(*a*) *every interval of the form $[\alpha, \beta)$ belongs to $\mathfrak{F}$;*

(*b*) *if E_1 and E_2 belong to $\mathfrak{F}$, and $E_2 \subset E_1$, then $E_1 - E_2$ belongs to $\mathfrak{F}$;*

(c′) if the sets $E_1, E_2 \cdots$ all belong to $\mathfrak{F}$, and are all contained in an interval $[\alpha, \beta)$, then $\bigcup E_n$ belongs to $\mathfrak{F}$;
then $\mathfrak{F}$ contains all bounded Borel sets.

Let $\mathfrak{F}^*$ be the family consisting of all sets E such that for every interval $I = [\alpha, \beta)$ the intersection EI belongs to $\mathfrak{F}$. Every interval $I_0 = [\alpha_0, \beta_0)$ belongs to $\mathfrak{F}^*$; for if $I = [\alpha, \beta)$ the intersection $I \cap I_0$ either is empty or is an interval $[\alpha', \beta')$, and in either case belongs to $\mathfrak{F}$. If E_1 and E_2 belong to $\mathfrak{F}^*$, and $E_2 \subset E_1$, then for every interval $I = [\alpha, \beta)$ the products E_1I and E_2I are in $\mathfrak{F}$, so by (b) $E_1I - E_2I \equiv (E_1 - E_2)I$ is in $\mathfrak{F}$. So $E_1 - E_2$ is in $\mathfrak{F}^*$. Finally, if $E_1, E_2 \cdots$ all belong to $\mathfrak{F}^*$, and I is any interval $[\alpha, \beta)$, all the sets E_nI belong to $\mathfrak{F}$ and are all contained in I. So by (c′) their sum $(\bigcup E_n)I$ belongs to $\mathfrak{F}$; whence $\bigcup E_n$ belongs to $\mathfrak{F}^*$.

Thus $\mathfrak{F}^*$ satisfies conditions (a), (b) and (c) of 49.1, and therefore contains all Borel sets. Let E be a bounded Borel set and I an interval $[\alpha, \beta)$ containing E. Since E is in $\mathfrak{F}^*$, the intersection EI is in $\mathfrak{F}$. But $EI = E$, so E is in the family $\mathfrak{F}$, and the proof is complete.

In the proof of 49.2 we showed that every interval $[\alpha, \beta)$ belongs to the family $\mathfrak{M}_g$ of sets of finite g-measure. Conditions (b) and (c′) are satisfied, by 19.6, 19.8 and 19.3. Hence by 49.3 $\mathfrak{M}_g$ contains every bounded Borel set.

Recovering the direction of our study, it is interesting to see how close the Borel sets come to filling up the family of g-measurable sets. We can prove

49.4. *If $g(x)$ is positively monotonic and E is g-measurable, there are Borel sets A and B such that $A \subset E \subset B$ and $m_g(B - A) = 0$.*

By 20.7, for each positive integer n there is an open set G_n containing E and a closed set F_n contained in E for which

$$m_g(G_n - E) < \frac{1}{n} \qquad \text{and} \qquad m_g(E - F_n) < \frac{1}{n}.$$

There is no loss of generality in supposing that $G_1 \supset G_2 \supset G_3 \supset \cdots$ and that $F_1 \subset F_2 \subset F_3 \subset \cdots$. Otherwise, we need only replace G_n by the (open) intersection $G_1G_2 \cdots G_n$ and F_n by the (closed) sum $F_1 \cup \cdots \cup F_n$. Define $A = \bigcup F_n$ and $B = \bigcap G_n$. These are Borel sets, and $A \subset E \subset B$. Also,

$$B - A = \bigcap [G_n - F_n]$$

and $m_g(G_n - F_n) < 2/n$, so by 19.9

$$m_g(B - A) = \lim_{n \to \infty} m_g(G_n - F_n) = 0.$$

REMARK. As the proof above shows, in 49.4 we may require that A be a sum of closed sets and that B and $B - A$ be products of open sets.

This theorem shows that if we can distinguish the Borel sets which have g-measure 0 we can construct the family of all g-measurable sets. For if E is g-measurable it is (by 49.4) the sum of a Borel set and a subset of a Borel set of g-measure 0. On the other hand, by 49.2 and 19.13 and 19.8, every such set is g-measurable.

EXERCISE. Let $a_j^{(m)}$ $(m, j = 1, 2, 3, \cdots)$ be a double sequence. If for each positive integer m the series $a_1^{(m)} + a_2^{(m)} + \cdots$ is absolutely convergent, and for each j the limit $a_j = \lim_{m\to\infty} a_j^{(m)}$ exists, then:

(i) if $|a_j^{(m)}| \leqq A_j$ for all m and j, and ΣA_j converges, then $\sum_{j=1}^{\infty} a_j$ is absolutely convergent, and

$$\sum_{j=1}^{\infty} a_j = \lim_{m\to\infty} \sum_{j=1}^{\infty} a_j^{(m)};$$

(ii) if $a_j^{(1)} \leqq a_j^{(2)} \leqq a_j^{(3)} \leqq \cdots$ for each j, then $\sum_{j=1}^{\infty} a_j$ is absolutely convergent if and only if $\lim_{m\to\infty} \sum_{j=1}^{\infty} a_j^{(m)}$ is finite, and in this case

$$\sum_{j=1}^{\infty} a_j = \lim_{m\to\infty} \sum_{j=1}^{\infty} a_j^{(m)}.$$

(In R_1, let P be the set of positive integers. Apply 18.4 and 29.3.)

EXERCISE. If $\{a_n\}$, $\{b_n\}$ are such that $\Sigma |a_n|^p$ and $\Sigma |b_n|^q$ converge, where p and q are greater than 1 and $1/p + 1/q = 1$, then $\Sigma(a_n b_n)$ is absolutely convergent, and

$$\left|\sum_{n=1}^{\infty} a_n b_n\right| \leqq \left\{\sum_{n=1}^{\infty} |a_n|^p\right\}^{1/p} \left\{\sum_{n=1}^{\infty} |b_n|^q\right\}^{1/q}.$$

(Apply 24.4.)

50. Although we have extended the theorems of the preceding chapters to Lebesgue-Stieltjes integrals with respect to positively monotonic functions $g(x)$, this by no means completes our task. There is a new sort of investigation needed, one which has no analogue in the theory of the integrals $\int f(x)\, dx$. This concerns the relationships between integrals involving different integrators $g(x)$. In this section we prove two such theorems.

50.1. *Let $g_0(x), g_1(x), g_2(x), \cdots$ be a finite or denumerable collection of positively monotonic functions; let E be a set in the space R_q, and let $f(x)$ be a function defined on E. Let the equation*

$$(A) \qquad \Delta_{g_0}I = \Delta_{g_1}I + \Delta_{g_2}I + \cdots$$

hold for every interval I. Then

(i) *E is g_0-measurable if and only if it is g_n-measurable for each n $(n = 1, 2, \cdots)$, and in that case*

$$(B) \qquad m_{g_0}E = m_{g_1}E + m_{g_2}E + \cdots;$$

(ii) *$f(x)$ is g_0-summable over E if and only if it is g_n-summable over E for each n $(n = 1, 2, \cdots)$ and the series*

$$(C) \qquad \sum \int_E |f(x)|\, dg_n(x)$$

converges; and in that case

$$(D) \qquad \int_E f(x)dg_0(x) = \int_E f(x)dg_1(x) + \int_E f(x)dg_2(x) + \cdots.$$

In the proof we shall several times make use of a limiting process which we state as a lemma.

Lemma. *Let $E_1, E_2, \cdots$ be a sequence of sets each of which is g_i-measurable $(i = 0, 1, \cdots)$ and for each of which equation (B) is satisfied. Let $\lim E_m$ exist, in the sense defined after* 19.11. *If either one of the conditions*

(i) *all the E_m are contained in an open interval I^*,*

(ii) *$E_1 \subset E_2 \subset E_3 \subset \cdots$*

is satisfied, then $\lim E_m$ is also g_i-measurable $(i = 0, 1, 2, \cdots)$ and for it equation (B) is satisfied.

For each i $(i = 0, 1, \cdots)$ the set $\lim E_m$ is g_i-measurable by 19.11, and by 19.12 we have

$$(E) \qquad m_{g_i}(\lim E_m) = \lim_{m\to\infty} m_{g_i}E_m \qquad (i = 0, 1, 2, \cdots).$$

If there are finitely many, say p, of the functions g_i, this and 19.8 imply

$$(F) \qquad \begin{aligned} m_{g_0}(\lim E_m) &= \lim_{m\to\infty} m_{g_0}E_m \\ &= \lim_{m\to\infty} \sum_{n=1}^{p} m_{g_n}E_m \\ &= \sum_{n=1}^{p} \lim_{m\to\infty} m_{g_n}E_m \\ &= \sum_{n=1}^{p} m_{g_n}(\lim E_m), \end{aligned}$$

so equation (B) holds. Suppose then that there are denumerably many of the functions $g_i(x)$. If hypothesis (i) holds, we have by 47.2

$$m_{g_i}E_m \leqq \Delta_{g_i}I^* \qquad (i = 0, 1, 2, \cdots),$$

and the right members have a finite sum $\Delta_{g_0}I^*$. If hypothesis (ii) holds, then

$$m_{g_i}E_1 \leqq m_{g_i}E_2 \leqq \cdots \qquad (i = 0, 1, 2, \cdots).$$

In either case, by a well-known theorem on infinite series (stated in the first exercise at the end of §49) we have (recalling (E))

$$\sum_{n=1}^{\infty} m_{g_n}(\lim E_m) = \lim_{m\to\infty} \sum_{n=1}^{\infty} m_{g_n}E_m.$$

But for each E_m equation (B) is satisfied, so

$$\begin{aligned}\sum_{n=1}^{\infty} m_{g_n}(\lim E_m) &= \lim_{m\to\infty} m_{g_0}E_m \\ &= m_{g_0}(\lim E_m).\end{aligned}$$

This completes the proof of the lemma.

Let I be a half-open interval $\alpha^{(i)} \leqq x^{(i)} < \beta^{(i)} (i = 1, \cdots, q)$. If we define I_m to be the interval $\alpha^{(i)} - \dfrac{2}{m} \leqq x^{(i)} \leqq \beta^{(i)} - \dfrac{2}{m}$, then $I = \lim I_m$. We cannot apply the lemma, since we know nothing about the measures of the I_m. But by the remark after 47.5 we have

$$\lim_{m\to\infty} \Delta_{g_i}I_m = m_{g_i}I \qquad (i = 0, 1, 2, \cdots).$$

If we use this in place of (E) and replace $m_{g_i}I_m$ by $\Delta_{g_i}I_m$, the proof of the lemma is still applicable; hypothesis (i) is clearly satisfied. We thus find that equation (B) holds for half-open intervals $I = [\alpha, \beta)$. By 19.8, it holds for sets E which are finite sums of disjoint intervals of type $[\alpha, \beta)$.

If E is an open set it is the sum of a denumerable collection I_1, $I_2, \cdots$ of disjoint half-open intervals $[\alpha_i, \beta_i)$, by 3.6. As we have just seen, for each finite sum $I_1 \cup \cdots \cup I_m$ equation (B) holds. So by the lemma equation (B) holds for all open sets E.

Suppose next that E is bounded and is the product of open sets $G_1, G_2, \cdots$. We may suppose $G_1 \supset G_2 \supset \cdots$; otherwise we could replace G_m by $G_1 \cap G_2 \cap \cdots \cap G_m$. Let I^* be an open interval containing E. We may suppose that each G_m is contained in I^*; otherwise we could replace it by $G_m I^*$. By the lemma, equation (B) holds for E.

Next let E be the product of open sets, but not bounded. Let W_m^0 be the interior of the interval W_m. By the preceding paragraph, equation (B) holds for each set EW_m^0. These expand as m increases, so by the lemma equation (B) holds for E also.

Suppose now that E is g_0-measurable. By 49.4 there are Borel sets A and B such that $A \subset E \subset B$ and $B - A$ has g_0-measure zero. By the remark after 49.4, we may suppose that B and $B - A$ are products of open sets. So by the preceding proof equation (B) holds for the sets B and $B - A$. Applying it to the latter set shows that

$$m_{g_i}(B - A) = 0 \qquad (i = 0, 1, 2, \cdots).$$

Since $B - E$ is contained in $B - A$, by 19.13 we have

$$m_{g_i}(B - E) = 0 \qquad (i = 0, 1, 2, \cdots).$$

Thus $B - E$ is g_i-measurable and has g_i-measure zero for each i, and B itself is g_i-measurable for each i by 49.2, so by 19.6 E is g_i-measurable for each i. Also

$$\begin{aligned} m_{g_0}E &= m_{g_0}B \\ &= m_{g_1}B + m_{g_2}B + \cdots \\ &= m_{g_1}E + m_{g_2}E + \cdots . \end{aligned}$$

Conversely, let E be g_n-measurable for $n = 1, 2, \cdots$. For each n there exist Borel sets A_n, B_n such that $A_n \subset E \subset B_n$ and

$$m_{g_n}(B_n - A_n) = 0.$$

By the remark after 49.4, we may suppose that B_n and $B_n - A_n$ are products of open sets. Define

$$A = \bigcup A_n, \qquad B = \bigcap B_n.$$

Then B is a product of open sets, and so is $B - A = \bigcap (B_n - A_n)$. By 19.13,

$$m_{g_n}(B - E) = m_{g_n}(B - A) = 0 \qquad (n = 1, 2, \cdots).$$

Since (B) holds for products of open sets, this implies

$$m_{g_0}(B - A) = 0,$$

and by 19.13 the subset $B - E$ also has g_0-measure zero. By 49.2 the set B is g_0-measurable, so by 19.6 so is the set $E = B - (B - E)$. As already shown this implies that equation (B) holds. This completes the proof of conclusion (i) of our theorem.

In the space R_{q+1} let us define functions

$$G_i(x^{(1)}, \cdots, x^{(q+1)}) \equiv x^{(q+1)} g_i(x^{(1)}, \cdots, x^{(q)}) \qquad (i = 0, 1, 2, \cdots).$$

If I^* is an interval in R_{q+1} defined by inequalities $\alpha^{(i)} \leqq x^{(i)} \leqq \beta^{(i)}$ $(i = 1, \cdots, q+1)$, and I is the interval in R_q defined by the first q of these inequalities, we readily compute

$$\Delta_{G_i} I^* = (\beta^{(q+1)} - \alpha^{(q+1)}) \Delta_{g_i} I.$$

Hence equation (A) still holds if we replace g_i by G_i. By the part of the proof already completed, conclusion (i) holds for sets E in R_{q+1} if we replace g_i by G_i.

In particular, let $f(x)$ be defined and non-negative on R_q, and let Ω be its ordinate-set. By 25.9, for each integer i the function $f(x)$ is g_i-summable over R_q if and only if the set Ω has finite G_i-measure, and in that case

$$\int_{R_q} f(x) dg_i(x) = m_{G_i} \Omega.$$

Applying conclusion (i) to the set Ω thus yields conclusion (ii) under the restriction $E = R_q$. To obtain the conclusion with arbitrary E, it is only necessary to apply what has just been proved to $f(x)K_E(x)$.

If $f(x)$ is defined on E, we write it in the form $f^+(x) - f^-(x)$ as in 15.4. Then by 18.3 the function $f(x)$ is g_0-summable if and only if both $f^+(x)$ and $f^-(x)$ are g_0-summable. By the preceding paragraph this is true if and only if both are g_n-summable for $n = 1, 2, \cdots$ and the sums

$$\sum_{n=1}^{\infty} \int_E f^+(x) dg_n(x) \qquad \text{and} \qquad \sum_{n=1}^{\infty} \int_E f^-(x) dg_n(x)$$

are finite. This is true if and only if $|f|$ is g_n-summable for $n = 1, 2, \cdots$ and

$$\sum_{n=1}^{\infty} \int_E | f(x) | \, dg_n(x)$$

is finite. In case $f(x)$ is g_0-summable, so are f^+ and f^-, and by the preceding paragraph equation (D) holds for each. By subtraction we obtain equation (D) for $f(x)$.

The next theorem is a corollary of 50.1.

50.2. *Let $g(x)$ and $h(x)$ be positively monotonic functions such that $\Delta_g I \geqq \Delta_h I$ for every interval I. Let E be a set in the space R_q and $f(x)$ a function defined on E. Then*

(i) *if E is g-measurable it is h-measurable, and*

$$m_h E \leqq m_g E;$$

(ii) *if $f(x)$ is g-summable over E it is h-summable over E; and if furthermore $f(x)$ is non-negative on E the inequality*

$$\int_E f(x)dh(x) \leqq \int_E f(x)dg(x)$$

is satisfied.

The interval function $\Delta_g I - \Delta_h I$ is finite, non-negative and additive, so by 45.5 there is a positively monotonic function $k(x)$ such that $\Delta_k I = \Delta_g I - \Delta_h I$ for every interval I. We now apply 50.1 with g_0, g_1, g_2 replaced by g, h, k respectively. By that theorem, if E is g-measurable it is h-measurable and k-measurable, and

$$m_g E = m_h E + m_k E.$$

The last term being non-negative, this yields our first conclusion. Also, if $f(x)$ is g-summable over E it is h-summable and k-summable, and

$$\int_E f(x)dg(x) = \int_E f(x)dh(x) + \int_E f(x)dk(x).$$

If $f(x)$ is non-negative the last integral is non-negative. This completes the proof.

51. We shall now remove the restriction that the "integrator" $g(x)$ is positively monotonic. Our definition of the general Lebesgue-Stieltjes integral is as follows.

51.1. *Let $g(x)$ be defined and finite on the space R_q, and of BV on every closed interval. Let $P(x)$, $N(x)$ be a minimum decomposition of $g(x)$. If $f(x)$ is defined on R_q, it is summable with respect to g, or g-summable, over R_q if it is summable with respect to both P and N; and in this case we define*

$$\int_{R_q} f(x)dg(x) \equiv \int_{R_q} f(x)dP(x) - \int_{R_q} f(x)dN(x).$$

There is an apparent ambiguity in this definition, since the functions $P(x)$ and $N(x)$ are not uniquely determined by $g(x)$. However, our next theorem will show that this causes no trouble, and that any minimum decomposition of $g(x)$ is in a sense the best possible way of representing $g(x)$ as a difference of positively monotonic functions.

51.2. *If $g(x)$ is of BV on every interval I, and $P(x)$, $N(x)$ are a minimum decomposition of $g(x)$, and $\pi(x)$, $\nu(x)$ are positively monotonic* *functions such that*

$$\Delta_\pi I - \Delta_\nu I = \Delta_g I \text{ for every interval } I,$$

then every function $f(x)$ which is summable with respect to both π and ν is also summable with respect to P and N and therefore with respect to g, and

$$\int_{R_q} f d\pi - \int_{R_q} f d\nu = \int_{R_q} f dP - \int_{R_q} f dN = \int_{R_q} f dg.$$

Define $h(x) = \pi(x) - P(x)$. Since $\Delta_\pi I - \Delta_\nu I = \Delta_P I - \Delta_N I$ we also have $\Delta_h I = \Delta_\nu I - \Delta_N I$. By 46.9*d*, $h(x)$ is positively monotonic, since $\Delta_h I = \Delta_\pi I - \Delta_P I \geqq 0$. Therefore if $f(x)$ is summable with respect to π and ν, we know from 50.1 that f is summable with respect to P, h, and N, and

$$\int_{R_q} f d\pi = \int_{R_q} f dP + \int_{R_q} f dh,$$
$$\int_{R_q} f d\nu = \int_{R_q} f dN + \int_{R_q} f dh.$$

Subtracting the second of these equations member by member from the first yields the conclusion of the theorem.

This enables us at once to extend theorem 50.1:

51.3. *If $f(x)$ is summable with respect to both g_1 and g_2, and $g(x)$ is a function such that $\Delta_g I = \Delta_{g_1} I + \Delta_{g_2} I$ for every interval I, then $f(x)$ is g-summable, and*

$$\int_{R_q} f dg = \int_{R_q} f dg_1 + \int_{R_q} f dg_2.$$

Let P_1, N_1 and P_2, N_2 be minimum decompositions of g_1 and g_2 respectively. Define $\pi(x) = P_1(x) + P_2(x)$, $\nu(x) = N_1(x) + N_2(x)$. Then π and ν are positively monotonic, and

$$\begin{aligned} \Delta_\pi I - \Delta_\nu I &= \Delta_{P_1} I + \Delta_{P_2} I - \Delta_{N_1} I - \Delta_{N_2} I \\ &= \Delta_{g_1} I + \Delta_{g_2} I = \Delta_g I \end{aligned}$$

for every interval I. If $f(x)$ is summable with respect to g_1 and g_2, by definition it is summable with respect to P_1, N_1, P_2 and N_2. By 50.1 it is summable with respect to π and ν. By 51.2 $f(x)$ is g-summable, and by 51.2 and 50.1

$$\begin{aligned} \int_{R_q} f(x)dg(x) &= \int_{R_q} f(x)d\pi(x) - \int_{R_q} f(x)d\nu(x) \\ &= \int_{R_q} f(x)dP_1(x) + \int_{R_q} f(x)dP_2(x) - \int_{R_q} f(x)dN_1(x) \\ &\qquad - \int_{R_q} f(x)dN_2(x) = \int_{R_q} f(x)dg_1(x) + \int_{R_q} f(x)dg_2(x). \end{aligned}$$

EXERCISE. Let $g_0(x)$, $g_1(x)$, $\cdots$ be a sequence of functions of BV on every interval such that

$$\sum_{n=1}^{\infty} \Delta_{\tau_n} I < \infty$$

for every I, where $\tau_n = P_n + N_n$ and P_n and N_n are a minimum decomposition of $g_n(x)$. Let $\Delta_{g_0} I = \Sigma \Delta_{g_n} I$ for every interval I. If $f(x)$ is defined on a set E in the space R_q, then

(i) if E is g_n-measurable ($n = 1, 2, \cdots$) it is g_0-measurable, and $m_{g_0} E = \Sigma m_{g_n} E$;

(ii) if $f(x)$ is g_n-summable over E ($n = 1, 2, \cdots$) and the series

$$\sum \int_E |f(x)| \, d\tau_n(x)$$

converges, conclusion (D) of 50.1 holds.

52. In so far as is possible, we shall now extend the principal theorems on Lebesgue integrals to the Lebesgue-Stieltjes integrals defined in §51. To avoid repetition, throughout this section the symbol $g(x)$ shall be used to denote a function which is of BV on every interval I; the functions $P(x)$ and $N(x)$ denote a minimum decomposition of $g(x)$; and $\tau(x) = P(x) + N(x)$, so that (by 46.9) $\Delta_\tau I = T_g[I]$ for every interval I. A brief indication of proof is sufficient for most of these theorems, since they are usually simple corollaries of the corresponding theorems for the integrals with respect to $P(x)$ and $N(x)$, together with 51.2. The list will not include lemmas superseded by later theorems.

First we state three theorems on elementary operations.

52.1. *If $f(x)$ is defined on E, then $f(x)$ is g-summable over E if and only if it is τ-summable over E.*

Let $f(x) = 0$ on CE. Then $f(x)$ is g-summable (over R_q) if and only if it is both P-summable and N-summable, by definition. But it is both P-summable and N-summable if and only if it is τ-summable, by 50.1.

52.2. *If $f_1(x)$ and $f_2(x)$ are both g-summable over E, and $f_1(x) \leqq f_2(x)$ on E, then*

$$\int_E f_1(x) d\tau(x) \leqq \int_E f_2(x) d\tau(x).$$

This follows from 18.2, since $\tau(x)$ is positively monotonic. Here we cannot replace $\tau(x)$ by $g(x)$; in fact, if $g(x)$ is negatively monotonic the inequality holds in the opposite direction.

52.3. *If $f_1(x)$ and $f_2(x)$ are both g-summable on E, then so is $f_1(x) + f_2(x)$ if it is defined on E; so is $cf_1(x)$ for every finite number c; and so are* $\sup\{f_1(x), f_2(x)\}$, $\inf\{f_1(x), f_2(x)\}$, $f_1^+(x)$, $f_1^-(x)$ *and* $|f_1(x)|$. *Moreover,*

$$\int_E f_1(x)dg(x) + \int_E f_2(x)dg(x) = \int_E [f_1(x) + f_2(x)]dg(x),$$

$$\int_E cf_1(x)dg(x) = c\int_E f_1(x)dg(x),$$

$$\int_E |f(x)|\,d\tau(x) \geqq \left|\int_E f(x)dg(x)\right|.$$

Except for the last inequality this follows at once if we apply 18.3 to $P(x)$ and $N(x)$ and then subtract. For the last inequality, by 18.3

$$\int_E |f(x)|\,dP(x) \geqq \left|\int_E f(x)dP(x)\right|,$$

$$\int_E |f(x)|\,dN(x) \geqq \left|\int_E f(x)dN(x)\right|;$$

so, adding and applying 51.1,

$$\int_E |f(x)|\,d\tau(x) \geqq \left|\int_E f(x)dP(x) - \int_E f(x)dN(x)\right| = \left|\int_E f(x)dg(x)\right|.$$

Next we establish some theorems on measurable sets and measurable functions.

52.4. Definition. *A set E is g-measurable if it is both P-measurable and N-measurable. If E is g-measurable and m_PE and m_NE are not both ∞, we define*

$$m_gE = m_PE - m_NE.$$

52.5. *A set E is g-measurable if and only if it is τ-measurable; it has finite g-measure if and only if it has finite τ-measure.*

The first statement follows from 52.4 and 50.1. For the second, we see by 52.4 that E has finite g-measure if and only if it has finite P-measure and finite N-measure, and by 50.1 this is true if and only if E has finite τ-measure.

Since g-measurability and τ-measurability are equivalent, we can take over most of §19 bodily:

52.6. *If E is g-measurable and m_gE is finite, then*

$$m_gE = \int_{R_q} K_E(x)dg(x).$$

(Apply 19.2 to P and N and subtract.)

52.7. *All open sets and all closed sets (in particular, the whole space and the empty set) are g-measurable. The empty set has g-measure zero.*

(From 19.4, 19.5 and 19.7.)

52.8. *If $E_1, E_2, \cdots$ are g-measurable, so are $E_1 - E_2$, $\bigcup E_i$ and $\bigcap E_i$. If the E_i are disjoint and the series $\Sigma m_T E_i$ converges, then*

$$m_g(\bigcup E_i) = \Sigma m_g E_i.$$

(From 19.6, 19.8 and 19.9.)

52.9. *If $E_1, E_2, \cdots$ is a sequence of g-measurable sets, then* lim sup E_n *and* lim inf E_n *are g-measurable. If furthermore all the sets E_n are contained in a set of finite g-measure and* lim E_n *exists (in particular, if $E_1 \subset E_2 \subset E_3 \subset \cdots$ or if $E_1 \supset E_2 \supset E_3 \supset \cdots$) then*

$$m_g(\lim E_n) = \lim_{n \to \infty} m_g E_n.$$

The measurability of lim sup E_n and lim inf E_n follow at once from 52.5 and 19.11, and the last conclusion is obtained by applying 19.12 to the P-measures and N-measures of the sets and subtracting.

Of course it is not in general true that $\Delta_g I = m_g I$ for intervals I. Nevertheless we can state a useful generalization of 16.1 and 47.5.

52.10. *If $g(x)$ is of BV on every interval and is left continuous, then for every half-open interval I of the type $[\alpha, \beta)$ the equation*

$$m_g I = \Delta_g I$$

is satisfied.

It is not immediately evident that $P(x)$ and $N(x)$ inherit left continuity from $g(x)$, so we use a slight artifice. We again define I_n to be the interval $\alpha^{(i)} - \frac{2}{n} \leqq x^{(i)} \leqq \beta^{(i)} - \frac{2}{n}$. By the remark after 47.5, the equations

$$\lim_{n \to \infty} \Delta_P I_n = m_P I,$$
$$\lim_{n \to \infty} \Delta_N I_n = m_N I$$

hold. Subtracting, we find

$$\lim_{n \to \infty} \Delta_g I_n = m_g I.$$

By the left continuity of $g(x)$, this implies

$$\Delta_g I = m_g I.$$

We re-phrase 21.1 in the following definition.

52.11. *A function $f(x)$ defined on a set E is g-measurable on E if the set $E[f \geqq a]$ is g-measurable for every number a.*

By 52.5 we see immediately that

52.12. *If $f(x)$ is defined on a set E, it is g-measurable on E if and only if it is τ-measurable on E.*

But $\tau(x)$ is positively monotonic, so we can state

52.13. *All theorems of §§21 and 22 remain valid if the words "measurable" and "summable" are replaced by "g-measurable" and "g-summable" respectively.*

The phrase "almost everywhere" will be defined to mean "except on a set of τ-measure zero." In general, we cannot regard a set of g-measure zero as negligible, since a set of g-measure zero may have subsets having positive or negative g-measure. With this convention, we can establish

52.14. *If $f(x)$ is g-summable over E, then $f(x)$ is finite almost everywhere in E.*

For then $f(x)$ is τ-summable, and the conclusion follows from 23.1.

52.15. *If $f(x)$ is defined on a set E of τ-measure 0, then $f(x)$ is g-measurable over E, and*

$$\int_E f(x)dg(x) = 0.$$

By 23.2, $f(x)$ is τ-summable over E, hence g-summable; and by 52.3 and 23.2

$$\left| \int_E f(x)dg(x) \right| \leqq \int_E |f(x)| d\tau(x) = 0.$$

52.16. *If $f(x)$ is g-summable over a set E and $h(x) = f(x)$ almost everywhere on E, then $h(x)$ is g-summable over E, and*

$$\int_E h(x)dg(x) = \int_E f(x)dg(x).$$

Since $f(x)$ is g-summable, it is τ-summable; so $h(x)$ is τ-summable by 23.3, hence g-summable. Also by 23.3,

$$\int_E h(x)dP(x) = \int_E f(x)dP(x), \int_E h(x)dN(x) = \int_E f(x)dN(x);$$

so by subtraction the remaining conclusion is established.

52.17. *If $f(x)$ is defined on a set $E_1 \cup E_2$, and $m_\tau(E_1 - E_2) = m_\tau(E_2 - E_1) = 0$, then $f(x)$ is g-summable over E_2 if and only if it is g-summable over E_1; and if it is g-summable over E_1 then*

$$\int_{E_1} f(x)dg(x) = \int_{E_2} f(x)dg(x).$$

The statement concerning summability follows at once from 52.1 and 23.4. The equality is established by applying 23.4 to the integrals with respect to P and N.

The inequalities of §24 already apply to τ-integrals, and we can obtain trivial extensions by combining them with the inequality of 52.3.

In §26 there is nothing needing comment except that in the extension of definition 26.8 and theorem 26.10 we shall use τ-measure, not g-measure. Thus the function of sets $F(E)$ is AC (with respect to $g(x)$) if for every positive number ϵ there is a positive number δ such that $|F(E)| < \epsilon$ for every set E with $m_\tau E < \delta$. With this convention we can establish

52.18. *If $f(x)$ is g-summable over the space R_q, and for all measurable sets E we define*

$$F(E) = \int_E f(x)dg(x),$$

then $F(E)$ is AC with respect to $g(x)$, is completely additive, of BV, and bounded.

By 52.1 and 52.3, $|f(x)|$ is τ-summable. By 27.1, the set function $\int_E |f(x)| d\tau(x)$ is AC, and by 52.3

$$|F(E)| \leqq \int_E |f(x)| d\tau(x).$$

This establishes the absolute continuity of $F(E)$. Each of the set functions

$$\int_E f(x)dP(x), \qquad \int_E f(x)dN(x)$$

is of BV, bounded and completely additive, by 27.5 and 27.3. By subtraction, the conclusion of the theorem is established.

52.19. *If $f(x)$ is defined and g-summable over the disjoint sets E_1, E_2, $\cdots$, and the sum*

$$\sum \int_{E_i} |f(x)| d\tau(x)$$

is finite, then $f(x)$ is g-summable over $\bigcup E_i$, and

$$\int_{\bigcup E_i} f(x)dg(x) = \sum \int_{E_i} f(x)dg(x).$$

The first part of the conclusion follows at once from 52.1 and 27.2. The remainder of the conclusion is established by applying 27.2 to the integrals of $f(x)$ with respect to P and to N and subtracting.

As with §26, we adopt §28 bodily, replacing measure by τ-measure throughout.

52.20. *If $\{f_n(x)\}$ is a sequence of functions defined and g-measurable on a set E, and there is a function h(x) defined and g-summable on E such that $|f_n(x)| \leqq h(x)$, and the functions $f_n(x)$ converge in measure or almost everywhere on E to a finite-valued limit function f(x), then f(x) is g-summable, and*

$$\lim_{n\to\infty} \int_E f_n(x)dg(x) = \int_E f(x)dg(x),$$

and

$$\lim_{n\to\infty} \int_E |f_n(x) - f(x)| d\tau(x) = 0.$$

The functions $f_n(x)$ are τ-summable. So by 29.3 the function $f(x)$ is τ-summable (hence g-summable), and

$$\lim_{n\to\infty} \int_E |f_n(x) - f(x)| d\tau(x) = 0.$$

By 52.3, the other inequality of our theorem follows at once.

52.21. *If*

(*a*) *the functions $f_n(x)$ $(n = 1, 2, \cdots)$ are defined, finite and g-summable on a set E^* of finite g-measure;*

(*b*) *the functions $f_n(x)$ converge in measure (or almost uniformly, or almost everywhere) on E^* to a limit function f(x);*

(*c*) *for every positive ϵ there is a $\delta > 0$ such that $\int_E |f_n(x)| d\tau(x) < \epsilon$ for all n and for every subset E of E^* with $m_\tau E < \delta$;*

(*d*) *either the integrals $\int_{E^*} |f_n(x)| d\tau(x)$ are bounded, or f(x) is finite almost everywhere;*

then f(x) is g-summable over E^*, and

$$\lim_{n\to\infty} \int_{E^*} f_n(x)dg(x) = \int_{E^*} f(x)dg(x),$$

and

$$\lim_{n\to\infty} \int_{E^*} |f_n(x) - f(x)| d\tau(x) = 0.$$

If in 29.6 we understand "summable" to mean τ-summable, we find that $f(x)$ is τ-summable (hence g-summable), and the last inequality holds. The other inequality follows at once.

A similar generalization of 29.7 is obvious.

The reader may note one somewhat unsatisfactory feature of this theorem. The hypotheses could all be stated in terms of integrals with respect to $g(x)$ if it were not for hypothesis (c). It would seem desirable to replace this by the weaker hypothesis

(c′) *for every positive* ϵ *there is a* $\delta > 0$ *such that* $\left| \int_E f_n(x)dg(x) \right| < \epsilon$ *for all* n *and all subsets* E *of* E^* *with* $m_\tau E < \delta$.

However, it can be shown by examples that this weakening is impossible.

For example, let $g(n^{-1}) = n^{-2}$, $n = 1, 2, \cdots$, and let $g(x) = 0$ elsewhere. We find readily that the point n^{-1} has τ-measure $2n^{-2}$, and more generally the set E has τ-measure $m_\tau E$ which is the sum of the numbers $2n^{-2}$ taken over the values of n such that n^{-1} is in E. Define $f(n^{-1}) = n$ $(n = 1, 2, \cdots)$, $f(x) = 0$ for all other x. Define $f_m(n^{-1}) = n$ for $n = 1, 2, \cdots, m$, $f_m(x) = 0$ elsewhere. Then

$$\int f_m(x)dg(x) = 0$$

for all m, hypotheses (a) and (b) hold (with E^* the whole space R_1) and $f(x)$ is finite everywhere. But $f(x)$ is not τ-summable (hence not g-summable) over E^*, for $\Sigma 2n^{-2} \cdot n$ is divergent.

Nevertheless, in §56 we shall show that if $g(x)$ happens to be left continuous, it is possible to replace by hypothesis (c) by hypothesis (c′).

53. In the preceding sections we have started with a function $g(x)$ which is of BV on every interval, and from it we have derived a measure function $m_g E$ and an integral. These last were not independent. We derived the measure function from the integral, and as we saw in 25.9 we could also have derived the integral from the measure function. However, the question arises: if we are given a measure function, can we always find a function $g(x)$ which generates it? The question as stated is too vague. Before we study it, we must specify what we shall mean by a measure function.

Clearly a measure function should be a function defined on a family $\mathfrak{F}$ of point sets. We wish $\mathfrak{F}$ to have the most elementary properties of sets of finite measure; so we shall ask that all the intervals $[\alpha, \beta)$ belong to $\mathfrak{F}$, and that the difference of sets $\mathfrak{F}$ shall be in $\mathfrak{F}$. It is too much to ask that the sums of infinite collections of sets of $\mathfrak{F}$ shall belong to $\mathfrak{F}$; the family of sets of finite Lebesgue measure lacks this property. But we *can* ask that if $\{E_n\}$ is a sequence of sets of $\mathfrak{F}$ all contained in a set of $\mathfrak{F}$, the sum $\bigcup E_n$ shall belong to $\mathfrak{F}$. On $\mathfrak{F}$, the set-function $\mu(E)$ should be finite.

Two properties of g-measure which we wish $\mu(E)$ to retain are the following. A set of finite measure shall not contain subsets of unbounded measure; and there should be some additivity of measures of disjoint sets of $\mathfrak{F}$. To make this first desideratum precise, we adopt the following definition.

53.1. *If $\mu(E)$ is defined and finite on a family $\mathfrak{F}$ of sets in the space R_q, and E_0 is a set in R_q, the total variation $\alpha(E_0)$ of $\mu(E)$ over E_0 is defined to be the* sup *of the sum $\sum_1^n |\mu(E_i)|$ for all finite aggregates $E_1, \cdots, E_n$ of the disjoint sets of the family $\mathfrak{F}$ contained in E_0. If $\alpha(E_0)$ is finite, we say that $\mu(E)$ is of bounded variation (BV) over E_0.*

We can now specify the exact requirements which we shall impose on a measure function.

53.2. *A function of sets $\mu(E)$ defined on a family $\mathfrak{F}$ of subsets of the space R_q is a measure function on $\mathfrak{F}$ if the following conditions are satisfied.*

(*a*) *The family $\mathfrak{F}$ contains all intervals of the type $[\alpha, \beta)$.*

(*b*) *If E_1 and E_2 belong to $\mathfrak{F}$, so does $E_1 - E_2$.*

(*c*) *If $E_1, E_2, \cdots$ is a finite or denumerable collection of sets of $\mathfrak{F}$ all contained in a set E_0 of $\mathfrak{F}$, the sum $\bigcup E_n$ belongs to $\mathfrak{F}$.*

(*d*) *For every set E of $\mathfrak{F}$, the function $\mu(E)$ is defined and finite.*

(*e*) *The function $\mu(E)$ is of BV on every set E of the family $\mathfrak{F}$.*

(*f*) *If $E_1, E_2, \cdots$ is a finite or denumerable collection of disjoint sets of $\mathfrak{F}$, all contained in a set E_0 of $\mathfrak{F}$, then*

$$\mu(\bigcup E_n) = \Sigma\mu(E_n).$$

In this last equation both members are defined. For $\bigcup E_n$ is in $\mathfrak{F}$ by (*c*), so the left member is defined by (*d*). On the right each term is defined; so the right member is either a finite sum or is an absolutely convergent series, since by (*e*)

$$\Sigma\,|\,\mu(E_n)\,| \leqq \alpha(E_0) < \infty.$$

By 49.3, hypotheses 53.2 (*a*, *b*, *c*) imply that all bounded Borel sets belong to the family $\mathfrak{F}$.

If $\mu(E)$ is defined, finite and non-negative for all sets of a family $\mathfrak{F}$ which satisfies (*a*, *b*, *c*) of 53.2 and (*f*) holds for finite sums, then (*e*) is necessarily satisfied, and in fact $\alpha(E) = \mu(E)$ for all sets E of $\mathfrak{F}$. On the one hand, if E belongs to $\mathfrak{F}$ then $\alpha(E) \geqq |\,\mu(E)\,| = \mu(E)$. On the other hand, for all collections $E_1, \cdots, E_n$ of disjoint subsets of E

which belong to $\mathfrak{F}$ the inequality

$$\sum_{j=1}^{n} |\mu(E_j)| = \sum_{j=1}^{n} \mu(E_j) \leqq \mu(\bigcup E_j) + \mu(E - \bigcup E_j) = \mu(E)$$

holds, so that $\alpha(E) \leqq \mu(E)$.

The definition 53.2 is unaltered in content if we replace (c) by the weaker assumption

(c') *If* $E_1, E_2, \cdots$ *is a finite or denumerable collection of disjoint sets of* $\mathfrak{F}$ *all contained in a set* E_0 *of* $\mathfrak{F}$, *the sum* $\bigcup E_n$ *also belongs to* $\mathfrak{F}$.
It is evident that if (c) is satisfied, so is (c'). Suppose that (c') holds. Let $E_1, E_2, \cdots$ be a collection as described in (c). Define successively

$$D_1 = E_1, \qquad D_2 = E_2 - D_1, \cdots,$$
$$D_n = E_n - (D_1 \cup \cdots \cup D_{n-1}), \cdots.$$

These sets are disjoint and are all contained in E_0. By induction, using (c') and (b), each D_n belongs to $\mathfrak{F}$. But $\bigcup D_n = \bigcup E_n$, so by ($c'$) the sum $\bigcup E_n$ belongs to $\mathfrak{F}$, and (c) is satisfied.

From the definition 53.2 we readily deduce the following corollaries.

53.3. *Let* $\mu(E)$ *be a measure function defined on* $\mathfrak{F}$, *and let* $E_1, E_2, \cdots$ *be a denumerable collection of sets of* $\mathfrak{F}$ *all contained in a set* E *of* $\mathfrak{F}$. *Then*

(a) *if* $E_1 \subset E_2 \subset \cdots$, *then* $\mu(\bigcup E_n) = \lim_{n\to\infty} \mu(E_n)$;

(b) *if* $E_1 \supset E_2 \supset E_3 \supset \cdots$, *then* $\mu(\bigcap E_n) = \lim_{n\to\infty} \mu(E_n)$;

(c) *if* $\lim_{n\to\infty} E_n$ *exists* (*that is, if* $\lim\sup E_n = \lim\inf E_n$) *and* $\mu(E)$ *is non-negative,* then*

$$\mu(\lim_{n\to\infty} E_n) = \lim_{n\to\infty} \mu(E_n).$$

To establish conclusion (a), we note that the sum $\bigcup E_n$ can also be written in the form $E_1 \cup (E_2 - E_1) \cup (E_3 - E_2) \cup \cdots$, and in this form the summands are disjoint sets of $\mathfrak{F}$ all contained in E. Therefore by (f) of 53.2 we have

$$\begin{aligned}\mu(\bigcup E_n) &= \mu(E_1) + \mu(E_2 - E_1) + \cdots \\ &= \mu(E_1) + \lim_{n\to\infty} \sum_{j=1}^{n-1} \mu(E_{j+1} - E_j) \\ &= \lim_{n\to\infty} \mu(E_n),\end{aligned}$$

* This restriction will be removed after 53.12 is established.

and (*a*) is established. From (*a*) we deduce the other conclusions just as in 19.9 and 19.12.

Although a set function $\mu(E)$ defined on a family and satisfying 53.2 will be called a measure function, it is not necessarily true that $\mu(E)$ will coincide with one of our previous measures $m_g E$. To ensure this we shall subject μ and $\mathfrak{F}$ to two further requirements, suggested by properties already established for functions $m_g E$. These two properties are specified in the following definitions.

53.4. *The measure function $\mu(E)$ is* regular *on $\mathfrak{F}$ if for every set E of $\mathfrak{F}$ there are Borel sets A, B such that $A \subset E \subset B$ and $\alpha(B - A) = 0$.*

53.5. *The family $\mathfrak{F}$ is the* natural range *of the measure function $\mu(E)$ if (a) $\mu(E)$ is regular on $\mathfrak{F}$ and (b) the family $\mathfrak{F}$ contains every set E for which there exist two Borel sets A, B such that $A \subset E \subset B$ and $\alpha(B - A) = 0$.*

In these definitions the family $\mathfrak{F}$ enters in a hidden way; for the definitions involve the total variation $\alpha(E)$, and the definition of $\alpha(E)$ involves sets $E_1, \cdots, E_n$ belonging to $\mathfrak{F}$. If $\mu(E)$ is regular this hidden occurrence of $\mathfrak{F}$ can be eliminated. We can in fact replace $\alpha(E)$ by $\alpha^*(E) =$ sup of the numbers $\sum_{j=1}^{n} |\mu(E_n)|$, where $E_1, \cdots, E_n$ are disjoint bounded Borel sets contained in E, thus:

53.6. *Let $\mu(E)$ be a regular measure function on the family $\mathfrak{F}$. Then*

(a) for every set E of $\mathfrak{F}$ and every $\epsilon > 0$ there is a bounded Borel set A contained in E such that $|\mu(E) - \mu(A)| < \epsilon$;

(b) for every set E the equation $\alpha^*(E) = \alpha(E)$ holds.

Let E belong to the family $\mathfrak{F}$, and let W_n be the interval

$$-n \leqq x^{(i)} \leqq n, \qquad i = 1, \cdots, q.$$

By 53.3

$$\lim_{n \to \infty} \mu(EW_n) = \mu(E).$$

Hence there is an n for which $|\mu(EW_n) - \mu(E)| < \epsilon$. The set EW_n belongs to $\mathfrak{F}$, so there are Borel sets A, B such that $A \subset EW_n \subset B$ and $\alpha(B - A) = 0$. By definition of $\alpha(E)$ we find $|\mu(EW_n - A)| \leqq \alpha(B - A) = 0$, so

$$\mu(A) = \mu(EW_n) - \mu(EW_n - A) = \mu(EW_n).$$

So the set A satisfies the requirements of part (*a*) of the theorem.

The sets used in calculating $\alpha^*(E)$ are a subset of the family of sets used in calculating $\alpha(E)$, so by 5.1 we see that $\alpha^*(E) \leqq \alpha(E)$. On the other hand, let h be any number less than $\alpha(E)$. There are disjoint sets $E_1, \cdots, E_n$ of the family contained in E for which

$$S = \sum_{j=1}^{n} | \mu(E_j) | > h.$$

By (*a*), each set E_j contains a bounded Borel subset A_j such that $| \mu(E_j) - \mu(A_j) | < (S - h)/n$, whence

$$\sum_{j=1}^{n} | \mu(A_j) | > h.$$

A fortiori, $\alpha^*(E) > h$. This holds for all numbers h less than $\alpha(E)$, so $\alpha^*(E) \geqq \alpha(E)$, and the proof of (*b*) is complete.

Before theorem 53.6 was proved, it was not evident that a function $\mu(E)$ might not have two natural ranges. That is, $\mu(E)$ might conceivably be defined on an aggregate $\mathfrak{G}$ of sets containing two distinct families $\mathfrak{F}_1$, $\mathfrak{F}_2$ each satisfying 53.4 and 53.5. Now, however, we see that as soon as $\mu(E)$ is specified for all bounded Borel sets we can single out a family $\mathfrak{F}$ which is the only family that can be the natural range of $\mu(E)$, and which is the natural range of $\mu(E)$ if $\mu(E)$ is defined for all sets of $\mathfrak{F}$. For then $\alpha^*(E)$ is uniquely determined; and in order to be the natural range of $\mu(E)$ the family $\mathfrak{F}$ must consist of those sets E such that there are Borel sets A, B satisfying the condition $A \subset E \subset B$, $\alpha^*(A) < \infty$, $\alpha^*(B - A) = 0$.

This suggests a question. Suppose that $\mu(E)$ is defined and regular on a family $\mathfrak{F}$, and that $\mathfrak{F}_1$ is the aggregate of sets just described which would be the natural range of $\mu(E)$ if $\mu(E)$ were defined on $\mathfrak{F}_1$. Is it always possible to find a measure function $\mu_1(E)$ on the natural range $\mathfrak{F}_1$ which is equal to $\mu(E)$ for all sets E of $\mathfrak{F}$? We could proceed to show directly that this extension is possible and is unique. But since this is a corollary of theorem 54.2 we prefer to postpone the study.

Theorem 53.6 also enables us to establish the following theorem.

53.7. *Let $\mu_1(E)$ and $\mu_2(E)$ be regular on the respective ranges $\mathfrak{F}_1$, $\mathfrak{F}_2$, and let $\mu_1(E) = \mu_2(E)$ for all bounded Borel sets E (which necessarily belong both to $\mathfrak{F}_1$ and to $\mathfrak{F}_2$). Then $\mu_1(E) = \mu_2(E)$ for all sets E belonging both to $\mathfrak{F}_1$ and to $\mathfrak{F}_2$; and if $\mathfrak{F}_1$ is the natural range of $\mu_1(E)$, then $\mathfrak{F}_2$ is contained in $\mathfrak{F}_1$.*

For all sets E we have $\alpha_1^*(E) = \alpha_2^*(E)$, where the definition of α_1^* and α_2^* are obvious; for in calculating α_i^* we use only bounded Borel sets $E_1, \cdots, E_n$, on which μ_1 and μ_2 coincide. So by 53.6 the respective total variations $\alpha_1(E)$, $\alpha_2(E)$ of $\mu_1(E)$ and $\mu_2(E)$ coincide for all sets E.

Now let E be a bounded set belonging both to $\mathfrak{F}_1$ and to $\mathfrak{F}_2$. We can find Borel sets A, B such that $A \subset E \subset B$ and $\alpha_1(B - A) = 0$, whence $\alpha_2(B - A) = 0$. By definition of α_i we see that $| \mu_i(E - A) | \leqq \alpha_i(B - A) = 0$, $i = 1, 2$. Hence

$$\mu_i(E) = \mu_i(E - A) + \mu_i(A) = \mu_i(A), \qquad i = 1, 2.$$

But A is a bounded Borel set, so $\mu_1(A) = \mu_2(A)$. Therefore $\mu_1(E) = \mu_2(E)$.

Next, let E be any set belonging to both families $\mathfrak{F}_1$ and $\mathfrak{F}_2$, and let W_n be the interval $-n \leqq x^{(i)} \leqq n$, $i = 1, \cdots, q$. By the preceding paragraph, $\mu_1(EW_n) = \mu_2(EW_n)$ for every n. So by 53.3,

$$\mu_1(E) = \lim_{n \to \infty} \mu_1(EW_n) = \lim_{n \to \infty} \mu_2(EW_n) = \mu_2(E).$$

Finally, let $\mathfrak{F}_1$ be the natural range of $\mu_1(E)$, and let E be a set of the family $\mathfrak{F}_2$. Since $\mu_2(E)$ is regular on $\mathfrak{F}_2$, there are Borel sets A, B such that $A \subset E \subset B$, $\alpha_2(A) \leqq \alpha_2(E) < \infty$, and $\alpha_2(B - A) = 0$. But $\alpha_1 = \alpha_2$, so $\alpha_1(A) < \infty$ and $\alpha_1(B - A) = 0$. Since $\mathfrak{F}_1$ is the natural range of $\mu_1(E)$, this implies that E belongs to $\mathfrak{F}_1$.

Closely related to this last theorem, but much deeper, is the following.

53.8. *If $\mu_1(E)$ and $\mu_2(E)$ are measure functions on the respective ranges $\mathfrak{F}_1$, $\mathfrak{F}_2$, and for all intervals of the form $I = [\alpha, \beta)$ the equation $\mu_1(I) = \mu_2(I)$ holds, then $\mu_1(E) = \mu_2(E)$ for every bounded Borel set.*

To avoid frequent repetition, in this proof the word "interval" shall always be understood to mean "interval of the type $[\alpha, \beta)$."

Let us first define an auxiliary class $\mathfrak{E}$ of sets. The class $\mathfrak{E}$ will consist of all sets E belonging both to $\mathfrak{F}_1$ and to $\mathfrak{F}_2$ and such that for every interval $I = [\alpha, \beta)$ the equation

$$\mu_1(IE) = \mu_2(IE)$$

holds. This class contains all intervals $I_1 = [\alpha_1, \beta_1)$; for $I_1 \cap I$ is again an interval of the type $[\alpha', \beta')$, for which the equality of μ_1 and μ_2 has been assumed. We need two properties of this class.

(i) *If E is in $\mathfrak{E}$ and I is an interval, then $I - E$ is in $\mathfrak{E}$.*

Let I_1 be any interval. Then (since I_1I is an interval)

$$\begin{aligned}\mu_1(I_1[I - E]) &= \mu_1(I_1I - I_1IE) = \mu_1(I_1I) - \mu_1(I_1IE) \\ &= \mu_2(I_1I) - \mu_2(I_1IE) = \mu_2(I_1[I - E]),\end{aligned}$$

so that $I - E$ is in $\mathfrak{E}$.

(ii) *If $E_1, E_2, \cdots$ is a finite or denumerable collection of disjoint sets of $\mathfrak{E}$, all contained in a set E_0 of $\mathfrak{E}$, then $\bigcup E_i$ is in $\mathfrak{E}$.*

The set E_0 belongs to $\mathfrak{E}$, therefore belongs both to $\mathfrak{F}_1$ and to $\mathfrak{F}_2$. Hence by 53.2(*c*) the sum $\bigcup E_j$ belongs both to $\mathfrak{F}_1$ and to $\mathfrak{F}_2$. Moreover, for each interval I the equation $\mu_1(IE_j) = \mu_2(IE_j)$, $j = 1, 2, \cdots$ all hold; so by 53.2(*f*) we have

$$\begin{aligned}\mu_1(I \cap [\bigcup E_j]) &= \mu_1(\bigcup IE_j) = \Sigma\mu_1(IE_j) \\ &= \Sigma\mu_2(IE_j) = \mu_2(\bigcup IE_j) = \mu_2(I \cap [\bigcup E_j]),\end{aligned}$$

and $\bigcup E_j$ belongs to $\mathfrak{E}$.

Next, let $\mathfrak{M}$ consist of all sets M of the class $\mathfrak{E}$ such that $M - E$ is in $\mathfrak{E}$ whenever E is in $\mathfrak{E}$.

(iii) *All intervals of type* $[\alpha, \beta)$ *belong to* $\mathfrak{M}$. This follows at once from (i).

(iv) *If* M_1 *and* M_2 *are in* $\mathfrak{M}$, *so is* $M_1 - M_2$. Since M_1 is in $\mathfrak{M}$ and M_2 is in $\mathfrak{E}$ (being in $\mathfrak{M}$), we know that $M_1 - M_2$ is in $\mathfrak{E}$. We must show that $(M_1 - M_2) - E$ is in $\mathfrak{E}$ whenever E is in $\mathfrak{E}$. By definition of the class $\mathfrak{M}$, the set $M_2 - E$ is in $\mathfrak{E}$. Now $M_2 - E$ and E are disjoint sets of $\mathfrak{E}$, so by (ii) the set $M_2 \cup E \equiv (M_2 - E) \cup E$ is in $\mathfrak{E}$. Since M_1 is in $\mathfrak{M}$, it follows that $M_1 - (M_2 \cup E)$ is in $\mathfrak{E}$. Since $M_1 - (M_2 \cup E) = (M_1 - M_2) - E$, this establishes statement (iv).

(v) *If* $M_1, M_2, \cdots, M_n$ *are in* $\mathfrak{M}$, *so is* $M_1 - (M_2 \cup \cdots \cup M_n)$. We need only apply (iv) $n - 1$ times.

(vi) *If* $M_1, M_2, \cdots$ *is a finite or denumerable collection of disjoint sets of* $\mathfrak{M}$, *all contained in a set* M_0 *of* $\mathfrak{M}$, *then* $\bigcup M_j$ *is in* $\mathfrak{M}$. Let E be any set of $\mathfrak{E}$; then $\bigcup M_j - E = \bigcup (M_j - E)$. The sets $M_j - E$ are each in $\mathfrak{E}$ and are disjoint, and they are all contained in the set M_0, which is in $\mathfrak{M}$ and therefore in $\mathfrak{E}$. So by (ii) their sum is in $\mathfrak{E}$. Thus for every set E of $\mathfrak{E}$ the difference $\bigcup M_j - E$ is in $\mathfrak{E}$, and by definition $\bigcup M_j$ is in $\mathfrak{M}$.

Statements (iii) and (iv) are hypotheses (*a*) and (*b*) of 49.3, and (vi) is hypothesis (*c'*). Hence by 49.3 the class $\mathfrak{M}$ contains all bounded Borel sets. But $\mathfrak{M}$ is itself contained in $\mathfrak{E}$, so from the definition of $\mathfrak{E}$ we see that for every bounded Borel set E the equation

$$\mu_1(IE) = \mu_2(IE)$$

holds for all intervals I. Letting I contain E, we find $\mu_1(E) = \mu_2(E)$ for all bounded Borel sets.

We now start toward the proof of a theorem which is interesting in itself besides being useful in later proofs. In general, a measure function $\mu(E)$ can assume both positive and negative values. The next theorem shows that in a sense the positive and negative values can be separated from each other; the space R_q can be split into two parts in such a way that $\mu(E)$ is positive for every set E of $\mathfrak{F}$ which is contained in one part and negative for every set E of $\mathfrak{F}$ contained in the other part. It is convenient first to define the positive and negative parts of the total variation of $\mu(E)$.

53.9. *If $\mu(E)$ is a measure function on a family $\mathfrak{F}$ of sets in the space R_q, and E_0 is a set in R_q, then $\pi(E_0)$ is defined to be* $\sup\{\mu(E) \mid E \subset E_0,$ *E in $\mathfrak{F}\}$; and $\nu(E_0)$ is defined to be* $\sup\{-\mu(E) \mid E \subset E_0, E$ *in* $\mathfrak{F}\}$.

From this definition we can deduce some trivial consequences.

53.10. *The functions $\pi(E)$, $\nu(E)$ satisfy the inequalities $0 \leqq \pi(E) \leqq \alpha(E)$, $0 \leqq \nu(E) \leqq \alpha(E)$. In particular, for all sets E of $\mathfrak{F}$ the numbers $\pi(E)$ and $\nu(E)$ are finite.*

The empty set Λ belongs to $\mathfrak{F}$, and $\mu(\Lambda) = 0$. Therefore 0 occurs in the collection of numbers whose sup is $\pi(E)$, and $\pi(E) \geqq 0$. A similar statement holds for $\nu(E)$.

If $h < \pi(E)$, there is a set E_0 contained in E and belonging to $\mathfrak{F}$ for which $\mu(E_0) > h$. Hence by 53.1, $\alpha(E) \geqq |\mu(E_0)| > h$. This holds for all h less than $\pi(E)$, so $\pi(E) \leqq \alpha(E)$. A similar proof holds for $\nu(E)$. In particular, if E belongs to $\mathfrak{F}$ the total variation $\alpha(E)$ is finite, so $\pi(E)$ and $\nu(E)$ are also finite.

Now we can proceed to establish our next theorem.

53.11. *If $\mu(E)$ is a measure function on a family $\mathfrak{F}$ of sets, there is a set P with the following properties. For every set E of $\mathfrak{F}$, the product EP and the difference $E - P$ belong to $\mathfrak{F}$. If E is a set of the family $\mathfrak{F}$ contained in P, then $\mu(E) \geqq 0$; and if E belongs to $\mathfrak{F}$ and is contained in the complement of P, then $\mu(E) \leqq 0$. Furthermore, if $\mu(E)$ is regular on $\mathfrak{F}$ we can take P to be a Borel set.*

Let I be an interval. For each n we can select a set E_n of the family $\mathfrak{F}$ which is contained in I and satisfies the inequality $\mu(E_n) > \pi(I) - 2^{-n}$. If $\mu(E)$ is regular, by 53.6 there is a Borel set A_n contained in E_n and such that $|\mu(E_n) - \mu(A_n)| < \mu(E_n) - [\pi(I) - 2^{-n}]$, so that $\mu(A_n)$ also exceeds $\pi(I) - 2^{-n}$. That is, if $\mu(E)$ is regular we may suppose that E_n is a Borel set.

Define $S_n = E_n \cup E_{n+1} \cup \cdots$. The sets $D_n = E_n, \cdots, D_m = E_m - [E_n \cup \cdots \cup E_{m-1}], \cdots$ are disjoint subsets of I and belong to $\mathfrak{F}$, so

$$\mu(S_n) = \mu(\bigcup_{j=n}^{\infty} E_j) = \mu(\bigcup_{j=n}^{\infty} D_j) = \sum_{j=n}^{\infty} \mu(D_j).$$

But D_j, being a subset of E_j, is contained in I, and by definition

$\pi(I) \geqq \mu(E_j - D_j)$. So

$$\mu(D_j) = \mu(E_j) - \mu(E_j - D_j) > [\pi(I) - 2^{-j}] - \pi(I) = -2^{-j}$$

Therefore

$$\pi(I) \geqq \mu(S_n) = \mu(D_n) + \sum_{j=n+1}^{\infty} \mu(D_j) > \mu(E_n) - \sum_{j=n+1}^{\infty} 2^{-j} > \pi(I) - 2^{1-n}.$$

By the definition of the sets S_n we see that $S_1 \supset S_2 \supset \cdots$. Hence by 53.3(*b*), if we write $P^* = \bigcap_{n=1}^{\infty} S_n$ we find that P^* is in $\mathfrak{F}$ and $\mu(P^*) = \lim_{n\to\infty} \mu(S_n) = \pi(I)$. Let E be a subset of P^* belonging to $\mathfrak{F}$. Then $\mu(E) = \mu(P^*) - \mu(P^* - E) = \pi(I) - \mu(P^* - E) \geqq 0$. On the other hand, let E be a subset of $I - P^*$ belonging to $\mathfrak{F}$. Then $P^* \cup E$ is in I, and

$$\mu(E) = \mu(P^* \cup E) - \mu(P^*) = \mu(P^* \cup E) - \pi(I) \leqq 0.$$

Now let $I_1, I_2, \cdots$ be a sequence of disjoint intervals whose sum is the whole space R_q. In each I_j we construct as above a set P_j such that $\mu(E) \geqq 0$ for every subset E of P_j belonging to $\mathfrak{F}$ and $\mu(E) \leqq 0$ for every subset E of $I_j - P_j$ belonging to $\mathfrak{F}$. We define $P = \bigcup P_j$. If E belongs to $\mathfrak{F}$, so does EP_j for every j; hence by 53.2*c* so does $EP = \bigcup EP_j$. Furthermore, if E is contained in P then $E = \bigcup EP_j$, and by 53.2*f*

$$\mu(E) = \Sigma\mu(EP_j) \geqq 0.$$

On the other hand, if E belongs to $\mathfrak{F}$ and is contained in CP, then $E = \bigcup (I_j - P_j)E$, and

$$\mu(E) = \Sigma\mu[(I_j - P_j)E] \leqq 0.$$

If each of the sets E_j is a Borel set, then each S_n is a Borel set, each P_j is a Borel set, and P is a Borel set.

This leads readily to the following theorem.

53.12. *If $\mu(E)$ is a measure function on a range $\mathfrak{F}$, then $\alpha(E)$, $\pi(E)$ and $\nu(E)$ are also measure functions on $\mathfrak{F}$, and the equations*

$$\pi(E) - \nu(E) = \mu(E), \qquad \pi(E) + \nu(E) = \alpha(E)$$

hold for every set E of $\mathfrak{F}$. If P is the set described in 53.11, *then for every set E of $\mathfrak{F}$*

$$\pi(E) = \mu(PE), \qquad \nu(E) = -\mu(E - P).$$

If $\mu(E)$ is regular on $\mathfrak{F}$, then $\pi(E)$, $\nu(E)$ and $\alpha(E)$ are regular on $\mathfrak{F}$.

The family $\mathfrak{F}$ is the natural range of $\mu(E)$ if and only if it is the natural range of $\alpha(E)$.

Let E be any set belonging to $\mathfrak{F}$. By definition of $\pi(E)$ we see that $\pi(E) \geqq \mu(EP)$. But if E_0 is a subset of E which belongs to $\mathfrak{F}$, by 53.11 the inequalities

$$\mu(E_0 - P) \leqq 0, \qquad \mu((E - E_0)P) \geqq 0$$

hold. Therefore

$$\begin{aligned}\mu(E_0) &= \mu(E_0P) + \mu(E_0 - P) \leqq \mu(E_0P)\\ &= \mu(EP - (E - E_0)P) = \mu(EP) - \mu((E - E_0)P)\\ &\leqq \mu(EP).\end{aligned}$$

Since this holds for all subsets E_0 of E which belong to $\mathfrak{F}$, it follows that $\pi(E) \leqq \mu(EP)$; and with the preceding inequality this shows that $\pi(E) = \mu(EP)$.

The equation $\nu(E) = -\mu(E - P)$ could be similarly proved. More easily, if we replace $\mu(E)$ by $-\mu(E)$ the rôles of π and ν are interchanged, and likewise for P and CP, so that $\nu(E) = -\mu(ECP) = -\mu(E - P)$. It follows at once that

$$\mu(E) = \mu(EP) + \mu(E - P) = \pi(E) - \nu(E).$$

By definition, if E belongs to $\mathfrak{F}$ then

$$\begin{aligned}\alpha(E) &\geqq |\mu(EP)| + |\mu(E - P)| = \mu(EP) - \mu(E - P)\\ &= \pi(E) + \nu(E).\end{aligned}$$

But if $E_1, \cdots, E_n$ are disjoint sets of the family $\mathfrak{F}$ contained in E, then

$$\begin{aligned}\sum_{j=1}^{n} |\mu(E_j)| &= \sum_{j=1}^{n} |\mu(E_jP) + \mu(E_j - P)|\\ &\leqq \sum_{j=1}^{n} |\mu(EP_j)| + \sum_{j=1}^{n} |\mu(E_j - P)|\\ &= \sum_{j=1}^{n} \mu(E_jP) - \sum_{j=1}^{n} \mu(E_j - P)\\ &= \mu(P \cap [\bigcup E_j]) - \mu(\bigcup E_j - P) = \pi(\bigcup E_j) + \nu(\bigcup E_j)\\ &\leqq \pi(E) + \nu(E).\end{aligned}$$

Hence

$$\alpha(E) \leqq \pi(E) + \nu(E),$$

which together with the preceding inequality proves $\alpha(E) = \pi(E) + \nu(E)$.

If E is in $\mathfrak{F}$, then EP and $E - P$ are in $\mathfrak{F}$ by 53.11. Hence by 53.10 $\pi(E)$, $\nu(E)$ and $\alpha(E)$ are all defined and finite. That is, they satisfy (*d*) of 53.2. From the equations $\pi(E) = \mu(PE)$ and $\nu(E) = -\mu(E - P) = -\mu(ECP)$ it follows that π and ν satisfy 53.2*f*. Since $\alpha(E) = \pi(E) + \nu(E)$, it too satisfies 53.2(*f*). Also all three set-functions α, π and ν are non-negative, so by the remark after 53.2 they satisfy 53.2*e*. Hence $\pi(E)$, $\nu(E)$ and $\alpha(E)$ all satisfy the definition of measure functions on $\mathfrak{F}$.

Suppose that $\mu(E)$ is regular. For each set E of $\mathfrak{F}$ there are Borel sets A, B of $\mathfrak{F}$ such that $A \subset E \subset B$ and $\alpha(B - A) = 0$. Since $\pi(E) + \nu(E) = \alpha(E)$ and $\pi(E)$ and $\nu(E)$ are non-negative, this implies that $\alpha(B - A) = \pi(B - A) = \nu(B - A) = 0$. But the total variations of the non-negative measure functions π, ν and α over $(B - A)$ are respectively equal to $\pi(B - A)$, $\nu(B - A)$ and $\alpha(B - A)$ (cf. remark after 53.2). Hence the sets A and B serve to show that π, ν and α are all regular.

Retaining the hypotheses that $\mu(E)$ is regular, the condition for $\mathfrak{F}$ to be the natural range of $\mu(E)$ is that it shall consist of the collection of sets E for which there exist Borel sets A, B such that $A \subset E \subset B$, $\alpha(A) < \infty$ and $\alpha(B - A) = 0$. The set-function $\alpha(E)$ is its own total variation, and is regular on $\mathfrak{F}$, so the above condition is exactly the condition for $\mathfrak{F}$ to be the natural range of $\alpha(E)$. Hence $\mathfrak{F}$ is the natural range of $\alpha(E)$ if and only if it is the natural range of $\mu(E)$.

We can now remove the restriction in 53.3(*c*) that $\mu(E)$ be non-negative. For since $\mu(E) = \pi(E) - \nu(E)$, and $\pi(E)$ and $\nu(E)$ are non-negative, by 53.3(*c*) the limit set $\lim E_n$ satisfies the equation

$$\begin{aligned}\lim_{n\to\infty} \mu(E_n) &= \lim_{n\to\infty} [\pi(E_n) - \nu(E_n)] = \lim_{n\to\infty} \pi(E_n) - \lim_{n\to\infty} \nu(E_n)\\ &= \pi(\lim_{n\to\infty} E_n) - \nu(\lim_{n\to\infty} E_n) = \mu(\lim_{n\to\infty} E_n).\end{aligned}$$

54. We now begin to relate these measure functions $\mu(E)$ to the functions m_gE of earlier sections. If we are given a $g(x)$ the relation is easy to establish.

54.1. *Let $g(x)$ be of BV on every interval, and let $\mathfrak{M}_g$ be the family of sets of finite g-measure. The set-function m_gE is a regular measure function on $\mathfrak{M}_g$. If $g(x)$ is positively monotonic, $\mathfrak{M}_g$ is the natural range of m_gE.*

(In 54.3 we shall find other conditions under which $\mathfrak{M}_g$ is the natural range of m_gE.)

The family $\mathfrak{M}_g$ satisfies conditions (*a, b, c*) of 53.2, as is shown by 52.8. Let $\tau(x)$ be defined as usual (§52). If E has finite g-measure, and $E_1, \cdots, E_n$ are disjoint subsets of E which also belong to $\mathfrak{M}_g$, then by 52.6 and 52.3

$$\sum_{j=1}^{n} |m_g E_j| = \sum_{j=1}^{n} \left| \int_{E_j} 1 dg(x) \right| \leqq \sum_{j=1}^{n} \int_{E_j} 1 d\tau(x) = \sum_{j=1}^{n} m_\tau(E_j)$$
$$= m_\tau(\bigcup E_j) \leqq m_\tau(E).$$

Hence if $\alpha_g(E)$ is the total variation of $m_g E$, we have $\alpha_g(E) \leqq m_\tau E$. This implies that $m_g E$ satisfies 53.2(*e*). Evidently 53.2(*d*) holds, and 53.2(*f*) follows from 52.8. Therefore $m_g E$ is a measure function on $\mathfrak{M}_g$, in the sense of §53.

If E has finite g-measure, it also has finite τ-measure. By 49.4, there are Borel sets A, B such that $A \subset E \subset B$ and $m_\tau(B - A) = 0$. A fortiori, $\alpha_g(B - A) = 0$, and $m_g E$ is regular.

If $g(x)$ is positively monotonic the function $m_g E$ is non-negative, and by the remark after 53.2 the equation $\alpha_g(E) = m_g E$ holds for every set E of the family $\mathfrak{M}_g$. In particular it holds for bounded Borel sets, since by the remark after 49.3 all such sets are in $\mathfrak{M}_g$. Let E be a set for which there are Borel sets, A, B such that $A \subset E \subset B$, $\alpha_g(B - A) = 0$ and $\alpha_g(A) < \infty$. For each interval W_n the set AW_n, BW_n and $(B - A)W_n$ are bounded Borel sets, so

$$m_g AW_n = \alpha_g(AW_n) \leqq \alpha_g(A)$$

and

$$m_g(B - A)W_n = \alpha_g[(B - A)W_n] \leqq \alpha_g(B - A) = 0.$$

Since $(B - A)W_n$ contains $(E - A)W_n$, by 19.13 this implies that $m_g(E - A)W_n = 0$. Since AW_n is a bounded Borel set it has finite g-measure, so that the sum $EW_n = AW_n \cup (E - A)W_n$ is g-measurable and

$$m_g EW_n = m_g AW_n = \alpha_g(AW_n) \leqq \alpha_g(A).$$

By 19.8, this implies that E has finite g-measure, so $\mathfrak{M}_g$ is the natural range of the g-measure function.

In the nature of a converse is the following more difficult theorem.

54.2. *If $\mu(E)$ is a regular measure function on a range $\mathfrak{F}$, there is a function $g(x)$ such that*

(i) *$g(x)$ is of BV on every interval,*

(ii) *$g(x) = 0$ if any one of the numbers $(x^{(1)}, \cdots, x^{(n)})$ is zero,*

(iii) *$g(x)$ is left continuous,*

(iv) *every set E of the family $\mathfrak{F}$ is g-measurable, and $m_g E = \mu(E)$.*

The class $\mathfrak{M}_g$ of sets of finite g-measure is the natural range of $m_g E$, and if $\mathfrak{F}$ is the natural range of $\mu(E)$ it coincides with $\mathfrak{M}_g$. There is only *one function* $g(x)$ *which satisfies* (i), (ii) *and* (iii) *and is such that* $m_g I = \mu(I)$ *for all intervals* I *of the type* $[\alpha, \beta)$.

Let the set P be defined as in 53.11, and let $\pi(E)$ and $\nu(E)$ be the positive and negative variations of $\mu(E)$, as in 53.9. Corresponding to each of these we define a function of BV as we did in 45.5. The process of 45.5 can be summarized as follows. Let x be any point of the space R_q. If any $x^{(i)}$ is 0, we define

$$p(x) = n(x) = a(x) = 0.$$

Otherwise, we define $\alpha^{(i)} = \inf \{x^{(i)}, 0\}$, $\beta^{(i)} = \sup \{x^{(i)}, 0\}$, let I be the interval $[\alpha, \beta)$, define $\nu(x)$ to be the number of superscripts i such that $x^{(i)} < 0$, and define

$$p(x) = (-1)^{\nu(x)}\pi(I), \qquad n(x) = (-1)^{\nu(x)}\nu(I),$$
$$g(x) = (-1)^{\nu(x)}\mu(I), \qquad a(x) = (-1)^{\nu(x)}\alpha(I).$$

Then by 53.12,

$$g(x) = p(x) - n(x), \qquad a(x) = p(x) + n(x).$$

As in 45.5, we find that for every interval I of type $[\alpha, \beta)$

$$(A)\ \Delta_g I = \mu(I),\ \Delta_p I = \pi(I),\ \Delta_n I = \nu(I),\ \Delta_a I = \alpha(I).$$

We now prove that the functions $g(x)$, $p(x)$, $n(x)$ and $a(x)$ are left continuous. First suppose that x is a point none of whose coordinates is 0. Let $\{x_n^{(i)}\}$ be a sequence of numbers such that $x_n^{(i)} < x^{(i)}$ and $\lim_{n\to\infty} x_n^{(i)} = x^{(i)}$, $i = 1, \cdots, q$. We wish to prove

$$\lim_{n\to\infty} g(x_n) = g(x),$$

and likewise for the other three functions. It is permissible to restrict our attention to those n large enough so that $x_n^{(i)} > 0$ if $x^{(i)} > 0$. As before, we define $\alpha_n^{(i)} = \inf \{x_n^{(i)}, 0\}$, $\beta_n^{(i)} = \sup \{x_n^{(i)}, 0\}$, $I_n = [\alpha_n, \beta_n)$. The number of superscripts i for which $x_n^{(i)} < 0$ is then the same as the number $\nu(x)$ of superscripts for which $x^{(i)} \leqq 0$. The interval I is the limit of the I_n, so by the remark after 53.12 we have $\mu(I_n) \to \mu(I)$. That is,

$$\lim_{n\to\infty} g(x_n) = (-1)^{\nu(x)} \lim_{n\to\infty} \mu(I_n) = (-1)^{\nu(x)}\mu(I) = g(x).$$

A like proof holds for the other three functions.

If any of the coordinates $x^{(i)}$ is 0, the proof in the preceding paragraph still holds if I is replaced by the empty set Λ.

From the definitions we have, for every interval $I = [\alpha, \beta)$,

$$0 \leqq \Delta_p(I) = \pi(I) \leqq \alpha(I) = \Delta_a I,$$
$$0 \leqq \Delta_n I = \nu(I) \leqq \alpha(I) = \Delta_a I,$$
$$\Delta_g I = \Delta_p I - \Delta_n I.$$

Hence by 50.2 every set of finite a-measure also has finite p-measure and finite n-measure, and by 52.4 it has finite g-measure. That is,

(*B*) *the class* $\mathfrak{M}_g$ *of sets of finite g-measure contains the class* $\mathfrak{M}_a$ *of sets of finite a-measure.*

Since $g(x)$ is left continuous, by 52.10 the equation $\Delta_g I = m_g I$ holds for all intervals of the type $[\alpha, \beta)$. But we already know that $\Delta_g I = \mu(I)$ for all such intervals, so $\mu(I) = m_g I$ for all intervals $[\alpha, \beta)$. By 53.8, the set-functions $\mu(E)$ and $m_g E$ are equal for all bounded Borel sets. In exactly the same way, the equations $\alpha(E) = m_a E$, $\pi(E) = m_p E$ and $\nu(E) = m_n E$ hold for all bounded Borel sets.

The function $a(x)$ is positively monotonic, so by 54.1 the family $\mathfrak{M}_a$ of sets of finite a-measure is the natural range of the measure function $m_a E$. By 53.7, every set of the family $\mathfrak{F}$ belongs to $\mathfrak{M}_a$, and by (*B*) belongs to $\mathfrak{M}_g$. That is, if E is in $\mathfrak{F}$ it has finite g-measure, and by 53.7

$$m_g E = \mu(E).$$

The natural range of a measure function is uniquely determined by its values on bounded Borel sets, as was remarked after 53.6. On such sets the functions $m_a E$ and $\alpha(E)$ coincide. Since $\mathfrak{M}_a$ is the natural range of $m_a E$, it is also the natural range of $\alpha(E)$.

The functions $\mu(E)$ and $m_g E$ coincide for all bounded Borel sets E, and they are regular, so by 53.6 we can calculate their total variations by using only bounded Borel sets. That is, for every set E_0 the total variation of $m_g E$ over E_0 is the same as the total variation $\alpha(E_0)$ of $\mu(E)$ over E_0.

The family $\mathfrak{M}_a$ satisfies 53.2(*a*, *b*, *c*), and since we have seen that it is contained in $\mathfrak{M}_g$, the function $m_g E$ is defined on it. Clearly $m_g E$ continues to satisfy conditions 53.2(*d*, *e*, *f*) on $\mathfrak{M}_a$. Since $\alpha(E)$ is the total variation of $m_g E$, and $\mathfrak{M}_a$ is the natural range of $\alpha(E)$, it is also the natural range of $m_g E$, by 53.12. Applying 53.7 to $m_g E$ as on $\mathfrak{M}_g$ and on $\mathfrak{M}_a$, we see that $\mathfrak{M}_g$ is contained in $\mathfrak{M}_a$. But by (*B*), $\mathfrak{M}_a$ is contained in $\mathfrak{M}_g$, so the two coincide. Hence $\mathfrak{M}_g$ is the natural range of $m_g E$.

If $\mathfrak{F}$ is the natural range of $\mu(E)$, by 53.7 (applied to $\mu(E)$ and $m_g E$) it contains $\mathfrak{M}_g$. We have already seen that $\mathfrak{F}$ is contained in $\mathfrak{M}_g$, so the two coincide.

It remains to prove the uniqueness of the function $g(x)$, subject to (i), (ii) (iii) and the equation $\Delta_g I = \mu(I)$ for all intervals $I = [\alpha, \beta)$. Let γ be a function which satisfies (i), (ii), (iii) and is such that $\Delta_\gamma I = \mu(I)$ for all intervals I of type $[\alpha, \beta)$. Let x be any point of the space; we shall show that $\gamma(x) = g(x)$. If any one of the $x^{(i)}$ is zero, this follows from (ii). Otherwise, we define as before $\alpha^{(i)} = \inf [x^{(i)}, 0]$, $\beta^{(i)} = \sup [x^{(i)}, 0]$, $I = [\alpha, \beta)$, $\nu(x) =$ number of superscripts i such that $x^{(i)} < 0$. If we form the sum which defines $\Delta_\gamma I$, all terms but one vanish because of (ii), and we find

$$\Delta_\gamma I = (-1)^{\nu(x)}\gamma(x).$$

But by 52.10, $\Delta_\gamma I = m_\gamma I$. By hypothesis, $m_\gamma I = \mu(I)$. Hence

$$\gamma(x) = (-1)^{\nu(x)}\Delta_\gamma I = (-1)^{\nu(x)}\mu(I).$$

This is exactly the definition of $g(x)$. Hence $\gamma(x) = g(x)$, and the function $g(x)$ is uniquely determined by the values of the measure function $\mu(E)$ on intervals $I = [\alpha, \beta)$. The proof of the theorem is complete.

One immediate consequence of this theorem is that we have an affirmative answer to the question raised in §53: if $\mu(E)$ is regular on a family $\mathfrak{F}$, can it be extended uniquely to its natural range? For $\mu(E)$ determines $g(x)$, which determines the measure function $m_g E$ on the natural range $\mathfrak{M}_g$; and $m_g E = \mu(E)$ for all sets E of $\mathfrak{F}$, so $m_g E$ is an extension of $\mu(E)$ to its natural range. If there were another such extension, say to a function $\mu_1(E)$ on a natural range $\mathfrak{F}_1$, then $\mu_1(E)$, $\mu(E)$ and $m_g E$ all coincide on all sets of $\mathfrak{F}$, and in particular on all bounded Borel sets; so $\mathfrak{F}_1 = \mathfrak{M}_g$ and $m_g E = \mu_1(E)$, by 53.7.

Another consequence is

54.3. *If $g(x)$ is of BV on every interval and is left continuous, the family $\mathfrak{M}_g$ of sets of finite g-measure is the natural range of $m_g E$. Furthermore, there is a function $\gamma(x)$ satisfying* (i), (ii), *and* (iii) *of theorem* 54.2 *such that if either one of the integrals*

$$\int_{R_q} f(x)dg(x), \qquad \int_{R_q} f(x)d\gamma(x)$$

exists so does the other, and the two are equal.

For every interval $I = [\alpha, \beta)$ we have

$$\Delta_g I = m_g I,$$

by 52.10. Since $m_g E$ is a regular measure function on the range $\mathfrak{M}_g$, by 54.2 there is a function $\gamma(x)$ satisfying conditions (i), (ii) and (iii) of 54.2 and such that $m_\gamma E = m_g E$ for all sets E of $\mathfrak{M}_g$, and the class $\mathfrak{M}_\gamma$ of sets of finite γ-measure is the natural range of $m_\gamma E$.

Let I be any interval containing an interior point. If we denote the closure of I by $[\alpha, \beta]$, and write I_0 for the corresponding right-open interval $[\alpha, \beta)$, then

$$\Delta_g I = \Delta_g I_0, \qquad \Delta_\gamma I = \Delta_\gamma I_0.$$

For in defining Δ_g only the values of $g(x)$ at the vertices of I were used, and I and I_0 have the same vertices. A like statement holds for $\Delta_\gamma I$. But by 52.10, for the interval I_0 we have

$$\Delta_\gamma I_0 = m_\gamma I_0 = m_g I_0 = \Delta_g I_0;$$

and from this and the preceding equation we conclude that $\Delta_g I = \Delta_\gamma I$. If I contains no interior point, we must have $\alpha^{(i)} = \beta^{(i)}$ for some i, and therefore $\Delta_g I = 0 = \Delta_\gamma I$. Thus the equation $\Delta_g I = \Delta_\gamma I$ holds for all intervals I.

Now in defining integral and measure the function $g(x)$ never entered except in the expression $\Delta_g I$. The functions $g(x)$ and $\gamma(x)$, though possibly different, have the same difference-function; $\Delta_g I = \Delta_\gamma I$. Hence they lead to the same integral and measure. This establishes the last statement of our theorem. Also, the classes $\mathfrak{M}_g$ and $\mathfrak{M}_\gamma$ are identical, and $m_g E = m_\gamma E$ for all sets E of $\mathfrak{M}_g$. Since $\mathfrak{M}_\gamma$ is the natural range of $m_\gamma E$ by 54.2, it follows that $\mathfrak{M}_g$ is the natural range of $m_g E$. This completes the proof.

Another corollary, similar to the last, informs us that we would have lost no generality if we had restricted our attention to left continuous functions from the start; for every $g(x)$ which is of BV on every interval can be replaced by another such function which is left continuous and gives the same value to the integral as $g(x)$ did.

54.4. *If $g(x)$ is of BV on every interval, there is a function $\gamma(x)$ which satisfies* (i), (ii), (iii) *of* 54.2 *and is such that*

(*a*) *every g-measurable set E is γ-measurable, and if $m_g E$ is finite the equation*

$$m_\gamma E = m_g E$$

holds;

(*b*) *every g-summable function $f(x)$ is also γ-summable, and*

$$\int_{R_q} f(x)d\gamma(x) = \int_{R_q} f(x)dg(x).$$

Here we shall only prove conclusion (*a*); we could go on to prove (*b*) also, but this will follow from (*a*) after 55.6 is established.

The set function m_gE is a regular measure function on the family $\mathfrak{M}_g$ of sets of finite g-measure. By 54.2, there is a function $\gamma(x)$ satisfying (i), (ii), (iii) of 54.2 and such that every set E of $\mathfrak{M}_g$ is γ-measurable, and $m_\gamma E = m_g E$. This establishes (a) for sets of finite g-measure. If E is g-measurable, but has not finite g-measure, we regard it as the sum for all n of the sets EW_n. Each of these has finite g-measure, and therefore is γ-measurable. Hence $E = \bigcup_n EW_n$ is γ-measurable.

It is interesting to observe that, with the same notation as in 54.2, the following theorem holds.

54.5. *Let $\mu(E)$ be a regular measure function on a range $\mathfrak{F}$, and let the functions $g(x)$, $p(x)$, $n(x)$, $a(x)$ be defined as in the proof of* 54.2. *Then $p(x)$ and $n(x)$ form a minimum decomposition of $g(x)$, and $a(x)$ is the same as $\tau(x)$.*

Let I be an interval of the type $[\alpha, \beta)$. Since $g(x) = p(x) - n(x)$, we know by 46.3 and 46.5 that $T_g[I] \leqq T_p[I] + T_\tau[I] = \Delta_p I + \Delta_n I$. To establish the reverse inequality let ϵ be a positive number, and let P be the set exhibited in 53.11. As stated in that theorem, we may suppose P to be a Borel set. In the proof of 54.2 it was shown that the equations $m_p E = \pi(E)$ and $m_n E = \nu(E)$ hold for all bounded Borel sets. Thus for every bounded Borel set E we have

$$m_p E = \pi(E) = \pi(EP) = m_p EP,$$
$$m_n E = \nu(E) = \nu(E - P) = m_n(E - P).$$

In particular, $m_n IP = \nu(IP) = 0$, so by 20.1 there is an open set U containing IP such that

$$(A) \qquad m_n U < \epsilon.$$

The set UI can by 3.6 be represented as the sum of intervals $\bigcup_{j=1}^{\infty} I_j$, where I_j has the form $[\alpha_j, \beta_j)$. Hence

$$m_p UI = \sum_{j=1}^{\infty} m_p I_j.$$

Let n be chosen large enough so that

$$(B) \qquad m_p UI - \sum_{j=1}^{n} m_p I_j < \epsilon.$$

The remainder $I - [I_1 \cup \cdots \cup I_n]$ can be represented as the sum of disjoint intervals $I_{n+1}^*, \cdots, I_m^*$ each of the type $[\alpha, \beta)$. Then

$$(C) \qquad m_p\left(\bigcup_{j=1}^{n} I_j\right) + m_p\left(\bigcup_{j=n+1}^{m} I_j^*\right) = m_p I,$$

$$(D) \qquad m_n\left(\bigcup_{j=1}^{n} I_j\right) + m_n\left(\bigcup_{j=n+1}^{m} I_j^*\right) = m_n I.$$

Since $I_1 \cup \cdots \cup I_n$ is contained in U, the first term in (D) is less than ϵ:

$$(E) \qquad m_n\left(\bigcup_{j=1}^{n} I_j\right) < \epsilon.$$

Since UI contains IP, we have $m_pI = m_pIP \leqq m_pUI$, and by (B) it follows that

$$(F) \qquad \sum_{j=1}^{n} m_pI_j > m_pI - \epsilon.$$

These inequalities, with (C) and (D), yield

$$m_p\left(\bigcup_{j=n+1}^{m} I_j^*\right) < \epsilon, \qquad m_n\left(\bigcup_{j=n+1}^{m} I_j^*\right) > m_nI - \epsilon.$$

Now

$$\begin{aligned} T_g[I] &\geqq \sum_{j=1}^{n} |\Delta_g I_j| + \sum_{j=n+1}^{m} |\Delta_g I_j^*| = \sum_{j=1}^{n} |m_g I_j| + \sum_{j=n+1}^{m} |m_g I_j^*| \\ &\geqq \left|\sum_{j=1}^{n} m_g I_j\right| + \left|\sum_{j=n+1}^{m} m_g I_j^*\right| \\ &= \left| m_p\left(\bigcup_{j=1}^{n} I_j\right) - m_n\left(\bigcup_{j=1}^{n} I_j\right)\right| \\ &\qquad + \left| m_n\left(\bigcup_{j=n+1}^{m} I_j^*\right) - m_p\left(\bigcup_{j=n+1}^{m} I_j^*\right)\right| \\ &> (m_pI - \epsilon - \epsilon) + (m_nI - \epsilon - \epsilon) \\ &= m_pI + m_nI - 4\epsilon = \Delta_pI + \Delta_nI - 4\epsilon. \end{aligned}$$

Since ϵ is arbitrarily small, this implies that

$$T_g[I] \geqq \Delta_pI + \Delta_nI.$$

Therefore $p(x)$, $n(x)$ is a minimum decomposition of $g(x)$.

It follows that for every interval I, the total variation $\Delta_\tau I$ is the sum of Δ_pI and Δ_nI. But $a(x) = p(x) + n(x)$, so $\Delta_aI = \Delta_pI + \Delta_nI$. Hence $\Delta_\tau I = \Delta_aI$ for every interval I. The functions $a(x)$, $\tau(x)$ were both so defined that they vanish whenever any one of the $x^{(i)}$ is zero. Hence, just as in proving the uniqueness of $g(x)$ in the last paragraph of proof of 54.2, we find that $\tau(x) = a(x)$. This completes the proof of the theorem.

55. If we are given a measure function $\mu(E)$ on a range $\mathfrak{F}$ we can at once utilize a process due to Lebesgue to define an integral. If $f(x)$ is defined on a set E of $\mathfrak{F}$, we say that $f(x)$ is $\mathfrak{F}$-measurable if the set $E[f \geqq a]$ belongs to $\mathfrak{F}$ for every number a. Suppose now that $f(x)$ is defined, bounded and $\mathfrak{F}$-measurable on a set E_0 of $\mathfrak{F}$, and that $\mu(E)$ is non-negative. Let a, b be numbers such that $a < f(x) < b$, and let $l_0, \cdots, l_k$ be numbers such that $a = l_0 < l_1 < \cdots < l_k = b$. This collection of numbers l_i we call a "ladder L." We now form the sums

$$(A) \qquad s(L) = \sum_{j=1}^{k} l_{j-1}\mu(E_j), \qquad S(L) = \sum_{j=1}^{k} l_j\mu(E_j),$$

where E_j is the set $E_0[l_{j-1} < f \leqq l_j]$. Clearly $s(L) \leqq S(L)$. Let L' be a ladder $\{l'_0, \cdots, l'_h\}$ which is "finer" than L; that is, each l_j is one of the numbers l'_i. If $l_{j-1} = l'_i < l'_{i+1} < \cdots < l'_{i+m} = l_j$, then the term $l_j\mu(E_j)$ in $S(L)$ is replaced by the sum $l'_{i+1}\mu(E'_{i+1}) + \cdots + l'_{i+m}\mu(E'_{i+m})$, where $E'_j = E_0[l'_{i-1} < f \leqq l'_i]$. Then $E'_{i+1} \cup \cdots \cup E'_{i+m} = E_j$, so

$$\begin{aligned} l'_{i+1}\mu(E'_{i+1}) + \cdots + l'_{i+m}\mu(E'_{i+m}) &\leqq l_j\mu(E'_{i+1}) + \cdots + l_j\mu(E'_{i+m}) \\ &= l_j\mu(E'_{i+1} \cup \cdots \cup E'_{i+m}) = l_j\mu(E_j). \end{aligned}$$

Therefore, adding for $j = 1, \cdots, k$, we find that $S(L') \leqq S(L)$. Similarly $s(L') \geqq s(L)$.

It follows that $s(L_1) \leqq S(L_2)$ for all ladders L_1, L_2. For if L' is the ladder consisting of all the numbers both of L_1 and of L_2, then $s(L_1) \leqq s(L') \leqq S(L') \leqq S(L_2)$. Therefore if we write $m = \sup s(L)$ and $M = \inf S(L)$ for all ladders L, we have $m \leqq M$. On the other hand, if all the steps $l_j - l_{j-1}$ of the ladder L have length less than ϵ, then

$$S(L) - s(L) = \Sigma(l_j - l_{j-1})\mu(E_j) \leqq \epsilon\Sigma\mu(E_j) = \epsilon\mu(E_0);$$

whence

$$s(L) \leqq m \leqq M \leqq S(L) \leqq s(L) + \epsilon\mu(E_0).$$

The positive number ϵ is arbitrarily small, so $m = M$. Incidentally, this proves that by making the length of the largest step of L arbitrarily close to zero we oblige both $s(L)$ and $S(L)$ to approach the common value $m = M$ as limit. We now define

$$\int_{E_0} f(x)d\mu(E) = M.$$

Next, let $f(x)$ be non-negative and $\mathfrak{F}$-measurable on E_0. Define $f_n(x)$ to be inf $\{f(x), n\}$. This function is also $\mathfrak{F}$-measurable and is bounded, so the integral

$$\int_{E_0} f_n(x)d\mu(E)$$

exists. It is rather easy to show that this integral increases monotonically with n. If it has a finite limit as $n \to \infty$, we say that $f(x)$ is μ-integrable on E_0, and we define

$$\int_{E_0} f(x)d\mu(E) \equiv \lim_{n\to\infty} \int_{E_0} f_n(x)d\mu(E).$$

Let W_n be the interval $-n \leqq x^{(i)} \leqq n$, $i = 1, \cdots, q$. If $f(x)$ is non-negative and is μ-integrable on each W_n, it is easy to show that

$$\int_{W_n} f(x)d\mu(E)$$

increases monotonically with n. If its limit is finite, we say that $f(x)$ is μ-integrable over the space R_q, and we define

$$\int_{R_q} f(x)d\mu(E) \equiv \lim_{n\to\infty} \int_{W_n} f(x)d\mu(E).$$

If $f(x)$ is defined over R_q, we define as usual $f^+(x) = \sup\{f(x), 0\}$, $f^-(x) = \sup\{-f(x), 0\}$. Then $f(x)$ is said to be μ-integrable over R_q if f^+ and f^- are both μ-integrable over R_q, and in that case we define

$$\int_{R_q} f(x)d\mu(E) \equiv \int_{R_q} f^+(x)d\mu(E) - \int_{R_q} f^-(x)d\mu(E).$$

Finally we remove the restriction that $\mu(E)$ be non-negative. With the notation of the preceding section, we have $\mu(E) = \pi(E) - \nu(E)$. Then we say that a function $f(x)$ defined on R_q is μ-integrable over R_q if it is both π-integrable and ν-integrable, and in that case we define

$$\int_{R_q} f(x)d\mu(E) = \int_{R_q} f(x)d\pi(E) - \int_{R_q} f(x)d\nu(E).$$

From these definitions it is easily seen that if $f(x)$ is defined on R_q and is $\mathfrak{F}$-measurable on every interval, a necessary and sufficient condition for $f(x)$ to be μ-integrable over R_q is that $|f(x)|$ be α-integrable over R_q.

We now prove the analogues of 50.1 and 50.2.

55.1. *Let $\mu_i(E)$ $(i = 0, 1, 2)$ be non-negative measure functions, all defined on the same family $\mathfrak{F}$, and such that $\mu_0(E) = \mu_1(E) + \mu_2(E)$ for all sets E of $\mathfrak{F}$. Let $f(x)$ be defined on R_q. Then $f(x)$ is μ_0-integrable if* *and only if it is both μ_1-integrable and μ_2-integrable, and in that case*

$$\int_{R_q} f(x)d\mu_0(x) = \sum_{i=1,2} \int_{R_q} f(x)d\mu_i(x).$$

Suppose $f(x)$ bounded and $\mathfrak{F}$-measurable on a set E_0 of $\mathfrak{F}$. If we define $s_i(L)$ and $S_i(L)$ $(i = 0, 1, 2)$ as in equation (A) at the beginning of the section, we easily see that

$$s_0(L) = s_1(L) + s_2(L), \qquad S_0(L) = S_1(L) + S_2(L)$$

for every ladder L. Hence we readily deduce

$$\int_{E_0} f(x)d\mu_0(x) = \sum_{i=1,2} \int_{E_0} f(x)d\mu_i(x).$$

The extension to functions $f(x)$ which are defined and μ_0-integrable (or both μ_1-integrable and μ_2-integrable) over R_q needs no detailed presentation; it follows by straightforward use of the definition.

Two corollaries of 55.1 can now be stated.

55.2. *If $\mu_0(E)$ and $\mu_1(E)$ are measure functions on $\mathfrak{F}$, and $0 \leqq \mu_1(E) \leqq \mu_0(E)$ for all E in $\mathfrak{F}$, then every function $f(x)$ which is μ_0-integrable over R_q is also μ_1-integrable.*

For all E in $\mathfrak{F}$, define $\mu_2(E) = \mu_0(E) - \mu_1(E)$. The requirements in 53.2 are easily verified, and so $\mu_2(E)$ is a non-negative measure function on $\mathfrak{F}$. So by 55.1, if $f(x)$ is μ_0-integrable it is both μ_1-integrable and μ_2-integrable.

55.3. *If $\mu_i(E)$ $(i = 0, 1, 2)$ are measure-functions on $\mathfrak{F}$ such that $\mu_0(E) = \mu_1(E) + \mu_2(E)$ for all sets E of the family $\mathfrak{F}$, then every function $f(x)$ which is both μ_1-integrable and μ_2-integrable over R_q is also μ_0-integrable, and*

$$\int_{R_q} f(x)d\mu_0(x) = \sum_{i=1,2} \int_{R_q} f(x)d\mu_i(x).$$

Let $\pi_i(E)$, $\nu_i(E)$ be the positive and negative variations of $\mu_i(E)$ $(i = 0, 1, 2)$ as defined in 53.9. If E is an arbitrary set and E_0 is a subset of E which belongs to $\mathfrak{F}$, then by the definition 53.9 we find

$$\begin{aligned} \mu_0(E_0) &= \pi_1(E_0) + \pi_2(E_0) - \nu_1(E_0) - \nu_2(E_0) \\ &\leqq \pi_1(E) + \pi_2(E). \end{aligned}$$

This holds for all subsets E_0 of E which belong to $\mathfrak{F}$, so

$$(A) \qquad \pi_0(E) = \sup \mu_0(E_0) \leqq \pi_1(E) + \pi_2(E).$$

In the same way we prove

$$(B) \qquad \nu_0(E) \leqq \nu_1(E) + \nu_2(E).$$

By 53.12, the set-functions $\pi_i(E)$ and $\nu_i(E)$ $(i = 0, 1, 2)$ are measure functions on $\mathfrak{F}$. If $f(x)$ is defined and both μ_1-integrable and μ_2-integrable over R_q, by definition it is integrable with respect to π_1, π_2, ν_1, and ν_2. By 55.2 with inequalities (A) and (B) it is π_0-integrable and ν_0-integrable; so by definition it is μ_0-integrable. Furthermore, for every set E of $\mathfrak{F}$ we have

$$\begin{aligned}\pi_0(E) - \nu_0(E) = \mu_0(E) &= \mu_1(E) + \mu_2(E)\\ &= \pi_1(E) + \pi_2(E) - \nu_1(E) - \nu_2(E).\end{aligned}$$

Transposing, we obtain

$$\pi_0(E) + \nu_1(E) + \nu_2(E) = \nu_0(E) + \pi_1(E) + \pi_2(E)$$

for all sets E of $\mathfrak{F}$. If $f(x)$ is both μ_1-integrable and μ_2-integrable, we already know that it is π_i-integrable and ν_i-integrable $(i = 0, 1, 2)$. By 55.1,

$$\begin{aligned}\int_{R_q} f(x)d\pi_0(E) + \int_{R_q} f(x)d\nu_1(E) + \int_{R_q} f(x)d\nu_2(E)&\\ = \int_{R_q} f(x)d\{\pi_0(E) + \nu_1(E) + \nu_2(E)\}&\\ = \int_{R_q} f(x)d\{\nu_0(E) + \pi_1(E) + \nu_2(E)\}&\\ = \int_{R_q} f(x)d\nu_0(E) + \int_{R_q} f(x)d\pi_1(E) + \int_{R_q} f(x)d\pi_2(E).&\end{aligned}$$

Transposing yields

$$\begin{aligned}\int_{R_q} f(x)d\mu_0(E) &= \int_{R_q} f(x)d\pi_0(E) - \int_{R_q} f(x)d\nu_0(E)\\ &= \sum_{i=1,2} \left\{\int_{R_q} f(x)d\pi_i(E) - \int_{R_q} f(x)d\nu_i(E)\right\}\\ &= \sum_{i=1,2} \int_{R_q} f(x)d\mu_i(E).\end{aligned}$$

Next we show that if the family $\mathfrak{F}$ is enlarged in such a way that the measure function remains regular, all functions $f(x)$ which were originally integrable remain integrable, and the values of their integrals do not change.

55.4. *Let $\mu_1(E)$ and $\mu_2(E)$ be measure functions defined and either non-negative or regular on the respective ranges $\mathfrak{F}_1$, $\mathfrak{F}_2$. Suppose that $\mathfrak{F}_1$ is contained in $\mathfrak{F}_2$, and that for each set E in $\mathfrak{F}_1$ the functions $\mu_1(E)$ and*

$\mu_2(E)$ are equal. Then every μ_1-integrable function $f(x)$ is also μ_2-integrable, and

$$\int_{R_q} f(x)d\mu_1(E) = \int_{R_q} f(x)d\mu_2(E).$$

Consider first the case of non-negative measure functions μ_1 and μ_2. Evidently every $\mathfrak{F}_1$-measurable function is $\mathfrak{F}_2$-measurable, since $\mathfrak{F}_1$ is contained in $\mathfrak{F}_2$. If $f(x)$ is bounded and $\mathfrak{F}_1$-measurable on a set E_0 of $\mathfrak{F}_1$, for every ladder L the sums $s(L)$, $S(L)$ have the same value whether μ_1 or μ_2 is used as measure function. Therefore

$$(A) \qquad \int_{E_0} f(x)d\mu_1(E) = \int_{E_0} f(x)d\mu_2(E).$$

If $f(x)$ is non-negative and μ_1-integrable on E_0, and

$$f_n(x) = \inf \{n, f(x)\},$$

then the limit

$$\lim_{n\to\infty} \int_{E_0} f_n(x)d\mu_1(x) = \lim_{n\to\infty} \int_{E_0} f_n(x)d\mu_2(x)$$

exists. The limits of the members of this equation are the respective members of (A), so (A) holds for such $f(x)$. The extension from integrals over E_0 to integrals over R_q is similarly made, and the restriction to non-negative $f(x)$ is removed by applying the result already proved to the functions f^+ and f^-.

If μ_1 and μ_2 are regular, though not necessarily non-negative, on the respective families $\mathfrak{F}_1$, $\mathfrak{F}_2$, they have the same positive and negative variations $\pi(E)$, $\nu(E)$; for by 53.6 we can compute $\pi(E)$ and $\nu(E)$ on bounded Borel sets, on which the two measure functions are surely equal. By the part of the proof already completed, if $f(x)$ is π- and ν-integrable when π and ν are regarded as measure functions on $\mathfrak{F}_1$, they are still π- and ν-integrable when π and ν are regarded as measure functions on $\mathfrak{F}_2$, and the integrals retain the same values. By subtracting, the conclusion of the theorem is established.

The integral defined in this section is very closely related to the integral defined in §51, and there called the Lebesgue-Stieltjes integral. (Historically, the form of definition used in this section has the better right to the name.) The next two theorems show this relationship.

55.5. *Let $g(x)$ be of BV on every interval. Let m_gE be the corresponding measure-function, and $\mathfrak{M}_g$ the family of sets of finite g-measure. Let $f(x)$ be defined over the space R_q. Then if $f(x)$ is g-summable in the sense of* §51 *over R_q it is also m_g-integrable in the sense of* §55 *over R_q, and*

$$\int_{R_q} f(x)dm_gE = \int_{R_q} f(x)dg(x).$$

If $g(x)$ is positively monotonic or left continuous the converse is also true: if $f(x)$ is m_g-integrable over R_q it is also g-summable over R_q.

Evidently $f(x)$ is g-measurable if and only if it is $\mathfrak{M}_g$-measurable, for there is only one concept of measure present. We first suppose that $g(x)$ is positively monotonic, so that m_gE is non-negative. Let $f(x)$ be g-measurable and bounded on a set E_0 of finite measure. Given a ladder L: $(l_0, l_1, \cdots, l_k)$, we define $\varphi(x)$ on E_0 to be the function whose value is l_{j-1} on the set $E_j \equiv E_0[l_{j-1} < f \leqq l_j]$, and we define $\Phi(x)$ on E_0 as the function whose value is l_j on E_j. These are g-measurable functions, and $\varphi(x) < f(x) \leqq \Phi(x)$. Hence

$$\begin{aligned} s(L) &= \sum_{j=1}^{k} l_{j-1}m_gE_j = \sum_{j=1}^{k} \int_{E_j} l_{j-1}dg(x) \\ &= \int_{E_0} \varphi(x)dg(x) \leqq \int_{E_0} f(x)dg(x) \\ &\leqq \int_{E_0} \Phi(x)dg(x) = \sum_{j=1}^{k} \int_{E_j} l_jdg(x) \\ &= \sum_{j=1}^{k} l_jm_gE_j = S(L). \end{aligned}$$

But by definition of the integral with respect to m_gE we have

$$s(L) \leqq \int_{E_0} f(x)dm_gE \leqq S(L).$$

Since $s(L)$ and $S(L)$ differ arbitrarily little, it must be true that

$$(*) \qquad \int_{E_0} f(x)dg(x) = \int_{E_0} f(x)dm_gE.$$

If $f(x)$ is non-negative and g-measurable on E_0, define $f_n(x) = \inf\{n, f(x)\}$. Then by the preceding proof

$$\int_{E_0} f_n(x)dg(x) = \int_{E_0} f_n(x)dm_gE$$

for every n. Now $f(x)$ is g-summable if and only if the limit of the right member is finite, by 18.4; this is true if and only if $f(x)$ is m_g-integrable. If the limits are finite, the limit of the left and right members are respectively the left and right members of (*), in the one case by definition and in the other by 18.4. Hence (*) holds in this case.

The extension to R_q is made by use of the intervals W_n, as before; the proof is like that in the preceding paragraph. The restriction that $f(x)$ be non-negative is removed by applying the part of the proof already established to f^+ and f^-. This completes the proof of the theorem for positively monotonic functions $g(x)$.

Suppose next that $g(x)$ is of BV on every interval, but is not necessarily positively monotonic. Let $P(x)$, $N(x)$ be a minimum decomposition of $g(x)$. If $f(x)$ is g-summable over R_q, it is also P-summable and N-summable. By the part of the proof already completed, it is both m_P-integrable and m_N-integrable, and

$$\begin{aligned}\int_{R_q} f(x)dg(x) &= \int_{R_q} f(x)dP(x) - \int_{R_q} f(x)dN(x)\\ &= \int_{R_q} f(x)dm_PE - \int_{R_q} f(x)dm_NE.\end{aligned}$$

But m_gE is the sum of the measure functions m_PE and $-m_NE$, so by 55.3 the function $f(x)$ is m_g-integrable, and

$$\int_{R_q} f(x)dg(x) = \int_{R_q} f(x)dm_gE.$$

All that remains is to establish the converse for left-continuous functions $g(x)$. Let $\pi(E)$ and $\nu(E)$ be respectively the positive and negative variations of m_gE, as defined in 53.9. By 54.1 and 53.12, these are regular measure functions on the family $\mathfrak{M}_g$ of sets of finite g-measure. If $p(x)$ and $n(x)$ are defined as in 54.2, they are left continuous, so for every interval I of the type $[\alpha, \beta)$ we know that $\Delta_pI = m_pI$ and $\Delta_nI = m_nI$, by 47.5. But $p(x)$ and $n(x)$ were so defined that $\Delta_pI = \pi(I)$ and $\Delta_nI = \nu(I)$ for all such intervals I (cf. 54.2, equation (A)). Hence, by 53.8 and 53.7, for all sets E of the family $\mathfrak{M}_g$ we have

$$m_pE = \pi(E), \qquad m_nE = \nu(E).$$

Now let $f(x)$ be a function defined and m_g-integrable over R_q. By definition, it is both π-integrable and ν-integrable. By the preceding equation, it is both m_p-integrable and m_n-integrable. But $p(x)$ and $n(x)$ are positively monotonic, so by the part of the proof already completed we know that $f(x)$ is p-summable and n-summable. From 54.5, this implies that $f(x)$ is g-summable, and the theorem is complete.

The next theorem is the analogue of 55.5 in which we start with a measure function $\mu(E)$ and seek a corresponding $g(x)$.

55.6. *Let $\mu(E)$ be a regular measure function on the range $\mathfrak{F}$, and let $g(x)$ be the function described in* 54.2. *If $f(x)$ is μ-integrable over R_q, it is g-summable over R_q, and*

$$\int_{R_q} f(x)dg(x) = \int_{R_q} f(x)d\mu(E).$$

If $\mathfrak{F}$ is the natural range of $\mu(E)$ and $f(x)$ is defined over R_q, then $f(x)$ is g-summable over R_q if and only if it is μ-integrable over R_q.

Let $\mathfrak{F}_1$ be the natural range of $\mu(E)$, and let $\mu_1(E)$ be the extension of $\mu(E)$ to its natural range $\mathfrak{F}_1$. This extension is possible and is uniquely determined, by the remark after 54.2. If $\mathfrak{F}$ is already the natural range of $\mu(E)$, then $\mathfrak{F}_1$ is $\mathfrak{F}$ and $\mu_1(E)$ is identically equal to $\mu(E)$ on $\mathfrak{F}$.

In defining the function $g(x)$ of 54.2 it is immaterial whether we use $\mu_1(x)$ or $\mu(x)$; for the definition of $g(x)$ involves only the values of $\mu(E)$ on intervals $[\alpha, \beta)$, and on such intervals the function μ and μ_1 are identical. Hence by 54.2 we conclude that the family $\mathfrak{M}_g$ of sets of finite g-measure is the same as $\mathfrak{F}_1$, and for all sets E of $\mathfrak{F}_1$ the equation $m_gE = \mu_1(E)$ holds. Thus if $f(x)$ is defined on R_q, it is μ_1-integrable if and only if it is m_g-integrable. Since $g(x)$ is left continuous, $f(x)$ is m_g-integrable if and only if it is g-summable by 55.5. Also, if the integrals exist,

$$\int_{R_q} f(x)dg(x) = \int_{R_q} f(x)dm_gE = \int_{R_q} f(x)d\mu_1(x).$$

If $\mathfrak{F}$ is the natural range of $\mu(E)$, so that $\mathfrak{F}_1 = \mathfrak{F}$ and $\mu_1 = \mu$, this completes the proof. Otherwise, by 55.4, if $f(x)$ is μ-integrable it is μ_1-integrable, and

$$\int_{R_q} f(x)d\mu(E) = \int_{R_q} f(x)d\mu_1(E).$$

This, with the preceding paragraph, completes the proof.

The postponed proof of conclusion (b) of 54.4 is now easily made. With the notation and hypotheses of that theorem, if $f(x)$ is g-summable it is m_g-integrable, and

$$\int_{R_q} f(x)dg(x) = \int_{R_q} f(x)dm_gE,$$

by 55.5. From conclusion (a) of 54.4, together with 55.4, we find that $f(x)$ is m_γ-integrable, and

$$\int_{R_q} f(x)dm_gE = \int_{R_q} f(x)dm_\gamma E.$$

Since γ is left continuous, by 55.5 the function $f(x)$ is γ-summable over R_q and

$$\int_{R_q} f(x)dm_\gamma E = \int_{R_q} f(x)d\gamma(x).$$

This completes the proof of 54.4.

Summing up, we have shown that there is a one-to-one correspondence between measure functions on their natural ranges and functions satisfying (i), (ii), and (iii) of 54.2. This correspondence is such that, if $\mu(E)$ on range $\mathfrak{F}$ corresponds to $g(x)$, then every set of $\mathfrak{F}$ has finite g-measure, and conversely; every μ-integrable function $f(x)$ is g-summable, and conversely; and if $f(x)$ is g-summable, the two integrals

$$\int_{R_q} f(x)dg(x) \qquad \text{and} \qquad \int_{R_q} f(x)d\mu(E)$$

are equal.

From this correspondence it is evident without further proof that all the theorems of §52 remain valid after mere notational changes for integrals

$$\int_{R_q} f(x)d\mu(E),$$

provided that $\mu(E)$ is a regular measure function on its natural range $\mathfrak{F}$. As always, the apparent restriction that the range of integration be R_q is actually no restriction, since every integral can be written in this manner.

EXERCISE. Let $g(x)$ be of BV on every interval, and let $\mu(E)$ be a regular measure function on a family $\mathfrak{F}$ of sets. Then there exists a set E^* with τ-measure zero, and a function $f(x)$ which is g-summable over every set E of the family $\mathfrak{F}$, such that for every set E of $\mathfrak{F}$ the equation

$$\mu(E) = \mu(EE^*) + \int_E f(x)dg(x)$$

holds. If $\mu(E) = 0$ whenever $m_\tau E = 0$, then $\mu(E)$ is absolutely continuous (with respect to $\tau(x)$) and

$$\mu(E) = \int_E f(x)dg(x)$$

for every set E of $\mathfrak{F}$.

(On bounded Borel sets consider the measure function $\mu(E) - r_i m_g E$, where $r_1, r_2 \cdots$ is an ordering of the rationals. Construct the set P of 53.11, and call it P_i. We may assume $P_i \subset P_j$ if $r_i > r_j$, for we can replace P_i by $\bigcup P_j$ for all j such that $r_j \geqq r_i$. Define E^* to be $(\bigcap P_i) \cup (C \bigcup P_i)$. Define $f(x)$ to be the sup of all r_j for which x is in P_j.) (Nikodym, Fundamenta Mathematicae vol. 15 (1930), p. 168. Also H. Wileński, Fund. Math. 16, p. 399.)

56. The set P of theorem 53.11 can be used in such a way as to permit us to replace integrals with respect to measure functions $\mu(E)$ by integrals with respect to non-negative measure-functions, and to replace integrals with respect to $g(x)$ by integrals with respect to $\tau(x)$.

56.1. *Let $\mu(E)$ be a measure function on its natural range $\mathfrak{F}$, and let P be the set defined in* 53.11. *Let $f(x)$ be defined on the space R_q. If we define $\sigma(x) = K_P(x) - K_{CP}(x)$, then $f(x)$ is μ-integrable if and only if $f(x)\sigma(x)$ is α-integrable, and in that case*

$$(A) \qquad \int_{R_q} f(x)d\mu(E) = \int_{R_q} f(x)\sigma(x)d\alpha(E).$$

Suppose first that $f(x)$ is μ-integrable; by definition, it is then π-integrable and ν-integrable. The functions $f(x)$ and $f(x)\sigma(x)$ differ only on the complement of P, which has π-measure zero. Therefore

$$\int_{R_q} f(x)d\pi(E) = \int_{R_q} f(x)\sigma(x)d\pi(E).$$

Likewise $-f(x) = f(x)\sigma(x)$ except on P, which has ν-measure zero, so

$$\int_{R_q} [-f(x)]d\nu(E) = \int_{R_q} f(x)\sigma(x)d\nu(E).$$

Since $\pi(E) + \nu(E) = \alpha(E)$, by 55.1 we see that $f(x)\sigma(x)$ is α-integrable and equation (A) holds.

Conversely, let $f(x)\sigma(x)$ be α-integrable. For each interval W_n the intersections PW_n and $(CP)W_n$ are bounded Borel sets, hence in $\mathfrak{F}$. By 22.5 (which we may use because of 52.13 and the remark at the end of §55) $f(x)\sigma(x)$ is α-integrable over W_n. Since $\sigma(x)$ is 1 on PW_n and -1 on $(CP)W_n$, it is bounded and α-measurable, so the product $[f(x)\sigma(x)]\sigma(x) \equiv f(x)$ is α-integrable over W_n. Now $f(x)\sigma(x)$ is α-integrable over R_q, so by 52.3 its absolute value is also α-integrable. But this is the same as $|f(x)|$, since $\sigma(x) = \pm 1$. From the definition of the integral with respect to α we readily find that the preceding three sentences imply that $f(x)$ is α-integrable over R_q. It is therefore π-integrable and ν-integrable by 55.1, so it is μ-integrable by definition. This completes the proof.

We deduce at once a corollary on g-summable functions.

56.2. *If $g(x)$ is left continuous and of BV on every interval, and P is the set determined by m_gE as in* 53.11, *and $\sigma(x) = K_P(x) - K_{CP}(x)$, then in order that a function $f(x)$ defined on R_q be g-summable over R_q it is necessary and sufficient that $f(x)\sigma(x)$ be τ-summable over R_q; and in that case the equation*

$$\int_{R_q} f(x)dg(x) = \int_{R_q} f(x)\sigma(x)d\tau(x)$$

holds.

By 55.5 and 55.6, this is merely 56.1 in another notation.

This theorem allows us to improve 52.21 slightly.

56.3. *If $g(x)$ is left continuous, hypothesis (c) of* 52.21 *can be replaced by*

(c′) *for every positive ϵ there is a positive δ such that*

$$\left| \int_E f_n(x)dg(x) \right| < \epsilon$$

for every g-measurable subset E of E^* with $m_\tau E < \delta$.

If we replace the integrator $g(x)$ by $\tau(x)$ and replace the functions $f(x)$, $f_n(x)$ by their respective products $f(x)\sigma(x)$, $f_n(x)\sigma(x)$ with the function $\sigma(x)$ of 56.1 we find that all the hypotheses of 52.21 are satisfied. Therefore by 52.21 the function $f(x)\sigma(x)$ is τ-summable, and

$$\lim_{n\to\infty} \int_{E^*} f_n(x)\sigma(x)d\tau(x) = \int_{E^*} f(x)\sigma(x)d\tau(x),$$

$$\lim_{n\to\infty} \int_{E^*} | f_n(x)\sigma(x) - f(x)\sigma(x) | \, d\tau(x) = 0.$$

The second conclusion is (omitting the factor $| \sigma(x) | \equiv 1$) exactly the conclusion of 52.21, and by 56.2 the first conclusion can be written in the form of the first conclusion of 52.21.

The next theorem concerns the reduction of Lebesgue-Stieltjes integrals in which the measure function is itself defined by means of an integral. As we saw in the preceding section, it does not matter which of our two forms of definition of the integral we use. So for convenience we use a hybrid form, involving both types of definition.

56.4. *Let $g(x)$ be defined and either positively monotonic or left continuous on R_q, and of BV on every interval. Let $h(x)$ be defined on R_q and g-summable over every set of finite g-measure. Let $\mathfrak{H}$ be the family of sets E over which the integral*

$$H(E) = \int_E h(x)dg(x)$$

is defined. Then $H(E)$ is a measure-function on the range $\mathfrak{H}$; and if $f(x)$ is defined on R_q, it is H-integrable if and only if $f(x)h(x)$ is g-summable, and in that case

$$(*) \qquad \int_{R_q} f(x)dH(E) = \int_{R_q} f(x)h(x)dg(x).$$

The class $\mathfrak{H}$ may well be larger than the class of sets of finite g-measure, for $h(x)$ may be g-summable over sets which do not have finite g-measure. It may even include sets which are not g-measurable, since $\mathfrak{H}$ will contain all sets on which h vanishes identically, even though they are not g-measurable.

For each n the function $h(x)$ is g-summable over W_n, so the subset $W_nR_q[\,|h| = \infty]$ of the interval W_n on which h is not finite has g-measure zero, by 23.1. Adding these sets for all n, we find that $R_q[\,|h| = \infty]$ has g-measure zero. On this set we can re-define $h(x)$ to be zero without changing any of the integrals mentioned in the theorem. That is, we may assume without loss of generality that $h(x)$ is finite-valued.

We must first show that $H(E)$ is a measure function on the family $\mathfrak{H}$. Condition 53.2(a) clearly holds, since every interval $[\alpha, \beta)$ has finite g-measure. Suppose $h(x)$ non-negative. If E_1, E_2 are in $\mathfrak{H}$, by definition the functions $h(x)K_{E_i}(x)$, $i = 1, 2$, are g-summable over the space R_q. By 52.3, so is $\sup\{0, h(x)K_{E_1}(x) - h(x)K_{E_2}(x)\}$. This is identical with $h(x)K_{E_1-E_2}(x)$, so $h(x)$ is summable over $E_1 - E_2$, and 53.2(b) holds. The restriction $h \geqq 0$ is readily removed by considering h^+ and h^- separately. Let $E_1, E_2, \cdots$ be disjoint sets of $\mathfrak{H}$, all contained in a set E_0 of $\mathfrak{H}$. If the E_i are finite in number, the function

$$h(x)K_{\bigcup E_i}(x) \equiv \Sigma h(x)K_{E_i}(x)$$

is g-summable by 52.3, and $\bigcup E_i$ is in $\mathfrak{H}$. If they are denumerably infinite, then

$$h(x)K_{\bigcup E_i}(x) = \lim_{n\to\infty} \sum_1^n h(x)K_{E_i}(x),$$

and

$$\left| \sum_1^n h(x)K_{E_i}(x) \right| \leqq |h(x)|\, K_{E_0}(x),$$

which is summable. By 52.20, conditions (c') (see remark after 53.2) and (f) are satisfied. For (e), we observe that if E is in $\mathfrak{H}$ and $E_1, \cdots, E_n$ are disjoint sets of $\mathfrak{H}$ contained in E, then by 52.3

$$\sum_1^n |H(E_i)| = \sum_1^n \left| \int_{E_i} h(x)dg(x) \right| \leqq \sum_1^n \int_{E_i} |h(x)|\, d\tau(x)$$
$$\leqq \int_E |h(x)|\, d\tau(x).$$

Thus $H(E)$ is actually a measure-function.

Suppose first that $h(x)$ is non-negative and $g(x)$ positively monotonic. Let $f(x)$ be a function defined on a closed interval I and having a finite number of distinct values $c_1, \cdots, c_k$, the value c_j being assumed on a set E_j of the family $\mathfrak{H}$. Then $f(x)$ is H-integrable, and

$$\int_I f(x)dH(E) = \sum_{j=1}^{k} c_j H(E_j) = \sum_{j=1}^{k} \int_{E_j} c_j h(x)dg(x)$$
$$= \int_I f(x)h(x)dg(x).$$

Hence $f(x)h(x)$ is g-summable over I, and the equation

$$(A) \qquad \int_I f(x)dH(E) = \int_I f(x)h(x)dg(x)$$

holds.

Suppose next that $f(x)$ is non-negative, bounded and H-integrable over I; say $0 \leqq f(x) < M$. We define the ladder L_n by subdividing the interval $[-M, M]$ into 2^{n+1} equal parts by points $l_j^{(n)} = j \cdot M \cdot 2^{-n}$, $j = -2^n, -2^n + 1, \cdots, 2^n$. As before, we define $E_j^{(n)}$ to be the set $I[l_{j-1}^{(n)} < f \leqq l_j^{(n)}]$. If $f_n(x)$ is defined to have the value $l_{j-1}^{(n)}$ on $E_j^{(n)}$, we see that

$$s(L_n) = \sum_{j=-2^n+1}^{2^n} l_{j-1} H(E_j^{(n)}) = \int_I f_n(x)dH(E).$$

Since $f_n(x)$ can also be regarded as the greatest one of the numbers $l_j^{(n)}$ which is less than $f(x)$, and the numbers $l_j^{(n)}$ are included among the $l_j^{(n+1)}$, it is immediate that $f_1(x) \leqq f_2(x) \leqq \cdots$. Moreover, it is easy to see that as $n \to \infty$ the functions $f_n(x)$ tend (in fact uniformly) to $f(x)$. Hence by the definition of the integral,

$$\lim_{n\to\infty} \int_I f_n(x)dH(E) = \lim_{n\to\infty} s(L_n)$$
$$= \int_I f(x)dH(E).$$

By the preceding paragraph, each product $f_n(x)h(x)$ is g-summable, and

$$(B) \qquad \int_I f_n(x)dH(E) = \int_I f_n(x)h(x)dg(x).$$

This, with the preceding equation, shows that both members of (B) tend to the left member of (A) as $n \to \infty$. But $h(x)$ being non-negative, we have $f_1(x)h(x) \leqq f_2(x)h(x) \leqq \cdots \to f(x)h(x)$. So by 18.4 the product $f(x)h(x)$ is g-summable and is the limit of the right member of (B), so that equation (A) is satisfied.

Conversely, let us suppose that $f(x)$ is non-negative and bounded and that $f(x)h(x)$ is g-summable over a closed interval I. If we can show that $f(x)$ is $\mathfrak{H}$-measurable on I, then since it is bounded it will be H-integrable, and by the preceding paragraph equation (A) will hold. Consider then the set $I[f > a]$. If $f(x) > a$, then $f(x)h(x)$ exceeds $ah(x)$ unless $h(x)$ vanishes. That is,

$$I[f > a] = I[fh > ah] \cup (I[h = 0] \cap I[f > a]).$$

The last summand in this equation is a subset of $I[h = 0]$; over this set $h(x)$ is g-summable, and its integral is zero. That is, the last summand has measure $H(I[h = 0] \cap I[f > a]) = 0$. It remains to show that the set

$$I_a \equiv I[fh > ah]$$

is $\mathfrak{H}$-measurable.

By hypothesis, the functions $f(x)h(x)$ and $ah(x)$ are g-summable over I. Hence $f(x)h(x) - ah(x)$ is also g-summable over I. By 21.3, the set $I_a \equiv I[fh - ah > 0]$ is g-measurable. By 22.5, $h(x)$ is g-summable over I_a. This is exactly the definition of the statement that I_a is $\mathfrak{H}$-measurable.

We now remove the restriction that $f(x)$ is bounded. Suppose that $f(x)$ is non-negative and H-integrable on I. By definition, for every positive integer n the function $f_n(x) = \inf \{n; f(x)\}$ is H-integrable, and

$$\lim_{n\to\infty} \int_I f_n(x)dH(E) = \int_I f(x)dH(E).$$

By the preceding proof, $f_n(x)h(x)$ is g-summable over I, and

$$\int_I f_n(x)h(x)dg(x) = \int_I f_n(x)dH(E).$$

This, with the preceding equation, shows that

$$\lim_{n\to\infty} \int_I f_n(x)h(x)dg(x) = \int_I f(x)dH(E).$$

But $f_n(x)h(x)$ increases monotonically with n and approaches $f(x)h(x)$; so by 18.4 the product $f(x)h(x)$ is g-summable over I, and equation (A) holds.

Conversely, let $f(x)$ be non-negative and let $f(x)h(x)$ be g-summable over I. Defining $f_n(x)$ as above, we see that

$$f_n(x)h(x) = \inf \{f(x)h(x), nh(x)\}.$$

Each of the functions $f(x)h(x)$, $nh(x)$ is g-summable over I, by hypothesis. Hence $f_n(x)h(x)$ is also g-summable over I. Since $f_n(x)$ is bounded, by a preceding part of the proof we know that $f_n(x)$ is

H-integrable and equation (B) holds. By 18.4, as $n \to \infty$ the right member of (B) tends to a finite limit, namely the right member of (A). Hence by definition the left member of (A) is defined and equation (A) holds.

Next, suppose that $f(x)$ is defined and non-negative on R_q. By using integrals over the intervals W_n instead of the integrals of the $f_n(x)$, we can prove as above that $f(x)$ is H-integrable over R_q if and only if $f(x)h(x)$ is g-summable over R_q, and in that case the equation (*) of the theorem is satisfied.

Let $f(x)$ be any function defined on R_q. By definition, it is H-integrable over R_q if and only if $f^+(x)$ and $f^-(x)$ are both H-integrable over R_q. Since f^+ and f^- are non-negative, the preceding proof shows that they are both H-integrable over R_q if and only if both $f^+(x)h(x)$ and $f^-(x)h(x)$ are g-summable over R_q. But

$$\begin{aligned} f^+(x)h(x) &= \sup\{f(x), 0\} \cdot h(x) = \sup\{f(x)h(x), 0\} \\ &= [f(x)h(x)]^+, \end{aligned}$$

and likewise

$$f^-(x)h(x) = [f(x)h(x)]^-.$$

Hence $f(x)$ is H-integrable over R_q if and only if both $[f(x)h(x)]^+$ and $[f(x)h(x)]^-$ are g-summable over R_q. This is the necessary and sufficient condition that $f(x)h(x)$ be g-summable over R_q. Moreover, if the integrals exist, then the analogues of (*) hold for f^+ and for f^-. By subtraction, equation (*) is satisfied.

We now remove the restriction that $h(x)$ be non-negative. The integral

$$\alpha(E) = \int_E |h(x)|\, dg(x)$$

is defined for all sets E of the family $\mathfrak{H}$, by 18.3. This set-function is in fact the total variation of $H(E)$ for all sets E of $\mathfrak{H}$, as we now prove. Let E_0 be any set of the family $\mathfrak{H}$ and let k be a number less than the total variation of $H(E)$ on E_0. Then there are disjoint sets $E_1, \cdots, E_p$ of the family $\mathfrak{H}$, all contained in E_0, and such that

$$\sum_{j=1}^{p} |H(E_j)| > k.$$

But then by 18.3

$$\begin{aligned} k < \sum_{j=1}^{p} \left| \int_{E_j} h(x)dg(x) \right| &\leq \sum_{j=1}^{p} \int_{E_j} |h(x)|\, dg(x) = \int_{\bigcup E_j} |h(x)|\, dg(x) \\ &\leq \int_{E_0} |h(x)|\, dg(x) = \alpha(E_0). \end{aligned}$$

This is true for all k less than the total variation of $H(E)$ over E_0, so that the total variation cannot exceed $\alpha(E_0)$. On the other hand, let P be the set of all x such that $h(x) > 0$. Then E_0P is measurable, as follows from 21.4 and 21.3; and

$$\begin{aligned}\textit{total variation of } H(E) \textit{ over } E_0 &\geqq | H(E_0P) | + | H(E_0 - P) | \\ &= \int_{E_0P} h(x)dg(x) - \int_{E_0-P} h(x)dg(x) \\ &= \int_{E_0} | h(x) | dg(x).\end{aligned}$$

Our statement about $\alpha(E)$ is thus established.

The set P on which h is positive serves as the P of 53.11, since $H(E)$ is non-negative if E is in P and non-positive if E is in the complement of P. Let $\sigma_0(x)$ be the function $K_P(x) - K_{CP}(x)$, which is 1 on P and -1 on CP. As we saw in 56.1, a function $f(x)$ is H-integrable if and only if $f(x)\sigma_0(x)$ is α-integrable, and if the integrals exist they are equal:

$$\int_{R_q} f(x)dH(E) = \int_{R_q} f(x)\sigma_0(x)d\alpha(E).$$

By the preceding part of the proof for non-negative functions $h(x)$, the function $f(x)\sigma_0(x)$ is α-integrable if and only if $f(x)\sigma_0(x) | h(x) |$ is g-summable, and in that case

$$\int_{R_q} f(x)\sigma_0(x)d\alpha(E) = \int_{R_q} f(x)\sigma_0(x) | h(x) | dg(x).$$

But $\sigma_0(x) | h(x) | = h(x)$, and we have therefore shown that $f(x)$ is H-integrable if and only if $f(x)h(x)$ is g-summable, and in that case equation (*) holds. The proof of the theorem for positively monotonic functions $g(x)$ is complete.

Finally, let $g(x)$ be left continuous on the whole space R_q and of BV on every interval. With the help of the set P of 53.11 we define the function $\sigma(x) = K_P(x) - K_{CP}(x)$. By 56.2, the existence and value of the integrals $H(E)$ and $\int f(x)h(x)dg(x)$ are unaffected if we simultaneously replace $g(x)$ by $\tau(x)$ and $h(x)$ by $h(x)\sigma(x)$. But $\tau(x)$ is positively monotonic, so by the preceding proof $f(x)$ is H-integrable if and only if $f(x)h(x)\sigma(x)$ is τ-summable. This in turn is true if and only if $f(x)h(x)$ is g-summable; and if the integrals exist, then

$$\begin{aligned}\int_{R_q} f(x)dH(E) &= \int_{R_q} f(x)h(x)\sigma(x)d\tau(x) \\ &= \int_{R_q} f(x)h(x)dg(x).\end{aligned}$$

The proof of the theorem is complete.

Let us now restrict our attention to simple integrals and choose $g(x)$ to be x. Let $\gamma(x)$ be a function defined on R_1 and AC on every interval, so that

$$\gamma(b) - \gamma(a) = \int_a^b \dot{\gamma}(x)\, dx.$$

Theorem 56.4 then yields the following corollary.

56.5. *If $\gamma(x)$ is defined on R_1 and AC on every interval, and $f(x)$ is defined on R_1, then $f(x)$ is γ-summable over R_1 if and only if $f(x)\dot{\gamma}(x)$ is Lebesgue summable, and in that case*

$$\int_{R_1} f(x) d\gamma(x) = \int_{R_1} f(x)\dot{\gamma}(x)\, dx.$$

CHAPTER VIII

The Perron Integral

57. The definitions of integral so far considered have all rested fundamentally on a limiting process performed on finite sums. But one of the oldest—if not *the* oldest—of all definitions of integral is based on the inversion of the process of differentiation. To Newton, an (indefinite) integral $F(x)$ of a function $f(x)$ meant a function, necessarily continuous, whose derivative $F'(x)$ is everywhere equal to $f(x)$. For us, this definition is too restrictive. It is possible to construct quite simple functions which cannot be *everywhere* the derivative of any function. For instance, define $f(x) = 0$ for $x \neq 0$, $f(0) = 1$. If $F'(x) = f(x)$, then by the theorem of the mean $F(x)$ is constant for $x < 0$ and for $x > 0$. Hence by continuity $F(x)$ is constant, and $F'(0) = 0 \neq f(0)$. On the other hand, if we weaken our requirements and ask only that $F'(x) = f(x)$ almost everywhere, we have gone too far. For example, if $f(x)$ is defined and identically zero on $0 \leqq x \leqq 1$, we wish $F(x)$ to be constant on $[0, 1]$. Yet in §9 we constructed a function which is continuous and is constant on each of a set of intervals in $[0, 1]$ whose total length is 1. If we take $F(x)$ to be this function we find $F'(x) = 0 = f(x)$ almost everywhere, while* $F(0) = 0$ and $F(1) = 1$. So if we are to ask only that $F' = f$ almost everywhere, we must impose some other sort of condition on F or else lose all meaning.

As an example, let $f(x)$ be defined on an interval (a, b), and let us say that $F(x)$ is an integral of $f(x)$ if $F(x)$ is AC and $F' = f$ almost everywhere. From our preceding theorems, we find at once that this definition coincides with the definition of the Lebesgue integral $\int_a^x f(x)\, dx$.

However, another path is suggested to us by the proof of 33.1. There we made use of continuous functions $U(x)$, $L(x)$ satisfying the relations $U(a) = L(a) = 0$, $-\infty \neq \underline{D}U(x) \geqq f(x)$, $+\infty \neq \bar{D}L(x) \leqq$

* It is even possible to show that if $f(x)$ is any measurable function on an interval $[a, b]$ and $\varphi(x)$ is continuous on $[a, b]$, and ϵ is an arbitrary positive number, there exists a function $F(x)$ such that $F'(x) = f(x)$ for almost all x in $[a, b]$ and $|F(x) - \varphi(x)| < \epsilon$ for all x in $[a, b]$. (Lusin, Integral and Trigonometric Series, Moscow (1915). Unfortunately this book has not been translated from the Russian and is quite rare.)

$f(x)$. For any function $f(x)$ we could seek all such functions $U(x)$ and $L(x)$, and if the inf $F(x)$ of all $U(x)$ was at the same time the sup of all $L(x)$, we could take $F(x)$ to be the (indefinite) integral of $f(x)$. This is the definition of integral due to O. Perron, and it is essentially this definition with which we shall now be concerned. However, we find it convenient to phrase the definition somewhat differently. Instead of using a pair of functions we use a set of four. The advantage is that this permits us to prove a very general theorem on integration by parts (65.1); Perron's definition does not seem to lead directly to any proof of this theorem. The disadvantage of the present definition is that it requires us to consider four functions instead of two, which sometimes lengthens the writing slightly but does not make the reasoning any more difficult.

Let $f(x)$ be defined on an interval $[a, b]$. We say that

57.1. *A set of four functions $\varphi^1(x)$, $\varphi^2(x)$, $\varphi^3(x)$, $\varphi^4(x)$ is a tetrad adjoined to f on $[a, b]$ if*

(*a*) *the $\varphi^i(x)$ are continuous on $[a, b]$, and $\varphi^i(a) = 0$;*

(*b*) *ofr all except at most a denumerable set of values of x the relations*

$$-\infty \neq D_+\varphi^1 \geqq f, \qquad -\infty \neq D_-\varphi^2 \geqq f,$$
$$+\infty \neq D^-\varphi^3 \leqq f, \qquad +\infty \neq D^+\varphi^4 \leqq f$$

are valid on $[a, b]$.

In case the functions $\varphi^i(x)$ also satisfy the condition

$$|\varphi^i(b) - \varphi^j(b)| < \epsilon \qquad (i, j = 1, 2, 3, 4)$$

the tetrad will be said to be *ϵ-adjoined to f.* Functions satisfying the conditions imposed on φ^1, φ^2, φ^3, φ^4 respectively will be called right majors, left majors, left minors, and right minors. (The superscripts are those suggested by the usual numbering of the four quadrants.)

In this terminology we give our definition of integrability.

57.2. *A function $f(x)$ is Perron integrable (or, more briefly, P^*-integrable) on an interval $[a, b]$ if for every positive ϵ there is a tetrad ϵ-adjoined to $f(x)$ on $[a, b]$.*

This defines the concept of integrability, but before we define the integral itself, it is convenient to establish a few preliminary theorems.

For brevity, throughout the rest of this chapter we shall understand the word "integrable" to mean P^*-integrable unless some other meaning is specifically stated.

57.3. *If $\psi(x)$ is any major of f and $\varphi(x)$ is any minor of f, then $\psi(x) - \varphi(x)$ is a monotonic increasing function.*

Suppose that ψ is a right major and φ a right minor; then except for a denumerable set of values of x the difference $D_+\psi - D^+\varphi$ is defined, since $D_+\psi$ and $D^+\varphi$ are not both infinite of the same sign. Also, except for a denumerable set of values of x this difference is non-negative, for $D_+\psi \geqq f$ and $D^+\varphi \leqq f$. Hence by 31.4 the inequalities $D_+(\psi - \varphi) \geqq D_+\psi - D^+\varphi \geqq 0$ hold except perhaps on a denumerable set, and $\psi - \varphi$ is monotonic increasing by 34.1. A like proof holds if ψ is a left major and φ a left minor.

In case ψ and φ are of opposite sides, the proof is rather more difficult; so we interrupt ourselves to establish a lemma.

57.4. *If $g(x)$ is any finite-valued function defined on an interval $[a, b]$, then except at most for a denumerable set E of values of x the inequalities*

$$D^+g(x) \geqq D_-g(x) \qquad \textit{and} \qquad D^-g(x) \geqq D_+g(x)$$

hold.

Let E_1 be the set on which $D^+g < D_-g$. For every x in E_1, there is a rational number ρ between $D^+g(x)$ and $D_-g(x)$; and if n be chosen large enough, the inequalities

$$(A) \qquad \frac{g(x') - g(x)}{x' - x} > \rho > \frac{g(x) - g(x'')}{x - x''}$$

hold whenever

$$(B) \qquad x - \frac{1}{n} < x' < x < x'' < x + \frac{1}{n}.$$

Thus if we define $E(n, \rho)$ to be the set of all points x for which the inequalities (A) hold whenever (B) holds, we see that every point x of E_1 belongs to some $E(n, \rho)$ for properly chosen n and ρ. Since n and ρ have denumerable ranges, there is only a denumerable number of sets $E(n, \rho)$; and when we show that each $E(n, \rho)$ contains only a finite number of points, the denumerability or finiteness of the set $E_1 \subset \bigcup E(n, \rho)$ will be established.

If two points x_1, x_2 belong to $E(n, \rho)$, their distance is at least $1/n$. Otherwise we would have, say, $x_1 < x_2 < x_1 + \frac{1}{n}$. In (A) we could take $x' = x_1$, $x = x_2$, and obtain

$$\frac{g(x_1) - g(x_2)}{x_1 - x_2} > \rho.$$

We could also take $x = x_1$, $x'' = x_2$, and obtain

$$\rho > \frac{g(x_1) - g(x_2)}{x_1 - x_2}.$$

This is a contradiction, so the distance between x_1 and x_2 must be at least $1/n$. Hence in the finite interval $[a, b]$ there can be only a finite number of points of $E(n, \rho)$, and so there can be at most a denumerable set of x at which $D^+g < D_-g$. The other statement of the lemma is similarly established.

Returning now to 57.3, suppose that ψ is a right major and φ a left minor of f. Then except on a denumerable set E the inequalities

$$\begin{aligned} D_+\varphi(x) &\leqq D^-\varphi(x) \leqq f(x), \\ D_+\varphi(x) &\leqq D^-\varphi(x) < \infty, \\ -\infty &\neq D_+\psi(x) \geqq f(x) \end{aligned}$$

all hold. Hence except on E the difference $D_+\psi(x) - D_+\varphi(x)$ is defined and non-negative, so by 31.4 we have

$$D^+(\psi(x) - \varphi(x)) \geqq D_+\psi(x) - D_+\varphi(x) \geqq 0$$

except on E. Applying 34.1, $\psi - \varphi$ is monotonic increasing. A like proof holds for the remaining case, in which ψ is a left major and φ a right minor.

A simple consequence is

57.5. *If the tetrad $[\varphi^i(x)]$ is ϵ-adjoined to f on $[a, b]$, it is ϵ-adjoined to f on $[a, x]$ for all x such that $a < x \leqq b$.*

The greatest of the four functional values at x is one of the numbers $\varphi^1(x)$, $\varphi^2(x)$, say to be specific $\varphi^1(x)$; and the least is one of the numbers $\varphi^3(x)$, $\varphi^4(x)$, say $\varphi^3(x)$. Then by 57.3, $0 \leqq \varphi^1(x) - \varphi^3(x) \leqq \varphi^1(b) - \varphi^3(b) < \epsilon$, so all four numbers $\varphi^i(x)$ differ by less than ϵ.

57.6. (**Corollary**). *If $f(x)$ is integrable over $[a, b]$, it is integrable over $[a, x]$ for $a < x \leqq b$.*

This enables us to prove

57.7. *If $f(x)$ is integrable over $[a, b]$, there is a function $F(x)$ which is the* inf *of all right majors $\varphi^1(x)$, the* inf *of all left majors $\varphi^2(x)$, the* sup *of all left minors $\varphi^3(x)$ and the* sup *of all right minors $\varphi^4(x)$. In this case we define*

$$\int_a^b f(x)\,dx \equiv F(b).$$

Let $\varphi_0(x)$ be any minor of f. For every right major $\varphi^1(x)$ we have $\varphi^1(x) \geqq \varphi_0(x)$, $a \leqq x \leqq b$; hence, defining $F(x)$ to be the inf of all right majors, $F(x) \geqq \varphi_0(x)$. Since this inequality holds for all minors $\varphi_0(x)$ the function $F(x)$ is *an* upper bound of $\varphi_0(x)$ for all minors. We now show that it is the sup of all right minors and also the sup of all left minors. By definition, for every $\epsilon > 0$ we can choose a right major $\varphi^1(x)$, a right minor $\varphi^4(x)$ and a left minor $\varphi^3(x)$ such that $0 \leqq \varphi^1(b) - \varphi^3(b) < \epsilon$ and $0 \leqq \varphi^1(b) - \varphi^4(b) < \epsilon$. Then

$$0 \leqq F(x) - \varphi^4(x) \leqq \varphi^1(x) - \varphi^4(x) \leqq \varphi^1(b) - \varphi^4(b) < \epsilon,$$
$$0 \leqq F(x) - \varphi^3(x) \leqq \varphi^1(x) - \varphi^3(x) \leqq \varphi^1(b) - \varphi^3(b) < \epsilon,$$

for all x in $[a, b]$. Hence $F(x)$ is the sup of all right minors $\varphi^4(x)$ and of all left minors $\varphi^3(x)$. Reversing the rôles of minors and majors, $F(x)$ is also the inf of all left majors $\varphi^2(x)$, as was to be proved.

The definition of the integral in 57.7 is actually equivalent to Perron's, but the proof of the equivalence is far from simple. Instead of proving it, we content ourselves with the remark that our definition reduces to Perron's if we require that for every ϵ there shall be tetrads φ^i which are ϵ-adjoined to $f(x)$ and satisfy the identities $\varphi^1(x) \equiv \varphi^2(x)$ and $\varphi^3(x) \equiv \varphi^4(x)$. Thus it is not surprising that many of our proofs—in fact, all except the proofs of 65.1 and 66.1—transform into the corresponding proofs for the usual Perron integral if we regard φ^1 and φ^2 as identical and also regard φ^3 and φ^4 as identical.

The next theorem is a corollary of 57.7.

57.8. *If $f(x)$ is integrable on an interval $[a, b]$ and $F(x)$ is the function described in 57.7, then*

$$\int_a^\xi f(x)\,dx = F(\xi)$$

for all ξ such that $a < \xi \leqq b$.

Let ϵ be a positive number, and let $[\varphi^i(x)]$ be a tetrad ϵ-adjoined to $f(x)$ on $[a, b]$. By 57.5, it is also ϵ-adjoined to $f(x)$ on $[a, \xi]$, so by the definition 57.7 of the integral

$$\varphi^1(\xi) \geqq \int_a^\xi f(x)\,dx \geqq \varphi^4(\xi) > \varphi^1(\xi) - \epsilon.$$

But by 57.7

$$\varphi^1(\xi) \geqq F(\xi) \geqq \varphi^4(\xi),$$

so

$$\left| F(\xi) - \int_a^\xi f(x)\,dx \right| < \epsilon.$$

Since ϵ is an arbitrary positive number, this establishes the theorem.

58. Let us now investigate the elementary properties of the P^*-integral.

58.1. *If $f(x)$ is integrable over $[a, b]$, the function*

$$F(x) = \int_a^x f(x)\,dx$$

is continuous.

For every positive integer n it is possible to find a right major $\varphi_n^1(x)$ and a right minor $\varphi_n^4(x)$ such that $\varphi_n^1(b) - \varphi_n^4(b) < 1/n$. Then by 57.3 and 57.7

$$0 \leqq F(x) - \varphi_n^4(x) \leqq \varphi_n^1(x) - \varphi_n^4(x)$$
$$\leqq \varphi_n^1(b) - \varphi_n^4(b) < \frac{1}{n}.$$

Hence the continuous functions $\varphi_n^4(x)$ converge uniformly to $F(x)$ on $[a, b]$, and $F(x)$ is therefore continuous.

Remark. From the proof of 58.1 it follows that if $[\varphi^i(x)]$ is any tetrad adjoined to $f(x)$ on $[a, b]$, the differences $\varphi^1(x) - F(x)$, $\varphi^2(x) - F(x)$, $F(x) - \varphi^3(x)$, $F(x) - \varphi^4(x)$ are monotonic increasing. Consider for example the function $\varphi^1(x)$. The $\varphi_n^i(x)$ being defined as in 58.1, the difference $\varphi^1(x) - \varphi_n^4(x)$ is monotonic increasing, by 57.3. But $\varphi^1(x) - F(x) = \lim_{n\to\infty} [\varphi^1(x) - \varphi_n^4(x)]$, so $\varphi^1(x) - F(x)$ is also monotonic increasing.

58.2. *If $f(x)$ is integrable on $[a, b]$, it is integrable over every subinterval $[\alpha, \beta]$ of $[a, b]$.*

Let the tetrad $[\varphi^i(x)]$ be $\epsilon/2$-adjoined to $f(x)$ on $[a, b]$. The functions $\Phi^i(x) \equiv \varphi^i(x) - \varphi^i(\alpha)$ are then ϵ-adjoined to $f(x)$ on $[\alpha, \beta]$. For they are continuous on $[\alpha, \beta]$ and satisfy the equations $\Phi^i(\alpha) = 0$. The conditions on the derivates are satisfied, since the $\Phi^i(x)$ have the same derivates as the $\varphi^i(x)$. And they are ϵ-adjoined, for

$$| \Phi^i(\beta) - \Phi^j(\beta) | \leqq | \varphi^i(\beta) - \varphi^j(\beta) | + | \varphi^i(\alpha) - \varphi^j(\alpha) | < \epsilon.$$

58.3. *If $a < b < c$, and $f(x)$ is integrable over $[a, b]$ and over $[b, c]$, then $f(x)$ is integrable over $[a, c]$, and*

$$\int_a^c f(x)\,dx = \int_a^b f(x)\,dx + \int_b^c f(x)\,dx.$$

Let $[\varphi_1^i(x)]$, $[\varphi_2^i(x)]$ be tetrads $\epsilon/2$-adjoined to $f(x)$ on the respective intervals $[a, b]$, $[b, c]$. Define

$$\Phi^i(x) = \varphi_1^i(x), \qquad a \leqq x \leqq b,$$
$$\Phi^i(x) = \varphi_1^i(b) + \varphi_2^i(x), \qquad b < x \leqq c.$$

Then $\Phi^i(a) = \varphi_1^i(a) = 0$, and the $\Phi^i(x)$ are continuous on $[a, c]$. The conditions on the derivates are satisfied, since the derivates of

$\Phi^i(x)$ are the same as those of $\varphi_1^i(x)$ for $a \leqq x < b$, and the same as those of $\varphi_2^i(x)$ for $b < x \leqq c$. Also,

$$| \Phi^i(c) - \Phi^j(c) | \equiv | \varphi_1^i(b) + \varphi_2^i(c) - \varphi_1^j(b) - \varphi_2^j(c) |$$
$$\leqq | \varphi_1^i(b) - \varphi_1^j(b) | + | \varphi_2^i(c) - \varphi_2^j(c) | < \epsilon.$$

Since ϵ is arbitrary, this proves that $f(x)$ is integrable on $[a, c]$.

By the definition of the integral,

$$\varphi_1^4(b) \leqq \int_a^b f(x)\,dx \leqq \varphi_1^1(b),$$
$$\varphi_2^4(c) \leqq \int_b^c f(x)\,dx \leqq \varphi_2^1(c).$$

Adding and recalling the definition of Φ^i, we obtain

$$\Phi^4(c) \leqq \int_a^b f(x)\,dx + \int_b^c f(x)\,dx \leqq \Phi^1(c).$$

But by the definition of the integral

$$\Phi^4(c) \leqq \int_a^c f(x)\,dx \leqq \Phi^1(c).$$

Hence the two members of the equation in 58.3 both lie in the same interval of length less than ϵ. Since ϵ is arbitrary, the two members are equal, and 58.3 is established.

58.4. *If $f(x)$ is integrable over $[a, b]$, and k is a finite constant, then $kf(x)$ is integrable over $[a, b]$, and*

$$\int_a^b kf(x)\,dx = k \int_a^b f(x)\,dx.$$

For $k = 0$ this is trivial. Suppose that k is negative. Let $[\varphi^i(x)]$ be $\epsilon/|k|$-adjoined to $f(x)$ on $[a, b]$, and define

$$\psi^1(x) = k\varphi^4(x), \qquad \psi^4(x) = k\varphi^1(x),$$
$$\psi^3(x) = k\varphi^2(x), \qquad \psi^2(x) = k\varphi^3(x).$$

The functions $\psi^i(x)$ then form a tetrad ϵ-adjoined to $kf(x)$ on $[a, b]$. For these functions are continuous and vanish at $x = a$. From the definition 31.1 of the derivates, together with 6.3, we find

$$D_+\psi^1(x) = D_+k\varphi^4(x) = kD^+\varphi^4(x).$$

Since $+\infty \neq D^+\varphi^4(x) \leqq f(x)$ save on a denumerable set, and k is negative, we find

$$-\infty \neq kD^+\varphi^4(x) \geqq kf(x)$$

except on the same denumerable set; that is, $\psi^1(x)$ satisfies the derivate conditions required of a right major. The other three functions can be discussed similarly. Finally,

$$|\psi^i(b) - \psi^j(b)| = |k| \cdot |\varphi^i(b) - \varphi^j(b)| < \epsilon,$$

so $[\psi^i(x)]$ is a tetrad ϵ-adjoined to $kf(x)$ on $[a, b]$.

From the inequality

$$\varphi^4(b) \leqq \int_a^b f(x)\,dx \leqq \varphi^1(b)$$

we deduce

$$\psi^1(b) = k\varphi^4(b) \geqq k \int_a^b f(x)\,dx \geqq k\varphi^1(b) = \psi^4(b).$$

But since $\psi^1(x)$ and $\psi^4(x)$ are respectively a right major and a right minor for $kf(x)$, we must have

$$\psi^1(b) \geqq \int_a^b kf(x)\,dx \geqq \psi^4(b).$$

Since the interval $[\psi^4(b), \psi^1(b)]$ has length less than the arbitrary positive number ϵ, and both members of the equation in 58.4 are in that interval, the equation must hold. This completes the proof if $k < 0$.

If k is positive, we apply the part of the theorem already proved first to $-kf(x)$ and then to $(-1)[(-k)f(x)]$. This completes the proof.

58.5. *If $f(x)$ is integrable over $[a, b]$, then $f^*(x) \equiv f(-x)$ is integrable over $[-b, -a]$, and*

$$(*) \qquad \int_{-b}^{-a} f^*(x)\,dx = \int_a^b f(x)\,dx.$$

Let $[\varphi^i(x)]$ be a tetrad ϵ-adjoined to $f(x)$ on $[a, b]$. For $-b \leqq x \leqq -a$, define

$$\begin{aligned}
\psi^1(x) &= \varphi^2(b) - \varphi^2(-x),\\
\psi^2(x) &= \varphi^1(b) - \varphi^1(-x),\\
\psi^3(x) &= \varphi^4(b) - \varphi^4(-x),\\
\psi^4(x) &= \varphi^3(b) - \varphi^3(-x).
\end{aligned}$$

These are continuous, and they all vanish at $x = -b$. By 31.7, for each x in $[-b, -a]$ the equations

$$D_+\psi^1(x) = D_-\varphi^2(-x), \qquad D_-\psi^2(x) = D_+\varphi^1(-x),$$
$$D^-\psi^3(x) = D^+\varphi^4(-x), \qquad D^+\psi^4(x) = D^-\varphi^3(-x)$$

all hold. It follows at once that the conditions in 57.1 on the derivates are satisfied; for instance, $D_+\psi^1(x) = D_-\varphi^2(-x) \neq -\infty$ and $D_+\psi^1(x) = D_-\varphi^2(-x) \geqq f(-x) = f^*(x)$ save on a denumerable set. So $[\psi^i(x)]$ is a tetrad adjoined to $f^*(x)$ on $[-b, -a]$. Also, each difference $|\psi^i(-a) - \psi^j(-a)|$ is one of the numbers $|\varphi^k(b) - \varphi^l(b)|$, and is therefore less than ϵ. Hence the tetrad is ϵ-adjoined to $f^*(x)$. Since ϵ is arbitrary, this proves that $f^*(x)$ is integrable over $[-b, -a]$.

To establish equation (*), we observe that the left member of (*) is in the interval $[\psi^4(-a), \psi^1(-a)]$, while the right member of (*) belongs to the interval $[\varphi^3(b), \varphi^2(b)]$, by the definition of integral. These intervals have length less than ϵ, and they coincide, since $\psi^4(-a) = \varphi^3(b)$ and $\psi^1(-a) = \varphi^2(b)$. Hence the two members of (*) differ by less than ϵ. But ϵ is an arbitrary positive number, so the two members of (*) are equal.

58.6. *If $f_1(x)$ and $f_2(x)$ are integrable over $[a, b]$, and the sum $f_1(x) + f_2(x)$ is defined on $[a, b]$, then $f_1(x) + f_2(x)$ is integrable on $[a, b]$, and*

$$\int_a^b [f_1(x) + f_2(x)]\,dx = \int_a^b f_1(x)\,dx + \int_a^b f_2(x)\,dx.$$

Let ϵ be a positive number, and let the tetrads $[\varphi_1^i(x)]$, $[\varphi_2^i(x)]$ be $\epsilon/2$-adjoined to $f_1(x)$ and $f_2(x)$ respectively. Define $\Phi^i(x) = \varphi_1^i(x) + \varphi_2^i(x)$, $i = 1, \cdots, 4$. It is clear that each $\Phi^i(x)$ is continuous and $\Phi^i(a) = 0$. Except on a denumerable set E the relations $-\infty \neq D_+\varphi_1^1(x) \geqq f_1(x)$, $-\infty \neq D_+\varphi_2^1(x) \geqq f_2(x)$ both hold. Hence by 31.4, except on E we have

$$D_+\Phi^1(x) \geqq D_+\varphi_1^1(x) + D_+\varphi_2^1(x) \geqq f_1(x) + f_2(x),$$
$$D_+\Phi^1(x) \geqq D_+\varphi_1^1(x) + D_+\varphi_2^1(x) > -\infty.$$

The other three functions $\Phi^i(x)$ can be shown similarly to satisfy the requirements in 57.1. Clearly

$$\begin{aligned}|\Phi^i(b) - \Phi^j(b)| &= |\varphi_1^i(b) + \varphi_2^i(b) - \varphi_1^j(b) - \varphi_2^j(b)| \\ &\leqq |\varphi_1^i(b) - \varphi_1^j(b)| + |\varphi_2^i(b) - \varphi_2^j(b)| < \epsilon.\end{aligned}$$

So $f_1(x) + f_2(x)$ is integrable.

To establish the equation, we have

$$\varphi_j^4(b) \leqq \int_a^b f_j(x)\,dx \leqq \varphi_j^1(b), \qquad j = 1, 2.$$

Adding, the right member of the equation in 58.6 lies in the interval $[\varphi_1^4(b) + \varphi_2^4(b), \varphi_1^1(b) + \varphi_2^1(b)]$; that is, in the interval $[\Phi^4(b), \Phi^1(b)]$. But the left member of the equation is also in this interval, which has length less than ϵ. So the two members of the equation differ by less than the arbitrary positive number ϵ, and must be equal.

58.7. Corollary. *If $f_1(x), \cdots, f_n(x)$ are integrable over $[a, b]$, and $\lambda_1, \cdots, \lambda_n$ are finite constants, then $\Sigma\lambda_i f_i(x)$ is integrable over $[a, b]$, and*

$$\int_a^b \sum_{i=1}^n \lambda_i f_i(x)\,dx = \sum_{i=1}^n \lambda_i \int_a^b f_i(x)\,dx.$$

This follows from 58.4 and 58.6.

58.8. *If $f_1(x)$ and $f_2(x)$ are integrable over $[a, b]$, and $f_1(x) \geqq f_2(x)$, then*

$$\int_a^b f_1(x)\,dx \geqq \int_a^b f_2(x)\,dx.$$

Let h be any number less than the right member of the inequality. There is a right minor function $\varphi^4(x)$ adjoined to $f_2(x)$ such that $\varphi^4(b) > h$. But this same function $\varphi^4(x)$ also serves as a right minor for $f_1(x)$; hence

$$\int_a^b f_1(x)\,dx \geqq \varphi^4(b) > h.$$

This holds for every h less than the integral of $f_2(x)$, and the inequality of the theorem follows at once.

This yields a simple corollary.

58.9. *If $f(x)$ and $|f(x)|$ are both integrable* on $[a, b]$, then*

$$\left|\int_a^b f(x)\,dx\right| \leqq \int_a^b |f(x)|\,dx.$$

This follows from 58.8 and the inequality

$$-|f(x)| \leqq f(x) \leqq |f(x)|.$$

58.10. *If the function $F(x)$ is defined and continuous on $[a, b]$, and for all except at most a denumerable set of values of x in $[a, b]$ the derivative $F'(x)$ is defined and finite, then $\dot{F}(x)$ is integrable, and*

$$\int_a^x \dot{F}(x)\,dx = F(x) - F(a).$$

If we take $\varphi^i(x) = F(x) - F(a)$, $i = 1, 2, 3, 4$, the tetrad $\varphi^i(x)$ satisfies the requirements in 57.1. It is ϵ-adjoined to $\dot{F}(x)$ on $[a, b]$ for every positive ϵ, so $\dot{F}(x)$ is integrable. Also,

$$F(x) - F(a) = \varphi^4(x) \leqq \int_a^x \dot{F}(x)\,dx \leqq \varphi^1(x) = F(x) - F(a),$$

establishing the equation in the statement of the theorem.

* We shall see that this is equivalent to requiring that $f(x)$ be Lebesgue summable.

59. We are now able to indicate the generality of the Perron integral. Suppose that $f(x)$ is Lebesgue summable over the interval $[a, b]$. By definition, for every positive ϵ there exists a U-function $u(x)$ and an L-function $l(x)$ such that $u(x) \geqq f(x) \geqq l(x)$ and

$$0 \leqq \int_a^b (u - l)\, dx < \epsilon.$$

Construct now the functions

$$U(x) = \int_a^x u(x)\, dx, \qquad L(x) = \int_a^x l(x)\, dx.$$

These functions are continuous, and $U(a) = L(a) = 0$. Also, by 33.1,

$$\underline{D}U(x) \geqq u(x), \qquad \text{and} \qquad -\infty \neq u(x) \geqq f(x);$$

and likewise

$$\bar{D}L(x) \leqq l(x), \qquad \text{and} \qquad +\infty \neq l(x) \leqq f(x).$$

Therefore $U(x)$ is a right major and a left major of f, (cf. 31.3), and $L(x)$ is a right minor and a left minor of f. We have thus obtained a tetrad

$$\varphi^1(x) = \varphi^2(x) = U(x), \qquad \varphi^3(x) = \varphi^4(x) = L(x)$$

which is ϵ-adjoined to f, and so f is Perron integrable. Moreover, both the Perron and Lebesgue integrals of f lie between $L(b)$ and $U(b)$, and so differ by less than ϵ. Since ϵ is arbitrary, the two integrals are equal. We have therefore proved the following theorem.

59.1. *If $f(x)$ is Lebesgue summable over $[a, b]$, it is P^*-integrable over $[a, b]$, and the Lebesgue integral of $f(x)$ over $[a, b]$ is equal to its Perron integral.*

This proves that the P^*-integral is at least as general as the Lebesgue integral, for functions defined on intervals in one-dimensional space. To prove that it is actually more general, we need only give an example of a function which is Perron integrable but not Lebesgue summable. For this purpose we define $F(x) = x^2 \cos \pi x^{-2}$ for $0 < x \leqq 1$, $F(0) = 0$. This function has a finite derivative for each value of x; for the differentiation is elementary for $x \neq 0$, while at $x = 0$ we have

$$|F'(0)| = \lim_{h \to 0} \left| \frac{h^2 \cos \pi h^{-2} - 0}{h - 0} \right| \leqq \lim_{h \to 0} \frac{h^2}{h} = 0.$$

But $F(x)$ is not of BV on the interval $[0, 1]$. For if we subdivide

[0, 1] into the intervals $[\alpha_i, \alpha_{i+1}]$, where $\alpha_0 = 0$, $\alpha_1 = n^{-\frac{1}{2}}$, $\alpha_2 = (n-1)^{-\frac{1}{2}}$, $\cdots$, $\alpha_{n-1} = 2^{-\frac{1}{2}}$, $\alpha_n = 1$, we find

$$\sum_{0}^{n-1} | F(\alpha_i) - F(\alpha_{i+1}) | > \sum_{1}^{n} j^{-1},$$

which is unbounded as $n \to \infty$.

The function $F'(x)$ is P^*-integrable on [0, 1], by 58.10; and

$$F(x) = \int_0^x F'(x)\, dx,$$

the integral being a P^*-integral. But $F'(x)$ is not Lebesgue summable on [0, 1]. If it were, by 59.1 the preceding equation would still be valid if the integral were understood to be a Lebesgue integral. By 27.7, $F(x)$ would be AC, hence by 9.1 it would be of BV. We have seen that it is not of BV.

60. Occasionally (for example, as we shall see, in the proof of 64.1) it is convenient to be able to relax condition (*b*) in the definition (57.1) of adjoined tetrads by requiring only that $D_+\varphi^1 \geqq f$, etc., almost everywhere. Hence we shall prove that this relaxation leaves the definition of integrability and of the P^*-integral unaltered.

60.1. *A necessary and sufficient condition that a function $f(x)$, defined on an interval $[a, b]$, shall be Perron integrable on $[a, b]$, is that for every $\epsilon > 0$ there shall exist a set of four functions $\psi^1, \psi^2, \psi^3, \psi^4$ such that*

(*a*) *$\psi^i(x)$ is continuous, and $\psi^i(a) = 0$;*

(*b*) *the conditions $D_+\psi^1 \neq -\infty \neq D_-\psi^2$, $D^-\psi^3 \neq +\infty \neq D^+\psi^4$ hold for all x in $[a, b]$ except at most those x belonging to a denumerable set E_0;*

(*c*) *the conditions $D_+\psi^1 \geqq f$, $D_-\psi^2 \geqq f$, $D^-\psi^3 \leqq f$, $D^+\psi^4 \leqq f$ hold for almost all x in $[a, b]$;*

(*d*) *the inequalities $| \psi^1(b) - \psi^4(b) | < \epsilon$, $| \psi^2(b) - \psi^3(b) | < \epsilon$ hold.*

If $f(x)$ is integrable, and the functions $\psi^i(x)$ satisfy (a), (b), (c), then

$$(*) \qquad \psi^i(b) \geqq \int_a^b f(x)\, dx \geqq \psi^j(b) \qquad (i = 1 \text{ or } 2,\ j = 3 \text{ or } 4).$$

The conditions are clearly necessary, for they are satisfied by every tetrad ϵ-adjoined to $f(x)$. We must prove them sufficient. Let E be the set of measure zero on which one or more of the inequalities in (*c*) fails to hold, and let $h(x)$ be $+\infty$ on E and 0 elsewhere. Then $h(x)$ is summable on $[a, b]$, and its Lebesgue integral is 0; so by 15.1, for every positive number δ there is a U-function $u(x)$ on $[a, b]$ such that $u(x) \geqq h(x)$ and the integral of $u(x)$ over $[a, b]$ is less than δ. Define

$$\gamma(x) = \int_a^x u(x)\, dx;$$

then

$$(A) \qquad 0 \leqq \gamma(b) < \delta,$$

and by 33.1 we have

$$(B) \qquad \underline{D}\gamma(x) \geqq u(x) \geqq h(x) \geqq 0.$$

We now define

$$\begin{aligned} \varphi^1(x) &= \psi^1(x) + \gamma(x), \\ \varphi^2(x) &= \psi^2(x) + \gamma(x), \\ \varphi^3(x) &= \psi^3(x) - \gamma(x), \\ \varphi^4(x) &= \psi^4(x) - \gamma(x), \end{aligned}$$

and we proceed to show that these functions form a tetrad $(2\epsilon + 4\delta)$-adjoined to $f(x)$ on $[a, b]$.

The four functions $\varphi^i(x)$ are clearly continuous and vanish at a. By 31.4 the inequalities

$$(C) \qquad \begin{aligned} D_+\varphi^1(x) &\geqq D_+\psi^1(x) + D_+\gamma(x), \\ D_-\varphi^2(x) &\geqq D_-\psi^2(x) + D_-\gamma(x), \\ D^-\varphi^3(x) &\leqq D^-\psi^3(x) - D_-\gamma(x), \\ D^+\varphi^4(x) &\leqq D^+\psi^4(x) - D_+\gamma(x) \end{aligned}$$

hold on $[a, b]$. This, with (B), shows that $D_+\varphi^1$ and $D_-\varphi^2$ are different from $-\infty$ except at most on the denumerable set E_0, and $D^-\varphi^3$ and $D^+\varphi^4$ are different from $+\infty$ except at most on E_0. Also, except on E_0 we can prove that

$$(D) \qquad \begin{aligned} D_+\varphi^1(x) &\geqq f(x) \geqq D^+\varphi^4(x), \\ D_-\varphi^2(x) &\geqq f(x) \geqq D^-\varphi^3(x). \end{aligned}$$

For except on E we know that $D_+\psi^1(x) \geqq f(x)$ by hypothesis and $D_+\gamma(x) \geqq 0$ by (B), so by (C) the first of inequalities (D) is satisfied. The others are similarly established. If x is in E but not in E_0, then $D_+\psi^1(x) > -\infty$ and $D_+\gamma(x) \geqq h(x) = \infty$, so by (C) we have

$$D_+\varphi^1(x) \geqq \infty \geqq f(x).$$

The other inequalities in (D) are established in a like manner. It is thus demonstrated that the tetrad $[\varphi^i(x)]$ is adjoined to $f(x)$ on $[a, b]$.

By hypothesis (d) and inequality (A),

$$\begin{aligned} |\,\varphi^1(b) - \varphi^4(b) &= |\,\psi^1(b) + \gamma(b) - \psi^4(b) + \gamma(b)\,| \\ &\leqq |\,\psi^1(b) - \psi^4(b)\,| + 2\,|\,\gamma(b)\,| \\ &< \epsilon + 2\delta. \end{aligned}$$

In the same way

$$| \varphi^2(b) - \varphi^3(b) | < \epsilon + 2\delta.$$

Since no minor can exceed any major, it is easy to deduce from these inequalities that

$$| \varphi^i(b) - \varphi^j(b) | < 2\epsilon + 4\delta \qquad (i, j = 1, 2, 3, 4).$$

So the tetrad $[\varphi^i(x)]$ is $(2\epsilon + 4\delta)$-adjoined to $f(x)$. Since ϵ and δ are arbitrary, this proves that $f(x)$ is P^*-integrable over $[a, b]$.

Finally, to establish inequality (*) in 60.1 we need only observe that if $i = 1$ or 2 and $j = 3$ or 4, the function $\varphi^i(x)$ is a major function for $f(x)$ and $\varphi^j(x)$ is a minor function. Hence

$$\begin{aligned} \psi^i(b) + \delta &\geqq \psi^i(b) + \gamma(b) = \varphi^i(b) \\ &\geqq \int_a^b f(x)\,dx \geqq \varphi^j(b) = \psi^j(b) - \gamma(b) \\ &\geqq \psi^j(b) - \delta. \end{aligned}$$

Since δ is an arbitrary positive number, this implies that inequality (*) is satisfied.

From 60.1 it follows that the usual extensions of the Riemann integral are also special cases of the Perron integral. Without entering too greatly into details, such an extension applies to functions $f(x)(a \leqq x \leqq b)$ which have infinite discontinuities on a closed denumerable set E_0, and yields as indefinite integral a continuous function $F(x)$ such that for every closed interval $[\alpha, \beta]$ in $[a, b]$ which contains no point of E_0 the difference $F(\beta) - F(\alpha)$ is the Riemann integral of $f(x)$ from α to β. If we define $\psi^1(x) = \psi^2(x) = \psi^3(x) = \psi^4(x) = F(x) - F(a)$, then the $\psi^i(x)$ are continuous, and $\psi^i(a) = 0$. Moreover, $\psi^{i\prime}(x) = F'(x) = f(x)$ almost everywhere, so that (c) of 60.1 is satisfied. If x_0 does not belong to the denumerable set E_0, then $f(x)$ is bounded on a neighborhood of x_0, so that $F(x)$ satisfies a Lipschitz condition and has finite derivatives on this neighborhood. Hence (b) of 60.1 is also satisfied. Consequently by 60.1 and 57.6 $f(x)$ is P^*-integrable, and its P^*-integral is $F(x) - F(a)$.

The next theorem follows readily from 60.1.

60.2. *If $f_1(x)$ is integrable on $[a, b]$, and $f_2(x) = f_1(x)$ almost everywhere on $[a, b]$, then $f_2(x)$ is also integrable, and*

$$(*) \qquad \int_a^b f_2(x)\,dx = \int_a^b f_1(x)\,dx.$$

For let $[\psi^i(x)]$ be a tetrad ϵ-adjoined to $f_1(x)$. Then with respect to $f_2(x)$, the tetrad $[\psi^i(x)]$ satisfies the conditions of 60.1. Hence $f_2(x)$

is integrable. By definition, the right member of equation (*) lies between $\psi^4(b)$ and $\psi^1(b)$, and by 60.1 the left member is also between those numbers. Hence the two integrals differ by at most

$$| \psi^1(b) - \psi^4(b) | ,$$

which is less than ϵ; and since ϵ is arbitrary, they are equal.

Theorem 60.1 permits us to generalize 58.10 somewhat:

60.3. *If $F(x)$ is defined and continuous on $[a, b]$, and all four derivates of $F(x)$ are finite except at most on a denumerable set, and $F'(x)$ is defined for almost all x in $[a, b]$, then $\dot{F}(x)$ is integrable over $[a, b]$, and*

$$\int_a^x \dot{F}(x)\,dx = F(x) - F(a).$$

If we choose $\psi^i(x) = F(x) - F(a)$, $i = 1, 2, 3, 4$, we find that the hypotheses of 60.1 are satisfied with $f(x) = \dot{F}(x)$. Hence by 60.1 and 57.6, $\dot{F}(x)$ is integrable over $[a, x]$, $a < x \leqq b$, and $F(x) - F(a) = \psi^1(x) \geqq \int_a^x \dot{F}(x)\,dx \geqq \psi^4(x) = F(x) - F(a)$. This establishes the theorem.

If two AC functions have the same derivative almost everywhere, they differ only by a constant, as 35.3 shows. The next theorem is an analogue for functions which are not necessarily AC.

60.4. *If $F_1(x)$ and $F_2(x)$ are continuous on an interval $[a, b]$, and except at most on a denumerable set all four derivates of each function are finite, and except on a set of measure 0 the derivatives $F_1'(x)$ and $F_2'(x)$ are defined and equal, then $F_1(x)$ and $F_2(x)$ differ only by a constant on $[a, b]$.*

By 60.3,

$$F_i(x) - F_i(a) = \int_a^x \dot{F}_i(x)\,dx, \qquad i = 1, 2.$$

But $\dot{F}_1(x) = \dot{F}_2(x)$ for almost all x, so by 60.2 we have

$$\int_a^x \dot{F}_1(x)\,dx = \int_a^x \dot{F}_2(x)\,dx.$$

This, with the preceding pair of equations, yields

$$F_1(x) - F_1(a) = F_2(x) - F_2(a).$$

61. The proof of the fundamental theorem for the Perron integral follows closely the corresponding proof for the Lebesgue integral (33.3).

61.1. *If $f(x)$ is Perron integrable on $[a, b]$ and*

$$F(x) = \int_a^x f(x)\,dx,$$

then for almost all x in $[a, b]$ the derivative $F'(x)$ exists and is finite and equal to $f(x)$.

Let ϵ and k be arbitrary positive numbers, and let $[\varphi^i(x)]$ be a tetrad which is $(k\epsilon/2)$-adjoined to $f(x)$ on $[a, b]$. Denote by E_0 the set (at most denumerable) on which one or more of the inequalities in 57.1(*b*) fails to hold. By the remark after 58.1, the functions $\varphi^i(x) - F(x)$ $(i = 1, 2)$ are monotonic increasing; they vanish at $x = a$, and $\varphi^i(b) - F(b) < k\epsilon/2$. Let E_k be the set on which

$$D^+[\varphi^1(x) - F(x)] \geqq k.$$

This set is measurable by 32.1, and by 32.3 we have

$$\frac{kmE_k}{2} \leqq \varphi^1(b) - F(b) < \frac{k\epsilon}{2}.$$

Hence $mE_k < \epsilon$.

By 31.4 the inequality

$$(A) \qquad D_+F(x) \geqq D_+\varphi^1(x) - D^+[\varphi^1(x) - F(x)]$$

holds wherever the right member is defined. But by 34.2 the function $\varphi^1(x) - F(x)$ has a finite derivative almost everywhere, and $D_+\varphi^1(x) > -\infty$ except on E_0, so

$$(B) \qquad D_+F(x) > -\infty \text{ almost everywhere.}$$

For all points of $[a, b] - (E_k \cup E_0)$ we find by (A) that

$$(C) \qquad D_+F(x) \geqq f(x) - k.$$

Thus the set on which this inequality fails is contained in a set $E_0 \cup E_k$ of measure less than ϵ, so by 20.3 its exterior measure is less than ϵ. But ϵ is arbitrary, so the set on which (C) fails has exterior measure 0. By 20.2 and 20.6, its measure is zero, and (C) holds almost everywhere in $[a, b]$.

In particular, if k is the reciprocal of a positive integer n we see that the inequality

$$D_+F(x) \geqq f(x) - \frac{1}{n}$$

holds except on a set N_n of measure 0. Except on $\bigcup N_n$, which has measure 0, this inequality holds for all n, so by 5.2

(D) $$D_+F(x) \geqq f(x) \text{ for almost all } x.$$

In a like manner, by using φ^2 instead of φ^1 we find

(E) $$D_-F(x) > -\infty \qquad \text{and} \qquad D_-F(x) \geqq f(x)$$

for almost all x. If we replace $f(x)$ by $-f(x)$, the indefinite integral $F(x)$ is replaced by $-F(x)$, and by 31.7 and (B), (D) and (E) we obtain

(F) $$\begin{aligned} D^+F(x) &< \infty, \qquad D^+F(x) \leqq f(x), \\ D^-F(x) &< \infty, \qquad D^-F(x) \leqq f(x) \end{aligned}$$

for almost all x. Now B, (D), (E) and (F), with 6.5, yield

$$D^+F(x) = D^-F(x) = D_+F(x) = D_-F(x) = f(x)$$

and

$$-\infty < D_+F(x) = f(x) = D^+f(x) < \infty$$

for almost all x. This completes the proof.

The next theorem is an immediate corollary.

61.2. *If $f(x)$ is Perron integrable on $[a, b]$, it is measurable and almost everywhere finite on $[a, b]$.*

For defining $F(x)$ as above, we have $f(x) = D^+F(x)$ almost everywhere; and $D^+F(x)$, being the derivate of a continuous function, is measurable by 32.1. Also, $D^+F(x)$ is almost everywhere finite, by 61.1.

62. In §59 we established the negative result that not every Perron integrable function is Lebesgue summable. It is however convenient to have conditions which, when added to Perron integrability, ensure the summability of a function $f(x)$.

62.1. *If $f(x)$ is P^*-integrable over $[a, b]$, and $f(x) \geqq 0$ for almost all x in $[a, b]$, then $f(x)$ is Lebesgue summable over $[a, b]$.*

The function $|f(x)|$ is almost everywhere equal to $f(x)$, so is P^*-integrable by 60.2. For $|f(x)|$, the function $\varphi^4(x) \equiv 0$ is a minor function. Let $\varphi^1(x)$ be any right major function of $|f(x)|$; then $\varphi^1(x) = \varphi^1(x) - \varphi^4(x)$ is monotonic increasing by 57.3. Hence by 34.2 $D_+\varphi^1(x)$ is summable. But $f(x)$ is measurable by 61.2, and $|f(x)| \leqq D_+\varphi^1(x)$; so by 22.2 $f(x)$ is summable.

62.2. *If $f_1(x)$ is Perron integrable on $[a, b]$, and $f_2(x)$ is Lebesgue summable on $[a, b]$, and $f_1(x) \geqq f_2(x)$ for almost all x in $[a, b]$, then $f_1(x)$ is also Lebesgue summable.*

There is no less of generality in assuming that $f_2(x)$ is finite-valued; for if we define $f_3(x)$ to be equal to $f_2(x)$ whenever $f_2(x)$ is finite and to be 0 on the set of measure zero on which $|f_2(x)| = \infty$, we see that $f_3(x)$ satisfies the hypotheses imposed on $f_2(x)$.

By 59.1, $f_2(x)$ is Perron integrable, and the difference $f_1(x) - f_2(x)$ is defined for all x in $[a, b]$; so by 58.7 the difference is integrable on $[a, b]$. But $f_1(x) - f_2(x) \geqq 0$ for almost all x, so by 62.1 $f_1(x) - f_2(x)$ is summable. Since $f_2(x)$ is summable by hypothesis, so is $f_2(x) + (f_1(x) - f_2(x)) = f_1(x)$.

63. The only convergence theorems which we are able to establish for the Perron integral are those already proved for the Lebesgue integral, modified by simply adding one and the same Perron integrable function to each summable function involved. For example:

63.1. *If a sequence of finite-valued Perron integrable functions $f_1(x), f_2(x), \cdots$ converges in measure on $[a, b]$ to a function $f(x)$, and if there is a summable function $g(x)$ such that $|f_1(x) - f_n(x)| \leqq g(x)$ for all n and all x, then $f(x)$ is Perron integrable, and*

$$\lim_{n\to\infty} \int_a^b f_n(x)\,dx = \int_a^b f(x)\,dx.$$

For then $f_n - f_1$ converges in measure to $f - f_1$, and each function $f_n - f_1$ is summable by 62.2. Hence by 29.3 the difference $f - f_1$ is summable, and

$$\lim_{n\to\infty} \int_a^b (f_n(x) - f_1(x))\,dx = \int_a^b (f(x) - f_1(x))\,dx.$$

The integrals can now be interpreted as P^*-integrals and the P^*-integral of f_1 added to both members to establish the theorem.

64. We now proceed to establish a theorem on integration by substitution.

64.1. *Let $f(x)$ be defined and integrable on an interval $[a, b]$. Let $g(y)$ be a function defined, continuous and monotonic increasing on an interval $[\alpha, \beta]$, and such that $g(\alpha) = a$, $g(\beta) = b$, and $\bar{D}g(y) \neq +\infty$ except at most for a denumerable set of values of y in $[\alpha, \beta]$. Then $f(g(y))\dot{g}(y)$ is integrable over $[\alpha, \beta]$, and*

$$\int_a^b f(x)\,dx = \int_\alpha^\beta f(g(y))\dot{g}(y)\,dy.$$

Let ϵ be an arbitrary positive number, and let $[\varphi^i(x)]$ be a tetrad ϵ-adjoined to $f(x)$ on $[a, b]$. We shall show that the functions

$$\psi^i(y) = \varphi^i(g(y)), \qquad i = 1, 2, 3, 4$$

satisfy the conditions of 60.1 with f replaced by $f(g(y))\dot{g}(y)$. It is immediate that the ψ^i are continuous and vanish at $y = \alpha$. For the rest we discuss $\psi^1(y)$ only; the rest can be similarly treated.

We divide the points of $[\alpha, \beta]$ into two classes; in class I we put all y for which there exists in $[\alpha, \beta]$ a number $y' > y$ with $g(y') = g(y)$; in class II we put all y such that $g(y') > g(y)$ for all $y' > y$. Consider first any y in class I. Since $g(y)$ is monotonic and $g(y) = g(y')$, we see that $g(y)$ is constant on $[y, y']$. Hence so is $\psi^1(y) \equiv \varphi^1(g(y))$. Also, $D_+g(y) = 0$, so $\dot{g}(y) = 0$; and then

$$-\infty \neq D_+\psi^1(y) = 0 = f(g(y))\dot{g}(y).$$

Thus at all points of class I conditions (b) and (c) of 60.1 are satisfied.

For each point y in class II we may write

$$(A) \qquad \frac{\psi^1(y') - \psi^1(y)}{y' - y} = \frac{\varphi^1(g(y')) - \varphi^1(g(y))}{g(y') - g(y)} \cdot \frac{g(y') - g(y)}{y' - y}.$$

Except at most on a denumerable set X_0 we have $D_+\varphi^1(x) \neq -\infty$. To each x of X_0 there corresponds at most one y of class II; for if two numbers y, y' both correspond to x, then $g(y') = g(y) = x$, so y and y' cannot both be in class II. Hence the class Y_0, consisting of those values of y in class II such that $g(y)$ is in X_0, is at most a denumerable set. We suppose that y belongs neither to this set Y_0 nor to the (at most denumerable) set Y_1 on which $\bar{D}g(y) = \infty$. Then if in (A) we let $y' > y$ tend to y, the first factor on the right is bounded below and the second factor is non-negative and bounded; so the product is bounded below, and $D_+\psi^1(y) \neq -\infty$. We have now proved this relation on all y of class I and all but a denumerable set of y of class II; so condition (b) of 60.1 holds.

Suppose in addition that $g'(y)$ exists; this supposition rejects only a set of measure 0. Now in (A) we let $y' > y$ tend to y. On the right, the last factor tends to $\dot{g}(y) = g'(y) \geqq 0$, and the first factor has a lower limit which cannot be less than the lower right derivate of φ^1 at $g(y)$; so

$$D_+\psi^1(y) \geqq D_+\varphi^1(g(y)) \cdot g'(y) \geqq f(g(y))g'(y).$$

This condition holds almost everywhere; so (c) of (60.1) is satisfied. (Of course we have considered only ψ^1 specifically, but the others can be similarly treated.) There remains only condition (d) of 60.1. But this is trivial, for

$$| \psi^1(\beta) - \psi^4(\beta) | = | \varphi^1(b) - \varphi^4(b) | < \epsilon,$$
$$| \psi^2(\beta) - \psi^3(\beta) | = | \varphi^2(b) - \varphi^3(b) | < \epsilon.$$

Hence by 60.1 we see that $f(g(y))\dot{g}(y)$ is integrable over $[\alpha, \beta]$. Also by 60.1,

$$\varphi^1(b) = \psi^1(\beta) \geqq \int_\alpha^\beta f(g(y))\dot{g}(y)\,dy \geqq \psi^4(\beta) = \varphi^4(b),$$

while

$$\varphi^1(b) \geqq \int_a^b f(x)\,dx \geqq \varphi^4(b).$$

So

$$\left| \int_a^b f(x)\,dx - \int_\alpha^\beta f(g(y))\dot{g}(y)\,dy \right| \leqq | \varphi^1(b) - \varphi^4(b) | < \epsilon.$$

Since ϵ is arbitrary, the two integrals are equal, completing the proof of the theorem.

REMARK. Although we did not explicitly require that $g(y)$ be AC, the other hypotheses imply the absolute continuity of $g(y)$. For let us take $f(x) \equiv 1$. By 64.1, if $\alpha \leqq \xi \leqq \beta$ we have

$$\int_\alpha^\xi 1 \cdot \dot{g}(y)\,dy = \int_a^{g(\xi)} 1\,dx = g(\xi) - a = g(\xi) - g(\alpha).$$

From 27.7 we see that this implies that $g(y)$ is AC.

As a corollary to 64.1 we have the following theorem.

64.2. *If $g(y)$ is continuous and monotonic decreasing for $\alpha \leqq y \leqq \beta$, and $g(\alpha) = b$, $g(\beta) = a$, and $\underline{D}g(y) \neq -\infty$ except at most for a denumerable set Y_1 of values of y; and if $f(x)$ is integrable on $[a, b]$; then $f(g(y))\dot{g}(y)$ is integrable on $[\alpha, \beta]$, and*

$$\int_a^b f(x)\,dx = -\int_\alpha^\beta f(g(y))\dot{g}(y)\,dy.$$

We define $\gamma(\eta)$ for $-\beta \leqq \eta \leqq -\alpha$ by the relation $\gamma(\eta) = g(-\eta)$. Then γ is monotonic increasing, and $\gamma(-\beta) = a$, $\gamma(-\alpha) = b$; and except when $-\eta$ is in Y_1 we have

$$\bar{D}\gamma(\eta) = -\underline{D}g(-\eta) \neq +\infty.$$

Also, $\dot{\gamma}(\eta) = -\dot{g}(-\eta)$. By 64.1, $f(\gamma(\eta))\dot{\gamma}(\eta)$ is integrable over $[-\beta, -\alpha]$, and by 64.1 and 58.5

$$\int_a^b f(x)\,dx = \int_{-\beta}^{-\alpha} f(\gamma(\eta))\dot{\gamma}(\eta)d\eta = -\int_\alpha^\beta f(g(y))\dot{g}(y)\,dy.$$

65. In order to have the theorem on integration by parts stated in generality adequate for the purposes of the next section, we utilize the notion of the Stieltjes integral also:

65.1. *If $f(x)$ is integrable on $[a, b]$, and $g(x)$ is of BV on $[a, b]$, then $f(x)g(x)$ is integrable on $[a, b]$ and*

$$(*)\qquad \int_a^b f(x)g(x)\,dx = F(b)g(b) - \int_a^b F(x)dg(x),$$

wherein

$$F(x) = \int_a^x f(x)\,dx.$$

The integral in equation (*) can be thought of as a Stieltjes integral, that is as the difference of two integrals like those in §11, taken with integrators $p(x)$, $n(x)$ of a minimum decomposition of $g(x)$. Or we can consider $g(x)$ as defined for all x by setting $g(x) = g(a)$ for $x < a$ and $g(x) = g(b)$ for $x > b$, and then regard the integral as a Lebesgue-Stieltjes integral. As we saw in §48, it is immaterial which point of view we choose; the value of the integral is the same in either case.

Since $g(x)$ is of BV on $[a, b]$, we can represent it in either of the forms

$$g(x) = g_1(x) - g_2(x) = g_3(x) - g_4(x),$$

where the g_i are non-negative and g_1 and g_2 are monotonic decreasing, while g_3 and g_4 are monotonic increasing. Let $[\varphi^i(x)]$ be a tetrad δ-adjoined to f on $[a, b]$, and let

$$\psi_j^i(x) = \varphi^i(x)g_j(x) - \int_a^x \varphi^i(x)dg_j(x), \qquad (i, j = 1, 2, 3, 4).$$

We shall now investigate the properties of these sixteen functions.

It is obvious that the functions $\psi_j^i(x)$ all vanish at $x = a$. To prove their continuity we observe that by 37.1 and 47.1 for any two points x and x' of $[a, b]$ the estimate

$$\int_x^{x'} \varphi^i(x)dg_j(x) = \varphi^i(\bar{x})[g_j(x') - g_j(x)]$$

holds, where $\bar{x}$ is between x and x' inclusive. Hence

$$(A)\quad \psi_j^i(x') - \psi_j^i(x) = \varphi^i(x')g_j(x') - \varphi^i(x)g_j(x) - \varphi^i(\bar{x})\{g_j(x') - g_j(x)\}$$
$$= g_j(x')[\varphi^i(x') - \varphi^i(x)] - (g_j(x') - g_j(x))[\varphi^i(\bar{x}) - \varphi^i(x)].$$

As x' tends to x, the quantities in square brackets tend to zero and their coefficients are bounded, hence $\psi_j^i(x')$ tends to $\psi_j^i(x)$.

Consider now a value of x for which the derivatives $g_j'(x)$ exist and are finite; this rules out only a set of measure zero, by 34.3. Equation (A) yields

$$\frac{\psi_j^i(x') - \psi_j^i(x)}{x' - x} = g_j(x)\,\frac{[\varphi^i(x') - \varphi^i(x)]}{x' - x} - \frac{g_j(x') - g_j(x)}{x' - x}\,[\varphi^i(\bar{x}) - \varphi^i(x')]$$

As x' tends to x, so does $\bar{x}$; the first factor in the last term approaches the finite limit $g_j'(x)$, the second factor approaches zero. Since $g_j(x)$ is non-negative, by 6.3 we find

$$D^+\psi_j^i(x) = g_j(x)D^+\varphi^i(x),$$

with like equations for the other three derivates. Thus we have shown that for almost all x in $[a, b]$ the following inequalities are satisfied.

$$(B) \qquad \begin{aligned} D_+\psi_j^1(x) &= g_j(x)D_+\varphi^1(x) \geqq g_j(x)f(x), \\ D_-\psi_j^2(x) &= g_j(x)D_-\varphi^2(x) \geqq g_j(x)f(x), \\ D^-\psi_j^3(x) &= g_j(x)D^-\varphi^3(x) \leqq g_j(x)f(x), \\ D^+\psi_j^4(x) &= g_j(x)D^+\varphi^4(x) \leqq g_j(x)f(x). \end{aligned}$$

We shall use these later.

Equation (A) can also be written in the form

$$(C) \quad \frac{\psi_j^i(x') - \psi_j^i(x)}{x' - x} \equiv g_j(x')\left[\frac{\varphi^i(x') - \varphi^i(x)}{x' - x}\right] + (g_j(x) - g_j(x'))\left[\frac{\varphi^i(\bar{x}) - \varphi^i(x)}{\bar{x} - x}\right]\cdot\left(\frac{\bar{x} - x}{x' - x}\right)$$

provided that we agree to define the expression in the last square bracket to have some value, say zero, when $\bar{x} = x$. We assume that x belongs to the set (consisting of all but a denumerable set of x) for which

$$D_+\varphi^1 \neq -\infty \neq D_-\varphi^2, \qquad D^-\varphi^3 \neq +\infty \neq D^+\varphi^4.$$

Let now j be 1 or 2, so that g_j is decreasing, and take $x' > x$. If $i = 1$, the quantities in square brackets are bounded below, since their lower limit is $D_+\varphi^1 \neq -\infty$ or possibly 0, and their coefficients are non-negative and bounded; hence the left member of (C) is bounded below. So $D_+\psi_j^1 \neq -\infty$ $(j = 1, 2)$ except at most on a denumerable set. If we repeat the whole argument, replacing φ^1 by φ^4 and "below" by "above," we find that $D^+\psi_j^4 \neq +\infty$ $(j = 1$ or $2)$ except at most on a denumerable set.

Moreover, we can repeat the whole argument with φ^2 in place of φ^1 and $j = 3$ or 4, if we take $x' < x$; and we can repeat this last with φ^3 in place of φ^2 if we replace "below" by "above." The result is that the following inequalities hold except for a denumerable collection of values of x.

$$(D) \qquad \begin{aligned} &D_+\psi_j^1 \neq -\infty, \qquad D^+\psi_j \neq +\infty, \qquad (j = 1 \text{ or } 2), \\ &D_-\psi_j^2 \neq -\infty, \qquad D^-\psi_j^3 \neq +\infty, \qquad (j = 3 \text{ or } 4). \end{aligned}$$

With a view to establishing (*d*) of 60.1, we observe that

$$(E)\quad \psi_j^1(b) - \psi_j^4(b) = [\varphi^1(b) - \varphi^4(b)]g_j(b) - \int_a^b [\varphi^1(x) - \varphi^4(x)]dg_j(x)$$
$$\leqq \delta \cdot g_j(b) + \delta \mid g_j(b) - g_j(a) \mid,$$

and if ϵ is an arbitrary positive number this last expression is less than $\epsilon/2$ provided that δ is small enough. In a like manner we prove

$$\psi_j^2(b) - \psi_j^3(b) < \frac{\epsilon}{2}.$$

Now let us define

$$\psi^1 = \psi_1^1 - \psi_2^4, \qquad \psi^2 = \psi_3^2 - \psi_4^3,$$
$$\psi^3 = \psi_3^3 - \psi_4^2, \qquad \psi^4 = \psi_1^4 - \psi_2^1.$$

For all x except at most those of a denumerable set we have, by 31.4 and (*D*),

$$D_+\psi^1 \geqq D_+\psi_1^1 - D^+\psi_2^4,$$
$$D_-\psi^2 \geqq D_-\psi_3^2 - D^-\psi_4^3,$$
$$D^-\psi^3 \leqq D^-\psi_3^3 - D_-\psi_4^2,$$
$$D^+\psi^4 \leqq D^+\psi_1^4 - D_+\psi_2^1.$$

Hence hypothesis (*b*) of 60.1 holds, while by (*B*) hypothesis (*c*) of 60.1 is satisfied. Also, from (*E*) we find

$$\psi^1(b) - \psi^4(b) = \psi_1^1(b) - \psi_1^4(b) + \psi_2^1(b) - \psi_2^4(b) < \epsilon,$$
$$\psi^2(b) - \psi^3(b) = \psi_3^2(b) - \psi_3^3(b) + \psi_4^2(b) - \psi_4^3(b) < \epsilon.$$

We have now verified all the hypotheses of 60.1. So we know that $f(x)g(x)$ is integrable over $[a, b]$, and

$$(F)\qquad \psi^4(b) \leqq \int_a^b f(x)g(x)\,dx \leqq \psi^1(b).$$

If we recall that $g_1(x)$ is positive and monotonic decreasing, so that $-g_1(x)$ is monotonic increasing, and write the definition of $\psi_1^i(x)$ in the form

$$\psi_1^i(x) = \varphi^i(x)g_1(x) + \int_a^x \varphi^i(x)d[-g_1(x)],$$

then from the inequality $\varphi^4(x) \leqq F(x) \leqq \varphi^1(x)$ (cf. 57.7) and 18.2 we conclude

$$\psi_1^4(b) \leqq F(b)g_1(b) + \int_a^b F(x)d[-g_1(x)] \leqq \psi_1^1(b).$$

In the same way,

$$\psi_2^1(b) \geqq F(b)g_2(b) + \int_a^b F(x)d[-g_2(x)] \geqq \psi_2^4(b).$$

These two inequalities, with the equation $g(x) = g_1(x) - g_2(x)$, imply

$$(G) \quad \psi^4(b) = \psi_1^4(b) - \psi_2^1(b) \leqq F(b)g(b) - \int_a^b F(x)dg(x) \leqq \psi_1^1(b) - \psi_2^4(b) = \psi^1(b).$$

Hence by (F) and (G) the two members of equation (*) of 65.1 both lie in the interval $[\psi^4(b), \psi^1(b)]$, whose length is less than ϵ. Since ϵ is an arbitrary positive number, equation (*) is established.

REMARK. If $f(x)$ happens to be summable, so is $f(x)g(x)$, and their Perron and Lebesgue integrals coincide. Hence 65.1 gives a generalization of the theorem on integration by parts for the Lebesgue integral. The conclusion that $f(x)g(x)$ is integrable, while trivial for Lebesgue integrals, is not trivial for Perron integrals. It is easy to construct Perron integrable functions $f(x)$ and continuous functions $g(x)$ (of course not of BV) such that $f(x)g(x)$ is not integrable.

66. The preceding theorem enables us to prove the second theorem of mean value for the Perron integral.

66.1. *If $f(x)$ is integrable on $[a, b]$ and $g(x)$ is monotonic, then there exists a ξ in $[a, b]$ such that*

$$\int_a^b f(x)g(x)\,dx = g(a)\int_a^\xi f(x)\,dx + g(b)\int_\xi^b f(x)\,dx.$$

By 65.1,

$$\int_a^b f(x)g(x)\,dx = F(b)g(b) - \int_a^b F(x)dg(x).$$

By 37.1 and 47.1 the last (Stieltjes) integral is equal to

$$F(\xi)(g(b) - g(a)),$$

where $a \leqq \xi \leqq b$. Hence, recalling the meaning of $F(x)$ and using 58.3,

$$\begin{aligned}\int_a^b fg\,dx &= F(b)g(b) - F(\xi)g(b) + F(\xi)g(a) \\ &= g(a)F(\xi) + g(b)[F(b) - F(\xi)] \\ &= g(a)\int_a^\xi f\,dx + g(b)\int_\xi^b f\,dx.\end{aligned}$$

CHAPTER IX

Differential Equations

67. In proving the fundamental existence theorem for solutions of systems of differential equations we need to know that a certain sequence of functions contains a uniformly convergent subsequence. It is evident that in general a sequence of continuous functions may fail to contain any uniformly convergent subsequence. For instance, the functions $|x|^{1/n}$, $-1 \leqq x \leqq 1$, $n = 1, 2, 3, \cdots$ are continuous, and they converge to a limit function which is 0 if x is 0 and $+1$ if $x \neq 0$. Every subsequence converges to this same discontinuous limit, and so the convergence cannot be uniform. Our next theorem (67.2) sets forth conditions which guarantee the existence of a uniformly convergent subsequence. Before stating this theorem it is convenient to define two new terms.

67.1. *Let $\mathfrak{F}$ be a family of functions $f(x)$, all defined and finite-valued on a set E. The functions of the family $\mathfrak{F}$ are called* equi-continuous *if for every positive number ϵ there is a positive number δ such that for every pair of points x_1, x_2 of E with distance $\| x_1, x_2 \| < \delta$ and for every function $f(x)$ of the family $\mathfrak{F}$ the inequality*

$$|f(x_1) - f(x_2)| < \epsilon$$

holds. The functions of $\mathfrak{F}$ are said to be uniformly bounded *if there is a finite number M such that $|f(x)| \leqq M$ for all functions $f(x)$ in $\mathfrak{F}$ and all x in E.*

Our theorem on the existence of uniformly convergent subsequences can now be stated as follows.

67.2. (Ascoli's Theorem). *If $\mathfrak{F}$ is a family of functions defined, equi-continuous and uniformly bounded on a bounded closed set E, then from every sequence $\{f_n(x)\}$ of functions of $\mathfrak{F}$ it is possible to select a uniformly convergent subsequence.*

It is convenient to separate the proof of this theorem into two lemmas, which will be available for separate use later.

First, let us observe that if E is a bounded closed set, it contains a finite or denumerable subset E_0 whose closure contains E. If E is finite, we take E_0 to be E. Otherwise, let x_1 be a point of E. We define $x_2, x_3, \cdots$ inductively; x_n is a point of E whose distance from the subset $\{x_1, x_2, \cdots, x_{n-1}\}$ is the greatest possible. The set E_0, consisting of $x_1, x_2, \cdots$, is denumerable, and we easily show that its closure contains E. Otherwise, there would be a point x^* of E and a positive ϵ such that no point x_n is in the ϵ-neighborhood of x^*. Thus the distance of x^* from $\{x_1, \cdots, x_{n-1}\}$ is at least ϵ, and by definition of x_n we must have $||x_m, x_n|| \geqq \epsilon$ if $m < n$. By the Cauchy criterion, no subsequence of the sequence $\{x_n\}$ can converge. But this contradicts the Bolzano-Weierstrass theorem, and our remark is established. (The use of the closedness of E is not at all essential.)

The first of our two lemmas concerns the behavior of the functions on the denumerable set E_0.

67.3. *Let $\{f_n(x)\}$ be a sequence of functions defined on a finite or denumerable set E_0. It is then possible to select a subsequence which converges to some limit (finite or infinite) at each point of E_0.*

Let us denote the points of E_0 be $x_1, x_2, \cdots$. The values of the functions $f_n(x)$ at $x = x_1$ form a sequence of numbers, so by 2.12 we can extract a subsequence (we denote it by $\{f_{1,n}(x)\}$) which converges at $x = x_1$. Again, the values of the functions $f_{1,n}(x)$ at $x = x_2$ form a sequence, from which by 2.12 we can extract a convergent subsequence. Let $\{f_{2,n}(x)\}$ be the chosen subsequence of $\{f_{1,n}(x)\}$, convergent at x_2. Thus we can proceed inductively; each sequence $\{f_{k,n}(x)\}$ converges at $x_1, x_2, \cdots, x_k$, and from it we choose a subsequence $\{f_{k+1,n}(x)\}$ which converges at x_{k+1}, and of course continues to converge at $x_1, \cdots, x_k$. If the set E_0 is finite the process stops when E_0 is exhausted, and the proof is then complete. If E_0 is denumerable, we proceed as follows.

Consider the sequence $\{f_{n,n}(x)\}$, $n = 1, 2, \cdots$. We shall prove that this converges at each point of E_0. Let x_k be a point of E_0. The function $f_{n,n}(x)$ is in the sequence $\{f_{n,j}(x)\}, j = 1, 2, \cdots$. This is a subsequence of $\{f_{n-1,j}\}$, and so on; hence if n is not less than k, $f_{n,n}(x)$ occurs in the sequence $\{f_{k,j}\}$, $j = 1, 2, \cdots$. Moreover, if $m > n \geqq k$ the term $f_{m,m}(x)$ follows the term $f_{n,n}(x)$ in the sequence $f_{k,1}(x), f_{k,2}(x), \cdots$. That is, except for its first $k - 1$ terms the sequence $\{f_{n,n}(x)\}$ is a subsequence of $\{f_{k,n}(x)\}$, and therefore converges at x_k. This completes the proof.

The procedure of proof used here is Cantor's "diagonal process."

67.4. *Let $\mathfrak{F}$ be a family of functions defined and equi-continuous on a bounded set E in q-dimensional space R_q. Let E_0 be a subset of E whose closure $\bar{E}_0$ contains E. If $\{f_n(x)\}$ is a sequence of functions of $\mathfrak{F}$ which converges to a finite limit at each point of E_0, this sequence converges uniformly on all of E.*

Let ϵ be a positive number. By the definition 67.1, there is a positive δ such that if x_1 and x_2 are in E and their distance $||x_1, x_2||$ is less than δ, then

$$|f_n(x_1) - f_n(x_2)| < \frac{\epsilon}{3}, \; n = 1, 2, \cdots.$$

Since E is contained in the closed set $\bar{E}_0$, so is $\bar{E}$. Thus if x is in $\bar{E}$, its δ-neighborhood contains a point x_0 of E_0; hence x is in the δ-neighborhood of x_0. The bounded closed set $\bar{E}$ is then covered by the open sets $N_\delta(x_0)$, x_0 in E_0, and by the Heine-Borel theorem (3.5) a finite number of these neighborhoods cover $\bar{E}$. We denote these by $N_\delta(x_m)$, $m = 1, \cdots, p$.

By hypothesis, the sequence $\{f_n(x)\}$ converges at each point x_m, $m = 1, \cdots, p$. So for each of these points there is an integer n_m such that

$$|f_i(x_m) - f_j(x_m)| < \frac{\epsilon}{3}$$

if i and j are both greater than n_m. Now let n_ϵ be the greatest of the numbers $n_1, \cdots, n_p$. For each point x of E there is an x_m such that $||x, x_m||$ is less than δ. Then by the inequalities already established we have

$$|f_i(x) - f_j(x)| \leqq |f_i(x) - f_i(x_m)| + |f_i(x_m) - f_j(x_m)| \\ + |f_j(x_m) - f_j(x)| < \epsilon,$$

provided that i and j both exceed n_ϵ. This is the necessary and sufficient condition for uniform convergence (6.17), and theorem 67.4 is established.

The proof of Theorem 67.2 can now be completed. By the remark after the statement of the theorem, there is a finite or denumerable subset E_0 of E whose closure contains E. Given any sequence $\{f_n(x)\}$ of functions of $\mathfrak{F}$, by 67.3 there is a subsequence which converges at each point of E_0. The limit cannot be $\pm\infty$, since the functions $\mathfrak{F}$ are uniformly bounded. Hence by 67.4 the chosen subsequence converges uniformly on E.

68. By a *solution* of a system of differential equations

$$(D) \qquad \dot{y}^{(i)}(x) = f^{(i)}(x, y(x)) \qquad (a \leqq x \leqq b; i = 1, \cdots, q),$$

we mean a system of q functions, absolutely continuous on the interval $[a, b]$, having values such that the point $(x, y^{(1)}(x), \cdots, y^{(q)}(x))$ lies in the range of definition of the function $f(x, y)$, and such also that equations (D) are satisfied for almost all x in $[a, b]$. An obviously equivalent formulation is

$$(I)\quad y^{(i)}(x) = y^{(i)}(\xi) + \int_\xi^x f^{(i)}(x, y(x))\,dx \qquad (a \leqq x \leqq b;\ i = 1, \cdots, q),$$

where ξ is in $[a, b]$.

For the rest of this section we shall be engaged in proving theorems concerning the existence and uniqueness of solutions of such equations.

68.1. *Let the functions* $f^{(i)}(x, y) \equiv f^{(i)}(x, y^{(1)}, \cdots, y^{(q)})$, $i = 1, \cdots, q$, *be defined for all x in an interval $[a, b]$ and all y, measurable in x for each fixed y, and continuous in y for each fixed x. Let there exist a function $S(x)$, summable over $[a, b]$, such that*

$$|f^{(i)}(x, y)| \leqq S(x)$$

for all x in $[a, b]$ and all y. Then for each ξ in the interval $[a, b]$ and each q-tuple $\eta \equiv (\eta^{(1)}, \cdots, \eta^{(q)})$ there is a function

$$y(x) = (y^{(1)}(x), \cdots, y^{(q)}(x))$$

satisfying D (or I) and such that $y^{(i)}(\xi) = \eta^{(i)}$, $i = 1, \cdots, q$.

In order that (I) shall hold, it is clearly essential that $f^{(i)}(x, y(x))$ shall be summable over $[a, b]$. Therefore we establish the following lemma.

68.2. *Let $f^{(i)}(x, y)$ satisfy the hypotheses of Theorem* 68.1. *Then if $y(x)$ is a step-function or a continuous function, $f^{(i)}(x, y(x))$ is summable over $[a, b]$.*

In order to prove this we need only show that $f^{(i)}(x, y(x))$ is measurable if $y(x)$ is a step-function or a continuous function; for then the hypothesis $|f^{(i)}(x, y)| \leqq S(x)$ will guarantee its summability. Let $y(x)$ be a step function, assuming the value c_j on the subinterval I_j where $I_1, \cdots, I_p$ is a partition of $[a, b]$. On I_j the function $f^{(i)}(x, y(x))$ coincides with $f^{(i)}(x, c_j)$, which is measurable by hypothesis. Now $f^{(i)}(x, y(x))$ can be written in the form

$$f^{(i)}(x, y(x)) = \sum_{j=1}^{p} f^{(i)}(x, c_j) K_{I_j}(x),$$

and is therefore measurable.

If $y(x)$ is continuous, it can be uniformly approximated by step-functions $y_n(x)$. Since $f^{(i)}(x, y)$ is continuous in y for each x, the function $f^{(i)}(x, y(x))$ is the limit of the functions $f^{(i)}(x, y_n(x))$, which have already been proved measurable. Hence it is itself measurable, and the lemma is established.

REMARK. If the $f^{(i)}(x, y)$ are defined on an open set V in $(q + 1) -$ dimensional space, and for fixed x are continuous for all y such that

(x, y) is in V, and for fixed y are measurable on the (necessarily open) set of x such that (x, y) is in V, then $f^{(i)}(x, y(x))$ is measurable whenever $y(x)$ is a step-function or continuous function such that $(x, y(x))$ is in V. The proof above applies without change to step-functions, and for continuous functions $y(x)$ needs only the additional remark that the approximating step-functions $y_n(x)$ can be so chosen that $(x, y_n(x))$ is in V.

We now return to the proof of Theorem 68.1. For each positive integer n, we define the function $u_n(x)$ as follows.

$$u_n(x) = x + \frac{1}{n} \qquad \left(x < \xi - \frac{1}{n}\right),$$

$$u_n(x) = \xi \qquad \left(\xi - \frac{1}{n} \leqq x < \xi + \frac{1}{n}\right),$$

$$u_n(x) = x - \frac{1}{n} \qquad \left(\xi + \frac{1}{n} \leqq x\right).$$

It is easy to verify that $u_n(x)$ is continuous, and that

$$| u_n(x_1) - u_n(x_2) | \leqq | x_1 - x_2 |,$$
$$| u_n(x) - x | \leqq \frac{1}{n}.$$

Next, we shall show that for each n there is a function $g_n(x)$, defined and continuous on $[a, b]$, such that

$$(A) \qquad g_n^{(i)}(x) = \eta^{(i)} + \int_\xi^{u_n(x)} f^{(i)}(x, g_n(x))\, dx.$$

At first glance, this equation seems no more manageable than (I). However, recalling the definition of $u_n(x)$, the equation is satisfied on $\left[\xi - \frac{1}{n}, \xi + \frac{1}{n}\right]$ by the function $g_n^{(i)}(x) \equiv \eta^{(i)}$. We can extend the range of definition of $g_n(x)$ inductively. If for some positive integer j the function $g_n(x)$ is already defined and satisfies equation (A) on a subinterval $\left[\xi, \xi + \frac{(j-1)}{n}\right]$ of $[a, b]$, it is necessarily continuous on the subinterval, being an integral. Hence $f^{(i)}(x, g_n(x))$ is summable over the subinterval. Now for all x in

$$\left[\xi + \frac{(j-1)}{n}, \xi + \frac{j}{n}\right]$$

the interval of integration $[\xi, u_n(x)]$ is contained in

$$\left[\xi, \xi + \frac{(j-1)}{n}\right],$$

so the integral in the right member of (A) is defined. Thus by successive steps we can extend the range of definition of g_n to the whole interval $[\xi, b]$. In a like manner we extend it to $[a, \xi]$.

The functions $g_n^{(i)}(x)$ have two properties important to us. They are uniformly bounded; for if x is in $[a, b]$, then

$$| g_n^{(i)}(x) | \leqq | \eta^{(i)} | + \left| \int_\xi^{u_n(x)} f^{(i)}(x, g_n(x))\, dx \right| \leqq | \eta^{(i)} | + \int_a^b S(x)\, dx.$$

They are equi-continuous. For let ϵ be a positive number. By 27.1, there is a positive number δ such that

$$\left| \int_\alpha^\beta S(x)\, dx \right| < \epsilon$$

if α and β are in $[a, b]$ and $| \beta - \alpha | < \delta$. Hence, for such α and β,

$$\begin{aligned} | g_n^{(i)}(\beta) - g_n^{(i)}(\alpha) | &= \left| \int_{u_n(\alpha)}^{u_n(\beta)} f^{(i)}(x, g_n(x))\, dx \right| \\ &\leqq \left| \int_{u_n(\alpha)}^{u_n(\beta)} S(x)\, dx \right| < \epsilon, \end{aligned}$$

since $| u_n(\beta) - u_n(\alpha) | \leqq | \beta - \alpha |$.

Now by Ascoli's theorem, from the sequence $g_n^{(1)}$ we can select a uniformly convergent subsequence $g_\alpha^{(1)}$; here α ranges over a certain subset of the positive integers. From the sequence $g_\alpha^{(2)}$ we can select a uniformly convergent subsequence $g_\beta^{(2)}$, and so on. After q steps, we obtain a subsequence $g_\kappa(x) (\kappa = n_1, n_2, \cdots)$ such that for each i $(i = 1, \cdots, q)$ the functions $g_\kappa^{(i)}(x)$ converge uniformly on $[a, b]$. We denote the limit function by $y(x)$; thus

$$(B) \qquad \lim_{\kappa \to \infty} g_\kappa^{(i)}(x) = y^{(i)}(x)$$

uniformly on $[a, b]$.

Equation (A) can be written in the form

$$(C) \quad g_\kappa^{(i)}(x) = \eta^{(i)} + \int_\xi^x f^{(i)}(x, g_\kappa(x))\, dx + \int_x^{u_\kappa(x)} f^{(i)}(x, g_\kappa(x))\, dx.$$

The last integral does not exceed

$$\left| \int_x^{u_\kappa(x)} S(x)\, dx \right|$$

in absolute value; and since $| u_\kappa(x) - x | \leqq 1/\kappa$, this tends to zero with $1/\kappa$. For each x, the functions $f^{(i)}(x, y)$ are continuous in y. Hence

$$\lim_{\kappa \to \infty} f^{(i)}(x, g_\kappa(x)) = f^{(i)}(x, y(x)).$$

Since $| f^{(i)}(x, g_\kappa(x)) | \leqq S(x)$, by 29.3 we see that

$$\lim_{\kappa \to \infty} \int_\xi^x f^{(i)}(x, g_\kappa(x))\, dx = \int_\xi^x f^{(i)}(x, y(x))\, dx.$$

From these, with (B), we see that

$$y^{(i)}(x) = \eta^{(i)} + \int_\xi^x f^{(i)}(x, y(x))\, dx \qquad (i = 1, \cdots, q;\ a \leqq x \leqq b).$$

By setting $x = \xi$ we obtain $y^{(i)}(\xi) = \eta^{(i)}$, so $y(x)$ is the function sought, and Theorem 68.1 is established.

The hypothesis in 68.1 that $|f^{(i)}(x, y)|$ shall have a summable bound $S(x)$ is unnecessarily stringent; it prevents us from applying 68.1 even to such simple equations as $y'(x) = y(x)$. In the next theorem we weaken this requirement. The symbol $||f||$ in the statement of the theorem is as usual an abbreviation for the square root of the sum of the squares of the $f^{(i)}$, $i = 1, \cdots, q$.

68.3. *Let the following hypotheses be satisfied.*

(i) *The functions $f^{(i)}(x, y)$, $i = 1, \cdots, q$, are defined for all x in an interval $[a, b]$ and all y, and are measurable functions of x for each fixed y and continuous functions of y for each fixed x in $[a, b]$.*

(ii) *There is a function $S(x)$ summable on $[a, b]$ and a function $\varphi(t)$, defined, positive and continuous for $t \geqq 0$ but not summable over $[0, \infty)$, such that*

$$||f(x, y)|| < \frac{S(x)}{\varphi(||y||)}$$

for all x in $[a, b]$ and all y.

Then for each ξ in $[a, b]$ and each q-tuple η there is a set of solutions $y^{(i)}(x)$ of equations D such that $y^{(i)}(\xi) = \eta^{(i)}$, $i = 1, \cdots, q$.

Evidently $S(x)$ has to be non-negative. The function $\varphi(t)$ is summable over $[0, ||\eta|| + 1]$, being continuous. It cannot be summable over $[||\eta|| + 1, \infty)$, for if it were it would be summable over $[0, \infty)$, contrary to hypothesis. Hence the integral of φ over $[||\eta|| + 1, N)$ increases without bound as N increases, and there is an N such that

$$(A) \qquad \int_{||\eta||+1}^N \varphi(t)\, dt > \int_a^b S(x)\, dx.$$

We define a set of q functions as follows.

$$g^{(i)}(x, y) = f^{(i)}(x, y) \quad \text{if} \quad a \leqq x \leqq b \quad \text{and} \quad ||y|| \leqq N,$$

$$g^{(i)}(x, y) = f^{(i)}\left(x, \frac{Ny}{||y||}\right) \quad \text{if} \quad a \leqq x \leqq b \quad \text{and} \quad ||y|| > N.$$

It is easy to see that these functions are continuous in y for fixed x and measurable in x for fixed y. Since for every y the value of $g^{(i)}(x, y)$ coincides with a value of $f^{(i)}(x, \bar{y})$ on the set $||\bar{y}|| \leqq N$, we also have

$$|g^{(i)}(x, y)| \leqq ||g(x, y)|| \leqq S(x) \cdot \sup \left\{ \frac{1}{\varphi(t)} \,\middle|\, t \leqq N \right\},$$

and the last expression is a summable function of x on $[a, b]$. Hence the functions $g^{(i)}(x, y)$ satisfy the hypotheses of 68.1, and there are AC functions $y^{(i)}(x)$ such that

$$(B) \qquad \dot{y}^i(x) = g^{(i)}(x, y(x))$$

for almost all x in $[a, b]$ and $y^{(i)}(\xi) = \eta^{(i)}$, $i = 1, \cdots, q$. All that remains is to show that $||y(x)|| \leqq N$ on $[a, b]$; for if this can be proved, then $g^{(i)}(x, y(x))$ is identically equal to $f^{(i)}(x, y(x))$ by the definition of $g^{(i)}$, so that equations (B) reduce to equations D.

Suppose this false, so that $||y(x)||$ sometimes exceeds N. To be specific, we suppose that $||y(x)||$ exceeds N somewhere on $[\xi, b]$. Let β be the least number greater than ξ such that $||y(\beta)|| = N$, and let α be the greatest number in $[\xi, \beta]$ such that $||y(x)|| = ||\eta|| + 1$. Then

$$||\eta|| + 1 \leqq ||y(x)|| \leqq N \qquad \text{for} \qquad \alpha \leqq x \leqq \beta,$$

and on $[\alpha, \beta]$ equations (B) are the same as equations D. Using the Cauchy inequality,

$$\begin{aligned} \sum_{i=1}^{q} y^{(i)}(x)\dot{y}^{(i)}(x) &= \sum_{i=1}^{q} y^{(i)}(x) f^{(i)}(x, y(x)) \\ &\leqq ||y(x)|| \cdot ||f(x, y(x))|| \\ &\leqq ||y(x)|| \cdot S(x)/\varphi(||y(x)||) \end{aligned}$$

for almost all x in $[\alpha, \beta]$. For such x,

$$\varphi(||y(x)||) \sum_{i=1}^{q} y^{(i)}(x) \frac{\dot{y}^{(i)}(x)}{||y(x)||} \leqq S(x).$$

But whenever the $y^{(i)}(x)$ are all differentiable, that is, for almost all

x in $[\alpha, \beta]$, this can be written in the form

$$\varphi(\|y(x)\|)\frac{d}{dx}\|y(x)\| \leqq S(x).$$

Hence

$$(C) \quad \int_\alpha^\beta \varphi(\|y(x)\|)\frac{d}{dx}\|y(x)\|\,dx \leqq \int_\alpha^\beta S(x)\,dx \leqq \int_a^b S(x)\,dx.$$

The function $\|y(x)\|$ is AC, since each $y^{(i)}$ is AC, so by 38.1

$$\int_{\|\eta\|+1}^N \varphi(t)\,dt = \int_\alpha^\beta \varphi(\|y(x)\|)\,\|y(x)\|^{\cdot}\,dx.$$

If we substitute this in (C), we find that (A) is contradicted, and the proof is complete.

EXERCISE. Suppose that the $f^{(i)}$ satisfy (i) of 68.3, and for each N there is a summable function $S_N(x)$ such that $\|f(x, y)\| \leqq S_N(x)$ if $\|y\| \leqq N$, and there is a positive AC function $M(x)$ such that

$$\dot{M}(x) \geqq \sup\left\{\sum_{i=1}^q \frac{y^{(i)}(x)f^{(i)}(x, y)}{M(x)}\middle|\, \|y\| \leqq M(x)\right\}$$

for almost all x in $[a, b]$. Show that if $\|\eta\| < M(a)$ there is a solution $y(x)$ of equations D such that $y^{(i)}(a) = \eta^{(i)}$, $i = 1, \cdots, q$. (Construct $g^{(i)}$ as in 68.3, with $N = \sup M(x)$. If $\|y(x)\|$ ever exceeds $M(x)$, let β be the first such x. Almost everywhere on $[a, \beta]$ the derivative of M^2 is at least equal to that of $\|y\|^2$. Use 35.1.)

The next theorem is a corollary of 68.1.

68.4. *Let the following hypotheses be satisfied.*

(i) *The functions $f^{(i)}(x, y)$ are defined on a set V in $(q + 1)$ – dimensional space, and (ξ, η) is a point interior to V.*

(ii) *There is a positive number ϵ and a function $S(x)$ summable on $[\xi - \epsilon, \xi + \epsilon]$ such that the set V_ϵ of points (x, y) with $|x - \xi| \leqq \epsilon$ and $\|y, \eta\| \leqq \epsilon$ is in V, and on V_ϵ the $f^{(i)}(x, y)$ are measurable in x for fixed y and continuous in y for fixed x, and*

$$|f^{(i)}(x, y)| \leqq S(x).$$

Then there is a solution $y^{(i)}(x)$ of equations D defined on an interval $[\alpha, \beta]$ containing ξ in its interior and such that

$$y^{(i)}(\xi) = \eta^{(i)}, \qquad (i = 1, \cdots, q).$$

Define

$$g^{(i)}(x, y) = f^{(i)}(x, y) \qquad ((x, y) \text{ in } V_\epsilon),$$

$$g^{(i)}(x, y) = f^{(i)}\left(x, \eta + \frac{\epsilon(y - \eta)}{||y, \eta||}\right) \qquad (|x - \xi| \leqq \epsilon, ||y, \eta|| > \epsilon).$$

As in 68.3, these functions satisfy the hypotheses of 68.1 with $a = \xi - \epsilon$ and $b = \xi + \epsilon$. Hence the equations

$$\dot{y}^{(i)} = g^{(i)}(x, y)$$

have a solution $y^{(i)}(x)$ defined in $[\xi - \epsilon, \xi + \epsilon]$ and reducing to $\eta^{(i)}$ at $x = \xi$. Since the $y^{(i)}(x)$ are continuous, on a subinterval $[\alpha, \beta]$ of $[\xi - \epsilon, \xi + \epsilon]$ containing ξ in its interior we have

$$||y(x), \eta|| < \epsilon.$$

But when this last condition holds the functions $g^{(i)}$ are identical with the $f^{(i)}$, so on $[\alpha, \beta]$ the functions $y^{(i)}(x)$ satisfy equations D.

Although the hypotheses of 68.1 and 68.3 guarantee the existence of at least one solution of equations D passing through (ξ, η), they are not strong enough to guarantee that there is only one such solution. For example, let $f(x, y) = 3y^{\frac{2}{3}}$. Both $y = x^3$ and $y = 0$ satisfy the equation $\dot{y} = f(x, y)$, and both have the value 0 at $x = 0$. Consequently we need to seek conditions which will guarantee the uniqueness of the solution.

68.5. *Let the functions $f^{(i)}(x, y)$ be defined and finite for all x in the interval $[a, b]$ and all y. Let $y_1^{(i)}(x)$ and $y_2^{(i)}(x)$ $(i = 1, \cdots, q)$ be two sets of solutions of equations D which coincide at a point ξ of $[a, b]$. Let any one of the following four conditions be satisfied.*

(i) *To each q-tuple $\bar{y}$ there corresponds a positive number ϵ and a function $M(x)$ summable over $[a, b]$ such that*

$$(A) \qquad \sum_{i=1}^{q} \{f^{(i)}(x, y + \eta) - f^{(i)}(x, y)\}\eta^{(i)} \leqq M(x)\,||\eta||^2$$

whenever x is in $[a, b]$ and $||\bar{y} - y||$ and $||\bar{y} - (y + \eta)||$ are both less than ϵ.

(ii) *The same as* (i) *except that inequality* (A) *is replaced by*

$$-\sum_{i=1}^{q} \{f^{(i)}(x, y + \eta) - f^{(i)}(x, y)\}\eta^i \leqq M(x)\,||\eta||^2.$$

(iii) *The same as* (i) *except that inequality* (A) *is replaced by*

$$\left|\sum_{i=1}^{q} \{f^{(i)}(x, y + \eta) - f^{(i)}(x, y)\}\eta^{(i)}\right| \leqq M(x)\,||\eta||^2.$$

(iv) *The same as* (i) *except that* (*A*) *is replaced by*

$$||f(x, y + \eta), f(x, y)|| \leqq M(x) ||\eta||.$$

Then the identity $y_1(x) \equiv y_2(x)$ *is satisfied on the corresponding interval:*

(i) $[\xi, b]$; (ii) $[a, \xi]$; (iii) $[a, b]$; (iv) $[a, b]$.

Suppose first that hypothesis (i) is satisfied. If y_1 and y_2 are not identical on $[\xi, b]$, there is a point x_0 in $[\xi, b]$ at which they differ. Let α be the greatest value of x in $[\xi, x_0]$ at which $y_1(x) = y_2(x)$; then $y_1(x) \neq y_2(x)$ for $\alpha < x \leqq x_0$. To the q-tuple $y_1(\alpha)$ corresponds a positive ϵ and a summable function $M(x)$ as in hypothesis (i). Since y_1 and y_2 are continuous, there is a number β such that $\alpha < \beta \leqq x_0$ and the inequalities $||y_1(x), y_1(\alpha)|| < \epsilon$, $||y_2(x), y_1(\alpha)|| < \epsilon$ both hold on $[\alpha, \beta]$. Hence, if we define

$$\eta^{(i)}(x) = y_2^{(i)}(x) - y_1^{(i)}(x), \qquad (i = 1, \cdots, q),$$

by hypothesis (i) the inequality

$$\sum_{i=1}^{q} \{f^{(i)}(x, y_1(x) + \eta(x)) - f^{(i)}(x, y_1(x))\}\eta^{(i)}(x) \leqq M(x) ||\eta(x)||^2$$

is satisfied.

Since both y_1 and y_2 satisfy equations D, for almost all x in $[\alpha, \beta]$ we have

$$\begin{aligned}\sum_{i=1}^{q} \eta^{(i)}(x)\dot{\eta}^{(i)}(x) &= \sum_{i=1}^{q} \eta^{(i)}(x)[\dot{y}_2^{(i)}(x) - \dot{y}_1^{(i)}(x)] \\ &= \sum_{i=1}^{q} \eta^{(i)}(x)\{f^{(i)}(x, y_2(x)) - f^{(i)}(x, y_1(x))\} \\ &= \sum_{i=1}^{q} \eta^{(i)}(x)\{f^{(i)}(x, y_1(x) + \eta(x)) - f^{(i)}(x, y_1(x))\} \\ &\leqq M(x) ||\eta(x)||^2.\end{aligned}$$

On $\alpha < x \leqq \beta$ the quantity $||\eta(x)||$ is positive, so this implies

$$||\eta(x)||^{-2} \sum_{i=1}^{q} \eta^{(i)}(x)\dot{\eta}^{(i)}(x) \leqq M(x)$$

for almost all x in $(\alpha, \beta]$. Hence if $\alpha < x \leqq \beta$ we have

$$\begin{aligned}\int_x^\beta ||\eta(x)||^{-2} \sum_{i=1}^{q} \eta^{(i)}(x)\dot{\eta}^{(i)}(x)\, dx &\leqq \int_x^\beta M(x)\, dx \\ &\leqq \int_a^b M(x)\, dx.\end{aligned}$$

Since $||\eta(x)||^2$ is AC, by 38.1 we find that

$$\begin{aligned}\log||\eta(\beta)||^2 - \log||\eta(x)||^2 &= \int_{||\eta(x)||^2}^{||\eta(\beta)||^2}\left(\frac{1}{t}\right)dt \\ &= \int_x^\beta \frac{1}{||\eta(x)||^2}(||\eta(x)||^2)^{\cdot}\,dx \\ &= \int_x^\beta ||\eta(x)||^{-2}\sum_{i=1}^{q} 2\eta^{(i)}(x)\dot{\eta}^{(i)}(x)\,dx.\end{aligned}$$

This, with the preceding inequality, yields

$$\log||\eta(x)||^2 \geqq \log||\eta(\beta)||^2 - 2\int_a^b M(x)\,dx,$$

or

$$\log||\eta(x)|| \geqq \log||\eta(\beta)|| - \int_a^b M(x)\,dx.$$

This inequality holds for all x in (α, β). But this is impossible; for as x approaches α the quantity $||\eta(x)||$ approaches 0, so its logarithm approaches $-\infty$. This contradiction establishes the theorem under hypothesis (i).

If hypothesis (ii) holds, we first change the independent variable from x to $\bar{x} = -x$. If we write

$$\begin{aligned}\bar{y}(\bar{x}) &= y(-\bar{x}), &\quad -b \leqq \bar{x} \leqq -a,\\ \bar{f}(\bar{x}, y) &= -f(-\bar{x}, y), &\quad -b \leqq \bar{x} \leqq -a,\end{aligned}$$

we see that equations D hold if and only if

$$\frac{d\bar{y}}{d\bar{x}} = \bar{f}(\bar{x}, \bar{y}(\bar{x}))$$

for almost all $\bar{x}$ in $[-b, -a]$. The system $\bar{f}$ satisfies hypothesis (i). So if the functions $\bar{y}_1$ and $\bar{y}_2$ coincide at $-\xi$ they are identical on $[-\xi, -a]$. That is, y_1 and y_2 are identical on $[a, \xi]$.

If hypothesis (iii) is satisfied, so are both (i) and (ii). Hence y_1 and y_2 are identical both on $[\xi, b]$ and on $[a, \xi]$; that is, they are identical on $[a, b]$.

If hypothesis (iv) holds, by the Cauchy inequality we have

$$\begin{aligned}&\left|\sum_{i=1}^{q}\{f^{(i)}(x, y+\eta) - f^{(i)}(x, y)\}\eta^{(i)}\right| \\ &\leqq ||f(x, y+\eta) - f(x, y)||\cdot||\eta|| \\ &\leqq M(x)||\eta||^2.\end{aligned}$$

Therefore hypothesis (iii) is satisfied.

EXERCISE. Let $g(x, y)$ be a function of two variables defined for $a \leqq x \leqq b$ and for all y, and monotonic decreasing in y for each fixed x. If $y_1(x)$ and $y_2(x)$ both satisfy the equation $\dot{y} = g(x, y)$ and coincide at a point ξ of $[a, b]$, they are identical on $[\xi, b]$.

EXERCISE. Let U be an open set in q-dimensional space, and define $\mu(x, \delta)$ to be the sup of $\sum_{i=1}^{q} \{f^{(i)}(x, y_1) - f^{(i)}(x, y_2)\}(y_1^{(i)} - y_2^{(i)})$ for all y_1 and y_2 in U such that $||y_1, y_2|| \leqq \delta$. Suppose that for each ξ in $[a, b]$ and each positive ϵ there is a function $\alpha(x)$ which is positive and AC on $[a, \xi]$, satisfies $\alpha(\xi) < \epsilon$, and has

$$\dot{\alpha}(x) \geqq \frac{\mu(x, \alpha(x))}{\alpha(x)}$$

for almost all x in $[a, \xi]$. Then if y_1 and y_2 are solutions of D lying in U and coinciding at ξ, they are identical on $[\xi, b]$. (From this and its analogue which guarantees identity on $[a, \xi]$ we can readily deduce 68.5.)

69. We now investigate the way in which the solutions of D depend on the initial values $\eta^{(i)}$ and on parameters which may appear in the equations. In order to keep the statements of our theorems from being inordinately verbose, we adopt a standard notation for this section. The functions $f^{(i)}(x, y, u)$ will be assumed to be defined for all x in an interval $[a, b]$, all y (unless a specific statement to the contrary is made) and all points $u = (u^{(1)}, \cdots, u^{(p)})$ of a set P in p-dimensional space R_p. The equations D with $f(x, y)$ replaced by $f(x, y, u)$ will be denoted by $D[u]$; these equations are then

$$D[u]: \qquad \dot{y}^{(i)}(x) = f^{(i)}(x, y, u) \qquad (i = 1, \cdots, q),$$

the differentiation in the left member being with respect to x. If ξ is in $[a, b]$ and η is a q-tuple and u is in P, and the equations $D[u]$ have exactly one solution which assumes the value η at $x = \xi$, we denote this solution by $y^{(i)}(x; \xi, \eta, u)$ $(i = 1, \cdots, q)$. The partial derivatives will be denoted by subscripts; thus for example

$$y_x^{(i)}(x; \xi, \eta, u) \equiv \frac{\partial}{\partial x} y^{(i)}(x; \xi, \eta, u),$$

$$f_{u^{(r)}}^{(i)}(x, y, u) = \frac{\partial}{\partial u^{(r)}} f^{(i)}(x, y, u).$$

69.1. *Let the following hypotheses be satisfied.*

(i) *The functions* $f^{(i)}(x, y, u)$ *are measurable functions of* x *on* $[a, b]$ *for all fixed* (y, u) *with* u *in* P, *and for each fixed* x *in* $[a, b]$ *they are continuous functions of* (y, u) *for all* y *and all* u *in* P.

(ii) *There are functions* $S(x)$, $\varphi(t)$ *as described in* 68.3 *such that*

$$||f(x, y, u)|| \leqq \frac{S(x)}{\varphi(||y||)}$$

for all x *in* $[a, b]$, *all* y *and all* u *in* P.

(iii) *To each point* (ξ, η, u) *in* R_{1+q+p} *with* ξ *in* $[a, b]$ *and* u *in* P *there corresponds exactly one set of solutions* $y^{(i)}(x; \xi, \eta, u)$ $(a \leqq x \leqq b)$ *of equations* $D[u]$.

Then the functions $y^{(i)}(x; \xi, \eta, u)$ *are continuous on the set*

$$\{(x, \xi, \eta, u) \mid x \text{ in } [a, b], \xi \text{ in } [a, b], \text{ all } \eta, u \text{ in } P\}.$$

Moreover, if the $f^{(i)}(x, y, u)$ *are continuous in all variables, the derivatives* $y_x^{(i)}(x; \xi, \eta, u)$ *are also continuous on the same set.*

Before beginning the proof, we observe that hypotheses (i) and (ii) guarantee the existence of at least one solution of equations $D[u]$ which passes through (ξ, η); hypothesis (iii) adds the requirement that there is not more than one solution. In 68.5 we have already listed conditions which insure that hypothesis (iii) is satisfied.

If the theorem were false there would exist a sequence of sets $[x_m, \xi_m, \eta_m, u_m]$, $m = 0, 1, 2, \cdots$, such that x_m and ξ_m are in $[a, b]$ and u_m is in P; x_m, ξ_m, η_m, u_m have the respective limits x_0, ξ_0, η_0, u_0 as m tends to ∞; but for some i the relation

$$\lim_{n\to\infty} y^{(i)}(x_n; \xi_n, \eta_n, u_n) = y^{(i)}(x_0; \xi_0, \eta_0, u_0)$$

fails to be satisfied. From this last statement it follows that we can select a subsequence (we may as well suppose it the original sequence) for which the limit on the left exists, but is different from the right member:

$$(A) \qquad \lim_{n\to\infty} y^{(i)}(x_n; \xi_n, \eta_n, u_n) = \bar{y}^{(i)} \neq y^{(i)}(x_0; \xi_0, \eta_0, u_0).$$

Since η_n approaches η_0, by rejecting a finite number of terms of the sequence we can bring it about that $||\eta_n, \eta_0|| < 1$. Now we select a number N as in the proof of 68.3, and define the functions $g^{(i)}(x, y, u)$ as we did there. As in that proof, we find that

$$||y(x; \xi_n, \eta_n, u_n)|| \leqq N \qquad (a \leqq x \leqq b, n = 0, 1, \cdots).$$

So the functions $y(x; \xi_n, \eta_n, u_n)$ satisfy the equations

$$\dot{y}^{(i)} = g^{(i)}(x, y, u),$$

and if we write

$$S_1(x) = S(x) \cdot \sup \left\{ \frac{1}{\varphi(t)} \middle| 0 \leqq t \leqq N \right\}$$

we have

$$| g^{(i)}(x, y, u) | \leqq S_1(x).$$

Now we can show as in 68.1 that the functions $y^{(i)}(x; \xi_n, \eta_n, u_n)$ are uniformly bounded and equi-continuous, and by 67.2 we can select a subsequence which converges uniformly on $[a, b]$:

$$\lim_{\alpha \to \infty} y^{(i)}(x; \xi_\alpha, \eta_\alpha, u_\alpha) = Y^{(i)}(x) \qquad (i = 1, \cdots, q)$$

uniformly on $[a, b]$, where α ranges over a sequence of values $n_1, n_2, \cdots$. Since the $g^{(i)}(x, y, u)$ have the summable bound $S_1(x)$ and are continuous in (y, u), this implies by 29.3 that

$$\begin{aligned} Y^{(i)}(x) &= \lim_{\alpha \to \infty} y^{(i)}(x; \xi_\alpha, \eta_\alpha, u_\alpha) \\ &= \lim_{\alpha \to \infty} \left\{ \eta^{(i)}_\alpha + \int_{\xi_\alpha}^x g^{(i)}(x, y(x; \xi_\alpha, \eta_\alpha, u_\alpha), u_\alpha)\, dx \right\} \\ &= \eta^{(i)}_0 + \int_{\xi_0}^x g^{(i)}(x, Y(x), u_0)\, dx. \end{aligned}$$

Thus by differentiating we see that $Y^{(i)}(x)$ satisfies equations $D[u]$ with $u = u_0$ and assumes the value $\eta^{(i)}_0$ *at* ξ_0. But by hypothesis (iii) there is only a single solution $y^{(i)}(x; \xi_0, \eta_0, u_0)$ of these equations with the given initial value, so

$$Y^{(i)}(x) \equiv y^{(i)}(x; \xi_0, \eta_0, u_0) \qquad (i = 1, \cdots, q; a \leqq x \leqq b).$$

In the inequality

$$\begin{aligned} &| y^{(i)}(x_\alpha; \xi_\alpha, \eta_\alpha, u_\alpha) - y^{(i)}(x_0; \xi_0, \eta_0, u_0) | \\ &\qquad \leqq | y^{(i)}(x_\alpha; \xi_\alpha, \eta_\alpha, u_\alpha) - y^{(i)}(x_\alpha; \xi_0, \eta_0, u_0) | \\ &\qquad\qquad + | y^{(i)}(x_\alpha; \xi_0, \eta_0, u_0) - y^{(i)}(x_0; \xi_0, \eta_0, u_0) | \end{aligned}$$

the first term on the right tends to zero because $y^{(i)}(x; \xi_\alpha, \eta_\alpha, u_\alpha)$ converges uniformly to $Y^{(i)}(x)$ (compare the preceding equation) and the second term tends to zero because $y^{(i)}(x; \xi_0, \eta_0, u_0)$ is a continuous function of x. Hence the left member approaches zero, and

$$\lim_{\alpha \to \infty} y^{(i)}(x_\alpha; \xi_\alpha, \eta_\alpha, u_\alpha) = y^{(i)}(x_0; \xi_0, \eta_0, u_0).$$

This contradicts (A), and the continuity of the functions $y^{(i)}(x, \xi, \eta, u)$ is established.

If it is true that the functions $f^{(i)}(x, y, u)$ are continuous on the set $\{(x, y, u) \mid x \text{ in } [a, b], \text{ all } y, u \text{ in } P\}$, the functions

$$f^{(i)}(x, y(x; \xi, \eta, u), u)$$

will be continuous for all x and ξ in $[a, b]$, all η, and all u in P, for they are then continuous functions of continuous functions. We write equation $D[u]$ in the integrated form; the equations

$$y^{(i)}(x; \xi, \eta, u) = \eta^{(i)} + \int_\xi^x f^{(i)}(x, y(x; \xi, \eta, u), u)\, dx$$

are then identities. The integrands on the right are continuous functions of x, so by 33.2 the left member has a derivative as to x on $[a, b]$, and this derivative is equal to the integrand on the right. That is, the equation

$$y_x^{(i)}(x; \xi, \eta, u) = f^{(i)}(x, y(x; \xi, \eta, u), u),$$

holds for all x and ξ in $[a, b]$, all u in P and all η, and the right member is continuous on the set of values specified in the theorem. This completes the proof.

As a special case of the preceding theorems, we consider the system of linear differential equations

$$L[u]: \quad \dot{y}^{(i)} \equiv \sum_{j=1}^{q} A_j^i(x, u) y^{(j)} + B^i(x, u) \qquad (i = 1, \cdots, q),$$

where as usual x has the range $[a, b]$ and u the range P.

69.2. *If the functions $A_j^i(x, u)$ and $B^i(x, u)$ are measurable functions of x on $[a, b]$ for each fixed u in P and are continuous functions of u on P for each fixed x in $[a, b]$, and there is a function $S(x)$ summable on $[a, b]$ such that*

$$| A_j^i(x, u) | \leqq S(x) \text{ and } | B^i(x, u) | \leqq S(x)$$

for all x in $[a, b]$ and all u in P, then for every ξ in $[a, b]$, every η and every u in P there is a unique solution $y^{(i)}(x; \xi, \eta, u)$ of equations $L[u]$ which has the value $\eta^{(i)}$ at $x = \xi$. The functions $y^{(i)}(x; \xi, \eta, u)$ are continuous for all x and ξ in $[a, b]$, all η, and all u in P. If the functions $A_j^i(x, u)$ and $B^i(x, u)$ are continuous in all variables, the partial derivatives $y_x^{(i)}(x; \xi, \eta, u)$ are also continuous on the same set.

Let us write $f^{(i)}(x, y, u)$ for the right member of $L[u]$. Then

$$|f^{(i)}(x, y, u)| \leqq \sum_{j=1}^{q} | A_j^i(x, u) | \cdot | y^{(j)} | + | B^i(x, u) | \leqq qS(x) \| y \| + S(x),$$

so

$$||f(x, y, u)|| \leqq q^2S(x)||y|| + qS(x) \leqq q^2S(x)(||y|| + 1).$$

Since the positive continuous function $\varphi(t) = \dfrac{1}{q^2(t+1)}$ is not summable over $[0, \infty)$, the hypotheses of 68.3 are satisfied, and so for each u, each η and each ξ in $[a, b]$ there is a solution of equations $L[u]$ reducing to η at $x = \xi$.

Since

$$|f^{(i)}(x, y + \eta, u) - f^{(i)}(x, y, u)| \leqq \sum_{j=1}^{q} |A_j^i(x, u)| \cdot |\eta^j| \leqq qS(x)||\eta||,$$

we have

$$||f(x, y + \eta, u), f(x, y, u)|| \leqq q^2S(x)||\eta||.$$

So hypothesis (iv) of 68.5 is satisfied, and the solution is unique.

The statements concerning continuity now follow from 69.1.

Our next theorem is closely related to 69.1, but is designed for application in situations in which we know only that the $f^{(i)}(x, y)$ have the desired properties in a neighborhood of a given solution of equations $D[u]$.

69.3. *Let the following hypotheses be satisfied.*

(i) *The functions $f^{(i)}(x, y, u)$, $i = 1, \cdots, q$, are defined and finite for all (x, y) in an open set V in $(q + 1)$-dimensional space and all u in a set P in p-dimensional space.*

(ii) *The functions $y^{(i)}(x)$, $i = 1, \cdots, q$, are AC on an interval $[a, b]$; for all x in $[a, b]$ the point $(x, y(x))$ lies in V, and the $y^{(i)}(x)$ satisfy equations $D[u]$ with $u = u_0$.*

(iii) *There is an interval $[a_0, b_0]$ containing $[a, b]$ in its interior and a function $S(x)$ summable over $[a_0, b_0]$ such that*

$$||f(x, y, u)|| \leqq S(x)$$

whenever (x, y) is in V, x is in $[a_0, b_0]$ and u is in P.

(iv) *For each fixed x in $[a_0, b_0]$ the functions $f^{(i)}(x, y, u)$ are continuous for all (y, u) such that (x, y) is in V and u is in P; and for each fixed (y, u) with u in P they are measurable functions of x on the (open) set of x such that (x, y) is in V.*

(v) *There is a positive number β and a function $M(x)$ summable over $[a_0, b_0]$ such that*

$$||f(x, y + \eta, u), f(x, y, u)|| \leqq M(x)||\eta||$$

whenever x is in $[a_0, b_0]$, u is in P and both (x, y) and $(x, y + \eta)$ are in V, and $||\eta|| < \beta$.

Then there are positive numbers ϵ, δ *such that for each* u *in* $N_\delta(u_0)P$ *and each* (ξ, η) *which has distance less than* δ *from some point of the curve* $(x, y(x))$, $a \leqq x \leqq b$ *the equations* $D[u]$ *have a unique solution* $y^{(i)}(x; \xi, \eta, u)$ $(i = 1, \cdots, q)$, *defined for* x *in* $[a - \epsilon, b + \epsilon]$, *and continuous in all variables for the values just indicated. If the* $f^{(i)}(x, y, u)$ *are continuous in all variables, so are the derivatives* $y_x^{(i)}(x; \xi, \eta, u)$ *for the indicated values of the variables.*

By 68.4 there are functions $y_1^{(i)}(x)$ and $y_2^{(i)}(x)$, defined and AC on intervals $[a - \epsilon, a + \epsilon]$, $[b - \epsilon, b + \epsilon]$ respectively, having $(x, y_1(x))$ and $(x, y_2(x))$ in V, satisfying equations $D[u]$ with $u = u_0$, and such that $y_1^{(i)}(a) = y^{(i)}(a)$ and $y_2^{(i)}(b) = y^{(i)}(b)$. We reduce the intervals $[a - \epsilon, a + \epsilon]$ and $[b - \epsilon, b + \epsilon]$ if necessary so that they are contained in $[a_0, b_0]$. If we define

$$\begin{aligned} y_0^{(i)}(x) &= y_1^{(i)}(x), & a - \epsilon \leqq x < a, \\ y_0^{(i)}(x) &= y^{(i)}(x), & a \leqq x \leqq b, \\ y_0^{(i)}(x) &= y_2^{(i)}(x), & b < x \leqq b + \epsilon, \end{aligned}$$

the functions $y_0^{(i)}(x)$ satisfy equations $D[u]$ with $u = u_0$ on $[a - \epsilon, b + \epsilon]$. They are easily seen to be AC on that interval, either directly from the definition of absolute continuity or else by observing that they satisfy the integrated form of equations $D[u]$.

Let E denote the set of all points $(x, y_0(x))$ with $a - \epsilon \leqq x \leqq b + \epsilon$. This set is bounded and closed, since the $y_0^{(i)}(x)$ are continuous. No point of E is in the (closed) complement CV of the set V, so each point of E has positive distance from CV. The distance of an arbitrary point (x, y) from CV being continuous, there is a point of E at which this distance assumes its minimum value on E. That is, the sup of the distance from $(x, y_0(x))$ to CV is positive, x ranging over $[a - \epsilon, b + \epsilon]$. We choose a positive number γ less than this sup and less also than $\beta/2$. Then the closed set E_γ, consisting of all (x, y) such that $a - \epsilon \leqq x \leqq b + \epsilon$ and $\| y, y_0(x) \| \leqq \gamma$, lies entirely in V.

Next we wish to extend the range of definition of the $f^{(i)}(x, y, u)$ in such a way that theorem 69.1 will be applicable. To do this we first define

$$\begin{aligned} \varphi(t) &= 1, & 0 \leqq t \leqq \gamma, \\ \varphi(t) &= \frac{\gamma}{t}, & t > \gamma. \end{aligned}$$

Then we define

$$g^{(i)}(x, y, u) = f^{(i)}(x, y_0(x) + \varphi(\| y, y_0(x) \|)(y - y_0(x)), u)$$

for $a - \epsilon \leqq x \leqq b + \epsilon$, all y, and all u in P. Since the vector

$$\varphi(\,||\,y, y_0(x)\,||\,)(y - y_0(x))$$

is the same as $y - y_0(x)$ if $||\,y, y_0(x)\,|| \leqq \gamma$ and has length γ otherwise, the right member of this equation is defined for all y, provided that x is in $[a - \epsilon, b + \epsilon]$ and u is in P, and it coincides with $f^{(i)}(x, y, u)$ if $||\,y, y_0(x)\,|| \leqq \gamma$. Also, for each fixed x in $[a - \epsilon, b + \epsilon]$ it is continuous for all y and all u in P, and if moreover the $f^{(i)}(x, y, u)$ are continuous for all (x, y) in V and all u in P the $g^{(i)}$ are continuous for x in $[a - \epsilon, b + \epsilon]$, all y and all u in P.

By the remark after 68.2, if $Y^{(i)}(x)$ are functions continuous on $[a - \epsilon, b + \epsilon]$ such that $(x, Y(x))$ is in V, the functions $f^{(i)}(x, Y(x), u)$ are summable over $[a - \epsilon, b + \epsilon]$ for all u in P. If y is fixed the functions $\varphi(\,||\,y, y_0(x)\,||\,)(y^{(i)} - y_0^{(i)}(x))$ are continuous on $[a - \epsilon, b + \epsilon]$ and the point $(x, y_0(x) + \varphi(\,||\,y, y_0(x)\,||\,)(y - y_0(x)))$ lies in E_γ, hence lies in V. Therefore by the definition of $g^{(i)}(x, y, u)$ it is measurable in x on $[a - \epsilon, b + \epsilon]$ for each fixed y and each fixed u in P. Also, the values of the $g^{(i)}$ are included among the values of the $f^{(i)}$ for (x, y) in V and u in P, so

$$||\,g(x, y, u)\,|| \leqq S(x).$$

In order to show that the $g^{(i)}$ also satisfy hypothesis (v) we need the trivial inequality

$$(A) \qquad ||\,\varphi(\,||\,y_1\,||\,)y_1, \varphi(\,||\,y_2\,||\,)y_2\,|| \leqq ||\,y_1, y_2\,||.$$

This is evident from a figure. To establish it analytically we suppose the notation chosen so that $||\,y_1\,|| \leqq ||\,y_2\,||$. Define

$$r = \frac{\varphi(\,||\,y_2\,||\,)}{\varphi(\,||\,y_1\,||\,)}.$$

Since φ is monotonic decreasing, this is at most 1. Since $t\varphi(t)$ is monotonic increasing, we have

$$||\,y_2\,||\,\varphi(\,||\,y_2\,||\,) \geqq ||\,y_1\,||\,\varphi(\,||\,y_1\,||\,),$$

so that $||\,y_1\,|| \leqq r\,||\,y_2\,||$. As we have assumed $||\,y_1\,|| \leqq ||\,y_2\,||$, this implies

$$||\,y_1\,|| \leqq \tfrac{1}{2}(1 + r)\,||\,y_2\,||.$$

Using the Cauchy inequality we find

$$\sum_{i=1}^{q} [\tfrac{1}{2}(1 + r)y_2^{(i)} - y_1^{(i)}]y_2^{(i)} \geqq \tfrac{1}{2}(1 + r)\,||\,y_2\,||^2 - ||\,y_1\,||\cdot||\,y_2\,||$$
$$\geqq 0.$$

From this, by a little algebraic manipulation, we find

$$||ry_2, y_1||^2 = ||y_2, y_1||^2 - 2(1 - r) \sum_{i=1}^{q} [\tfrac{1}{2}(1 + r)y_2^{(i)} - y_1^{(i)}]y_2^{(i)}$$
$$\leqq ||y_2, y_1||^2.$$

But

$$||\varphi(||y_1||)y_1, \varphi(||y_2||)y_2||$$
$$= \varphi(||y_1||)\,||y_1, ry_2||$$
$$\leqq ||y_1, ry_2||,$$

and this and the preceding inequality establish (A).

Now let x be in $[a - \epsilon, b + \epsilon]$ and u in P. If y_1 and y_2 are arbitrary vectors, we write Y_1, Y_2 for $y_1 - y_0(x)$ and $y_2 - y_0(x)$ respectively. Using inequality (A) and hypothesis (v),

$$||g(x, y_1, u), g(x, y_2, u)||$$
$$= ||f(x, y_0(x) + \varphi(||Y_1||)Y_1, u), f(x, y_0(x) + \varphi(||Y_2||)Y_2, u)||$$
$$\leqq M(x)\,||\varphi(||Y_1||)Y_1, \varphi(||Y_2||)Y_2||$$
$$\leqq M(x)\,||Y_1, Y_2||$$
$$= M(x)\,||y_1, y_2||.$$

Hence the $g^{(i)}(x, y, u)$ satisfy (v) (that is, satisfy (iv) of Theorem 68.5) for all y_1 and y_2.

By 68.1 and 68.5, the equations

$$G[u]: \qquad \dot{y}^{(i)} = g^{(i)}(x, y, u)$$

have unique solutions

$$y^{(i)}(x; \xi, \eta, u) \qquad (i = 1, \cdots, q; a - \epsilon \leqq x \leqq b + \epsilon)$$

defined for all ξ in $[a - \epsilon, b + \epsilon]$, all η, and all u in P. By 69.1, these functions are continuous in all variables on the range indicated, and so also are the derivatives $y_x^{(i)}$ if the $f^{(i)}(x, y, u)$ are continuous in all variables. It remains to show that they satisfy the original equations $D[u]$ for properly restricted values of the variables.

Denote by Q the closed set consisting of the points (x, y) such that $a - \epsilon \leqq x \leqq b + \epsilon$ and $||y, y(x)|| \geqq \gamma$. The coordinates of the point $(x, y(x; \xi, \eta, u))$ are continuous functions of all variables for x and ξ in $[a - \epsilon, b + \epsilon]$ and u in P, hence the distance d from this point to Q is also continuous. If the variables are in the set

$$(B) \qquad \{(x, \xi, \eta, u) \mid x \text{ in } [a - \epsilon, b + \epsilon], (\xi, \eta) \text{ in } E, u = u_0\}$$

the point $(x, y(x; \xi, \eta, u))$ lies on the solution of equations $G[u]$ which has value η at $x = \xi$. But $y_0^{(i)}(x)$ is the only such solution; so for the set of values (B) the point $(x, y(x; \xi, \eta, u))$ lies in the set E and has positive distance d from Q. Since d is a continuous function, on a neighborhood of the set (B) d remains positive. That is, there is a positive number δ (we may take it less than ϵ) such that on the set $\{(x, \xi, \eta, u) \mid x$ *in* $[a - \epsilon, b + \epsilon]$, (ξ, η) *in the* δ*-neighborhood of* E, u *in* $N_\delta(u_0)P\}$ the point $(x, y(x; \xi, \eta, u))$ is not in Q. It then must lie in E_γ. But then the right members of equations $G[u]$ are identical with $f^{(i)}(x, y, u)$, and so the solutions $y^{(i)}(x; \xi, \eta, u)$ of $G[u]$ also are solutions of equations $D[u]$. This completes the proof.

REMARK. If the $f^{(i)}(x, y, u)$ are continuous in all variables they are bounded on a neighborhood of E, so in this case hypothesis (iii) is superfluous.

We now investigate the differentiability of the solutions of equations $D[u]$. The amazing length of the statement of the next theorem is partly accounted for by the fact that it really is five theorems rolled into one, all five theorems having most of their hypotheses in common. Roughly stated, we are going to prove that the differentiability properties of the solutions $y^{(i)}(x; \xi, \eta, u)$ with respect to ξ, η, and u are the same as those of the functions $f^{(i)}(x, y, u)$ with respect to x, y, and u respectively.

69.4. *Let the following hypotheses be satisfied.*

(i) *The functions* $f^{(i)}(x, y, u)$ *are defined and finite for all* (x, y) *in an open set* V *in* $(q + 1)$*-dimensional space and all* u *in an open set* P *in* p*-dimensional space.*

(ii) *For each fixed* u *in* P *and each fixed* y, *the functions* $f^{(i)}(x, y, u)$ *are measurable functions on the (open) set of values of* x *such that* (x, y) *is in* V.

(iii) *For each fixed* x, *the functions* $f^{(i)}(x, y, u)$ *are continuous functions of* (y, u) *for all* y *such that* (x, y) *is in* V *and all* u *in* P, *and for all such* (y, u) *they possess continuous partial derivatives with respect to the variables* $y^{(i)}$ *of all orders* $\leqq m$ *(*m *a positive integer).*

(iv) *There is a function* $S(x)$ *summable over an interval* $[a_0, b_0]$ *containing* $[a, b]$ *in its interior and such that all the* $f^{(i)}(x, y, u)$ *and all the partial derivatives mentioned in* (iii) *have absolute values at most equal to* $S(x)$ *whenever* x *is in* $[a_0, b_0]$, (x, y) *is in* V *and* u *is in* P.

(v) *The functions* $y^{(i)}(x)$ $(i = 1, \cdots, q; a \leqq x \leqq b)$ *are solutions of equations* $D[u]$ *with* $u = u_0$ *in* P, *and the set* E *of points* $(x, y(x))$ $(a \leqq x \leqq b)$ *is in* V.

Then

(C_1) *there exist positive numbers* ϵ, δ *such that for each* (ξ, η) *in the* δ*-neighborhood of* E *and each* u *in the* δ*-neighborhood of* u_0 *there is a* *unique solution*

$$y^{(i)}(x;\ \xi,\ \eta,\ u) \qquad (a - \epsilon \leqq x \leqq b + \epsilon;\ i = 1, \cdots, q)$$

of equations $D[u]$ *such that* $y^{(i)}(\xi;\ \xi,\ \eta,\ u) = \eta^{(i)}$. *These functions are continuous in all variables on the set of values indicated, and so are their partial derivatives with respect to the* $\eta^{(i)}$ *of all orders* $\leqq m$.

(C_2) *If hypothesis* (iv) *is strengthened by assuming that for each fixed* x *the* $f^{(i)}$ *possess continuous partial derivatives of all orders* $\leqq m$ *with respect to the variables* $y^{(i)}$ *and* $u^{(r)}$, *and all these partial derivatives have absolute values at most equal to the summable function* $S(x)$, *then the solutions* $y^{(i)}(x;\ \xi,\ \eta,\ u)$ *have continuous partial derivatives of all orders* $\leqq m$ *with respect to the variables* $\eta^{(i)}$ *and* $u^{(r)}$ *on the set of values indicated in* (C_1).

(C_3) *If in addition to hypotheses* (i) *to* (v) *we assume that the functions* $f^{(i)}(x,\ y,\ u)$ *are continuous in all variables on the set* $\{(x,\ y,\ u) \mid (x,\ y)$ *in* V, u *in* $P\}$, *then on the set of values indicated in* (C_1) *the derivatives* $y_x^{(i)}(x;\ \xi,\ \eta,\ u)$ *are also continuous and possess continuous partial derivatives of all orders* $\leqq m$ *with respect to the variables* $\eta^{(i)}$.

(C_4) *If the functions* $f^{(i)}(x,\ y,\ u)$ *are continuous in all variables on the set* $\{(x,\ y,\ u) | (x,\ y)$ *in* V, u *in* $P]$, *and hypothesis* (iv) *is strengthened as in* (C_2), *then on the set indicated in* (C_1) *the derivatives* $y_x^{(i)}(x;\ \xi,\ \eta,\ u)$ *are also continuous and possess continuous partial derivatives of all orders* $\leqq m$ *with respect to the variables* $\eta^{(i)}$ *and* $u^{(r)}$.

(C_5) *If in addition to hypotheses* (i) *to* (v) *we assume that the* $f^{(i)}(x,\ y,\ u)$ *possess continuous partial derivatives of all orders* $\leqq m$ *with respect to all the variables* x, $y^{(i)}$, $u^{(r)}$, *then on the set of values of* $(x,\ \xi,\ \eta,\ u)$ *indicated the functions* $y^{(i)}(x;\ \xi,\ \eta,\ u)$ *and* $y_x^{(i)}(x;\ \xi,\ \eta,\ u)$ *possess continuous partial derivatives with respect to all the variables* x, ξ, $\eta^{(i)}$, $u^{(r)}$ *of all orders* $\leqq m$.

In proving this theorem it is convenient to observe a certain interrelation between the various conclusions, which we express in the following lemma.

Lemma. *If for a given* m *the hypotheses of* 69.4 *imply conclusion* (C_1), *then for the same* m *they also imply* (C_2), (C_3) *and* (C_4).

Suppose that the functions $f^{(i)}(x,\ y,\ u)$ are continuous in all variables. As we saw in proving 69.3, the equations

$$y_x^{(i)}(x;\ \xi,\ \eta,\ u) = f^{(i)}(x,\ y(x;\ \xi,\ \eta,\ u),\ u)$$

are identities in x, ξ, η, u on the set of variables indicated in (C_1). On this same set the functions $y^{(i)}(x;\ \xi,\ \eta,\ u)$ have continuous partial derivatives of all orders $\leqq m$ with respect to the variables $\eta^{(i)}$. Therefore from the preceding equation we see that the derivatives $y_x^{(i)}$ have continuous partial derivatives of all orders $\leq m$ with respect to $\eta^{(i)}$. This establishes conclusion (C_3).

We introduce p new variables $y^{(q+r)}$ $(r = 1, \cdots, p)$ whose range is P, and we define

$$f^{(i)}(x, y^{(1)}, \cdots, y^{(q+p)}) = f^{(i)}(x, y^{(1)}, \cdots, y^{(q)}, u^{(1)}, \cdots, u^{(p)})$$
$$(i = 1, \cdots, q),$$
$$f^{(i)}(x, y^{(1)}, \cdots, y^{(q+p)}) \equiv 0 \qquad (i = q + 1, \cdots, q + p).$$

Suppose now that for each fixed x the functions $f^{(i)}(x, y, u)$ have partial derivatives of all orders $\leq m$ with respect to the variables $y^{(i)}$ and $u^{(r)}$ whenever (x, y) is in V and u in P, and that none of these derivatives exceed $S(x)$ in absolute value. The equations $D[u]$, with parameters u and initial conditions $y^{(i)}(\xi) = \eta^{(i)}$, are equivalent to the set of $q + p$ equations

$$(D^*) \qquad \dot{y}^{(i)} = f^{(i)}(x, y), \qquad i = 1, \cdots, q + p$$

with the initial conditions

$$y^{(i)}(\xi) = \eta^{(i)} \qquad (i = 1, \cdots, q),$$
$$y^{(q+r)}(\xi) = u^{(r)} \qquad (r = 1, \cdots, p).$$

For the last p of the differential equations D^* force $y^{(q+r)}$ to be constant, and by the initial conditions this constant value is $u^{(r)}$. This new system satisfies all the hypotheses of 69.4 if we replace V by the (open) set of all (x, y, u) with (x, y) in V and u in P. Since we have already established conclusions (C_1) and (C_3) we know that the solutions

$$y^{(i)}(x; \xi, \eta^{(1)}, \cdots, \eta^{(q)}, u^{(1)}, \cdots, u^{(p)})$$

are continuous in all variables and so are their partial derivatives of all orders $\leq m$ with respect to the $\eta^{(i)}$ and $u^{(r)}$ for all (x, ξ, η, u) such that x and ξ are in $[a - \epsilon, b + \epsilon]$ and (ξ, η, u) is in the δ-neighborhood of the set of points

$$\{(x, y, u) \mid a - \epsilon \leqq x \leqq b + \epsilon, y = y(x), u = u_0\}.$$

In particular, the solutions have the desired continuity and differentia bility properties if (ξ, η) is in the $\delta/2$-neighborhood of E and u is in the $\delta/2$-neighborhood of u_0, which is conclusion (C_2) except for the trivial replacing of δ by $\delta/2$. Moreover, by (C_3) the derivatives

$$y_x^{(i)}(x; \xi, \eta, u)$$

also have these same differentiability properties if the functions $f^{(i)}(x, y, u)$ are continuous in all variables. This is conclusion (C_4).

The lemma is therefore established, and we return to the proof of the main theorem.

If (x, y) is in V and u is in P, and β is a positive number so small that the β-neighborhood of (x, y) is in V, we compute

$$(A)\quad f^{(i)}(x, y + \eta, u) - f^{(i)}(x, y, u) = \int_0^1 \frac{d}{dt} f^{(i)}(x, y + t\eta, u)\, dt$$
$$= \int_0^1 \sum_{j=1}^{q} f^{(i)}_{y^{(j)}}(x, y + t\eta, u)\eta^{(j)}\, dt$$
$$(i = 1, \cdots, q),$$

provided that $\| \eta \| < \beta$. From this, with hypothesis (iv) and the Cauchy inequality, we deduce

$$| f^{(i)}(x, y + \eta, u) - f^{(i)}(x, y, u) |$$
$$\leqq \int_0^1 \left\{ \sum_{j=1}^{q} [f^{(i)}_{y^{(j)}}(x, y + t\eta, u)]^2 \right\}^{\frac{1}{2}} \| \eta \|\, dt$$
$$\leqq q^{\frac{1}{2}} S(x) \| \eta \|,$$

whence

$$\| f(x, y + \eta, u), f(x, y, u) \| \leqq qS(x) \| \eta \|.$$

Thus hypothesis (v) of 69.3 is satisfied. The other hypotheses of 69.3 have been assumed satisfied, so there are positive numbers ϵ, δ such that equations $D[u]$ have unique solutions $y^{(i)}(x;\ \xi, \eta, u)$ $(i = 1, \cdots, q;\ a - \epsilon \leqq x \leqq b + \epsilon)$ for all (ξ, η, u) in the set

U: $\{ (\xi, \eta, u) \mid (\xi, \eta)$ *in the* δ*-neighborhood of* E, u *in the* δ*-neighborhood of* $u_0 \}$,

and these solutions are continuous for all (x, ξ, η, u) with x in $[a - \epsilon, b + \epsilon]$ and (ξ, η, u) in U.

Let j be any particular one of the numbers $1, \cdots, q$. If u is in P and the functions $y^{(i)}(x)$, $a - \epsilon \leqq x \leqq b + \epsilon$, are continuous and $(x, y(x))$ is in V, the functions $f^{(i)}(x, y(x), u)$ and

$$f^{(i)}(x, y^{(1)}(x), \cdots, y^{(j-1)}(x), y^{(j)}(x) + t, \quad y^{(j+1)}(x), \cdots,$$
$$y^{(q)}(x), u), \qquad (i = 1, \cdots, q)$$

are measurable if $| t |$ is small enough, by the remark after 68.2. If we subtract the first of these from the second and divide by t we obtain a measurable function. If we now let t tend to zero through a sequence of values, we obtain a sequence of measurable functions converging to

$$f^{(i)}_{y^{(j)}}(x, y(x), u),$$

which is therefore measurable on $[a - \epsilon, b + \epsilon]$.

If (ξ, η) is in the δ-neighborhood of E, u is in the δ-neighborhood of u_0 and x is in $[a - \epsilon, b + \epsilon]$, the point $(x, y(x; \xi, \eta, u))$ is in V. Because of the continuity of the solutions, the line-segment joining this point to the point $(x, y(x; \xi, \eta + h, u))$ will also lie in V if h is any q-tuple lying in a certain neighborhood of $(0, \cdots, 0)$. Henceforth we suppose that h is in that neighborhood. If h is in the given neighborhood of $(0, \cdots, 0)$ and t is in the interval $[0, 1]$ the functions

$$f^{(i)}_{y^{(j)}}(x, y(x; \xi, \eta, u) + t\{y(x; \xi, \eta + h, u) - y(x; \xi, \eta, u)\}, u)$$

are continuous in t and h for fixed x. By the preceding paragraph they are measurable in x for fixed (t, h), since the arguments are continuous functions of x on $[a - \epsilon, b + \epsilon]$. So for each x in $[a - \epsilon, b + \epsilon]$ we have

$$\begin{aligned} A^i_j(x, \eta, h, u) &\equiv \int_0^1 f^{(i)}_{y^{(j)}}(x, y(x; \xi, \eta, u) \\ &\qquad + t\{y(x; \xi, \eta + h, u) - y(x; \xi, \eta, u)\}, u)\, dt \\ &= \lim_{n\to\infty} \frac{1}{n} \sum_{m=0}^{n-1} f^{(i)}_{y^{(j)}}(x, y(x; \xi, \eta, u) \\ &\qquad + (m/n)\{y(x; \xi, \eta + h, u) - y(x; \xi, \eta, u)\}, u). \end{aligned}$$

For each n all the summands in the right member are measurable functions of x on $[a - \epsilon, b + \epsilon]$, so $A^i_j(x, \eta, h, u)$ is measurable on that interval for each fixed h. Since the first partials of $f^{(i)}$ are continuous functions of the $y^{(i)}$, and $y^{(i)}(x; \xi, \eta + h, u)$ and $y^{(i)}(x; \xi, \eta, u)$ are continuous functions of η and h, it is easily seen that $A^i_j(x, \eta, h, u)$ is also continuous in η and h for fixed x.

Next let k be any one of the numbers $1, \cdots, q$, and let h be such that all its components except $h^{(k)}$ are zero. We define

$$z^{(i)}(x; \eta, h, u) = \frac{y^{(i)}(x; \xi, \eta + h, u) - y^{(i)}(x; \xi, \eta, u)}{h^{(k)}}, \quad i = 1, \cdots, q.$$

Using equation (A), we find that for almost all x in $[a - \epsilon, b + \epsilon]$ this satisfies the equation

$$\begin{aligned} \dot{z}^{(i)}(x; \eta, h, u) &= (h^{(k)})^{-1}\{f^{(i)}(x, y(x; \xi, \eta + h, u), u) \\ &\qquad - f^{(i)}(x, y(x; \xi, \eta, u), u)\} \\ &= (h^{(k)})^{-1} \int_0^1 \sum_{j=1}^{q} f^{(i)}_{y^{(j)}}(x, y(x; \xi, \eta, u) + t\{y(x; \xi, \eta + h, u) \\ &\qquad - y(x; \xi, \eta, u)\}, u)(y^{(j)}(x; \xi, \eta + h, u) - y^{(j)}(x; \xi, \eta, u))dt \\ &= \sum_{j=1}^{q} A^i_j(x; \eta, h, u) z^{(j)}(x; \eta, h, u). \end{aligned}$$

Also, since $y^{(i)}(x;\ \xi,\ \eta,\ u)$ has the value $\eta^{(i)}$ at $x = \xi$ we see that

$$z^{(i)}(\xi;\ \eta,\ h,\ u) = \frac{\{(\eta^{(i)} + h^{(i)}) - \eta^{(i)}\}}{h^{(k)}}$$
$$= \delta_{ik},$$

where δ_{ik} is 1 if $i = k$ and is 0 otherwise.

Consider the differential equations

$$(L) \qquad \dot{v}^{(i)} = \sum_{j=1}^{q} A_j^i(x,\ \eta,\ h,\ u)v^j.$$

By 69.2, for each q-tuple $\bar{\eta}$ these equations have unique solutions $v^{(i)}(x;\ \xi,\ \bar{\eta},\ \eta,\ h,\ u)$ $(i = 1, \cdots, q;\ a - \epsilon \leqq x \leqq b + \epsilon)$ reducing to $\bar{\eta}^{(i)}$ at $x = \xi$ and continuous in all variables. Hence in particular the limit

$$(B) \qquad \lim_{||h|| \to 0} v^{(i)}(x;\ \xi,\ \bar{\eta},\ \eta,\ h,\ u) = v^{(i)}(x;\ \xi,\ \bar{\eta},\ \eta,\ 0,\ u)$$

exists and is continuous in x, ξ, $\bar{\eta}$, η and u. But for $h^{(k)} \neq 0$ the functions $z^{(i)}(x;\ \eta,\ h,\ u)$ satisfy these same differential equations L. If $\bar{\eta}$ is the vector whose k-th component is 1 and whose other components are 0, by the uniqueness of the solutions of L we have

$$\frac{y^{(i)}(x;\ \xi,\ \eta + h,\ u) - y^{(i)}(x;\ \xi,\ \eta,\ u)}{h^{(k)}} = z^{(i)}(x;\ \eta,\ h,\ u)$$
$$= v^{(i)}(x;\ \xi,\ \bar{\eta},\ \eta,\ h,\ u).$$

Thus equation (B) informs us that the partial derivative of $y^{(i)}$ with respect to $\eta^{(k)}$ exists, and

$$(C) \qquad y^{((i))}_{\eta^{(k)}}(x;\ \xi,\ \eta,\ u) = v^{(i)}(x;\ \xi,\ \bar{\eta},\ \eta,\ 0,\ u),$$

which is a continuous function of x, ξ, η and u.

We have thus obtained conclusion (C_1) of our theorem subject to the restriction that $m = 1$. By the lemma, conclusions (C_2), (C_3) and (C_4) also hold for $m = 1$.

To remove the restriction $m = 1$ we use induction. Suppose that the hypotheses of 69.4 imply (C_1), (and by the lemma imply (C_2), (C_3) and (C_4)) whenever m is less than a certain m' greater than 1; we show that the hypotheses still imply conclusion (C_1) when $m = m'$. The equations

$$y^{(i)}(x;\ \xi,\ \eta,\ u) = \eta^{(i)} + \int_\xi^x f^{(i)}(x,\ y(x;\ \xi,\ \eta,\ u),\ u)\ dx$$

are identities for $a - \epsilon \leqq x \leqq b + \epsilon$ and (ξ, η) in the δ-neighborhood of E and u in the δ-neighborhood of u_0, and the $y^{(i)}$ have continuous first-order partial derivatives with respect to the variables $\eta^{(k)}$. If for some k we denote the partial derivative of $y^{(i)}$ with respect to $\eta^{(k)}$ by $v^{(i)}$, we see by 39.2 that the equations

$$v^{(i)} = \delta_{ik} + \int_\xi^x \left\{ \sum_{j=1}^q f^{(i)}_{y^{(j)}}(x, y(x;\, \xi, \eta, u), u) v^{(j)} \right\} dx$$

are satisfied, where δ_{ik} is 1 if $i = k$ and is 0 otherwise. These equations are equivalent to the differential equations

$$(P) \qquad \dot{v}^{(i)} = \sum_{j=1}^q f^{(i)}_{y^{(j)}}(x, y(x;\, \xi, \eta, u), u) v^{(j)}$$

with the initial conditions $v^{(i)}(\xi) = \delta_{ik}$. Let us consider u fixed somewhere in the δ-neighborhood of u_0 and ξ also as fixed; η we will permit to vary, subject to the requirement that (ξ, η) shall remain in the δ-neighborhood of the set E. Now equations (P) are of the type of equations $D[u]$, with certain notational changes; y is replaced by v, the parameters are $\eta^{(j)}$ instead of $u^{(h)}$, the initial values are fixed at δ_{ik} instead of being the (arbitrary) $\eta^{(j)}$, and the numbers u in equations (P) are to be temporarily disregarded as parameters, since we are keeping them fixed. By hypothesis the functions $f^{(i)}$ have continuous partial derivatives of all orders $\leqq m'$ with respect to the variables $y^{(j)}$, and by the induction hypothesis and conclusion (C_1) the functions $y^{(i)}(x;\, \xi, \eta, u)$ have continuous partial derivatives of all orders $\leqq m' - 1$ with respect to the variables $\eta^{(j)}$. Hence for each fixed x in $[a - \epsilon, b + \epsilon]$ the right members of equations (P) have continuous partial derivatives of all orders $\leqq m' - 1$ with respect to the variables $v^{(j)}$ and $\eta^{(j)}$. We have assumed (C_1), (C_2), (C_3) and (C_4) all valid if m is less than m', so by (C_2) we find the equations (P) have solutions

$$v^{(i)} = v^{(i)}(x;\, \xi, v, \eta)$$

assuming the initial values $v^{(i)}$ at $x = \xi$ and having continuous partial derivatives of all orders $\leqq m' - 1$ with respect to the variables $v^{(i)}$ and $\eta^{(j)}$. In particular, if we fix $v^{(i)}$ at the value δ_{ik} this solution is

$$y^{(i)}_{\eta^{(k)}}(x;\, \xi, \eta, u),$$

as we saw in deriving equations (P); so these functions have continuous partial derivatives of all orders $\leqq m' - 1$ with respect to the variables $\eta^{(j)}$. That is, the functions $y^{(i)}(x;\ \xi,\ \eta,\ u)$ have continuous partial derivatives of all orders $\leqq m'$ with respect to the variables $\eta^{(j)}$, and conclusion (C_1) holds for $m = m'$. By the lemma, the same is true of (C_2), (C_3) and (C_4), so by induction these conclusions all hold for all integers m.

All that remains is to establish conclusion (C_5). We therefore assume that the functions $f^{(i)}(x, y, u)$ have continuous partial derivatives of all order $\leqq m$ with respect to all variables on the range (x, y) in V, u in P. If (ξ, η) is in the δ-neighborhood of E, so is $(\xi + h, \eta)$ for all sufficiently small values of h. We assume h to be such that $(\xi + h, \eta)$ is in the δ-neighborhood of E and $\xi + h$ is in $[a - \epsilon, b + \epsilon]$ and we define $Y^{(i)}(h)$ to be the value of $y^{(i)}(x;\ \xi + h,\ \eta,\ u)$ at $x = \xi$, that is,

$$Y^{(i)}(h) = y^{(i)}(\xi;\ \xi + h,\ \eta,\ u).$$

The functions $y^{(i)}(x;\ \xi + h,\ \eta,\ u)$ satisfy equations $D[u]$ and have values $Y^{(i)}(h)$ *at* $x = \xi$. But the only solutions of equations $D[u]$ with those initial values are the functions $y^{(i)}(x;\ \xi,\ Y(h),\ u)$. That is,

$$y^{(i)}(x;\ \xi + h,\ \eta,\ u) = y^{(i)}(x;\ \xi,\ Y(h),\ u) \qquad (i = 1, \cdots, q)$$

for all x in $[a - \epsilon,\ b + \epsilon]$, and all h near 0 such that $\xi + h$ is in $[a - \epsilon, b + \epsilon]$. Since the functions $y^{(i)}(x;\ \xi + h,\ \eta,\ u)$ satisfy equations $[u]$ and have the values $\eta^{(i)}$ at $x = \xi + h$, by integration we find that;

$$y^{(i)}(\xi;\ \xi + h,\ \eta,\ u) = \eta^{(i)} + \int_{\xi+h}^{\xi} f^{(i)}(x,\ y(x;\ \xi + h,\ \eta,\ u),\ u)\,dx,$$

or

$$Y^{(i)}(h) - \eta^{(i)} = -\int_{\xi}^{\xi+h} f^{(i)}(x,\ y(x;\ \xi + h,\ \eta,\ u),\ u)\,dx.$$

We divide by h and use the theorem of mean value, obtaining

$$\frac{Y^{(i)}(h) - \eta^{(i)}}{h} = -f^{(i)}(\bar{x},\ y(\bar{x};\ \xi + h,\ \eta,\ u),\ u),$$

where $\bar{x}$ is between ξ and $\xi + h$ inclusive. The right member of this equation is continuous in all variables, so

$$\lim_{h\to 0} \frac{Y^{(i)}(h) - \eta^{(i)}}{h} = -f^{(i)}(\xi,\ y(\xi;\ \xi,\ \eta,\ u),\ u) = -f^{(i)}(\xi,\ \eta,\ u) \qquad (i = 1, \cdots, q).$$

The first-order partial derivatives of the $y^{(i)}$ with respect to the $\eta^{(j)}$ have already been shown to exist and to be continuous. So by Taylor's theorem and the preceding equations,

$$\frac{y^{(i)}(x;\,\xi+h,\,\eta,\,u)-y^{(i)}(x;\,\xi,\,\eta,\,u)}{h}$$
$$=\frac{y^{(i)}(x;\,\xi,\,Y(h),\,u)-y^{(i)}(x;\,\xi,\,\eta,\,u)}{h}$$
$$=\sum_{j=1}^{q} y^{(i)}_{\eta^{(j)}}(x;\,\xi,\,\eta+\vartheta(Y(h)-\eta),\,u)\,\frac{[Y^{(j)}(h)-\eta^{(j)}]}{h},$$

where ϑ is between 0 and 1. As h tends to zero $Y(h)$ tends to $Y(0)$, which is η; so the first factor in the right member tends to a limit. The second factor has already been shown to approach a limit. Hence the left member approaches a limit. That is, $y^{(i)}$ has a partial derivative with respect to ξ, and

$$(C)\qquad y^{(i)}_{\xi}(x;\,\xi,\,\eta,\,u)=-\sum_{j=1}^{q} y^{(i)}_{\eta^{(j)}}(x;\,\xi,\,\eta,\,u)f^{(j)}(\xi,\,\eta,\,u).$$

This proves also that the left member is a continuous function of all variables.

From equations $D[u]$ themselves we see that the partial derivatives of the $y^{(i)}$ with respect to x are continuous functions of all variables, being identically equal to $f^{(i)}(x,\,y(x;\,\xi,\,\eta,\,u),\,u)$. So we have shown that the functions $y^{(i)}(x;\,\xi,\,\eta,\,u)$ have continuous first-order partial derivatives with respect to all variables on the range indicated, and conclusion (C_5) is established for $m=1$.

Suppose next that m is greater than 1. The functions $y^{(i)}(x;\,\xi,\,\eta,\,u)$ have been shown in (C_4) to have continuous partial derivatives of all orders $\leqq m$ with respect to the $\eta^{(j)}$ and $u^{(r)}$, and the $f^{(i)}(x,\,y,\,u)$ have been assumed to have continuous partial derivatives of all orders $\leqq m$ with respect to all variables. From equations (C) it now follows that the first partial derivatives of the $y^{(i)}$ with respect to ξ have continuous partial derivatives of all orders $\leqq m-1$ with respect to the $\eta^{(j)}$ and $u^{(r)}$, for the differentiation of the right member will introduce only derivatives known to exist and be continuous. In a like manner, from equations $D[u]$ we see that the partial derivatives of the $y^{(i)}$ with respect to x have continuous partial derivatives of all orders $\leqq m-1$ with respect to the variables $\eta^{(j)}$ and $u^{(r)}$. Returning to (C), we now see that if $m\geqq 2$ the partial differentiation of the right member with respect to x or ξ will involve only partial derivatives now known to exist and be continuous, and likewise for the right member of $D[u]$. That is, if

$m \geqq 2$ the partial derivatives

$$y^{(i)}_{\xi\xi},\ y^{(i)}_{\xi x},\ y^{(i)}_{xx}$$

all exist, are continuous and have continuous partial derivatives of all orders $\leqq m - 2$ with respect to the $\eta^{(j)}$ and $u^{(r)}$. Returning with this information to (C) and $D[u]$, we find that if $m \geqq 3$ the third-order partial derivatives of $y^{(i)}$ with respect to x and ξ possess continuous partial derivatives of all orders $\leqq m - 3$ with respect to the $\eta^{(j)}$ and $u^{(r)}$. We continue the process; the outcome is that all partial derivatives of $y^{(i)}$ of order $\leqq m$ exist and are continuous for the indicated values of the variables. Finally from equations $D[u]$ themselves it follows that $y^{(i)}_x$ also possesses continuous partial derivatives of all orders $\leqq m$ with respect to all the variables, and the proof of the theorem is complete.

CHAPTER X

Differentiation of Multiple Integrals

70. The theory of differentiation for multiple integrals rests upon the covering theorem of Vitali, to which we devote this section. Before stating the theorem, it is convenient to define regular convergence of a sequence of sets to a point.

70.1. *A sequence of measurable sets $E_1, E_2, \cdots$ converges regularly to a point x_0 if there exists a set $\{Q_n\}$ of cubes (that is, intervals all of whose edges are equal) with centers at x_0 and possessing the following properties. For each n, the cube Q_n contains the set E_n. As n tends to ∞, the side of Q_n tends to zero. There is a positive number α such that*

$$mE_n \geqq \alpha m Q_n$$

for each n.

A number α with which the last inequality is satisfied is called a *modulus of regularity* of the sequence $\{E_n\}$.

We can now state Vitali's theorem.

70.2. *Let E be a set in q-dimensional space R_q, contained in an open set U. Let $\mathfrak{F}$ be a family of closed sets such that for each x in E there is a sequence $\{F_n\}$ of sets of the family $\mathfrak{F}$ which converges regularly to x. Then there exists a finite or denumerable collection $F_1, F_2, \cdots$ of sets of the family $\mathfrak{F}$, no two having any points in common, such that $\bigcup F_i$ is contained in U and contains all of E except at most a subset of measure 0:*

$$m(E - \bigcup F_i) = 0.$$

The proof which we shall give is due to S. Banach.

We first prove the theorem under two auxiliary hypotheses.

(I) *The set U is bounded.*

(II) *There is a positive number α such that for each x_0 of E there is a sequence of sets of $\mathfrak{F}$ converging to x_0 with modulus of regularity α, each set of the sequence containing the point x_0.*

Our first step is to reject certain sets from $\mathfrak{F}$. Every set F of $\mathfrak{F}$ not contained in U can be rejected without disturbing the hypotheses; for if $\{F_i\}$ is a sequence of sets of $\mathfrak{F}$ converging regularly to a point x_0 of E, all but a finite number of the F_i are contained in U, and the rejection of these leaves us with a sequence of sets of $\mathfrak{F}$ contained in U and converging regularly to x_0. Next we reject from $\mathfrak{F}$ all sets F

which are not contained in any cube Q with $mQ \leqq \alpha^{-1}mF$. This too leaves our hypotheses undisturbed; for if $\{F_i\}$ converges to x_0 with modulus of regularity α, by definition each set F_i is contained in a cube Q_i with center at x_0 and such that $mF_i \geqq \alpha m Q_i$.

It is convenient to introduce a new symbol. For each set F of the family $\mathfrak{F}$, we define $e(F)$ to be half the edge of the least cube containing F. Thus for some x_0 the set F is contained in the cube

$$\{x \mid x_0^{(i)} - e(F) \leqq x^{(i)} \leqq x_0^{(i)} + e(F), (i = 1, \cdots, q)\},$$

but it is not contained in any cube with smaller edge.

The sequence $\{F_n\}$ will be defined inductively. For F_1 we choose an arbitrary set of the family $\mathfrak{F}$. The sets $F_1, \cdots F_p$ having been chosen, it is possible that $F_1 \cup \cdots \cup F_p$ contains all of E. In this case we already have the collection of sets F_n of the theorem, and the proof is complete. Otherwise, there will be a point x_0 of E not in $F_1 \cup \cdots \cup F_p$. This last set, being the sum of finitely many closed sets, is closed. Thus x_0 is in the open complement of $F_1 \cup \cdots \cup F_p$, so a neighborhood $N_\delta(x_0)$ of the point x_0 is in that complement. By hypothesis, there is a sequence $\{F_i^*\}$ of sets of $\mathfrak{F}$ converging regularly to x_0. All but a finite number of these are contained in $N_\delta(x_0)$, so infinitely many sets of the sequence $\{F_i^*\}$ have no points in common with $F_1 \cup \cdots \cup F_p$. Now let ϵ_{p+1} be the sup of $e(F)$ for all those sets F of the family $\mathfrak{F}$ which have no point in common with $F_1 \cup \cdots \cup F_p$. We choose F_{p+1} to be a set of $\mathfrak{F}$ having no points in common with $F_1 \cup \cdots \cup F_p$ and having $e(F)$ greater than $\epsilon_{p+1}/2$.

As previously remarked, if the defining process stops at some finite stage, the theorem is established. Suppose then that it does not stop, so that $F_1, F_2, \cdots$ is an infinite sequence chosen in the manner just described; we must show that it satisfies the conditions of the theorem.

Clearly $\bigcup F_n$ is contained in U; we had already rejected from $\mathfrak{F}$ all sets not contained in U. Clearly no two sets $F_m, F_n (m \neq n)$ have points in common; the method of choice provides this. It remains to show that $E - \bigcup F_n$ is a set of measure zero. Suppose this false; then the outer measure $m_e(E - \bigcup F_n)$ is positive.

To each set F_n of the sequence there corresponds a point x_n such that the cube

$$Q_n: \{x \mid x_n^{(i)} - e(F_n) \leqq x^{(i)} \leqq x_n^{(i)} + e(F_n), (i = 1, \cdots, q)\}$$

contains F_n. Let Q_n^* be the cube

$$Q_n^*: \{x \mid x_n^{(i)} - 5e(F_n) \leqq x^{(i)} \leqq x_n^{(i)} + 5e(F_n), (i = 1, \cdots, q)\}.$$

The edge of Q_n^* is five times that of Q_n, so

$$mQ_n^* = 5^q mQ_n.$$

Each set F of the family $\mathfrak{F}$ is contained in some cube Q such that $mQ \leqq \alpha^{-1}mF$, because of the rejection process at the beginning of this proof. Since Q_n is the smallest cube containing F_n, we have

$$mQ_n^* = 5^q mQ_n \leqq 5^q\alpha^{-1}mF_n.$$

The series ΣmF_n converges, for $F_1, F_2, \cdots$ is a sequence of disjoint closed sets all contained in a set U of finite measure. By the last inequality, the series ΣmQ_n^* also converges. Since $m_e(E - \bigcup F_n)$ is positive by hypothesis, there is an integer N so large that

$$\sum_{n=N+1}^{\infty} mQ_n^* < m_e(E - \bigcup F_n).$$

By 19.8 and 20.3 the set $Q_{N+1}^* \cup Q_{N+2}^* \cup \cdots$ cannot contain all of $E - \bigcup F_n$. Let x_0 be a point of $E - \bigcup F_n$ which does not belong to $Q_{N+1}^* \cup Q_{N+2}^* \cup \cdots$. We then have the following situation.

(*A*) *There is a point* x_0 *of* E *which does not belong to any set* $F_n (n = 1, 2, \cdots)$ *nor to any set* Q_n^* $(n = N + 1, N + 2, \cdots)$.

The point x_0 does not belong to any set F_n, so it does not belong to the (closed) set $F_1 \cup \cdots \cup F_N$. As in defining the sequence $\{F_n\}$, we see that there is a neighborhood $N_\delta(x_0)$ of the point x_0 which is free of points of $F_1 \cup \cdots \cup F_N$. There is a sequence of sets of $\mathfrak{F}$ containing x_0 and converging regularly to x_0; and all but a finite number of these must be contained in $N_\delta(x_0)$. We choose such a set, and call it F. Thus we have established

(*B*) *There is a set* F *of the family* $\mathfrak{F}$ *which contains* x_0 *but contains no point of* $F_1 \cup \cdots \cup F_N$.

This leaves two possibilities; either F has a point in common with some F_n, $n > N$, or it has not. We shall show that each of these leads to a contradiction. This will prove that the assumption $m_e(E - \bigcup F_n) > 0$ leads to a contradiction, and will complete the proof of the theorem (with the additional hypotheses (I) and (II)).

CASE I. *The set* F *has no point in common with any* F_n.

For each integer p, the set F is a set of $\mathfrak{F}$ containing no point of $F_1 \cup \cdots \cup F_p$. The supremum ϵ_{p+1} of the function $e(F^*)$ for all such sets F^* is at least equal to $e(F)$. By the law of formation of the sequence, $e(F_{p+1})$ is at least $e(F)/2$. Therefore the edge of the cube Q_{p+1} is at least $e(F)$, the edge of Q^*_{p+1} is at least $5e(F)$, and the measure of Q^*_{p+1} is at least $[5e(F)]^q$, which is a positive number independent of p. But this is impossible; we have already seen that the series ΣmQ^*_p converges.

CASE II. *There is an n such that F_n has a point in common with F.*

Let $p + 1$ be the least integer such that F_{p+1} has a point in common with F, and let $\bar{x}$ be a point belonging to both F_{p+1} and F. By (B), the number $p + 1$ cannot be one of the numbers $1, \cdots, N$, so $p \geqq N$. Since F is a set of the family $\mathfrak{F}$ having no point in common with $F_1 \cup \cdots \cup F_p$, the number ϵ_{p+1} is at least equal to $e(F)$, so by the law of formation of the sequence we know that

$$e(F_{p+1}) > \frac{e(F)}{2}.$$

The points $\bar{x}$ and x_0 both belong to F, which can be enclosed in a cube of side $2e(F)$. Hence,

$$| x_0^{(i)} - \bar{x}^{(i)} | \leqq 2e(F), \qquad i = 1, \cdots, q.$$

The point $\bar{x}$ also belongs to F_{p+1}, which is contained in Q_{p+1}. So by definition of Q_{p+1} there is a point x_{p+1} such that

$$x_{p+1}^{(i)} - e(F_{p+1}) \leqq \bar{x}^{(i)} \leqq x_{p+1}^{(i)} + e(F_{p+1}).$$

Combining the last three inequalities, we obtain

$$x_{p+1}^{(i)} - 5e(F_{p+1}) \leqq x_0^{(i)} \leqq x_{p+1}^{(i)} + 5e(F_{p+1}), \qquad i = 1, \cdots, q.$$

That is, x_0 belongs to the cube Q^*_{p+1}. But $p + 1 > N$, so this contradicts statement (A).

Thus our theorem has been established subject to the additional hypotheses (I) and (II). It is not difficult to replace (II) by the weaker hypothesis

(II′). *There is a positive number α such that for each x_0 of E there is a sequence of sets of $\mathfrak{F}$ converging to x_0 with modulus of regularity α.*

Let $\mathfrak{F}_0$ be the family of sets each consisting of a set of $\mathfrak{F}$ plus a single point of E. If $\{F_j\}$ is a sequence of sets of $\mathfrak{F}$ converging to x_0 with modulus of regularity α, we define F_j^0 to be F_j plus the single point x_0. Then the sets F_j^0 converge to x_0 with modulus of regularity α, and each of them contains x_0. That is, if the family $\mathfrak{F}$ satisfies the hypotheses of the theorem and also satisfies (I) and (II′) the family $\mathfrak{F}_0$ satisfies the hypotheses of the theorem and also satisfies (I) and (II). By the previous proof, a finite or denumerable collection $[F_n^0]$ of disjoint sets of $\mathfrak{F}_0$ covers all of E save a set of measure zero. Each F_n^0 consists of a set F_n plus a single point x_n. Therefore the sets F_n are disjoint, and their sum contains all of E except $E - \bigcup F_n$, which consists of $(E - \bigcup F_n^0)$ plus some or all of the individual points $x_1, x_2, \cdots$ and so has measure zero.

The removal of hypothesis (II′) requires a little more effort. Let E_n be the subset of E consisting of those points x_0 of E for which there is a sequence $\{F_j\}$ of sets of $\mathfrak{F}$ converging to x_0 with modulus of regularity $1/n$. Clearly $E_1 \subset E_2 \subset E_3 \subset \cdots$; and since for each point x_0 of E there is a sequence $\{F_j\}$ converging to x_0 with some positive modulus of regularity, each x_0 of E belongs to some set E_n.

Now we define inductively a sequence of positive integers $n_1 < n_2 < n_3 < \cdots$ and a sequence of disjoint sets $F_1, F_2, F_3, \cdots$, each one of which either belongs to $\mathfrak{F}$ or is empty, such that

$$(C) \qquad m_e\left(E_k - \bigcup_1^{n_k} F_n\right) < \frac{1}{k}.$$

The set E_1 satisfies the hypotheses of the theorem and also satisfies (I) and (II′) with $\alpha = 1$, so it can be covered, save for a subset of measure zero, by a finite or denumerable sequence $\{F_i\}$ of sets of $\mathfrak{F}$. If the number of these sets is finite, we denote it by n_1; then for $k = 1$ the left member of (C) vanishes. If the sequence is denumerable, the series ΣmF_i converges, for the sets F_i are disjoint and are contained in the bounded set U. We then choose an integer n_1 large enough so that

$$\sum_{n_1+1}^{\infty} mF_i < 1.$$

Then the set $E_1 - (F_1 \cup \cdots \cup F_{n_1})$ is contained in the sum of $E_1 - \bigcup F_i$ and $(F_{n_1+1} \cup \cdots)$. Hence by 20.3, 20.4, 20.8 and 19.8

$$m_e\left(E_1 - \bigcup_1^{n_1} F_i\right) \leqq m_e\left(E_1 - \bigcup F_i\right) + \sum_{n_1+1}^{\infty} mF_i < 1.$$

Suppose that the sequence has been chosen up to n_k and F_{n_k}. The set $(F_1 \cup \cdots \cup F_{n_k})$ is closed, so the set

$$U_k = U - (F_1 \cup \cdots \cup F_{n_k})$$

is open. The set $E_{k+1}U_k$ satisfies the hypotheses of the theorem and also the additional hypotheses (I) and (II)′, with $\alpha = 1/(k+1)$. If it is empty, we take $n_{k+1} = n_k + 1$ and let F_{n_k+1} be the empty set. Otherwise, by the first part of this proof there is a finite or denumerable sequence $\{F_i^*\}$ of disjoint sets of $\mathfrak{F}$ such that $\bigcup F_i^*$ is contained in U_k and contains all of $E_{k+1}U_k$ save at most a subset of measure zero. Since the F_i^* are contained in U_k, they cannot have any points in common with F_i, $i = 1, \cdots, n_k$. If there are finitely many sets F_i^*, we make the series formally infinite by writing infinitely many repetitions of the empty set Λ after them. In any case, ΣmF_i^* converges, so there is a number m such that

$$\sum_{m+1}^{\infty} mF_i^* < \frac{1}{(k+1)}.$$

Then the set $E_{k+1}U_k - (F_1^* \cup \cdots \cup F_m^*)$ is contained in the set $[E_{k+1}U_k - \bigcup F_i^*] \cup (F_{m+1}^* \cup F_{m+2}^* \cup \cdots)$, and so has exterior measure less than $1/(k+1)$. We define $n_{k+1} = n_k + m$, and write

$$F_{n_k+i} = F_i^*, \qquad i = 1, \cdots, m.$$

The set $E_{k+1} - (F_1 \cup \cdots \cup F_{n_{k+1}})$ is in U and contains no point of $(F_1 \cup \cdots \cup F_{n_k})$, so it is in U_k. Hence

$$E_{k+1} - \bigcup_1^{n_{k+1}} F_i = U_k[E_{k+1} - \bigcup_1^{n_{k+1}} F_i] = E_{k+1}U_k - \bigcup_{n_k+1}^{n_{k+1}} F_i$$
$$= E_{k+1}U_k - \bigcup_1^m F_i^*,$$

and this last has outer measure less than $1/(k+1)$. This completes the inductive definition of the sets F_n.

The sets F_n thus defined are disjoint and are contained in U. Let k be any positive integer. If p is greater than k, the set $E_k - \bigcup F_n$ is contained in $E_p - \bigcup F_n$, which in turn is contained in $E_p - \bigcup_1^{n_p} F_n$. This last has measure less than $1/p$. So the exterior measure of $E_k - \bigcup F_n$ is less than $1/p$ for arbitrarily large p, and must be zero. By 20.2 and 20.6, $E_k - \bigcup F_n$ is measurable and its measure is 0:

$$m(E_k - \bigcup F_n) = 0, \qquad k = 1, 2, \cdots.$$

But

$$E - \bigcup F_n = \bigcup_{k=1}^{\infty} (E_k - \bigcup_1^{\infty} F_n),$$

so $E - \bigcup F_n$ has measure zero. If we now reject the empty sets written among the F_n, the remainder are sets of $\mathfrak{F}$ with all the required properties.

All that remains is to remove the additional hypothesis (I). For each set $(n_1, \cdots, n_q)$ of integers let $U(n_1, \cdots, n_q)$ be the open cube

$$\{x \mid n_i < x^{(i)} < n_i + 1, \quad i = 1, \cdots, q\}$$

These cubes cover the whole space except a set of measure zero. By the previous proof, for each set $(n_1, \cdots, n_q)$ there is a finite or denumerable collection of disjoint sets F_i belonging to $\mathfrak{F}$, contained in $U \cap U(n_1, \cdots, n_q)$, and covering all of $EU(n_1, \cdots, n_q)$ save a subset of measure zero. If we take all these sets together, we have a finite or denumerable collection of sets of $\mathfrak{F}$ contained in U and covering all of E but a part of measure zero. The sets are disjoint; for if two sets F_i, F_j are contained in the same cube they were chosen disjoint, and if they are contained in different cubes they clearly can have no common points. This completes the proof.

71. Let us recall the definition of the upper derivate of a function $f(x)$ at x_0. If $f(x)$ is defined on an interval I_0: $\{x \mid a \leqq x \leqq b\}$, for each subinterval $I = [\alpha, \beta]$ we can define $\varphi(I) = f(\beta) - f(\alpha)$. By the definition 31.1, together with 6.4, we can state that $\bar{D}f(x_0)$ is the largest number l for which there exists a sequence of closed intervals $\{I_n\}$ contained in I_0, having one end at x_0, with lengths tending to zero, and such that $\lim [\varphi(I_n)/\Delta I_n] = l$. In fact, by the exercise after 31.1 we can replace the requirement that x_0 be an end of I_n by the weaker requirement that x_0 belong to I_n.

This suggests defining the upper derivate of a function of intervals $\Phi(I)$ to be the greatest number l for which there exists a sequence $\{I_n\}$ of closed intervals converging to x_0 and such that $\Phi(I_n)/\Delta I_n$ tends to l. But in the light of the preceding section, we see that in space of two or more dimensions there is a distinction which is absent from the one-dimensional case. A sequence $\{I_n\}$ of intervals containing x_0 may have edges tending to zero, and yet not converge regularly to x_0 in the sense of definition 70.1; for instance, in two-space the intervals I_n: $\{x \mid x_0^{(1)} \leqq x^{(1)} \leqq x_0^{(1)} + n^{-1}, x_0^{(2)} \leqq x^{(2)} \leqq x_0^{(2)} + n^{-2}\}$ fail to converge regularly to x_0. We therefore have a choice in the above definition. Either we can demand merely that the I_n contain x_0 and have edges tending to zero; the derivate thus defined is called the *strong upper derivate* of Φ at x_0, and we shall not discuss it further. Or else we can demand that the I_n converge regularly to x_0; the derivate thus defined is the *ordinary upper derivate* of Φ at x_0. Precisely,

71.1. *Let $\Phi(E)$ be a function of sets which is defined for all closed intervals contained in an interval I_0. Let x_0 be a point of I_0. The ordinary upper derivate $\bar{D}\Phi(x_0)$ and the ordinary lower derivate $\underline{D}\Phi(x_0)$*

of Φ at x_0 are respectively the greatest and the least of all numbers l for which there exists a sequence $\{I_n\}$ of closed subintervals of I_0 containing x_0, converging regularly to x_0 and having $\lim_{n\to\infty} [\Phi(I_n)/\Delta I_n] = l$. If the ordinary upper derivate and the ordinary lower derivate of Φ at x_0 are equal, we define the ordinary derivative of Φ at x_0 to be their common value, and denote it by $D\Phi(x_0)$.

Several variants of this definition suggest themselves. We could require in 71.1 that the I_n be cubes. More stringently, we could ask that the I_n be cubes with center at x_0; the derivates thus defined are sometimes called central derivates. Obviously, if the ordinary derivate exists so does the central derivate, and the two are equal.

In the opposite direction, we can relax the conditions by replacing the intervals I_n by closed sets E_n not necessarily containing x_0. The derivates thus defined are called general derivates.

71.2. *Let $\Phi(E)$ be a function of sets defined for all closed subsets E of the interval I_0, and let x_0 be a point of I_0. The general upper derivate $\bar{D}^*\Phi(x_0)$ and the general lower derivate $\underline{D}_*\Phi(x_0)$ of Φ at x_0 are respectively the greatest and the least of all numbers l for which there exists a sequence $\{E_n\}$ of closed subsets of I_0 converging regularly to x_0 and having $\lim_{n\to\infty} [\Phi(E_n)/mE_n] = l$. If these two general derivates are equal, we define the general derivative $D^*\Phi(x_0)$ of Φ at x_0 to be their common value,*

$$D^*\Phi(x_0) = \bar{D}^*\Phi(x_0) = \underline{D}_*\Phi(x_0).$$

Since all closed intervals are closed sets, the numbers l of definition 71.1 are included among the numbers l of 71.2. Hence

$$\underline{D}_*\Phi(x_0) \leqq \underline{D}\Phi(x_0) \leqq \bar{D}\Phi(x_0) \leqq \bar{D}^*\Phi(x_0).$$

In particular, if the general derivative exists so does the ordinary derivative, and the two are equal.

These upper and lower derivates satisfy the same set of inequalities as the derivates in §31.

71.3. *If $\Phi_1(E)$ and $\Phi_2(E)$ are finite-valued functions of intervals defined for all subintervals of I_0, then at each point x_0 of I_0 the following relations hold, insofar as the sums and differences are defined:*

$$\begin{aligned} \underline{D}\Phi_1 + \underline{D}\Phi_2 &\leqq \underline{D}[\Phi_1 + \Phi_2] \leqq \underline{D}\Phi_1 + \bar{D}\Phi_2 \leqq \bar{D}[\Phi_1 + \Phi_2] \\ &\leqq \bar{D}\Phi_1 + \bar{D}\Phi_2, \\ \underline{D}\Phi_1 - \bar{D}\Phi_2 &\leqq \underline{D}[\Phi_1 - \Phi_2] \leqq \begin{Bmatrix} \bar{D}\Phi_1 - \bar{D}\Phi_2 \\ \underline{D}\Phi_1 - \underline{D}\Phi_2 \end{Bmatrix} \\ &\leqq \bar{D}[\Phi_1 - \Phi_2] \leqq \bar{D}\Phi_1 - \underline{D}\Phi_2. \end{aligned}$$

If $\Phi(E)$ is a function of sets defined and finite for all closed subsets of I_0, the analogous relations hold between the general upper and lower derivates at x_0.

Let $\{I_n\}$ be a sequence of subintervals of I_0 converging regularly to x_0 and such that $[\Phi_1(I_n) + \Phi_2(I_n)]/\Delta I_n$ tends to $\bar{D}[\Phi_1 + \Phi_2]$. For a properly chosen subsequence $\{I_\alpha\}$, $\alpha = n_1, n_2, \cdots$ the sequences $\Phi_1(I_\alpha)/\Delta I_\alpha$ and $\Phi_2(I_\alpha)/\Delta I_\alpha$ converge to limits, finite or infinite. Then

$$(A) \qquad \bar{D}[\Phi_1 + \Phi_2] = \lim_{\alpha\to\infty} \frac{\Phi_1(I_\alpha)}{\Delta I_\alpha} + \lim_{\alpha\to\infty} \frac{\Phi_2(I_\alpha)}{\Delta I_\alpha},$$

provided that the sum on the right is defined. By the definition of ordinary upper derivate, we have

$$(B) \qquad \lim_{\alpha\to\infty} \frac{\Phi_s(I_\alpha)}{\Delta I_\alpha} \leqq \bar{D}\Phi_s, \qquad s = 1, 2.$$

So if the sum on the right in (A) is defined, (A) and (B) imply

$$(C) \qquad \bar{D}[\Phi_1 + \Phi_2] \leqq \bar{D}\Phi_1 + \bar{D}\Phi_2.$$

If the sum on the right in (A) is undefined, one of the summands is $+\infty$. By (B), one of the terms on the right in (C) is also $+\infty$, so if the right member of (C) has a meaning it means $+\infty$. So in either case (C) is established. As in 31.4, all the other relations are merely transformations of (C).

The same discussion applies to general derivates; we have only to use closed sets E_n instead of intervals I_n.

As a corollary we have the analogue of 31.5, both for ordinary derivates and for general derivates. The proof is the same as that of 31.5, and we shall not repeat it.

72. The principal theorem of this section will be the analogue of 33.3: the derivative of an indefinite integral is almost everywhere defined and equal to the integrand. In fact, except for needing the Vitali theorem in proving the analogue of 32.3, we can follow the proof of 33.3 with only trifling changes.

72.1. *Let $u(x)$ be a U-function summable over the closed interval I_0. For every measurable subset E of I_0, define $U(E) = \int_E u(x)\,dx$. Then at all points x_0 of I_0 the inequality*

$$\underline{D}_* U(x_0) \geqq u(x_0)$$

is satisfied.

Let h be an arbitrary number less than $u(x_0)$. By 7.4, there is a neighborhood $N_\epsilon(x_0)$ such that $u(x)$ exceeds h for all x in $I_0 N_\epsilon(x_0)$.

Let $E_1, E_2, \cdots$ be a sequence of closed subsets of I_0, converging regularly to x_0, such that $\lim [U(E_n)/mE_n] = \underline{D}_* U(x_0)$. All but a finite number of the sets E_n are contained in $I_0 N_\epsilon(x_0)$, so for all large n we have

$$\int_{E_n} u(x)\, dx > \int_{E_n} h\, dx = hmE_n.$$

That is, for all large n,

$$\frac{U(E_n)}{mE_n} > h.$$

Hence,

$$\underline{D}_* U(x_0) \equiv \lim_{n\to\infty} \frac{U(E_n)}{mE_n} \geqq h.$$

This holds for all h less than $u(x_0)$, so by 5.2 the theorem is established.

72.2. *Let $\Phi(I)$ be a function of intervals which is defined, additive and non-negative for all subintervals of an interval I_0, and let h be an arbitrary number. If the inequality $\bar{D}\Phi(x_0) \geqq h$ holds on a subset E of I_0, then*

$$\Phi(I_0) \geqq hm_eE.$$

This is trivial if h is negative or zero. If h is positive, let ϵ be a positive number less than h. For each point x_0 of E there is a sequence of intervals $\{I_n\}$ contained in I_0, containing x_0, converging regularly to x_0 and having $\lim \Phi(I_n)/\Delta I_n = \bar{D}\Phi(x_0) \geqq h$. By discarding a finite number of terms of this sequence, we are left with a sequence such that $\Phi(I_n)/\Delta I_n > h - \epsilon$. By the Vitali theorem (70.2) a finite or denumerable set of these intervals cover almost all of E and are disjoint. That is, there are disjoint subintervals $I'_1, I'_2, \cdots$ of I_0, each satisfying the inequality $\Phi(I'_j) > (h - \epsilon)\Delta I'_j$, and together covering almost all of E, so that

$$\sum \Delta I'_n \geqq m_eE.$$

We select a finite number of these I'_j, say $I'_1, \cdots, I'_k$, such that

$$\sum_1^k \Delta I'_j > m_eE - \epsilon.$$

Hence, using a preceding inequality,

$$\sum_1^k \Phi(I'_j) > \sum_1^k (h - \epsilon)\Delta I'_j > (h - \epsilon)(m_eE - \epsilon).$$

The remainder of I_0 can be subdivided into non-overlapping subintervals $I_1'', \cdots, I_p''$. By hypothesis,

$$\Phi(I_0) = \sum_1^k \Phi(I_j') + \sum_1^p \Phi(I_n'') \geqq \sum_1^k \Phi(I_j') > (h - \epsilon)(m_eE - \epsilon).$$

Since ϵ is an arbitrary positive number, this implies $\Phi(I_0) \geqq hm_eE$, establishing the theorem.

72.3. *Let $\Phi(E)$ be an additive function of sets, defined and non-negative for all bounded Borel sets. Let h be an arbitrary number. If $\bar{D}^*\Phi(x_0) \geqq h$ on a subset E of a bounded Borel set B, then*

$$\Phi(B) \geqq hm_eE.$$

We may think of $\Phi(E)$ as a measure function on its natural range; the extension to the natural range is possible, by the remark after 54.2, but we really have no need for Φ except on Borel sets. It is now possible to regard Φ as a measure to which all s- theorems apply. Therefore, by 20.7, for every positive ϵ there is an open set U containing B and such that $\Phi(U) < \Phi(B) + \epsilon$. Now, as in proving 72.2, we obtain a set of disjoint closed sets $E_1, \cdots, E_n$ contained in U, of total measure at least $m_eE - \epsilon$, and satisfying the inequality $\Phi(E_j)/mE_j > h - \epsilon$. This yields

$$\Phi(B) > \Phi(U) - \epsilon > \Phi(\bigcup_1^n E_j) - \epsilon = \sum_1^n \Phi(E_j) - \epsilon$$
$$> (h - \epsilon) \sum_1^n mE_j - \epsilon \geqq h \sum_1^n mE_j - \epsilon(1 + mU) > h(m_eE - \epsilon)$$
$$- \epsilon(1 + mU).$$

Since ϵ is an arbitrary positive number, the theorem is proved.

72.4. *If $f(x)$ is summable over R_q, and for all measurable sets E we define $F(E) = \int_E f(x)\,dx$, then for almost all x the ordinary derivative and the general derivative of $F(E)$ at x are defined, finite and equal to $f(x)$.*

We need prove only the statement about the general derivative; as we saw in §71, if this is defined so is the ordinary derivative, and the two derivatives are then equal.

Let I_0 be a closed interval, and let ϵ and k be positive numbers. Since $f(x)$ is summable over I_0, there is a U-function $u(x) \geqq f(x)$ on I_0 such that

$$\int_{I_0} u(x)\,dx < \int_{I_0} f(x)\,dx + \epsilon k.$$

For each measurable set E contained in I_0 we define

$$U(E) = \int_E u(x)\,dx,$$
$$G(E) = U(E) - F(E).$$

Let E_k be the subset of I_0 on which $\bar{D}^*G(x) \geqq k$. By 72.3,

$$\begin{aligned} \epsilon k &> U(I_0) - F(I_0) \\ &= G(I_0) \\ &\geqq k m_e E_k, \end{aligned}$$

so E_k has exterior measure less than ϵ. Except on E_k we have by 71.3 and 72.1

$$\begin{aligned} \underline{D}_* F(x) &\geqq \underline{D}_* U(x) - \bar{D}^* G(x) \\ &> u(x) - k \\ &\geqq f(x) - k. \end{aligned}$$

Thus for each positive k the set on which the inequality

$$\underline{D}_* F(x) > f(x) - k$$

fails to hold has exterior measure less than ϵ, which is an arbitrary positive number. So its exterior measure is zero; that is, its measure is zero.

In particular, if k is the reciprocal of a positive integer n the set on which the inequality

$$\underline{D}_* F(x) > f(x) - 1/n$$

fails to hold is a set N_n of measure zero. Except on $\bigcup N_n$, which has measure zero, the above inequality holds for every n. Hence by 5.2

$$\underline{D}_* F(x) \geqq f(x)$$

almost everywhere in I_0.

This holds true for $-f(x)$ as well as for $f(x)$, so for almost all x we have

$$\underline{D}_*[-F(x)] \geqq -f(x),$$

that is,

$$\bar{D}^* F(x) \leqq f(x).$$

The two inequalities together prove that the general derivative of $F(E)$ exists almost everywhere in I_0 and has the value $f(x)$. If we reject the set of measure zero on which $f(x)$ is infinite, we find that

$$D^*F(x) = f(x) \neq \pm\infty$$

almost everywhere in I_0.

To complete the proof, we use the cubes

$$W_n: \{x \mid -n \leqq x^{(i)} \leqq n, i = 1, \cdots, q\}.$$

Let E_0 be the set of x at which it is false that $D^*F(x)$ exists, is finite, and equals $f(x)$. The part of E_0 interior to W_n has measure zero, by the preceding proof. This holds for every n; so on adding we find $mE_0 = 0$, and the proof is complete.

73. Although we have not defined the concept of total variation for a function of intervals, the definition is self-suggesting; we merely paraphrase 46.2. Another way of stating it is this. If $\Phi(I)$ is a function of intervals defined and additive for all subintervals of I_0, there is by 45.5 a finite-valued function $g(x)$ on I_0 such that $\Delta_g I = \Phi(I)$ for every sub-interval I of I_0. We then say that $\Phi(I)$ is of BV on I_0 if $g(x)$ is of BV on I_0, as defined in 46.2. If this is so, by 46.9 there are non-negative interval functions $\Delta_P I$ and $\Delta_N I$ whose difference is $\Phi(I)$.

We now prove a theorem analogous to 34.3.

73.1. *Let $\Phi(I)$ be an additive function of intervals, defined for all subintervals of an interval I_0, and of BV on I_0. Then at almost all points x of I_0 the ordinary derivative $D\Phi(x)$ exists, and it is summable over the set E on which it exists. Moreover, if $\Phi(I)$ is non-negative the inequality*

$$\int_{I_0} D\Phi(x)\,dx \leqq \Phi(I_0)$$

is satisfied.

By the preceding paragraph, there is no loss of generality in supposing $\Phi(I)$ non-negative; for $\Phi(I)$ is the difference of two non-negative functions $\Delta_P I$ and $\Delta_N I$, and if the theorem holds for each of these it holds for their difference $\Phi(I)$.

For each pair h, k of positive integers we define $E_{h,k}$ to be the set of all points x of I_0 at which

$$\underline{D}\Phi(x) < \frac{h}{k} < \frac{h+1}{k} \leqq \bar{D}\Phi(x).$$

At each point of $E_{h,k}$ the derivative $D\Phi(x)$ fails to exist. Conversely, if the derivative fails to exist at x, the difference $\bar{D}\Phi(x) - \underline{D}\Phi(x)$ is positive. If k is an integer large enough so that $2/k$ is less than that difference, and h is the smallest integer such that h/k exceeds $\underline{D}\Phi(x)$, then $(h + 1)/k$ cannot exceed $\bar{D}\Phi(x)$, and so x is in $E_{h,k}$. Therefore

(A) *the set of points at which the derivative fails to exist is* $\bigcup_{h,k} E_{h,k}$.

Consider any one of these sets $E_{h,k}$. For an arbitrary positive number ϵ it is possible by 20.7 to find an open set U containing $E_{h,k}$ and having measure $mU < m_eE_{h,k} + \epsilon$. At each point x_0 of $E_{h,k}$ the lower derivate is less than h/k, so there exists a sequence of intervals I'_n converging regularly to x_0 and having $\Phi(I'_n)/\Delta I'_n < h/k$. By the Vitali theorem (70.2), there is a finite or denumerable set $I_1, I_2, \cdots$ of such intervals which are disjoint, lie in U and cover almost all of $E_{h,k}$. Hence

$$(B) \qquad \sum_n \Phi(I_n) \leqq \left(\frac{h}{k}\right) \sum_n \Delta I_n \leqq \left(\frac{h}{k}\right) mU < \left(\frac{h}{k}\right) (m_eE_{h,k} + \epsilon).$$

Consider now the part of $E_{h,k}$ in an interval I_n. On this set the upper derivate of Φ is at least $(h + 1)/k$, by hypothesis; so by 72.2 we have

$$\Phi(I_n) \geqq \left[\frac{h+1}{k}\right] \cdot m_e(E_{h,k}I_n).$$

By 20.4, this yields

$$(C) \qquad \begin{aligned} \sum_n \Phi(I_n) &\geqq \left[\frac{h+1}{k}\right] \cdot \sum_n m_e(E_{h,k}I_n) \\ &\geqq \left[\frac{h+1}{k}\right] m_e(E_{h,k} \cap [\bigcup I_n]) \\ &= \left[\frac{h+1}{k}\right] m_eE_{h,k}, \end{aligned}$$

since almost all of $E_{h,k}$ is in $\bigcup I_n$. Comparing (B) and (C) yields

$$\left[\frac{h+1}{k}\right] m_eE_{h,k} < \left(\frac{h}{k}\right) (m_eE_{h,k} + \epsilon),$$

or

$$m_eE_{h,k} < h\epsilon.$$

Since ϵ is arbitrary, $E_{h,k}$ has measure zero.

Returning to statement (A), we now see that the set on which the derivative fails to exist is the sum of a denumerable collection of sets of measure zero, so it has measure zero. If we denote by E the set on which $D\Phi(x)$ exists, we have shown that E constitutes almost all of I_0. It remains to show that $D\Phi(x)$ is summable over I_0.

We first effect a partition of I_0 into 2^q disjoint subintervals $I_{1,j}$, $j = 1, \cdots, 2^q$ by bisecting each side of I_0. Again, by bisecting each side of each $I_{1,j}$ we split I_0 into 2^{2q} disjoint intervals $I_{2,j}$, $j = 1, \cdots, 2^{2q}$. Continuing thus, for each integer n we split I_0 into 2^{nq} intervals $I_{n,j}$, $j = 1, \cdots, 2^{nq}$. For each n, we define $f_n(x) = \Phi(I_{n,j})/\Delta I_{n,j}$ for x in $I_{n,j}$. These are non-negative step functions, and it is evident that

$$(D) \qquad \int_{I_0} f_n(x)\, dx = \sum_{j=1}^{2^{nq}} \Phi(I_{n,j}) = \Phi(I_0).$$

Let x_0 be a point of E. For each n the point x_0 belongs to a certain one $I_{n,j(n)}$ of the intervals $I_{n,j}$. These intervals converge regularly to x_0, and $D\Phi(x_0)$ exists, so

$$\lim_{n\to\infty} \frac{\Phi(I_{n,j(n)})}{\Delta I_{n,j(n)}} = D\Phi(x_0).$$

But by definition the fraction on the left is the value of $f_n(x_0)$. So we have shown that $f_n(x_0)$ converges to $D\Phi(x_0)$ for each x_0 of E. Now by 29.1 the function $D\Phi(x)$ is summable over E, and

$$\int_E D\Phi(x)\, dx \leqq \lim_{n\to\infty} \int_E f_n(x)\, dx = \Phi(I_0).$$

73.2. *Let $\Phi(E)$ be a completely additive function of sets (or a non-negative additive function of sets), defined and finite on a family containing all bounded Borel sets. Then the general derivative $D^*\Phi(x)$ is defined and finite for almost all x, and is summable over every bounded measurable set on which it is defined. Moreover, if Φ is non-negative and E is a bounded Borel set on which $D^*\Phi(x)$ is defined, the inequality*

$$\int_E D^*\Phi(x)\, dx \leqq \Phi(E)$$

is satisfied.

If $\Phi(E)$ is completely additive, it is the difference of two non-negative completely additive functions of sets, by 53.12. So there is no loss of generality in supposing $\Phi(E)$ non-negative.

Let E be the set on which the general derivative $D^*\Phi(x)$ is defined. If we restrict our attention to subsets of an interval I_0, we can repeat the proof of 73.1, replacing ordinary derivatives by general derivatives and intervals by closed sets. We find that $I_0 - E$ has measure zero, and that $D^*\Phi(x)$ is summable over EI_0. It is therefore finite on almost all of EI_0 and is summable over every measurable subset of EI_0, by 22.5.

Since this is true for each of the cubes

$$W_n: \{x \mid -n \leqq x^{(i)} \leqq n, i = 1, \cdots, q\},$$

by letting n tend to ∞ we obtain the conclusion of the theorem.

In this last theorem it is clearly not necessary that $\Phi(E)$ should be the integral of its derivative; we did not assume that $\Phi(E)$ is absolutely continuous. We now shall show that the difference between Φ and the integral of its derivative is a *singular* function, in the following sense.

73.3. *A set function $\Phi(E)$, defined and completely additive on a completely additive family $\mathfrak{F}$ of sets, is a* singular function *if there is a Borel set B_0 of measure zero such that for every set E of the family $\mathfrak{F}$ contained in the complement of B_0 the equation $\Phi(E) = 0$ holds.*

It follows readily that $\Phi(E) = \Phi(EB_0)$ for every set E of $\mathfrak{F}$. Moreover, from 53.9 the positive and negative variations of Φ vanish on every set E in CB_0, so they too are singular functions. So also is the total variation of Φ.

73.4. *Let $\Phi(E)$ be defined, non-negative and completely additive on a completely additive family $\mathfrak{F}$ of sets. Then $\Phi(E)$ is singular if and only if $D^*\Phi(x) = 0$ for almost all points x.*

Suppose that Φ is singular, and that B_0 is the set specified in the definition.

For each interval I, the set $I - B_0$ is a Borel set. If E_n is the part of it on which $\bar{D}^*\Phi(x) \geqq 1/n$, then by 72.3 and 73.3

$$0 = \Phi(I - B_0) \geqq \left(\frac{1}{n}\right) m_e E_n,$$

so E_n has measure 0. The subset of $I - B_0$ in which $\bar{D}^*\Phi$ is positive is the set $\bigcup E_n$, hence has measure zero. Thus $0 \leqq \underline{D}^*\Phi \leqq \bar{D}^*\Phi \leqq 0$ for almost all x in $I - B_0$, and $D^*\Phi(x)$ exists and is zero almost everywhere in $I - B_0$. But B_0 has measure zero, so $D^*\Phi(x) = 0$ almost everywhere in I. If we now replace I by W_n and let n tend to ∞, we see that $D^*\Phi(x)$ is defined and equal to zero for almost all x.

Conversely, suppose $\Phi(x)$ has a vanishing general derivative for almost all x; say for all x in E_0, when CE_0 has measure 0. For each point x_0 of E_0 and each positive ϵ, there is a sequence $\{E_n'\}$ of closed sets converging regularly to x_0 and having $\Phi(E_n')/mE_n' < \epsilon$. If I_0 is an interval, we can choose a finite or denumerable collection of such closed sets $E_1, E_2, \cdots$ which are disjoint and cover almost all of

I_0E_0. Then $\Phi(\bigcup E_n) = \Sigma\Phi(E_n) < \epsilon\Sigma mE_n = \epsilon mI_0$. We write S for the union $\bigcup E_n$.

If we choose ϵ to have the successive values $1, \frac{1}{2}, \cdots, 1/n, \cdots$, we obtain corresponding sets $S_1, S_2, \cdots$, for each of which $m(I_0 - S_n) = 0$ and $\Phi(S_n) < mI_0/n$. The intersection $\bigcap S_n$ is contained in each S_n, so $\Phi(\bigcap S_n) \leqq \Phi(S_n) < mI_0/n$ for each n. Therefore $\Phi(\bigcap S_n)$ must be zero. The complement in I_0 of $\bigcap S_n$ is

$$I_0 - \bigcap S_n = \bigcup (I_0 - S_n),$$

which has measure zero. We can therefore enclose $I_0 - \bigcap S_n$ in a Borel set B which also has measure zero. For every set E of $\mathfrak{F}$ contained in $I_0 - B$ we have $0 \leqq \Phi(E) \leqq \Phi(I_0 - B) \leqq \Phi(I_0 - [I_0 - \bigcap S_n]) = \Phi(\bigcap S_n) = 0$.

For each cube W_n we thus find a Borel set B_n of measure zero, as described. Define $B_0 = \bigcup B_n$. This is a Borel set. If E is a set of the family $\mathfrak{F}$ contained in CB_0, then for each n the product EW_n is contained in $W_n - B_0$, which is contained in $W_n - B_n$. Its subset $EW_n - EW_{n-1}$ is also contained in $W_n - B_0$. Hence $\Phi(EW_n - EW_{n-1}) = 0$, whence by adding we obtain $\Phi(E) = \Sigma\Phi(EW_n - EW_{n-1}) = 0$.

73.5. *Let $\Phi(E)$ be a completely additive function of sets, defined and finite on the family of all bounded Borel sets. Let $f(x)$ be $D^*\Phi(E)$ where $D^*\Phi(E)$ is defined and finite, while $f(x) = 0$ elsewhere. Then $\Phi(E)$ is the sum of $\int_E f(x)\,dx$ and a singular function $\Psi(E)$; if $\Phi(E)$ is non-negative, so is $\Psi(E)$.*

Suppose first that Φ is non-negative. By 73.2, the function $f(x)$ is summable over every bounded measurable set. For each bounded Borel set E we define

$$\Psi(E) = \Phi(E) - \int_E f(x)\,dx.$$

By 73.2, the function Ψ is non-negative. For almost all x, the equation

$$D^*\Psi(x) = D^*\Phi(E) - f(x)$$

holds, by 73.1. This vanishes almost everywhere, by definition of $f(x)$; so by 73.4, $\Psi(E)$ is singular.

To remove the restriction that $\Phi(E)$ be non-negative, we need only observe that if $\Phi(E)$ is completely additive it is the difference of two non-negative functions. Each is expressible as the sum of the integral of its derivative and a singular function. By subtracting and observing that the difference of singular functions is singular, we complete the proof.

Appendix

In the preceding pages we have defined three important kinds of integral, and named them the Riemann, Lesbegue and Lebesgue-Stieltjes integrals. There is a sense in which these names are inadequate. Riemann's method of defining an integral was as follows. Let $f(x)$ be defined and bounded on a closed interval I. Let Π be a partition on I into nonoverlapping intervals $I_1, \cdots, I_n$, and let $x_1, \cdots, x_n$ be points in $I_1, \cdots, I_n$ respectively. Let $N(\Pi)$ be the greatest of the diagonals of $I_1, \cdots, I_n$. If the sum $\Sigma f(x_i)\Delta I_i$ tends to a limit as $N(\Pi) \to 0$, this limit is the Riemann integral of $f(x)$ over I. The definition in (11.1) is due to Darboux, so the integral there defined could justly be called the "Darboux-Riemann" integral. It is equivalent to the Riemann integral as just defined, in the sense that if $f(x)$ is integrable according to either definition it is integrable according to the other, the integrals being equal.

The concept of upper and lower integral introduced in 14.1 is essentially due to Daniell, so that we have called these the upper and lower Daniell integrals. Accordingly, we could justly have given the name "Daniell-Lebesgue" integral to the integral defined in 15.1. Lebesgue's original method of defining his integral presupposed a theory of measure. For bounded measurable functions on bounded sets the Lebesgue integral is defined by exactly the process used in proving 22.1, so for such functions the Lebesgue and Daniell-Lebesgue integrals are equivalent. The extension to unbounded functions and unbounded sets is made by means of the limiting process used in 22.2. It is easily seen that the two concepts of integral are equivalent.

Another scheme of definition, leading to another equivalent concept of integral, is due to F. Riesz. An exposition of it can be found in *Theory of Functions of a Real Variable* by L. M. Graves (McGraw-Hill, 1946).

In 47.1 and 51.1 an integral was defined and named the Lebesgue-Stieltjes integral. For the case of continuous functions of one variable, the concept of the integral of $f(x)$ with respect to $g(x)$ was introduced by Stieltjes. J. Radon extended such integration processes in the manner of Lebesgue. Our definition is still of Daniell's type; so the integral defined in 51.1 could justly be given the rather cumbersome name of "Daniell-Radon-Lebesgue-Stieltjes" integral. It is equivalent to that defined by Radon, in the Lebesgue pattern. It is also equivalent to the integral, which could be called the "Riesz-Radon-Lebesgue-Stieltjes" integral, defined in Graves's *Theory of Functions of a Real Variable.*

List of Special Symbols and Abbreviations

(In a few cases notation used by other authors has been given in parentheses)

Index

Lightning Source UK Ltd.
Milton Keynes UK
UKHW021822220123
415737UK00006B/306

9 780691 625751